H. Ibbeken R. Schleyer

Source and Sediment

A Case Study of Provenance and Mass Balance at an Active Plate Margin (Calabria, Southern Italy)

With a Contribution by R. Valloni

Supported by Research of P. Ergenzinger, G. Mezzadri, J. Mouton, and J. Rumohr

With 161 Figures

Springer-Verlag
Berlin Heidelberg New York London Paris
Tokyo Hong Kong Barcelona Budapest

Professor Dr. HILLERT IBBEKEN
Dr. RUPRECHT SCHLEYER

Institut für Geologie
Freie Universität Berlin
Altensteinstraße 34a
W-1000 Berlin 33, FRG

Cover picture: Calabrian Massif and Ionian Shelf. The scenery is topped by Mt. Scorda, 1572 m, 17 km from the coast. The Buonamico and Careri river basins feed the submarine canyon which reaches a depth of 750 m in the foreground, 8.5 km off the coast.

ISBN-13:978-3-642-76167-6 e-ISBN-13:978-3-642-76165-2
DOI: 10.1007/978-3-642-76165-2

Library of Congress Cataloging-in-Publication Data. Ibbeken, H. (Hillert), 1935– Source and sediment: a case study of provenance and mass balance at an active plate margin (Calabria, Southern Italy) / H. Ibbeken, R. Schleyer with a contribution by R. Valloni; supported by research of P. Ergenzinger, G. Mezzadri, J. Mouton, and J. Rumohr. p. cm. Includes bibliographical references and index. ISBN-13:978-3-642-76167-6 (U.S.)1.Geology–Italy–Calabria Region.2. Rocks, Sedimentary – Italy – Calabria Region. 3. Sediments (Geology) – Italy – Calabria Region – Analysis. 4. Mass budget (Geophysics) I. Schleyer, R. (Ruprecht), 1955– . II. Valloni, R. III. Title. QE272.I23 1991 554.5'78–dc20 91-11067

Typesetting: K+V Fotosatz GmbH, Beerfelden
32/3145-543210 – Printed on acid-free paper

Contents

Introduction

Why Provenance ...?

Why write a provenance study, when the main interest on terrigenous sediments is being shifted towards facies interpretation, basin fill, seismic stratigraphy and the like? We know that provenance work has yielded many results until now, but the main emphasis of modern studies has been on big rivers at passive margins, the results being mostly qualitative, with little regard to mass balances. Moreover, almost as a rule, only a limited number of rivers of a given area, if not only one, were usually investigated. As a consequence, no statistical tests to prove the reliability of the findings could be undertaken.

We try in this book to give a quantitative picture of provenance and mass balance of a series of small mountain rivers at an active margin – 19 small rivers, but powerful sediment contributors, which may also account for larger basin fills.

There are two more arguments in favour of resuming and refining provenance research. First, our quantitative knowledge of the source area via digitized maps and primary and derived data makes it possible to follow up the differentiation of certain properties through a sequence of environments from the source to the river-mouth sediment, in some cases even to the beach or to the shelf, and to answer the question why and how certain properties become lost and others are acquired. Second, this is made possible mainly by our concept of predicting sediment composition. Rock properties, areal distribution and erosion rates allow a prediction, a quantitative estimate of the composition of the possible river-mouth sediment to be formed and of the mass balances involved under certain circumstances. The discussion of the differences between the prognosticated and the real river-mouth sediment seems to us a meaningful goal of provenance research in modern environments, and one not yet attempted.

Why Calabria ...?

In Europe, Calabria, at the southern tip of Italy, may be one of the best-suited places to study provenance and to calculate mass balances in modern high energy environments.

The Calabrian arc can be seen as a small plate element with an active external margin, so that its plate tectonic position is well definable. It is a powerful source of sediments because of its strong Pleistocene uplift (1000 m) accompanied by strong erosion until today.

The many remainders of a Mid-Pleistocene peneplain allow the calculation of erosion rates over the entire research area of 2000 km^2. The relief is high and differentiated, steep slopes create extended landslides triggered by excessive rainfall or earthquakes. The debris is transported by short, steep-graded mountain rivers through the temporary reservoirs of large river-mouth areas into the Ionian Sea. The rock distribution is both simple enough to remain typical and differentiated enough not to be trivial: granitic, metamorphic and sedimentary rocks participate in more or less equal amounts; the ophiolitic suite, however, or acid volcanic rocks do not appear.

The climatic differentiation is strong, mean precipitation ranging between <600 mm/a close to the coast and >2000 mm/a in the central mountains. Heavy rain storms with precipitations of >400 mm/day hit the area and provoke disastrous floods.

Finally, man's impact is low, so that the original, natural conditions are almost conserved. This implies another advantage: at least until the 1960's, Calabria was a kind of under-developed region of Europe. The Italian government spent enormous amounts on developing the area. Among others, air photo-based topographic maps 1:10000 and 1:25000 were published and, a real treasure for our purposes, a geological map with the amazing scale of 1:25000 covering the entire area was edited. Consequently, quantifying source area conditions, we could deal with an almost unique data base.

Last but not least, Calabria is 2300 freeway-kilometers from Berlin; far, but not too far for effective travelling with heavy equipment but without troublesome and expensive sea and airways. All this, together with the kind support of local authorities and friendly acceptance by the population, made Calabria a great place for this provenance study.

... *"daß man sich von einer gebirgigen Gegend nicht schneller einen Begriff machen kann, als wenn man die Gesteinarten untersucht, die in den Bächen herabgeschoben werden, und daß hier auch die Aufgabe sei, durch Trümmer sich eine Vorstellung von jenen ewig klassischen Höhen des Erdaltertums zu verschaffen".*

Johann Wolfgang von Goethe, Palermo, 4. April 1787, Italienische Reise

... *"that there is no readier way of forming an idea of a mountainous district, than by examining the nature of the stones which are washed down by the streams and that in so doing, the purpose was to acquire a right notion of those eternally classic heights of the ancient world".*

Johann Wolfgang von Goethe, Palermo, 4 April 1787, Travels in Italy,
Translation by the Rev. A. J. W. Morrison, 1849

Problem and Concept

The goal of provenance studies is to reconstruct source area geology. Terrigenous sediments are often the only witnesses of former mountains eroded away and buried under higher tectonic units or younger sediments. The goal of actuogeologic provenance studies is to develop the tools and to test the modern relationships between sources and sediments which may be applied to the past.

Nobody expects a terrigenous sediment to represent faithfully and quantitatively the geological features of its feeder area. Weathering, transportation and deposition strongly filter the source area signals, and the transient or final deposits contain only poor and mutilated information about their origin. Much research has been done to refine the methodology and to deepen our understanding of the geological meaning of grain-size distributions, the petrography of gravels, of light and heavy minerals and of their geochemistry, with much success but probably even more frustration. So why attack again this barely vincible enemy? There are two reasons.

Actuogeologic provenance studies usually provide very detailed data about the products, the sediments, but little about the producers, the source areas. Important source area features such as geological setting, relief or geomorphology and climate are mostly taken into consideration, but never really thoroughly and quantitatively. At least we are not aware of a provenance study where the quality of the source area data corresponds to that of the modern sediment analysis usually applied. Many broad scattered papers treat single aspects of provenance problems: only one drainage area or one grain-size interval, one rock type or one degree of relief. This limitation is quite understandable, because the number of variables should be kept as low as possible. But the question remains whether such a limitation does not have the opposite effect, in so far as the complexity of the origin of a terrigenous sediment cannot be captured by a noncomplex research. So the first reason is to better balance the level of knowledge of both source area and sediment, to bridge this gap and to aim at more comprehensiveness.

The other reason is that we feel the need for a new concept for this kind of provenance research. The line of sight of the classical geologist is directed backwards, in our case from the sediment to the source area. The actuogeological approach, however, makes it possible to turn this line of sight and to look forwards, from the producer to the product, and to pursue a kind of prognostic geology. Janus with his two opposite faces, the Roman god of doors and gates and all beginnings, may serve as a symbol for this situation. If we gather adequate data about a mountainous area being eroded, we should be able to predict certain features of the resulting river-mouth sediments – by certain assumptions. Again, nobody expects these predictions to be verified by nature, but the interpretation of the differences between the predicted and the real sediment should improve our understanding of provenance.

This attempt to predict or prognosticate a kind of "would-be" sediment has the advantage that we are not primarily programmed by the shortcomings of the sediment, i.e. by the limited information of the product itself. From such a product point of view we are not able to ask why the remainders of certain rocks do not appear in the sediment. Seen from the producer, however, we have all the information about the hinterland, the source area, and we can try to explain why some materials are transmitted to the river-mouth sediment and why others are not. Besides this, it mostly makes more sense to follow the genetic development, and by doing so, the source area comes first and then the sediment.

Seen as a whole, this study is deliberately empirical and data-oriented. The overwhelming complexity of both the data base and the results hindered us from theorizing and developing "provenance models", but we learned to increase the probability of certain conclusions by the exclusion of others. A multivariate comparison made it possible to rank the variables of source and sediment which separate the four provinces of the study area.

Contributors

Our results are the sum of many geo-projects during the last 20 years in Calabria. Many diploma and doctorate theses of geologists, geomorphologists and geophysicists were carried out in Calabria, many students were employed to do all those thousand little things, among them the most terrible, digitizing maps.

Peter Ergenzinger, professor of Geomorphology at the Freie Universität Berlin, did the research on landslides and sedimentary budget in the Buonamico basin (Chap. 15).

Giovanni Mezzadri, professor of Petrography at the University of Parma, Italy; the treatment of heavy and light minerals of the soils and river sands is his work, together with Renzo Valloni (Chaps. 8 and 14).

Jean Mouton is director of Italpros, Rome, a consulting office for applied geology. He contributed hydrological data, geophysical data of slopes and paleovalleys and calculated the volumes of alluvial deposits (Chaps. 7 and 15).

Jan Rumohr, research geologist at the University of Kiel, Germany, did fieldwork in Calabria for many years. He did the X-ray analysis of the river's silt and clay fraction and the evaluation of the marine geology (Chaps. 2, 14 and 16).

Renzo Valloni, professor of Sedimentology at the University of Parma, Italy, shared the treatment of heavy and light minerals of the soils and river sands with Mezzadri and wrote the respective sections of the Chapters 8 and 14.

Acknowledgements. It was quite impossible to make a "two-man" business of the Calabrian geology and the Calabrian data set; the spectrum of topics was simply too broad. Thus help was needed and given from many persons and institutions.

Financial support for the many Calabrian projects came from the Deutsche Forschungsgemeinschaft and the Freie Universität Berlin; the Stiftung Volkswagenwerk granted an Akademie-Stipendium to H. I.

Data were provided by the Aeronautica Militare, Rome; Cassa per il Mezzogiorno, Rome; Cementir D'Agostino, Siderno; Istituto Geografico Militare, Firenze; Istituto Idrografico Marina, Genova and Italpros, Rome; submarine sampling was done by the Scientific Diving Group Clausthal.

The Azienda Medici, Bianco, served as long-standing field base for coastal and marine research; various help came from the local authorities, the Carabinieri, Bianco, the Corpo Forestale dello Stato, Reggio Calabria, from Gaetano and Valerio Costarella, Mario and Fedele Stranges, Toto Galluccio and, last but not least, from the local inhabitants, the friendly Calabrian population.

Hans Füchtbauer, Bochum (FRG) and Paul E. Potter, Cincinnati (USA) read an earlier draft of the book.

Alexander Bartholomä (geology), Wilhelm Büsch (mineralogy), Heinz Burger (mathemathics), Michael Diepenbroek (data processing), Ruth Drescher-Schneider (botany), Walter Franke (chemistry), Konrad Görler (geology), Evelin Hilberg (geology), Otto Hirsch (geodesy), Bernhard Mädger (technician), Barry Maynard (geochemistry), Peter Obenauf (geomorphology), Birgit Ott (geology), Klaus J. Reutter (geology), Lou Schermerhorn (mineralogy), Martin Schreiber (mineralogy), Günter Schultz (civil engineering), Mike Shulman (data processing), Christian Sommer (data processing) and Detlef Warnke (geology) helped in many ways and endless discussions.

Anouchka Aslanidis printed the photos and Rita Mensching mounted the computer plots.

Michaela Mechler and Ralf Sauer assisted in organizing the manuscript and compiled the reference section.

Last but not least, we were fortunate in having friendly and cooperative publishers.

We thank them all.

July 1990

Hillert Ibbeken
Ruprecht Schleyer

Chapter 1

Position and Geological Framework

The goal of this chapter is to give a brief summary of the tectonic setting of Calabria before the background of our problem, of provenance and mass balance. We have to present the possible reasons for the tectonic activity of the area which give rise to uplift, erosion and sediment production, and we have to make comparison possible with other, similar regions. For example, sand and sandstone of different plate tectonic realms are known to display different and typical chemical compositions; consequently, we have to check if the Calabrian data fit into this picture. It is not the goal of this chapter to give a geotectonic analysis as such, for which the reader is referred to the most recent comprehensive study of Boccaletti et al. (1984) and the papers of Rehault et al. (1986) and Aifa et al. (1988).

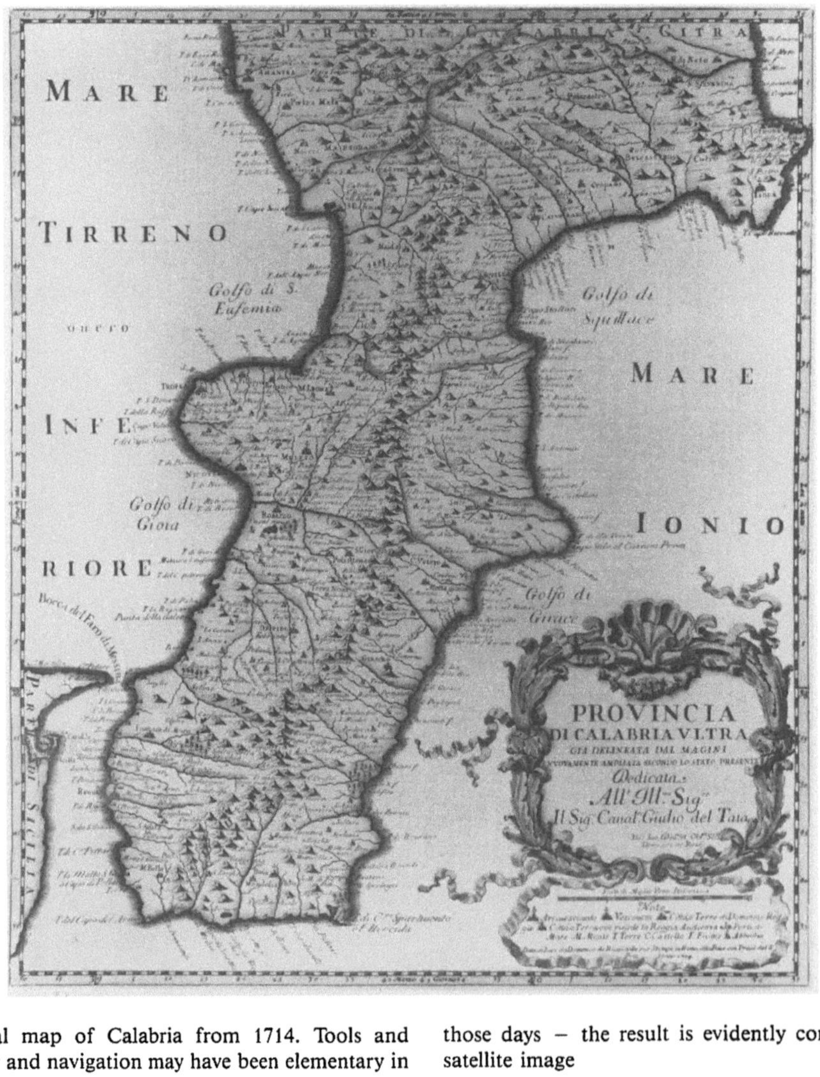

Fig. 1.1. Historical map of Calabria from 1714. Tools and methods of survey and navigation may have been elementary in those days – the result is evidently comparable with today's satellite image

Part I

The Source

Part I is intended to describe the Calabrian source area quantitatively in order to present an equal-ranking counterpart for the comparison with the river-mouth sediment. Main topics are the geological framework, the configuration of the basins and their channel network, the altitudes and the relief. Climate and vegetation, evidently factors of minor weight at this active plate margin, are treated only briefly, more emphasis is given to the single rock types, their regional distribution and especially their erosion rates. At first glance, the determination of the water budget of the rivers and their dissolved load does not fit in a provenance study of terrigenous sediments, but we needed this in order to obtain at least an estimation of the entire erosional budget of the area.

We are not interested, in the context of this book, in describing Calabria as a product of its geological history; we simply want to define it as a present-day producer of sediments.

Summary. Calabria forms part of the African deformed margin; the Ionian slope of the Calabrian arc is interpreted as an active plate margin, characterized by strong neotectonics, 1000 m uplift during 1 Ma, and strong seismic activity. So we are dealing, for the purpose of this study, with a well-definable plate tectonic setting.

1.1 Position of the Research Area

Calabria lies in the center of the Mediterranean Sea (Figs. 1.3A) at the southern tip of Italy, just opposite Sicily; the latitude is about 38 °N. Here only 50 km wide, the Italian main land is symmetrically divided by the Apenninic main divide into a Thyrrhenian west slope and an Ionian east slope. Our area under investigation (Figs. 1.1 and 1.2) comprises the Ionian east slope and encompasses nearly 2000 km^2, forming an elongated rectangle about 20 km wide and 100 km long. Maximum elevations reach nearly 2000 m; consequently short but high-gradient mountain rivers make this area a powerful source of sediment.

1.2 Tectonic Setting

There is no agreement among geoscientists about the tectonic setting of Calabria. The complex pattern of the interaction between the African and the European plate makes different models possible. Figure 1.4 sketches the general situation, the European foreland and its deformed margin and the African foreland and its deformed margin. Calabria forms the center of a strongly curved structure, the Calabrian arc. This arc makes part of

Fig. 1.2. NASA ERTS image of Calabria, 05 Nov 1972. Number E-1105-09065-6. The *dotted line* marks the Apenninic main divide, the *solid one* is the northern divide of the Assi basin, these are the limits of the area under investigation. The study area covers 1832 km^2

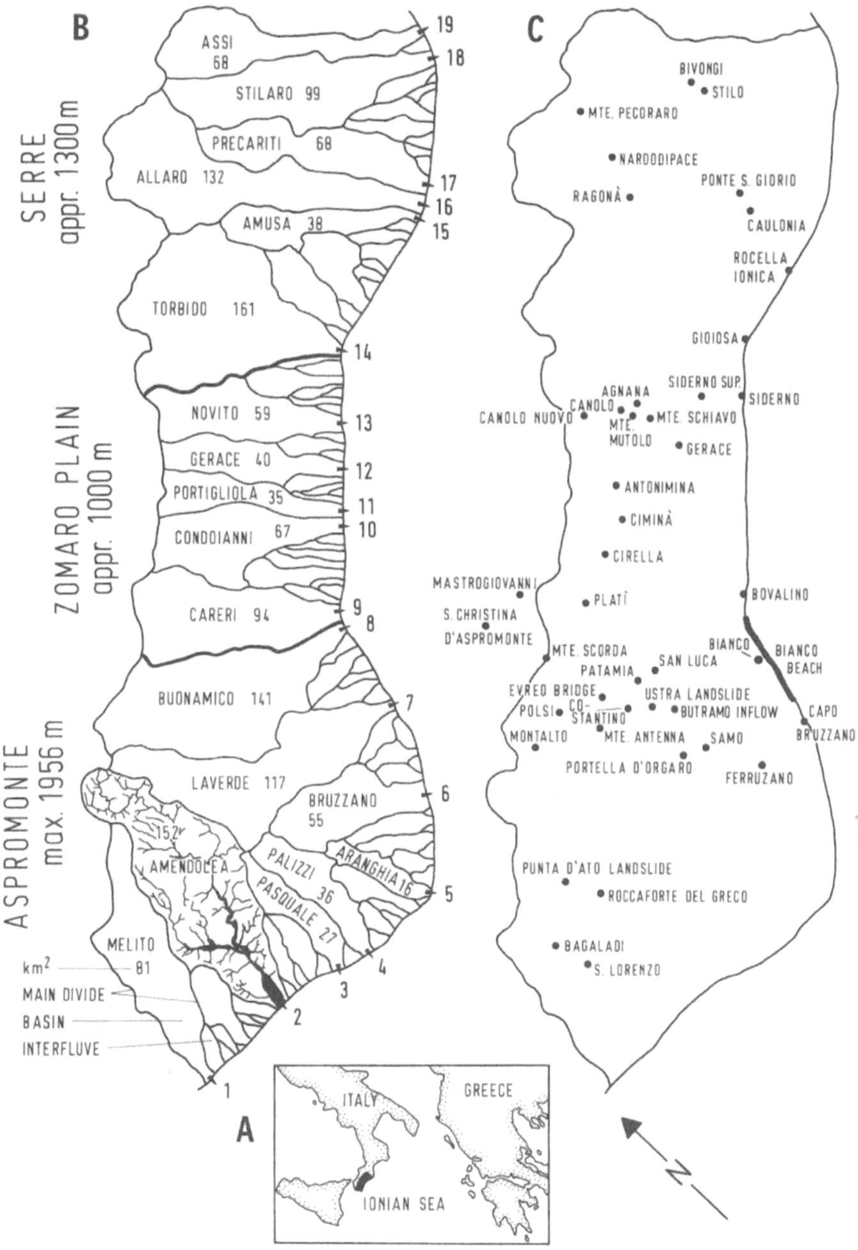

Fig. 1.3. A Position of the study area. **B** The 103 basins and interfluves, some data of the 19 basins studied in more detail and the physiographic provinces, Aspromonte Province (basin 01–08), Zomaro Province (high plain, basin 09–13) and Serre Province (basin 14–19). See Table 2.1 and Appendix B 01–20 for more detail. **C** Location of all villages and mountains cited in this study

the African deformed margin. Figure 1.5 gives a more detailed sketch of the Calabrian arc (Ghisetti 1981).

The most eastward and external element of the arc is the Apulian foreland, which extends from the Apulian platform, the "heel" of Italy, over the Ionian sea to the Hyblean and Ragusa platform of southeast Sicily. It is followed westward by a foredeep, the Bradano trough with its Sicilian equivalent, the Catania-Gela trough. The "folded arc" is made up by the Apenninic chain, most parts of the Sila, the Serre and the Aspromonte on the Italian mainland and the Peloritan massif on the northeastern tip of Sicily. The Eolian islands are part of a calc-alkaline volcanic arc. The Tyrrhenian Sea, finally, can be seen as a backarc basin.

According to Boccaletti et al. (1984), the folded Calabrian-Peloritan chain is composed of two sec-

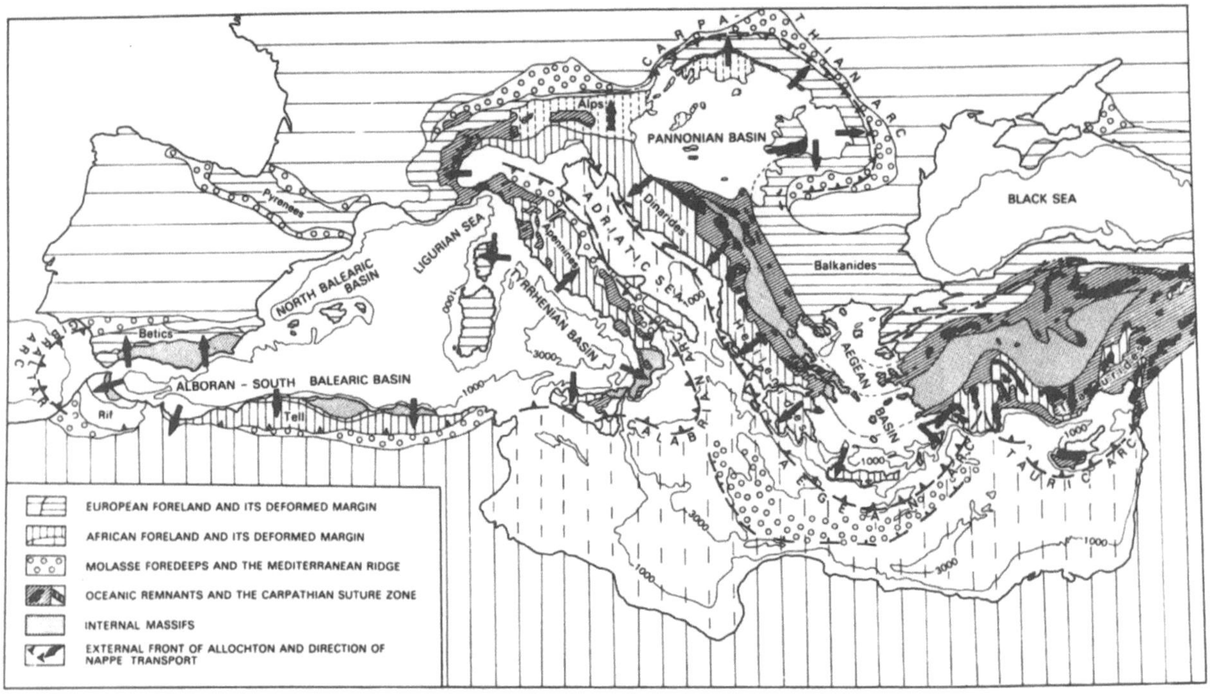

Fig. 1.4. Geotectonic setting of Mediterranean back-arc basins between the African and European forelands (from Horvarth and Berckhemer 1982, Fig. 1). Note the strong curvature of the Calabrian arc and the exposed position of its Ionian slope, whose flux of sediment is directed towards the external Calabrian arc, especially the Crotone-Spartivento basin (Fig. 1.10)

tors, differing in neotectonics and chain-building phases and separated by the Soverato-Capo Vaticano Line (Fig. 1.5). Both sectors contain extended areas of Hercynian granitic and metamorphic rocks and are covered by younger sediments. The southern sector, however, which comprises the entire study area, is characterized by the absence of ophiolitic units and Alpine metamorphism. This southern sector resembles much more the evolution pattern of the Sicilian Maghrebides chain, whereas the northern sector forms part of the Apenninic chain (op. cit., p. 223).

Chapter 8 gives a description of the rock suite of the research area in the southern sector.

The crustal thickness of the area (Schütte 1978) increases from about 30 km beneath the Apulian platform to about 45 km beneath the Sila and then strongly decreases and thins further to the West to 25 km. It reaches its lowest value of 16–18 km in the region of the Eolian islands. Figure 1.6 shows the contour map of the crust-mantle boundary in Southern Italy. The zone of maximum thickness forms an accentuated arc which is followed by the folded chain. Thus maximum crustal thickness and Apenninic chain do not coincide (Schütte 1978, p. 320).

The crustal thickening and its geometry gave rise to differing explanations: several subduction models, a thrust model and a "krikogenesis" model were developed. The subduction models interpret the Calabrian arc and the Tyrrhenian Sea as parts of a system arc-marginal sea (Ritsema 1972; Giese and Görler 1978; Gasparini et al. 1982; Rehault et al. 1986; Aifa et al. 1988). The subduction models refer to the distribution of earthquakes of the area: "The area of major continuity in the occurrence of intermediate and deep earthquakes closely corresponds to the area of maximum concavity, i.e. with the zone of maximum distortion of the arc. This zone may be interpreted as a strongly deformed remnant of a previously continuous Benioff zone extending during Oligocene time from the Northern Apennines to Gibraltar (Gasparini et al. 1982, p. 283).

Figure 1.7A shows the distribution of the epicenters (1910–1975) of the area and Fig. 1.7B the distribution of the hypocenters along a profile normal to the Calabrian arc crossing our research area.

Rehault et al. (1986) interpret the Tyrrhenian basin as a marginal sea developed through drifting of the Calabrian-Sicilian arc system towards the

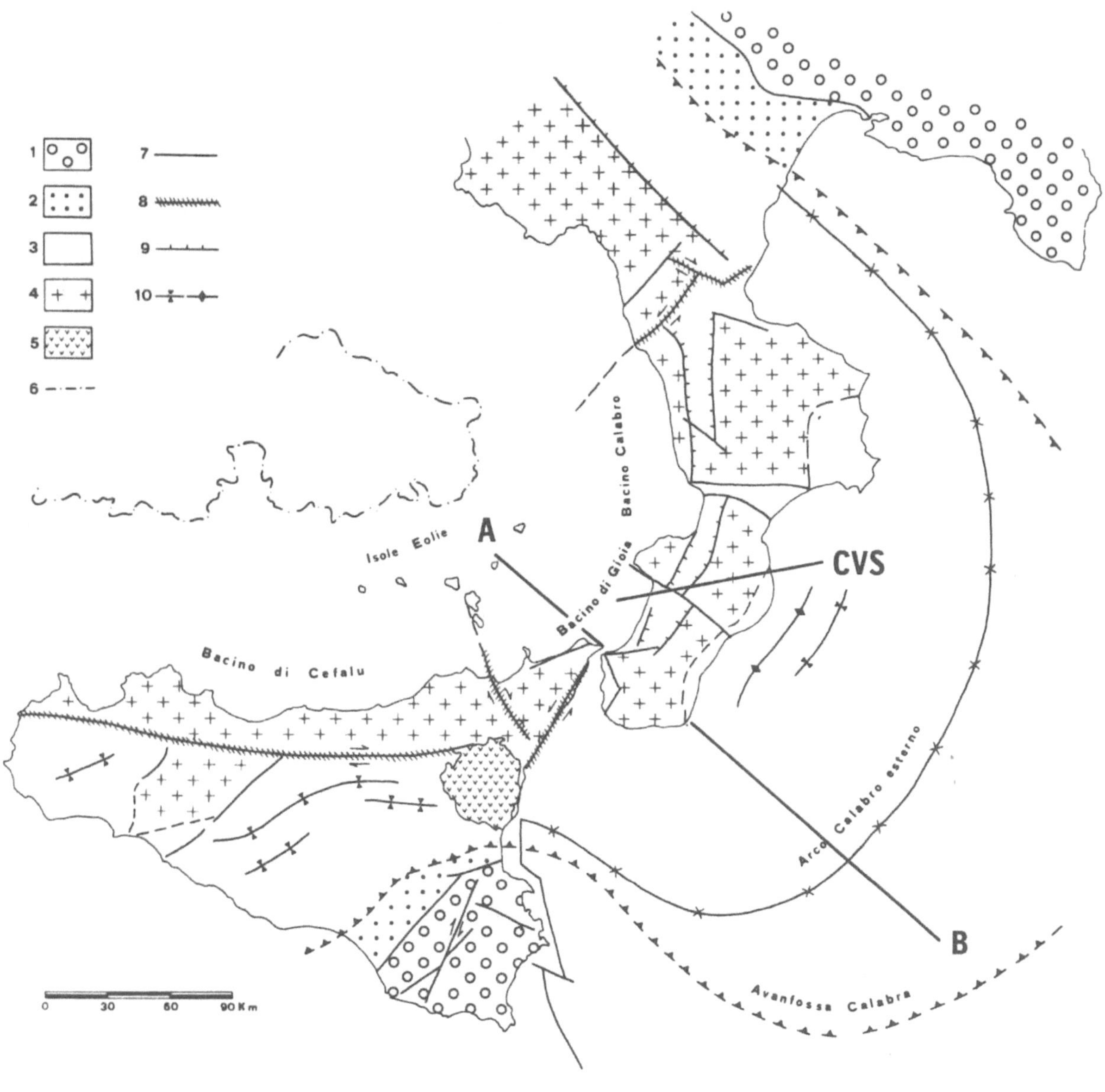

Fig. 1.5. The structure of the Calabrian arc. *1* Foreland (Apulia and Hyblea); *2* Plio-Pleistocene troughs developed along the border of the foreland (Bradano and Catania-Gela trough); *3* Plio-Pleistocene basins developed within the chain (S. Arcangelo, Mesima, Caltanisetta); *4* chain (Sila, Serre, Aspromonte, Peloritani); *5* Mount Etna; *6* limit of the Tyrrhenian deep sea plain; *7* faults; *8* surficial equivalents of the deep shear zone with transcurrent movements prevailing; *9* surficial equivalents of the deep shear zone with normal movements prevailing; *10* axes of synclines and anticlines. *AB* Position of the cross-section given in Fig. 1.10 (Ghisetti 1981, Fig. 12). *CVS* Capo Vaticano-Soverato Line (added)

east-southeast (Fig. 1.8). Aifa et al. 1988, p. 150) emphasize the simultaneous occurrence, during the Quaternary, of uplifts and rotational deformations (Fig. 1.9).

According to these subduction models, the Ionian slope of the Calabrian Massif has to be seen at least as a remnant of an active plate margin.

Ghisetti and Vezzani (1982, p. 153) explain the crustal thickening at the inner margin of the Calabrian arc as a deep shear zone which determines a crustal shortening of 40–60 km (Fig. 1.10). According to this "thrust" model the Ionian slope of the chain could also be regarded as an active margin.

The krikogenesis model (Wezel 1982), finally, postulates a post-orogenic ring-like system (kriko = ring, Greek) concentric round the Marsili backarc basin in the Tyrrhenian Sea. "The present shape of the arc has largely resulted from Plio-Quaternary vertical tectonic movements, superim-

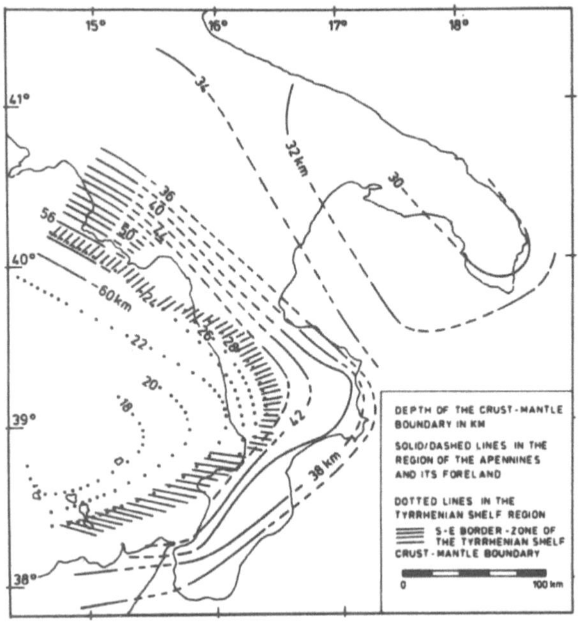

Fig. 1.6. Contour map of the crust/mantle boundary in southern Italy (Schütte 1978, Fig. 6). Maximum crustal thickness and Apenninic chain do not coincide, demonstrating an isostatic non-equilibrium

posed on compressive structures caused by previous phases of tectonic deformation" (op. cit., p. 345). In contrast to the foregoing models based on the presumed interaction and horizontal convergence of two lithospheric plates, the krikogenesis model relates the neotectonics and the volcanism of the Calabro-Sicilian region to intra-plate deformations. According to this model, the present-day structure of the mountain arc, the Serre and the Aspromonte, results from intene Plio-Quaternary vertical tectonic movements.

Inter- or intraplate deformation, subduction or deep-seated shear zone – the Ionian slope of the Calabrian Massif as part of the Calabro-Peloritan arc is a site of highest tectonic activity bordered by a deep marine basin, conditions which favour high erosion rates and a high flux of sediment from the source to the receiving basin. For the purposes of this study, it seems to be justified to consider the Ionian border, the external part of the central Calabrian arc, as an active plate margin. In terms of global dispersal systems and tectonics (Dickinson 1988) Calabria forms part of the suture belt and collision orogens of the "Eurotype" composite continent.

To go into more detail, two elements of striking importance for the sediment production still have to be presented, the neotectonics of the area, which creates relief, and the seismicity, which causes fragmentation and triggers land slides.

Fig. 1.7. A Epicenter map of deep and intermediate earthquakes which occurred in southern Italy in the period 1910–1979. **B** Distribution of intermediate and deep hypocenters along the line *A'A*. Only earthquakes with horizontal distance less than 50 km from the line are included (Gasparini et al. 1982, Figs. 1 and 2). Graph **B** is plotted conversely to obtain an appropriate E-W dip of the presumed Benioff plane

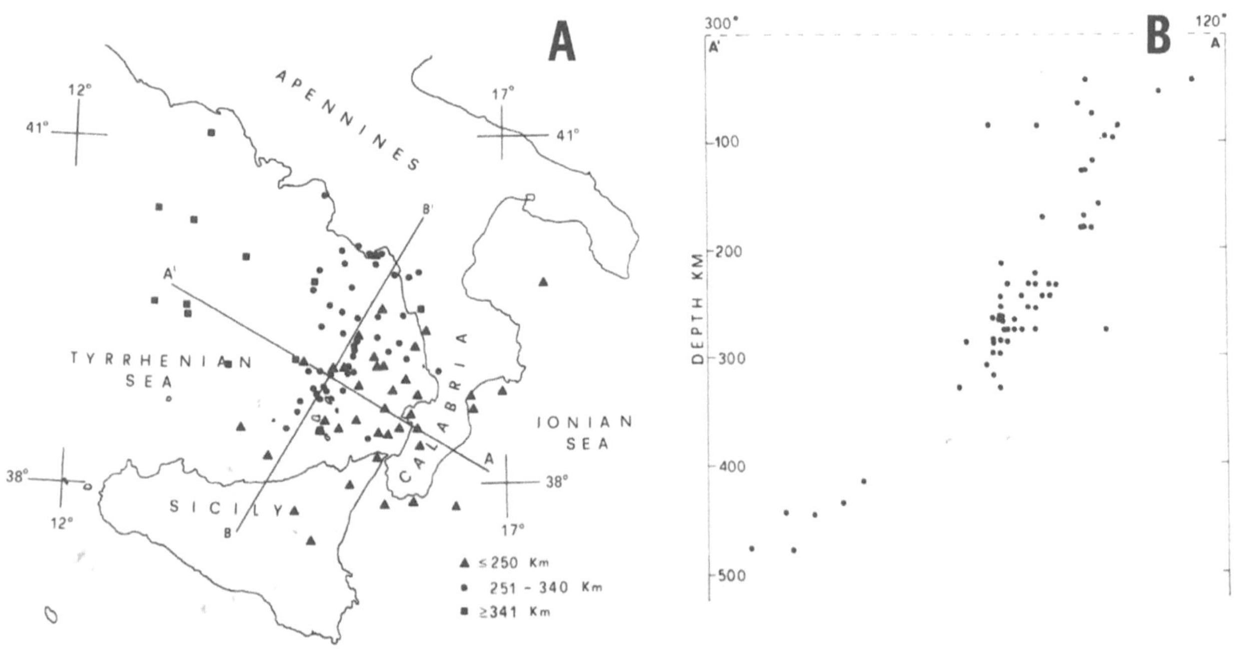

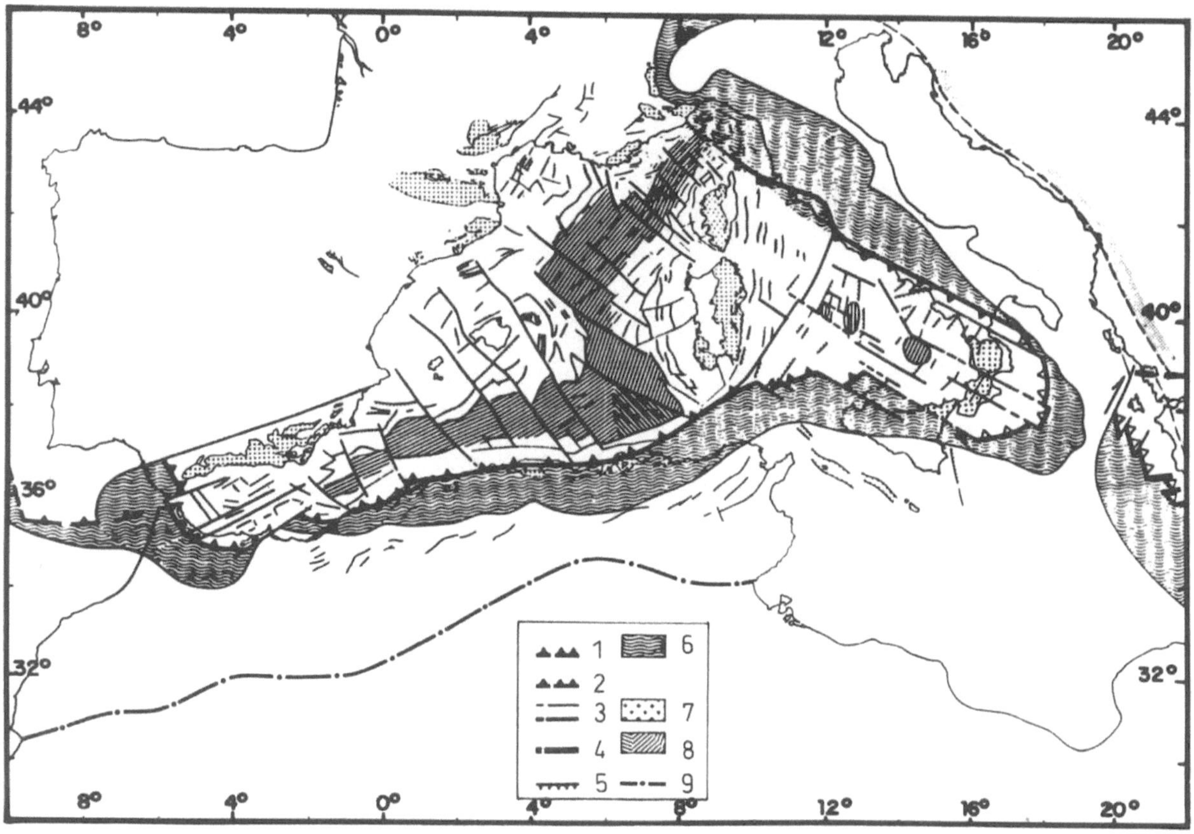

Fig. 1.8. Present situation of the Tyrrhenian basin and the Calabrian arc. *1* Suture zone and paleo-subduction; *2* active subduction zone or overthrust; *3* and *4* transform faults and strike-slip structures; *5* main distensive structures; *6* deformation zone in subduction and/or collisional area; *7* Pre-Mesozoic basement; *8* Neogene oceanic crust; *9* northern limit of the stable African craton (Rehault et al. 1986, Fig. 16)

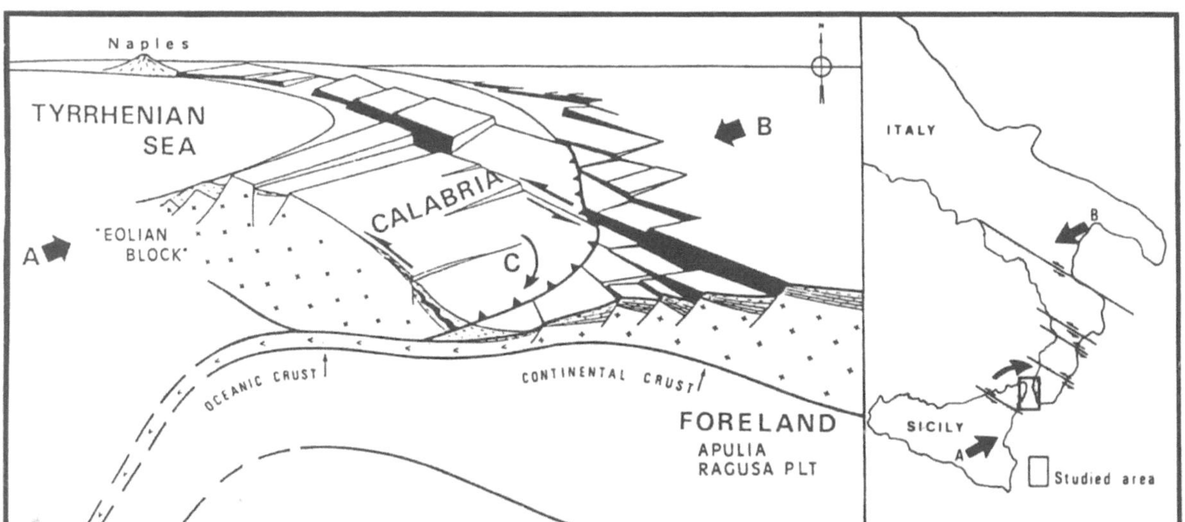

Fig. 1.9. Uplift and rotational movement in the Tyrrhenian arc. *A, B* Quaternary compression. *C* Clockwise rotation, *opposite parallel arrows* correspond to the horizontal component of the normal wrench fault (Aifa et al. 1988, Fig. 11)

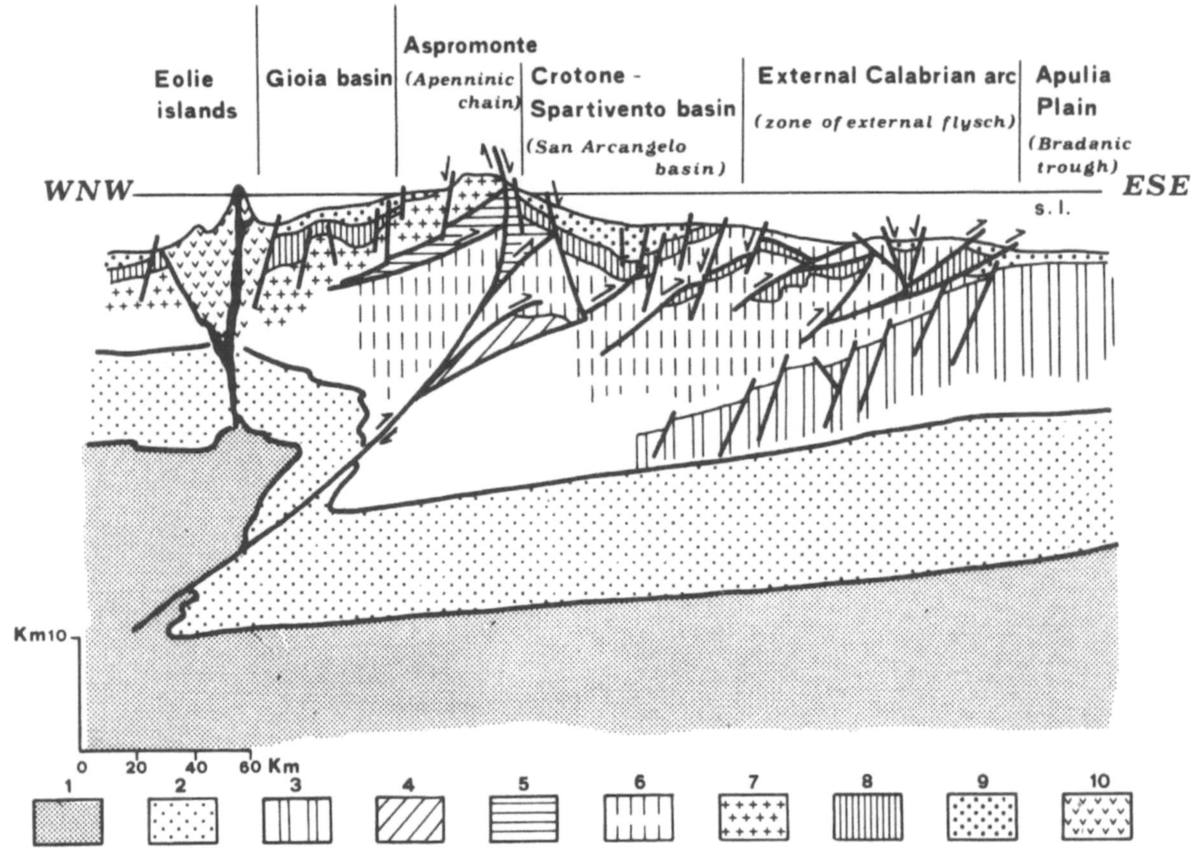

Fig. 1.10. Geological cross-section in southern Calabria along the line *AB* of Fig. 1.5, illustrating the connection between the structures at depth and at the surface. *1* Upper mantle; *2* lower crust; *3* Apulia plain sedimentary covers; *4* Imerese unit; *5* Panormide unit; *6* Sicilide unit; *7* Calabride unit; *8* Middle Miocene-Lower Pliocene terrigenous sequences; *9* Upper Pliocene-Pleistocene terrigenous sequences; *10* volcanics (Ghisetti and Vezzani 1982, Fig. 5)

1.3 Neotectonics

The neotectonics, the post-Miocene structural history of the area is described by Ciaranfi et al. (1983). Here the time span of about 5.2 Ma from the Pliocene until today is divided into four intervals characterized by different phases of uplift and subsidence. The areal distribution of uplift and depression and of the linear neotectonic elements is based on stratigraphic-structural and morpho-tectonic mapping, not taking into account eustatical sea level changes during Pliocene and Lower Pleistocene (op. cit., p. 10).

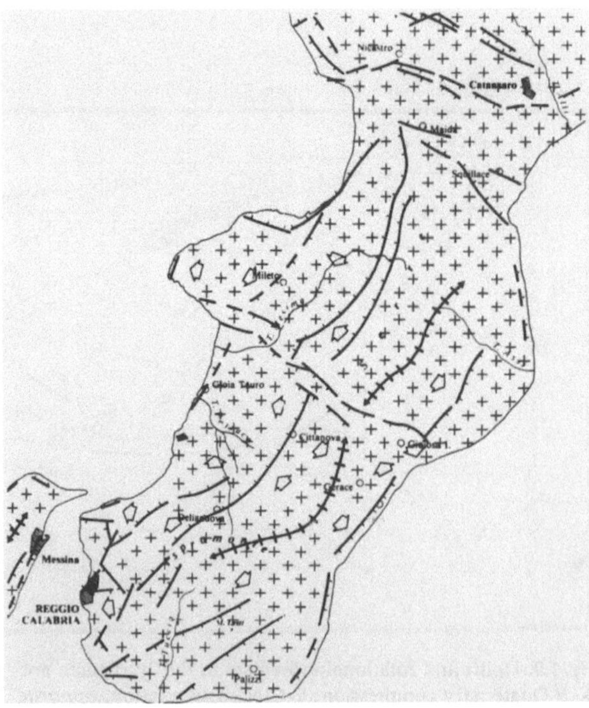

Fig. 1.11. Neotectonic map of Calabria, situation of the period from ca. 0.7 Ma until today, Intervallo IV–V (Ciaranfi et al. 1983, Foglio II). Note that the Ionian slope as a whole is tectonically less differentiated or more homogeneous than the Tyrrhenian one. ─┬─┬─ fault; + + + uplift; ┼┼┼┼ axis of uplift; ← direction towards less uplifted areas

Figure 1.11 sketches the situation during the last interval, from 0.7 Ma until today. The entire area is uplifted, with some faulting along the coastal areas. A central axis divides areas of different uplift rates.

The high plain in the center of the massif, between the Aspromonte and the Serre, is uplifted by about 1000 m in about 1 Ma or even less – this means a mean uplift rate of 1000 mm/ka. This uplift, combined with maximum river lengths of only about 20 km, creates an enormous relief and, consequently, strong erosion and sediment production.

The dating of this uplift is a key problem for the calculation of erosion rates and mass balances. This will be discussed in Chapter 6.

1.4 Seismic Activity

Figure 1.7 shows the epicenter map and the distribution of hypocenters of deep and intermediate earthquakes. This distribution gave rise to the assumption of a Benioff plane beneath the Calabrian arc. Shallow crustal earthquakes, which mainly cause catastrophic surficial events, are not plotted. These are restricted to the external parts of the folded chain itself.

According to Ghisetti and Vezzani (1982), the Ionian slope of the Calabrian Massif, our study area, is characterized by earthquakes of only lower intensities. The Crati and Mesima troughs at the Tyrrhenian slope (Fig. 1.5) are the site of epicenters with the highest magnitudes. The focal depth of these crustal earthquakes is 20–30 km. Ghisetti and Vezzani link the seismotectonic zoning of the Calabrian arc with the crustal pattern described in the foregoing section, especially with the deep-seated shear zone (Fig. 1.10). Because the position of this shear zone is internal with respect to the Ionian border of the Calabrian Massif, the historical seismicity is not elevated (op. cit., p. 162). The most important earthquakes are those of Ferruzano (see location map, Fig. 1.3) (Oct. 23, 1907), of Catanzaro about 50 km north of our area (April 13, 1973) and again Ferruzano (March 11, 1978; Bottari et al. 1981–82). However, the recurrence interval of only about 100 years of earthquakes with highest magnitudes of ≥ 8 in the area of the Crati and Mesima valleys (Fig. 1.5) and the strait of Messina also implies a strong seismic impact on the Ionian slope. The distance between the two areas measures between only 20 and 40 km. Thus it is not only the high relief but its combination with seismic fragmentation and the triggering of landslides which creates the high erosion rates of the Ionian slope of the Calabrian Massif, assisted by extreme precipitations (Chap. 4).

1.5 Conclusion

Most provenance studies refer to big rivers at passive margins; this implies relatively large drainage areas and a relatively mature sediment delivery. These rivers are commonly believed to be main contributors of sediment to the oceans, and geologists tend to extrapolate this into the past.

From this point of view, the small Calabrian peninsula, evidently an active plate margin delivering immature sediments, appears to be not very representative for provenance studies. However, even smaller source areas may account for high rates of sediment production, in both relative and absolute numbers. The sediment yield, the sediment load per square kilometer and year of a drainage basin are known to increase with decreasing basin size. This tells little about the total sediment discharge of a river, because a small source area with high sediment yield may be surpassed by a larger one with a lower sediment yield. There are, however, small but tectonically active areas which contribute tremendous amounts of sediment to the oceans. Taiwan, for example, which covers an area of only 0.46% of that of the conterminous United States, produces practically the same annual sediment load, nearly 300×10^6 t/a (Milliman and Meade 1983, p. 17).

This seems important to us, because a small source area, which nonetheless serves as a powerful source of sediment and which is responsible for the formation of thick and/or extended deposits, may be extinguished much more easily by erosion or later tectonic events than a larger one, giving rise to many speculations about its former state.

For these reasons, our provenance study may be helpful for source area reconstructions of orogenic sediments. The geographic extent of Calabria is very limited, its power in producing sediments, however, is impressive.

Chapter 2

Basins and Channel Network

The objective of our geomorphological analysis is to determine if there are definable and quantifiable geomorphological properties which are transmitted as distinguishable signals into the river-mouth sediments of the basins. In other words, we want to understand if and how the geological process of sediment formation may be influenced by geomorphology.

We deduced the classical geomorphological variables for each drainage area quantitatively and compared them with the data of their respective river-mouth sediments. In this Chapter 2, the drainage divides, drainage basins and channel networks are treated; Chapter 3 deals with the altitudes and the slopes.

In addition to this sediment-related interest on geomorphology, we wanted to find support for our concept calculating the erosional budget (Chap. 6). We believe that the emerging massif, 1 Ma ago, was roofed by a more or less homogeneous emergence plain, and we use the remnants of this emergence plain as reference level for the calculation of valley volumes. A random development of this geomorphological feature would help to make this concept acceptable.

Summary. We digitized the topographic map 1:25000 using a 250-m geomatrix and quantified nearly all geomorphological properties of the area which comprises 103 basins and interfluves. We subdivided three main physiographic provinces, Aspromonte, Zomaro and Serre Provinces; 19 larger rivers were investigated more closely.

An order and size classification of the drainage divides and the 103 basins and interfluves did not yield meaningful results, remaining purely descriptive. Most basins are distinctly elongated; basin shapes (Horton's formula) do not correlate with properties of the river-mouth sediments. The arrangement of the drainage basins and the basin shapes makes us assume that uplift started with a continuous Mid-Pleistocene emergence plain (Chap. 6).

The size of the drainage basins obviously has an effect upon shelf width; the larger the hinterland, the smaller the shelf. The rivers are highly braided multi-channel systems. Despite maximum basin lengths of only 20 km, four rivers of the 19 main basins reach the seventh order, the other 15 rivers are all of the sixth order. Drainage density ranges between 5.1 1/km and 10.36 1/km; bifurcation ratios vary between 3.8 and 5.8.

The architecture of the channel network of some rivers allows us at least to speculate upon possible former river-mouth positions which may have been situated at the former shelf break.

Differences of basin configuration and channel network do not correlate with distinguishable differences of the river-mouth sediments.

2.1 Digitization of Maps

We follow Evans (1972, p. 24), who discusses the differences between digitized contours or planimetric measurements and a digitized point grid, strongly favouring the point system. Irregardless of today's state of affairs, the relatively simple point or geomatrix method was the only one possible for us. The "geomatrix" is a matrix in which both rows and columns are in order according to their position in geographic space (op. cit. 1972, p. 24).

A 250-m point grid was laid over the entire research area. Each point thus represents the midpoint of a square of $250 \times 250\ \text{m} = 62500\ \text{m}^2 = 0.0625\ \text{km}^2 = 1/16\ \text{km}^2$. This corresponds to a 1-cm net on a 1:25000 map. Reducing the point distance to only 200 m would have been an increase from 16 to 25 points per square kilometer or an additional expenditure of more than 50% of work.

The total net comprises 290 squares in west-east direction (72.5 km) and 298 in north-south direction (74.5 km). The resulting 86420 squares were numbered linewise beginning lower left, the numbers being used as square identifiers (Appendix A, data management). This could also have been done by the geographical coordinates. We preferred to use an internal, independent coordinate system with the coordinates 0/0, by definition, for the lower left corner and ten units for one square edge.

Consequently, one unit corresponds to 25 m in nature and 1 mm on a 1:25000 map. The midpoint coordinates of the first square of the geomatrix are 5/5, and those of the last one 2895/2975. From the bulk of 86420 squares only 29296 squares fit the research area (Fig. 2.1) due to its elongated, NE-SW-striking shape. The angular pattern of the drainage divides of Fig. 2.1 reflects the solution of the square net applied. In addition, 7950 squares fit onto an off-shore area covering 500 km^2 of shelf and slope (Fig. 3.7).

For the evaluation procedure, the point net was laid over the maps and at each point the information needed was read off, listed and stored in the data bank. Thus each item of point information is enlarged to stand for an area of 1/16 km^2.

2.2 Drainage Divides and Drainage Basins

Data Base. Topographic map 1:25000 from the Italian Istituto Geografico Militare with 25 m contour lines. Sea chart No. 23, 1:100000, C° Milazzo-Roccella Ionica from the Italian Istituto Idrografico della Marina, 100 m and 200 m depth contour lines. Geomatrix, data bank.

Terminology. The drainage divide or watershed (Fig. 2.1) is the line of separation between two adjacent drainage basins. A drainage basin (drainage area or catchment of other authors) is the area enclosed by the drainage divides and the coastline. We outlined 103 drainage basins. However, we call the drainage basins simply basin and we delimit the term basin to the 19 larger rivers of SE-Calabria under closer investigation, that is, with a treatment of their fluvial sediments, and we designate all smaller drainage basins without such a treatment interfluves. These are mainly the real interbasin areas along the coast between the river mouths of the larger basins (Fig. 2.1 B, C). Most of these interfluves have no clear or only barely distinguishable networks or outflows to the sea. However, some of them, the larger ones, do have differentiated river networks, but these interfluves are geologically of so little importance that for our objectives they did not merit closer investigation.

The basins investigated more closely, the 19 rivers, were numbered separately from 01–19 and named individually. Thus the complete ID of such a basin is, for example, Allaro [16] (Fig. 1.3). Usually the basins 01–19 will be treated individually, whereas the characteristics of the remaining 84 interfluves will be given as a whole. Appendix B summarizes all graphic plots of the basins 01–19 and the "average" basin 20.

Method. The drainage divides were delineated by hand on the topographic map 1:25000, their lengths were measured by means of a curve wheel. The basins and interfluves outlined by the drainage divides were numbered continuously from south to north, from 01 to 103. Digitizing the maps, the basin or interfluve number of each point of the geomatrix was registered.

The geomorphological analysis differentiates between the linear aspects of channel systems (stream orders, stream length) and the areal aspects of drainage basins, for example the arrangement of areal elements and the size frequency distributions of basin areas (Strahler 1964).

One procedure to obtain information about the architecture of a basin is the statistical treatment of the elements of one or more subbasins. Usually these elements display geometric sequences if their logarithms are plotted against their order. However,

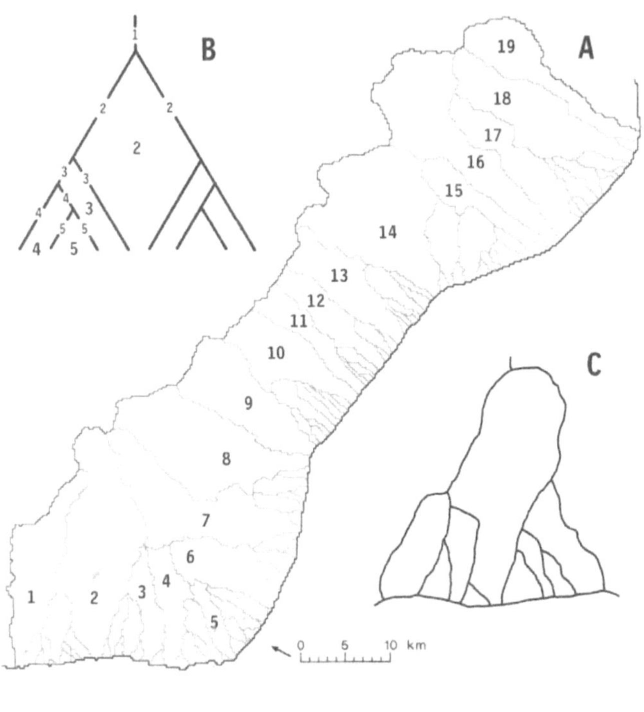

Fig. 2.1. A Drainage divides bordering the 103 basins and interfluves; *heavy lines* indicate the Apenninic main divide and the coastline. *Numbers* refer to the 19 basins investigated, see Table 2.1 and Appendix B 01–19 for more details. *Steps of the plotted lines* indicate the resolution of 250-m geomatrix. **B** Classification of the orders of divides and basins, schematic; *small numbers* refer to divides, *large numbers* to basins bordered by the divides. **C** Example of the symmetric configuration of interfluves and subinterfluves, blowup of the interfluve complex between Melito and Amendolea (basins 1 and 2). Note the similarity to the schematic pattern of partition (**B**) and the resulting basin shapes, which are either subparallel, opening or closing towards the sea

in this chapter we shall not treat these basin-internal analyses, but shall compare the 103 basins and interfluves of the Ionian slope of the Calabrian Massif, which, side by side, individually drain into the Ionian Sea and which, as a consequence, are expected to be less dependent on each other than the connected subbasins of a larger stream.

To systematize the basins and interfluves from this point of view, we first consider the linear aspects of the drainage divides and then the areal aspects of the individual drainage basins. The question is if it is possible even in an assemblage of drainage divides and drainage basins to find a system or hierarchy analogous to the basin-internal ones. If yes, it should be determined whether, for example, drainage areas of different order somehow produce different river-mouth sediments.

The basin pattern, however, is of interest not only because of its possible role for sedimentological provenance studies. The emerging massif, about 1 Ma ago, is believed to have been roofed by a kind of homogeneous emergence plain. Today the remnants of this surface serve as reference plain for the calculation of erosion rates (Chap. 6). As a prerequisite for this procedure, both the homogeneity of this possible plain and a more or less steady uplift and tilting have to be proved. A randomly developed drainage pattern not too much disturbed by neotectonics and geological structures would point in this direction.

2.2.1 The Classification of Drainage Divides

2.2.1.1 Order Classification

At first glance a classification of drainage divides and basins into orders and classes appears to be academic in the bad sense of the word and thus superfluous. However, if we want to understand or at least to describe the Calabrian Massif as an actuogeological complex, we have to treat its basins, which are mostly independent of each other, by means of a quantitative procedure which should correspond to classical basin analysis, applied to the more or less connected elements of interbasin relationships. Consequently, if we want to characterize the partition of the Ionian slope of the massif by its drainage divides, why not apply the order system of the classical network analysis?

Hence we call the central Apenninic drainage divide (Fig. 2.1 A) 0 order (zero order) and each divide branching off this zero order is a first order divide. If a further divide branches from this first order divide, both divides are second order divides from this bifurcation point and so on (Fig. 2.1 B). As in network analysis, counting of the drainage divide orders starts with the internal elements distant from the coast. However, whereas the single channel elements of a basin combine towards the main trunk and become longer, forming higher orders, the higher orders of divides ramify and become shorter towards the coast.

Segment Number. We counted 184 elements of drainage divides, not taking into account the elements of the central Apenninic divide (zero order), from which all first order members branch. The frequency of the drainage divide elements, plotted against their order number, displays no linear regression as is frequently the case when plotting the stream segments, but forms a nearly perfect normal distribution (Fig. 2.2 A), the fourth order being the most frequent, the modal class.

Segment Length. The mean length of the drainage divide segments, plotted against their order, decreases with increasing order (Fig. 2.2 B); the correlation coefficient is $r = -0.72^*$[1]. The mean length of the first order divides is 15.3 km, that of the ninth order is 2.3 km.

2.2.1.2 Length Classification

We classified the 184 divide segments using a geometric scale with the class limits 0.5, 1, 2, 4, 8, 16 and 32 km.

Segment Number. The segment numbers of the segment classes are more or less regularly distributed (Fig. 2.2 C), with the segment class 2–4 km forming the mode.

Segment Length. The segment lengths of the segment classes are not regularly distributed, the mode lies in the class 4–8 km (Fig. 2.2 D).

The classification of the drainage divides makes it possible to classify even the basins themselves and to look for possible rules or laws of their size distribution and spatial arrangement.

---

[1] Correlation coefficients marked by an asterix indicate significant correlations of a 95% confidence level.

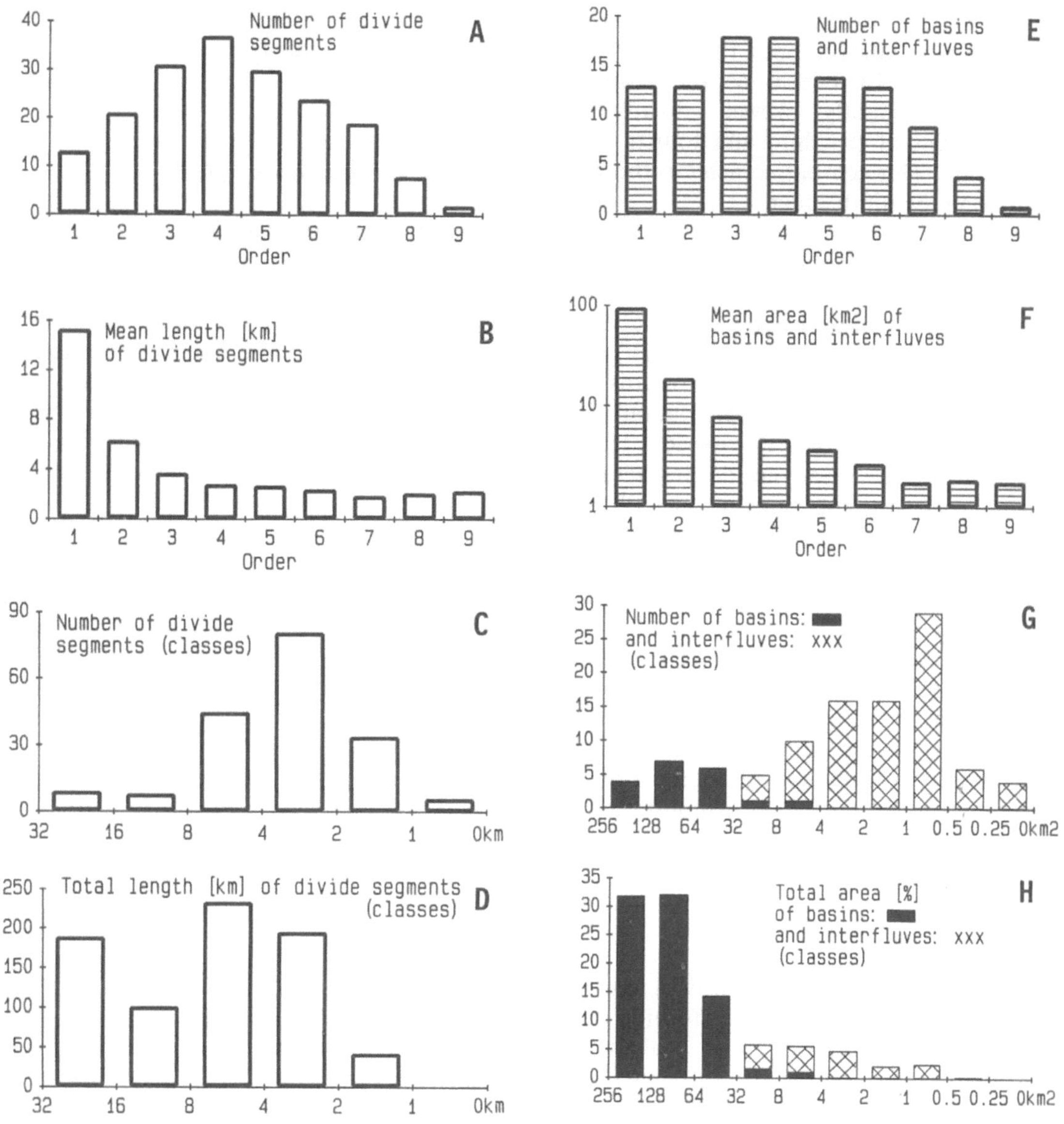

Fig. 2.2 A–H. Histograms of drainage divide segments, basins and interfluves. Within the range of the single geomorphic variables no direct influence on the formation of river-mouth sediments is detectable. Drainage divide segments. **A** Number of the drainage divide segments of the orders 1–9. **B** Mean length [km] of the drainage divide segments of the orders 1–9. **C** Number of drainage divide segments of the single segment length classes. **D** Total length [km] of drainage divide segments of the single segment length classes. Basins and interfluves. **E** Number of the basins and interfluves of the orders 1–9. **F** Mean area [km^2] of the basins and interfluves of the orders 1–9 (log-scale). **G** Number of the basins and interfluves of the single area classes. **H** Total area of the basins and interfluves of the single area classes. ***Black areas*** of the plots **G** and **H** indicate the portion of the 19 basins under closer investigation

2.2.2 The Classification of Drainage Basins

2.2.2.1 Order Classification

The drainage divides include 103 basins plus interfluves which are classified analogously to the drainage divides. The order of a basin is determined by the lowest order bifurcation of the drainage divides in its hinterland (Fig. 2.1 B), whereby the central Apenninic main divide remains disregarded, so that the large main basins are included in the first order. The highest class achieved in this area is the ninth order, necessarily with a very small area, by our definition.

Basin Number. The distribution of the 103 basins and interfluves by number onto the nine orders yields no regular distribution as the drainage divides, but a slightly skewed distribution (Fig. 2.2E). Basins of the third and fourth order make the mode. Thirteen basins achieve the first order, only one the ninth order.

Basin Size. The mean size of the basins of the single orders, plotted against the order number, displays, quite similarly to the mean lengths of the divides, a regular, but non-linear regression (Fig. 2.2F). The mean basin size decreases quickly towards the higher orders, passing into a relatively flat tail. The correlation coefficient is only r = −0.66*. This is not surprising, because we are not dealing with the size distribution of the subbasins of one coherent superbasin, a case in which we may expect stronger internal relationships, but we are confronted with 103 individual basins and interfluves, each having its own access to the sea.

Nevertheless, these 103 basins also have to share in the Ionian slope of the Calabrian Massif and, consequently, have to have a kind of "regard" for each other. The classification of the basins into pure size classes, without regard to their order, makes this even more evident, as demonstrated in the next section.

2.2.2.2 Size Classification

The 103 basins plus interfluves form a unit of 1832 km^2. The largest basin, the Torbido [14] (Fig. 2.1 A), covers 160.2 km^2, the smallest interfluve only 0.31 km^2, so the size difference corresponds to a factor of 517. We classified this with a geometric scale, with class limits by factor 2, from 0.125 to 256 km^2.

Basin Number. Figure 2.2G shows the described geometric class sequence of the basins by number percentage. This class distribution is irregular, nearly unimodal and shows a negative skewness. There is a distinct mode in the 0.5−1 km^2 class. About two thirds of all basins, in this case mostly interfluves, fall in the 0.5−4 km^2 classes. The 19 main basins are marked black.

Basin Size. The distribution of the area percentages over the single classes is given in Fig. 2.2H. Here the same tendency results as is shown by the distribution of the mean order sizes, that is, the majority of the total area is taken by the larger basins. About two thirds of the entire area are covered by basins < 64 km^2. Again, the 19 main basins are marked black.

2.2.3 The Physiographic Provinces

It is no problem to divide the 103 basins and interfluves into three physiographic provinces (Fig. 1.3, Table 2.1). The gneiss dome of the Aspromonte Province in the south with the Montalto, 1956 m; the granite dome of the Serre Province in the north with the Mte. Pecoraro, 1423 m and, in between, the Zomaro Province with the high plain at about 1000 m. The multivariate comparison of the 19 basins (Chap. 17) made it

Table 2.1. The physiographic provinces, their basins and interfluves. The 19 basins which yield the data basis for the provenance study cover 81% of the entire area. Italics refers to the abbreviations used on the graphs and plots. Underlined basins make part of the large Aspromonte Province resulting from the multivariate comparison (Chap. 17)

Physiographic province	Basin		km^2
Aspromonte	*Me*lito	[01]	80.31
	*Am*endolea	[02]	151.75
	*Pa*squale	[03]	28.00
	*Pa*lizzi	[04]	36.44
	*Ar*anghia	[05]	16.25
	*Br*uzzano	[06]	54.13
	*La*verde	[07]	117.25
	*Bu*onamico	[08]	138.81
Sum investigated basins			622.94 km^2 = 80%
Sum interfluves			157.38 km^2 = 20%
Sum Aspromonte Province			780.32 km^2 = 100%
Zomaro (High Plain)	*Ca*reri	[09]	92.94
	*Co*ndoianni	[10]	67.13
	*Po*rtigliola	[11]	35.44
	*Ge*race	[12]	40.56
	*No*vito	[13]	60.38
Sum investigated basins			296.45 km^2 = 78%
Sum interfluves			82.84 km^2 = 22%
Sum Zomaro Province			379.29 km^2 = 100%
Serre	*To*rbido	[14]	160.19
	*A*mu*sa	[15]	37.63
	*Al*laro	[16]	134.56
	*Pr*ecariti	[17]	65.88
	*St*ilaro	[18]	98.69
	*As*si	[19]	67.06
Sum investigated basins			563.01 km^2 = 84%
Sum interfluves			109.04 km^2 = 16%
Sum Serre Province			672.05 km^2 = 100%
Total			
Sum all investigated basins			1482.40 km^2 = 81%
Sum all interfluves			349.26 km^2 = 19%
Sum total			1831.66 km^2 = 100%

necessary to subdivide the Aspromonte Province into a large one whose basins reach the main divide and a small one whose basins do not.

Some trouble is caused by the assignment of the Torbido [14], which seems to play a special role, not only from the purely geological point of view.

2.2.4 The Selection of the "19 Rivers"

It was not possible and even not necessary to investigate the entire area of 103 basins and interfluves to the same degree. Starting the project in 1969, we decided to concentrate on 18 rivers and especially their river-mouth sediments; later one more river, overlooked during the first campaign, was included, so that 19 rivers were investigated in some detail.

This means that there are complete analyses of the river-mouth sediments which make direct comparison possible between the producing source area and the produced river-mouth sediment. In addition, complete morphometric analyses of all 19 basins were carried out. The digitization of the regional data set, on the other hand, concerning geology, geomorphology, precipitation and so on, always covers the entire area, the 19 basins and also the 84 interfluves.

It may be of interest to follow briefly the criteria of our river selection. The basis of our decision at that time was in fact only common sense: we tried to find the big rivers reaching up to the central divide (12 rivers) or at least far into the crystalline hinterland (7 rivers). Today we know that three of the basins not chosen are larger than the smallest river that we selected.

Moreover, the big oeuvre of the Carta Geologica Della Calabria 1:25000 had not yet been published, we did not even know about it – so it was obviously only a certain impression of the river mouths in the vicinity of highway and railway bridges that made one river and its drainage basin more attractive to us than another. We were really amazed – and recount it for this reason – that our decision on these 19 rivers was confirmed independently by a cluster analysis. We reduced the entire data set of the digitized Calabrian maps onto 69 variables, the frequency distributions of altitude, gradient and erosion rates, the lithology and the precipitation, and subjected the 103 drainage basins and interfluves to the cluster analysis (Chap. 17.1): only two main groups were formed, "our" 19 rivers and the remaining 84 interfluves.

2.2.5 Size, Arrangement and Shape of the Basins and Interfluves

The Size. The largest basin of the "19" is Torbido [14], 160.2 km^2, the smallest one is Aranghia [05], 16.3 km^2. Thus, the 19 basins comprise a full order of magnitude. This is important, because it enables us to answer the question whether the mere size of a basin, under otherwise similar conditions, has an influence upon the properties of the river-mouth sediments; obviously there is no influence.

Figure 2.3 A and B show the 19 basins following their geographic arrangement from south to north with their absolute areas [km^2] and area percentages [% km^2]. In Fig. 2.3 C the same 19 basins are ranked by size. Surprisingly, there is no preferred basin size; the sizes decrease so regularly that their correlation against the rank number is nearly perfect, $r = -0.98$*.

The Spatial Arrangement. The main drainage divide (zero order), the central Apeneninic divide, strikes NE-SW and divides the Calabrian Massif into a Tyrrhenian half or slope in the northwest and an Ionian half or slope in the southeast (Fig. 1.2). Thus the shape of the Ionian slope corresponds roughly to an elongated rectangle of approximately 20×100 km (Fig. 2.1 A). This rectangle is drained from NW to SE by a range of basins lying side by side and more or less normal to its long axis.

Thus mean river lengths are fixed at about 20 km, the coast is about 100 km long. Only the almost circular gneiss dome of the Aspromonte develops a radial drainage pattern (Fig. 1.3 B); the other provinces, even the Serre, are drained by subparallel basins. As shown in Chapter 3, this arrangement surprisingly does not influence the distribution of slope aspects, the slope directions.

Only two drainage divides succeed in passing straight through from the central Apenninic main divide to the coast without any branching or ramification, the one between Buonamico [08] and Careri [09] and that between Condoianni [10] and Portigliola [11] (Figs. 1.3 B and 2.1 A). Moreover, most drainage divides of the Zomaro Province are surprisingly straight despite the broad valley which follows the strike of the siltstone or olistostrome subparallel to the coast (Fig. 5.8), sandwiched by more resistant sandstone formations. The resistance of this siltstone body is very low. Thus a longitudinal or strike valley is developing but not yet completed. The valley is crossed by low but straight valley floor divides, a kind of "consequent divides" still bearing witness to the original slope

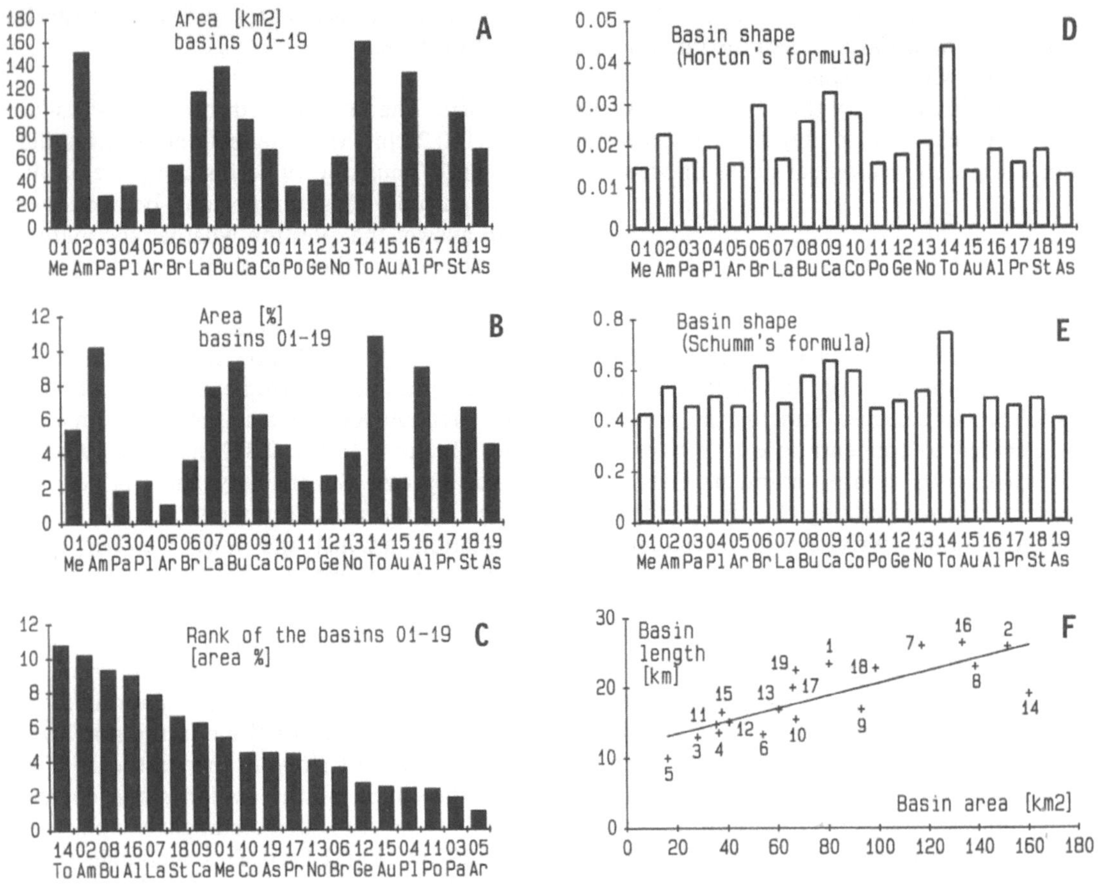

Fig. 2.3 A–F. Sizes and shapes of the 19 basins, grouped following the geographical arrangement from south (*left*) to north (*right*). Within this range of sizes and shapes no direct influence on the formation of river-mouth sediments is detectable. **A** Area [km^2]. **B** Area percent, without interfluves. **C** Size rank. **D** Basin shape according to Horton's formula. **E** Basin shape according to Schumm's formula. **F** Basin size versus basin length, r = 0.79*, *numbers* refer to the single basins. *01 Me*lito; *02 Am*endolea; *03 Pa*squale; *04 Pa*lizzi; *05 Ar*anghia; *06 Br*uzzano; *07 La*verde; *08 Bu*onamico; *09 Ca*reri; *10 Co*ndoianni; *11 Po*rtigliola; *12 Ge*race; *13 No*vito; *14 To*rbido; *15 A*musa; *16 Al*laro; *17 Pr*ecariti; *18 St*ilaro; *19 As*si

of the emerging massif (Figs. 2.4 and 2.5). An equivalent example is given by the Mte. Schiavo in the Novito [13] basin, an isolated mountain cut by a classical superimposed valley (Fig. 3.6).

The drainage divides of the smaller basins or interfluves are grouped following a very simple pattern, a bilateral symmetric triangle (Fig. 2.1 B). This feature may be created artificially by the regular branching of the sides of an isosceles triangle.

The Shapes of the Basins and Interfluves. As an inevitable consequence, three basic shapes of interfluves result from the triangular pattern of nearshore drainage divides, a rhomboidal one, whose drainage divides narrow towards the sea, a triangular one, whose drainage divides open towards the sea, and a third type bordered by subparallel drainage divides (Fig. 2.1 B). The opening and the subparallel type are the proper interfluves or interbasin areas with no distinct drainage network. Of course, this classification could be followed in more detail along the coast; however, we stopped at the level of Fig. 2.1, because the scale of 1:25000 provided no more information and further classification.

The Shape of the 19 Basins. The 19 larger rivers may be grouped into five basic basin shapes, as shown by Fig. 2.6.

Type A. Rectangular, subparallel basins. The basins are practically slices of the rectangular cake of the Ionian slope, the thalweg following the long axis. Aranghia [05] and Portigliola [11] are good examples (Fig. 2.6 A), but most basins belong to this type.

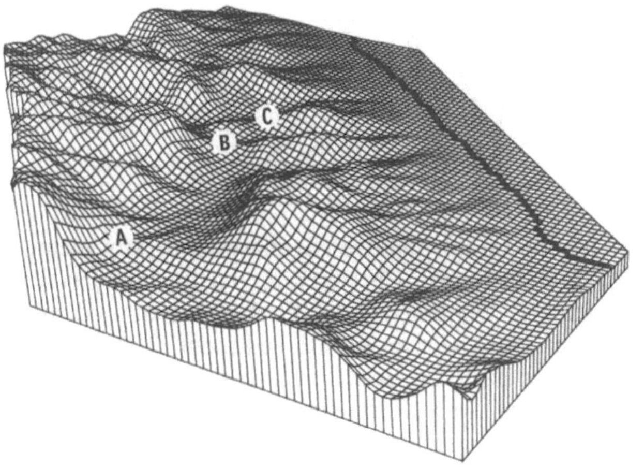

Fig. 2.4. View towards NE into the strike valley of the siltstone. The straight valley floor divides *A*, *B* and *C* between Careri [09], Condoianni [10] and Portigliola [11] are a kind of "consequent divide". The *heavy line* indicates the coast line. Compare the more or less corresponding photography, Fig. 2.5

Type B. Rhomboidal basins. These basins start and end relatively acute-angled. They may derive from type A by an enlargement of the central area normal to the main axis. The Amendolea [02] is a typical representative (Fig. 2.6B).

Type C. Pear-shaped basins. This is the classical basin type characterized by a broad hinterland and a relatively narrow river-mouth area (Horton 1941).

Fig. 2.5. View from the Zomaro plain above Platì towards ENE into the strike valley of the siltstone; valley floor divide between Careri [09] in the foreground and Condoianni [10]. The village in the *center* is Cirella, *left in the rear* lies Ciminà. Compare the more or less corresponding topographic computer plot, Fig. 2.4

Representative of this type is the Torbido [14] (Fig. 2.6C). However, for the elongated Calabrian peninsula this basin type is not typical. It simply stems from its position at the junction between the rotund dome of the Serre and the elongated Zomaro Province. A corresponding pear-shaped basin emerges if we consider the Buonamico [08] and the Careri [09] between Aspromonte and Zomaro Provinces as only one system. These rivers join within the head of the submarine canyon.

Type D. Bottle-shaped basins. This type may be seen as a modification of type C, the pear-shaped one, because the narrow river-mouth area, like a bottle neck, runs far into the hinterland. The best example is the Allaro [16] (Fig. 2.6D).

Type E. Angular basins. These basins are hard to assign to the basin types discussed above, because the basin axis is kinked halfway to the sea, 135° in the case of the Laverde [07], 165° in the case of the Buonamico [08] (Fig. 2.6E).

Based on such a qualitative description and classification of types of basin shape in Calabria, it is impossible to ascertain the possible influence of basin shape upon river-mouth sediment. The reason for the assumption of possible relationships is the fact that elongated basins with high bifurcation ratios (Chap. 2.3.1) should yield low but extended peak flows, whereas rotund basins with low bifurcation ratios produce sharp peak flows (Strahler 1964, p. 4–45; Gregory and Walling 1973, p. 269).

At least the tails of the grain-size distributions of river-mouth sediments, boulder and gravels on one side and silt and clay on the other, should be sen-

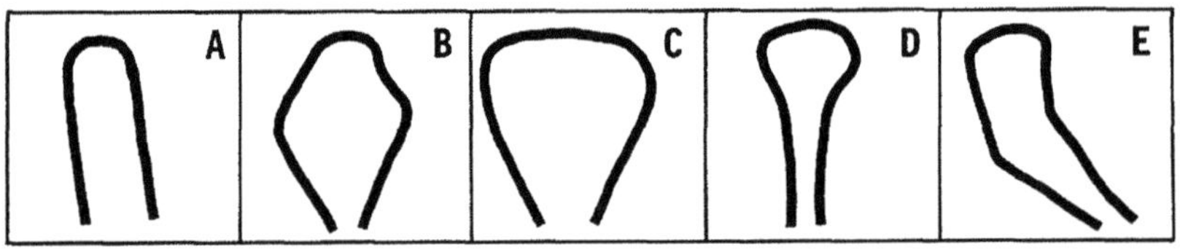

Fig. 2.6 A–E. Basic shapes of Calabrian basins (drainage areas). **A** Rectangular, subparallel basins. **B** Rhomboidal basins. **C** Pear-shaped basins. **D** Bottle-shaped basins. **E** Angular basins. Elongated basins of type **A, D** and **E** are by far the most common in the study area

sitive to such a hydrodynamic differentiation, the rotund basin type possibly favouring coarser river-mouth sediments and the elongated basin type finer ones.

Thus, in order to successfully correlate basin shape and river-mouth sediment, the basin shape was also quantified. The simplest procedure is to relate length and area of the basins (Hack 1957; Mueller 1973; Mesa and Gupta 1987).

The plot basin area versus basin length (Fig. 2.3 F) shows that in Calabria elongated basins from the subparallel A-type prevail due to a good correlation ($r = 0.79^*$) between basin area and basin length. Consequently, more rotund basins fall relatively far from the regression line, as demonstrated by the Torbido [14].

In addition, we used the form factor R_f from Horton (1932, p. 351) and the elongation ratio R_e from Schumm (1956, p. 612).

$$R_f = \frac{A_u}{L_b^2}$$

R_f is the dimensionless ratio of basin area A_u and the square of the basin length L_b.

R_e is the ratio of the diameter of a circle of the same area as the basin to the maximum basin length.

Figure 2.3 D and E show the ratios of the 19 basins according to Horton's and Schumm's formulae. The results are identical, the solution of Horton's method is somewhat better due to the squaring of L_b. So it is hardly necessary to discuss the preference of one of the methods.

The R_f-values run from the explicitly pear-shaped Torbido [14], $R_f = 0.44$ to the extremely elongated Assi [19], $R_f = 0.13$. However, it is impossible to assign certain types of basin shape to the three physiographic provinces Aspromonte, Zomaro and Serre. Only the Serre rivers, with ratios between $R_f = 0.13$ and 0.19, are relatively homogeneous.

There are no positive correlations or any meaningful relationships between the basin shape, as far as this is expressed by the ratios cited, and any property of the river-mouth sediment. The same is true for both the relief of the single basins formulated as the altitude standard deviation and the dissection calculated as altitude skewness (Chap. 3.1.1.2), variables which fail to show any dependence on basin shape.

2.2.6 Basin Size and Shelf Width

The morphological features of Calabria extend beyond the coast line. As will be pointed out, there are no specific properties of the river-mouth sediments which can be linked with specific properties of the source area morphology. However, the sediment production of the source areas obviously shapes the shelf, predominantly by cutting or deepening submarine canyons. In this respect the mere size of the source basins obviously models the first marine successor reservoirs beyond the alluvial plains of the river-mouth sediments.

Figure 2.7 A shows that the shelf width of our area under investigation varies between some 100 meters and some kilometers. In the Amendolea area [02] the distance of the 100-m and 200-m depth contour off the coast is only 150 m and 500 m respectively, in the Torbido area [14] the distance is 400 m (800 m). Maximum values off the Aspromonte coast are 4 km (5.5 km) and off the Serre the shelf break occurs at 5 km (6 km). It becomes evident that the shelf narrows off the river mouths of larger basins and that it widens in between (Rumohr and Ibbeken 1975).

The existence of large submarine canyons makes us suppose that a former more or less homogeneous shelf could have been eroded off the river mouths. Theoretically, the opposite is conceivable as well, a widening of the shelf off large rivers due to delta formation. The following section attempts to settle this question.

Data Base. Size distribution of the 103 basins and interfluves, distance of the 100-m depth contour line from the coast according to the nautical map

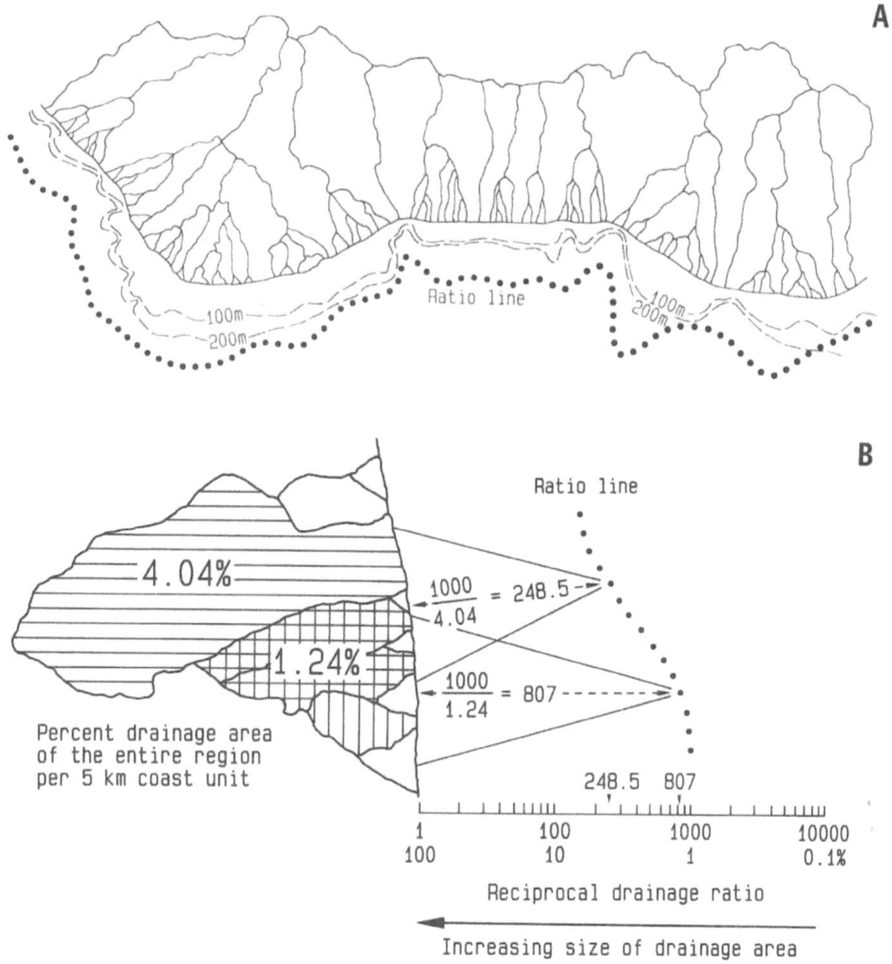

Fig. 2.7 A, B. Size of drainage basins and shelf configuration. **A** The 100-m and 200-m depth contour lines come close to the coast where large rivers flow into the sea and vice versa, the shelf widens off smaller source areas. *Hatched line* the line of the reciprocal drainage ratio is a calculated data line. Its distance to the coast is conversely proportional to the relative size of the respective hinterland. This ratio line mostly fits the shelf break, the correlation coefficient is r = 0.66*. **B** Construction of the reciprocal drainage ratio line. The reciprocal distances are calculated per 5-km coast unit, the ratio line is smoothed by a sliding mean of 2-km steps

No. 23, 1:100000, C° Milazzo-Rocella Ionica, Istituto Idrografico della Marina Italiana.

Terminology. Drainage ratio, D_r

$$D_r = \log \frac{S_{da}\ [\%]}{C_{5\,km}} .$$

The drainage ratio D_r is the logarithm of the area percentage of the hinterland S_{da} draining at a 5-km section $C_{5\,km}$ along the coast.

Reciprocal drainage ratio, R_{dr}

$$R_{dr} = \log \frac{1000}{S_{da}\ [\%]/C_{5\,km}} .$$

Method. We developed a data line, which graphically represents the size of the respective hinterland per unit coastline and which makes it possible to compare this with the actual shelf width. The area percentage of the basins per unit 5 km coast line was calculated. These values were plotted, on an appropriate scale, normal to the coast; to span the broad range a logarithmic scale was necessary. The connection of all these points is the drainage ratio line. Based only on 5-km steps, this line is very irregular, so it was smoothed using a sliding mean based on 2-km steps. The resulting drainage ratio line runs exactly opposite to the actual shelf break because it leaps seawards off the larger feeder areas and joins the coast in the areas between the larger river mouths. Plotting the reciprocal drainage ratio R_{dr} (Fig. 2.7 B), under the same procedure, yields the reciprocal ratio line, which fits the configuration of the present-day shelf line (Fig. 2.7A).

The graphical fit between the reciprocal drainage ratio line and the shelf break is supported by the correlation between the numerical values; the coefficient is r = 0.66*.

We are convinced that this agreement is causal and not an artefact. The river-mouth areas of all 19 rivers, but even those of the remaining larger basins or interfluves, are underlain by paleovalleys sharply cut into the basement, sometimes 20–40 m deep, very often 60–80 m deep and in the case of the Torbido [14] even 150 m deep (Chap. 15.2).

We cannot discuss the question as to when these paleovalleys currently filled with sediment were formed, whether during low stages of the sea level during the Messinian or during the Pleistocene. We only want to emphasize, in the context of this chapter, that the shelf widths of our Calabrian area are obviously controlled by erosion and that there is good fit between the degree of this erosion and the size of the present-day hinterland. There is one remarkable exception: the Laverde [07], one of the most important and most powerful rivers, possesses a 70-m-deep paleovalley, but today there is no trace of a submarine canyon, and, on the contrary, this river enters one of the broadest shelf areas of the region, and to crown this exceptional position it develops a delta, the only one of the area (Chap. 16.4.1, Fig. 16.8).

Fig. 2.8. The middle and lower course of the Amendolea [02] between Roccaforte del Greco and the coast, view towards S. This river section is 13 km long, the altitude of the river bed in the foreground is 370 m. Note the steep slopes and the blanket of debris at the *left* "ready to go" (cf. Figs. 15.3 and 15.4)

2.3 Channel Network

Classification of the Channel Systems. The short and youthful Calabrian rivers, the so-called Fiumara, had no time or space to develop through the classic stages of a river course. Near the heads of the valleys, frequently non-alluvial, straight channels occur. If, however, the deep and steep-walled valleys of the upper courses are left at the fall line, sometimes even earlier, braided rivers develop, highly braided multi-channel systems according to Rust 1978 (Figs. 2.8 and 2.9).

Data Base. Topographic map 1:10000 of the Italian Istituto Geografico Militare.

Terminology. We follow Strahler 1964 (pp. 4–44):

Order. The smallest fingertip channels of a network are designated order 1. Where two first order channels join, a channel segment of order 2 is formed; where two of order 2 join, a segment of order 3 is formed; and so forth (Fig. 2.10A).

Fig. 2.9. Channelized lower course of the Melito [01] towards the coast, shortly after a minor flooding. The entire width of the river bed is activated. View towards S. All lower courses of Calabrian rivers are confined by levees. Today most of these environments are destroyed by continuous gravel extraction and waste disposal

N_u: Number of segments of the order u

R_b: Bifurcation ratio, $R_b = \dfrac{N_u}{N_u + 1}$

Horton's law of stream numbers: $N_u = R_b^{k-u}$
k: order of the trunk segment
b: Regression coefficient
average bifurcation ratio $R_b = \log^{-1} b$
L_u: mean length of the segments of the order u
D: Drainage density, ratio of total channel segment lengths cumulated for all orders of a basin to the basin area [1/km].

Method. To remain comparable with most literature, we follow the classical method of network analysis (Strahler 1964). Gregory and Walling (1973, p. 44) prefer more refined methods.

The channel segments were drawn by hand from the map following the outlined channels and disregarding the uppermost crenulations of the contour lines. On the copy, the segments were assigned to their orders, measured and counted by means of a digitizing table.

2.3.1 Stream Orders

The seventh order is reached by only four rivers, a small one, Bruzzano [06] and three large ones,

Table 2.2. Drainage properties of the 19 basins. Highest stream order, correlation coefficients of stream-segment number and mean stream-segment length versus the order of the 19 basins, mean bifurcation ratio and drainage density

		Max. order	Corr. coeff.		Mean bifurcation (arith.)	Drainage density 1/km
			Segment number	Mean length		
Melito	[01]	6	−0.79	0.88*	5.1	6.91
Amendolea	[02]	6	−0.78	0.73	5.2	6.40
Pasquale	[03]	6	−0.77	0.84*	4.3	8.36
Palizzi	[04]	6	−0.76	0.72	4.5	10.26
Aranghia	[05]	6	−0.81*	0.74	5.2	8.85
Bruzzano	[06]	7	−0.74	0.77*	3.8	10.36
Laverde	[07]	7	−0.72	0.82*	4.0	7.90
Buonamico	[08]	7	−0.72	0.92*	4.3	7.54
Careri	[09]	6	−0.76	0.86*	5.0	8.03
Condoianni	[10]	6	−0.76	0.74	4.9	8.65
Portigliola	[11]	6	−0.78	0.78	4.0	7.61
Gerace	[12]	6	−0.78	0.73	4.3	7.69
Novito	[13]	6	−0.77	0.79	4.3	6.48
Torbido	[14]	7	−0.73	0.97*	4.2	6.09
Amusa	[15]	6	−0.78	0.82*	4.7	8.36
Allaro	[16]	6	−0.78	0.80	5.8	5.10
Precariti	[17]	6	−0.78	0.82*	4.4	7.28
Stilaro	[18]	6	−0.77	0.82*	4.7	6.01
Assi	[19]	6	−0.77	0.68	4.6	6.36

* Significant correlation (95% level).

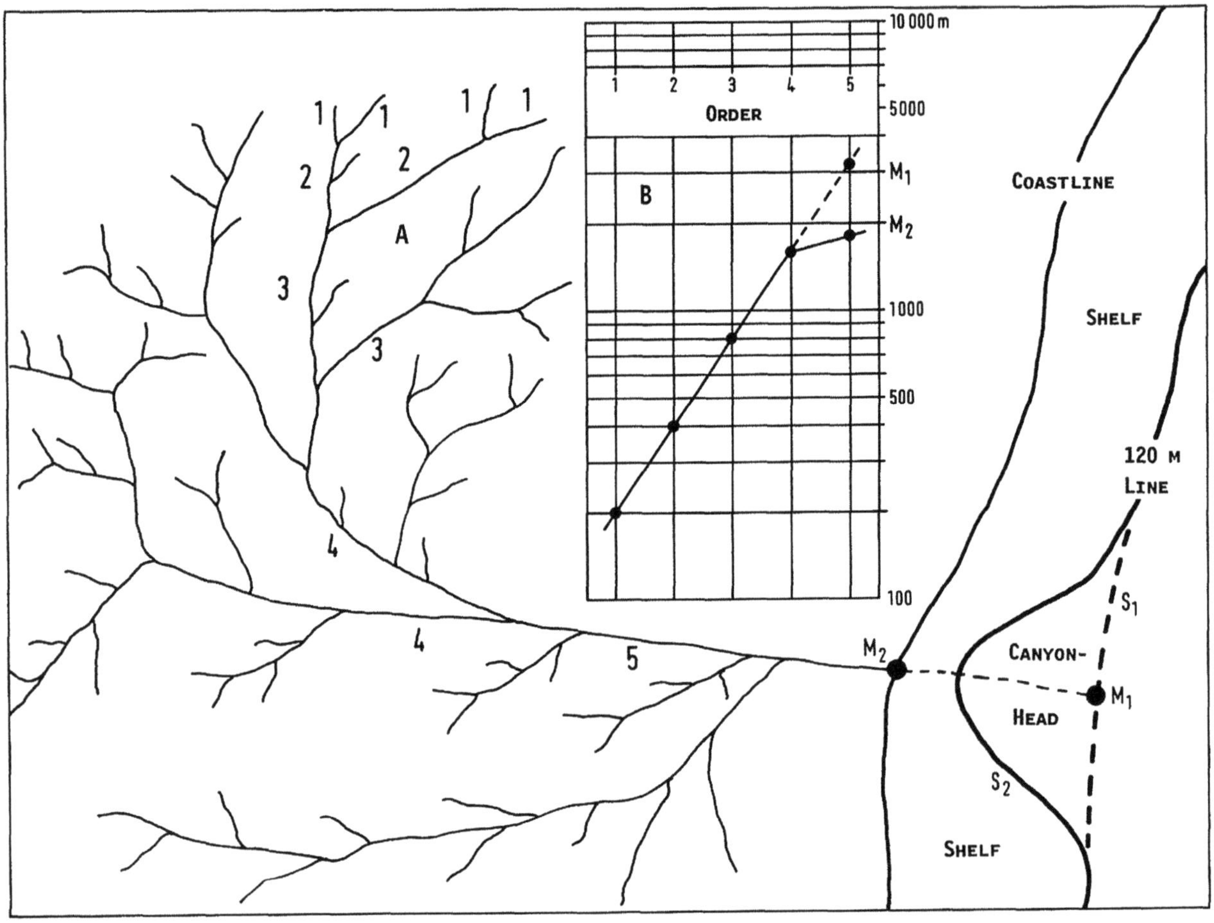

Fig. 2.10 A, B. River network analysis (schematic). **A** Definition of segment orders. Only if two elements of the same order n join, is an element of the order n+1 formed. **B** Logarithm of mean order length [m] plotted versus the order. See Appendix B 01–19 for the regressions of the 19 basins. The modern river-mouth position, M_2, terminates a trunk segment which is "too short" compared to the regression line of the lower orders. A former river mouth, M_1, seems appropriate, which may have been situated at a former shelf break S_1. The present-day 120-m line of the canyon head is S_2

Laverde [07], Buonamico [08] and Torbido [14]; all other 15 rivers yield the sixth order (Table 2.2). The regression of the number of segments versus the order number displays the usual and therefore expected linearity (Appendix B 01–19); however, this almost perfect linearity is disturbed in the case of the highest and the penultimate order of some rivers. Producing the regression line given by the lower orders towards the higher ones, one sometimes produces a misfit of the penultimate order (for example Melito [01], Pasquale [03] and Amusa [15]), or the highest order seems somewhat dislocated, as in the case of the Bruzzano [06], Laverde [07] and Buonamico [08]. This is hard to explain because the highest order usually can occur only once. It certainly makes no sense to measure this too critically and to expect perfect linearity from nature. Therefore, this would not have been commented on if the regression of the mean stream length of some rivers did not display much stronger irregularities, as will be pointed out in the next section.

Bifurcation. Strahler (1964, Fig. 4.II.4) draws different hypothetical basins of extreme and moderate bifurcation ratios, an assumption which should result, applied to Calabrian conditions, in a certain relationship between bifurcation and basin shapes (Fig. 2.6). This is definitely not the case. Strahler (1964, p. 4–45) states that bifurcation ratios of 3:5, which is more or less the Calabrian range (Table 2.2), are characteristic for watersheds in which the geologic structures do not distort the drainage pattern. In Calabria the independence of morphology and geology has to be underlined again and again.

2.3.2 Stream Length

The most striking feature of some Calabrian river networks is the misfit of the highest order or main trunk to the regression given by the sequence of the lower orders, a deviation which occurs in both directions, signalizing "too short" and "too long" main trunks (Appendix B 01 – 19). Based on the concept that a normally developed river displays a more or less linear relationship, when the logarithms of the mean segment lengths are plotted against their orders (Strahler 1964, p. 4–44), a deviation from this linearity should indicate a kind of anomaly. It makes little sense to overstress this principle and to over-interpret smallest deviations. However, we wonder about the fact that only ten of our 19 rivers are "normal" in the sense of the linearity concept. The main trunk of five rivers may be interpreted as "too long", but four rivers, Pasquale [03], Buonamico [08], Torbido [14] and Amusa [15] display really strong deviations of their highest order. All these rivers characterized by "too short" trunks are rivers with active submarine canyons, where the 100-m depth contour comes close to the coast.

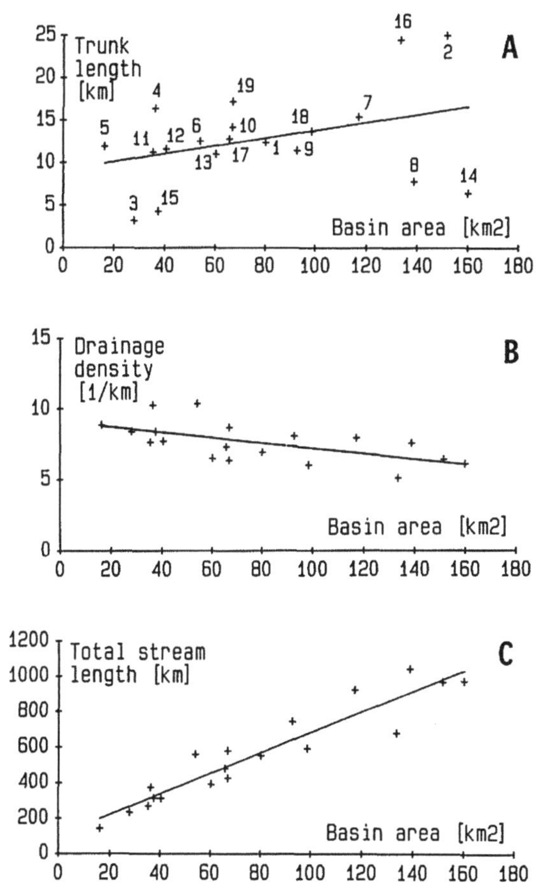

Strangely enough, this misfit is supported independently if we simply plot the main trunk length, the length of the highest order segment, against the area of the respective basin (Fig. 2.11 A). The correlation is very low, r = 0.37, but exactly the same four rivers which did not fit into their length/order regression, Pasquale [03], Buonamico [08], Torbido [14] and Amusa [15], prove to possess "too short" trunk segments compared to the other 15 basins. There is at least a tendency of the "too long" trunk segments to behave in an opposite manner.

The philosophy of this finding is that in the case of the four rivers presumed to possess "too short" trunks the extrapolation of the straight element given by the lower orders towards the trunk segment yields an ideal or theoretical length and river-mouth position which accords with the undisturbed shelf break or shelf line (Fig. 2.10, M_1). Nearly all rivers with "too short" trunk segments flow into submarine canyon heads (Fig. 2.7 A). This could mean that the modern river network conserves a kind of memory of a former stage during stages of lower sea level. However, this does not explain the situation of the basins with "normal" or even "too long" trunks. Theoretically they could have joined on the shelf to form trunks of higher orders.

However, not to become lost in speculations, it might suffice to state that the misfit of the trunk length may be seen as a distortion of the channel network close to the coast. Because the pattern of the interior basins is more or less well balanced, this distortion may reflect the influence of sea level changes.

Drainage Density. The drainage densities of the Ionian slope of our Calabrian area are not very high (Table 2.2). They are, according to Strahler (1964, p. 4–52), comparable to large areas of the humid central and eastern United States on rocks of moderate resistance under a deciduous forest cover or parts of the Rocky Mountain region. The drainage densi-

Fig. 2.11. A Length of the trunks (segment of highest order) versus area of the basins. The four rivers characterized by "too short" trunks compared to their regressions of mean segment length versus segment order, namely Pasquale [03], Amusa [15], Buonamico [08] and Torbido [14] (Appendix B), surprisingly turn out to be also "too short" compared to their basin area and the other Calabrian basins. **B** Drainage density versus drainage area: there is a strong tendency of larger basins to develop lower drainage densities (r = −0.61*). **C** Total stream length correlates strongly with basin size (r = 0.94*). Both data show that Calabrian rivers behave as most other rivers of the world ty decreases with increasing basin area (Fig. 2.11 B), an unexpected relation confirmed by Gregory and Walling (1973, Fig. 2.1). Plotting the total stream length of the basins against their areas (Fig. 2.11 C), we found exactly the relationships reported by Gregory and Walling from various rivers of the world (1973, Fig. 5.12).

2.4 Conclusion

Order, size and shape of the Calabrian basins have no visible influence on the river-mouth sediments; the same is true for all channel network characteristics. This is disappointing, because the digitization of these properties and their quantitative treatment should have made it possible to detect such relationships. We expected at least maximum grain sizes and properties such as bimodality to be possibly dependent on geomorphology. This is discussed in Chapter 10. In short, the broad range of grain sizes of the river-mouth sediments should be sensitive to the hydrodynamic conditions of the rivers and these in turn should be sensitive to or shaped by the most important basin characteristics. The size differs by one magnitude, 16 and 160 km^2, the shape varies from very elongated ($R_f = 0.13$) to pear-shaped ($R_f = 0.44$) and the drainage density varies between 5.1 and 10.36 1/km. So if these properties shape the river-mouth sediments, it should be quite evident; but this is not the case. Consequently this expectation is falsified in Calabria.

However, it might be concluded from our analysis of drainage basins and channel networks that the Ionian slope of the Calabrian Massif obviously underwent a homogeneous, very uniform development far removed from being shaped significantly by geological factors. For this reason, it might be justified to rely on the homogeneity of the Mid-Pleistocene plain as reference level of erosion rate calculations.

Chapter 3

Altitude and Slopes

Relief probably is one of the most frequently and most uncritically used terms in provenance discussions. It is the goal of this chapter to treat the altitude frequency distribution of the area and the statistical parameters: one of them, the altitude standard deviation, is defined as relief, another one, the altitude skewness, is interpreted as dissection or hypsometric integral, the "opening" of an area by erosion. The data bank makes it possible to calculate these variables, to decide if there is a discernible influence of relief and dissection upon properties of the river-mouth sediments such as grain-size distribution or petrographic composition, and to prove this statistically. Moreover, it becomes possible to play with the data to obtain correlations whose derivation by conventional methods would have been unjustifiably time-consuming; for example, diagrams in which the altitude of each point of the geomatrix of a basin is plotted against its distance from the river mouth, a procedure which leads to very indicative graphs.

The following is deliberately a morphometric study. The provenance concept offers little room for questions such as why and how certain properties of the source area developed. We simply accept these properties as such, but we have to quantify them as a prerequisite to the understanding of their possible influence on sediment formation.

Summary. The altitude standard deviation, the measure of relief, varies considerably between the single basins (213 to 481 m), as does the altitude skewness, the measure of dissection (−0.144 to 1.156). However, both variables exert no influence on properties of the river-mouth sediments such as uni- or bimodality of grain-size distributions.

We found many interesting byproducts but no striking correlation between a specific geomorphological parameter of the source area and a specific sedimentological feature of the river-mouth sediment, without regard to the fact that we are dealing with steep-graded mountain rivers.

Plots of altitude versus distance from the river mouth yield insight into internal basin differentiation and the river's longitudinal profiles. The main trunk slopes, the highest-order segments, vary between 1 : 30 and 1 : 110.

The regional distribution of the gradients makes it possible to delineate low- and high-gradient belts of strengthened slope erosion subparallel to the coast. The frequency distribution of the aspects, of the directions in which the slopes are facing, reveals a surprising symmetry and signalizes an almost perfect independence of the slope directions from geology or even altitude.

3.1 Altitude

Data Base. Onshore. Topographic map 1 : 25000 of the Istituto Geografico Militare Italiano with 25-m contour lines. There also exists a generated topographic map 1 : 10000 based on aerial photos, the evaluation of which, however, would have involved six times as much work.

Offshore. Echosounding map 1 : 10000 with 25-m contour lines (Rumohr and Ibbeken 1975). Geomatrix, data bank.

Terminology. The altitude matrix is a point grid with altitude values. The relief is expressed by the standard deviation of the altitude frequency distribution. The hypsometric integral (or dissection or regional convexity) is given by the skewness of the altitude frequency distribution.

Method. The altitude values were read off at the grid points of the geomatrix by visual interpolation, listed and transferred into the data bank. The error was ±5 m. For the statistical treatment the altitude values were classified into 100-m intervals, from sea level (0 m) down to −900 m in the submarine canyon and up to 2000 m in the Aspromonte.

3.1.1 Statistics

3.1.1.1 Altitude Frequency Distribution

The histograms of the altitude frequency distribution provide a useful base for the morphometric discussion of the scenery. On the one hand, they contain the correctly obtained but condensed data set and, on the other hand, they are not yet so abstract that the contact to nature is lost. Appendix B 01 – 20 shows the altitude frequency distributions of the 19 basins and the "average basin". The corresponding standardized cumulative curves are also plotted.

The average altitude frequency distribution of the 19 basins (Appendix B 20) is broad and relatively balanced, the correlation between the percentage of altitudes and altitude classes is very good, r = −0.97*. A primary mode in the 100 – 200-m class is followed by a weak secondary mode in the 1000 – 1300-m class. The first mode stems from the extended alluvial plains close to the coast, the second is caused by the elevations of the Serre.

We will take the mean altitude frequency distribution of the 19 rivers as a standard, against which the single 19 altitude frequency distributions may be measured. Appendix B 01 – 19 demonstrates the deviation of the altitude frequency distributions of the 19 basins against this standard. All basins are so different from this standard, that no one of them can claim to be the "standard river" or basin. This means that the mean or standard distribution has to be understood as a mixtum compositum of 19 morphologically very differently structured basins. This is surprising because personal field experience suggests strong similarities between the 19 basins. At the same time, it is a gain for the philosophy of this study: if relief and dissection had a distinct influence upon the formation of the river-mouth sediments, then this influence should be recognizable due to these differences between the basins, or, if not, such an influence may not exist. We will follow up and test this argument when we discuss the problem of bimodality in chapter 10. Basically, there are all kinds of deviations from the standard. They regard all altitude classes, all directions of the deviation, positive or negative, and their strength. The qualitative, synoptical comparison, however, is misleading and we will turn consequently to the more precise statistical parameters to describe the basins.

3.1.1.2 Statistical Parameters, Relief and Hypsometric Integral

Figure 3.1 gives the regional distribution of some derivatives of the altitude frequency distribution, of the mean, standard deviation and skewness; it also displays the median and the most frequently used geomorphical parameter of an altitude distribution, the hypsometric integral (Strahler 1952).

Mean. Convincing as it would be to favour such an important and simple parameter as the altitude mean (Fig. 3.1 A) as a characteristic value for altitude frequency distributions, it would be incorrect, because the mean, at least in Calabria, is strongly dependent on the size of the basins. This is quite logical, because the larger a basin, the higher its reach into the mountains and consequently the higher the altitude mean.

Thus, size and altitude mean of the Calabrian basins consistently correlate with r = 0.73* (Table 3.1) and the mean as a characteristic value for the comparison of differently sized basins becomes superfluous.

Standard Deviation and Relief. The conventional geological terminology uses the term relief only qualitatively and not very critically in the simple sense of "high" relief in steep, high mountains, of "low" relief in the plains and perhaps of "mean" relief in between. Unfortunately, there is a strong discrepancy between the implicitly accepted meaning of "relief" as such, stressed in almost all geologic discussions as an expression of tectonics or even plate tectonics, and the fact that this term has never been conceptually defined and exactly quantified in geology.

Geomorphologists basically discuss two methods to determine relief, the difference in altitude between the highest and the lowest point of a given area and the altitude standard deviation (Evans 1972, p. 31). The first method is criticized because it is based on only two extreme values instead of all values of a distribution like the second one, the altitude standard deviation, which consequently is statistically much more stable (Fig. 3.1 C).

It is clear that an area with a small dispersion or standard deviation of its altitude values has a smaller relief than one having a higher dispersion. In this book we use the term relief exclusively in the sense of altitude standard deviation.

Table 3.1. Correlation coefficients between geomorphological variables of the 19 basins

	Altitude frequency distribution			Area
	Mean	St. dev. (Relief)	Skewness (Dissection)	
Altitude:				
Maximum	0.74*	0.90*	−0.15	0.67*
Median	0.79*	0.77*	−0.76*	0.67*
Hyps. integral	0.65*	0.33	−0.83*	0.38
Area	0.73*	0.69*	−0.26	

* Significant correlation (95% level).

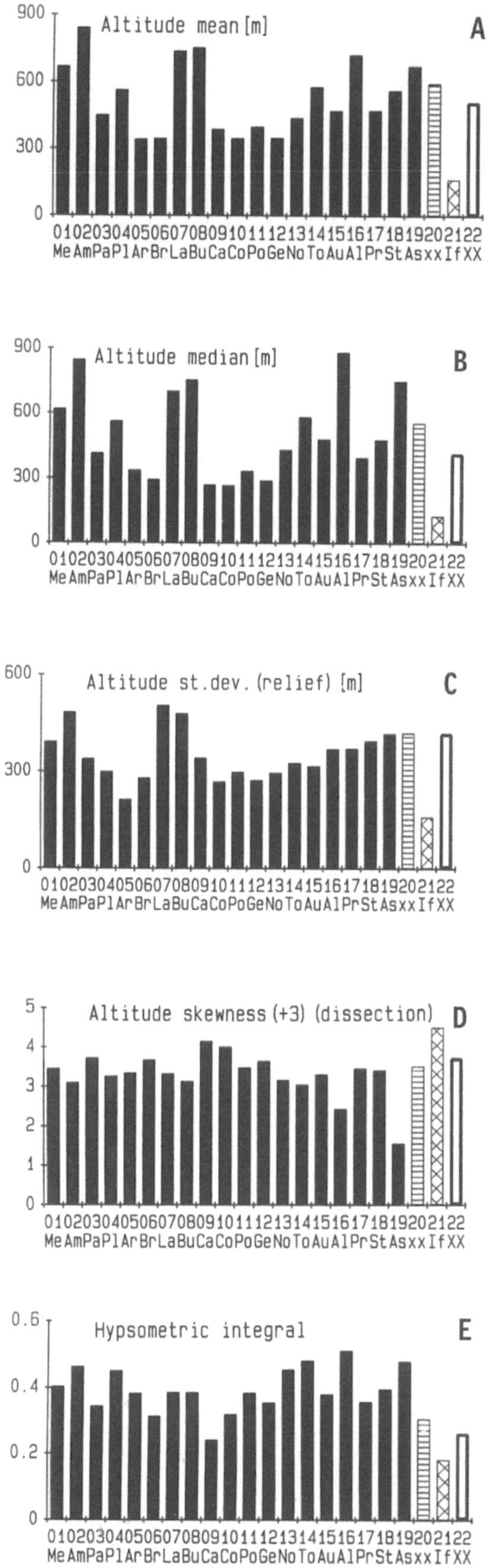

Fig. 3.1 A–E. Statistical parameters of the altitude frequency distribution. **A** Mean. **B** Median. **C** Standard deviation or relief. **D** Skewness or dissection; for better synoptical comparison, skewness values are made positive by adding 3. **E** Hypsometric integral. *01 Me*lito; *02 Am*endolea; *03 Pa*squale; *04 Pa*lizzi; *05 Ar*anghia; *06 Br*uzzano; *07 La*verde; *08 Bu*onamico; *09 Ca*reri; *10 Co*ndoianni; *11 P*ortigliola; *12 Ge*race; *13 No*vito; *14 To*rbido; *15 A*m*u*sa; *16 Al*laro; *17 Pr*ecariti; *18 St*ilaro; *19 As*si; *20 xx* mean of the basins 01–19; *21 If* mean of the 84 smaller basins and interfluves and *22 XX* total mean (103 basins and interfluves)

Skewness and Hypsometric Integral. A further frequently used geomorphological parameter besides relief is the "hypsometric integral" (Strahler 1952) (Fig. 3.1 E). This is the degree of dissection of an area, of its "opening", especially by erosion. This parameter may be determined, among others, from the cumulative altitude frequency distribution. Appendix B 01–20 gives the standardized cumulative altitude frequency curves of the area. The hypsometric integral, in this case, corresponds to the area beneath the curves, or, in nature, to the mountain volume above the deepest point in consideration. This is the "base level", in our case the sea level, and the volume is the "volume to be consumed" (Evans 1972, p. 45). Figure 3.1 E shows the regional distribution of the hypsometric integrals of the Calabrian basins.

Evans (op. cit.) criticizes the dependence of the hypsometric integral on the heterogeneity and the extreme values of an area and proposes to substitute the hypsometric integral by the skewness of the altitude frequency distribution (Fig. 3.1 D). In Calabria, both values, hypsometric integral and skewness, correlate nicely, the correlation coefficient being $r = -0.83$*. None of them displays a dependence on the size of the basins (Hypsometric Integral: $r = 0.38$; skewness: $r = -0.26$, Table 3.1), a circumstance that enhances the value of these parameters.

Maximum. The maximum altitude plays an important role, limiting the hypsometric integral as the "upper enveloping surface", referring to Strahler's original interpretation; however, its value as a single point in a basin is very limited. A locally restricted but high maximum may yield a statistically high importance not justified for the judgement of the entire area.

3.1.1.3 Relationships of the Variables

The relationships between these variables are summarized in Table 3.1, expressed by the correlation coefficients. The geomorphologically most important parameters are the altitude standard deviation as a measure of relief and the altitude skewness as a measure of dissection. The altitude standard deviation, the relief, correlates very well with the maximum, r = 0.90*, less so with the median, r = 0.77* and the size of the basins, r = 0.69*.

It is impossible to overlook the internal links of these relations. If the size of the basins correlates with both the altitude maximum and the altitude standard deviation, the relief, a similar correlation has to be expected for the maximum and the standard deviation; they correlate with r = 0.90*.

The regional distributions of altitude standard deviation and median are surprisingly similar (Figs. 3.1 B, C). However, this does not revaluate the median as a useful statistical parameter, because there are substantial exceptions. The Allaro [16], for example, Serre Province, yields the highest median with 883 m, but its relief is not very strong, not to be compared with the most spectacular rivers of the Aspromonte Province, Amendolea [02], Laverde [07] and Buonamico [08], whose median values are even less. The reason for this is simple. The Allaro [16] displays an internal, nearly closed and very elevated drainage area (Fig. 2.3 B), connected with the sea by only a very narrow, elongated valley, a kind of corridor. Thus, the intermediate altitude values are under-represented and no high relief value is yielded due to the missing dispersion.

The altitude skewness, the measure of dissection, correlates very well with the hypsometric integral (r = −0.83*, Figs. 3.1 D, E). The different meaning of dissection and relief may be emphasized using the Palizzi [04], the Buonamico [08] and the Torbido [14]: their altitude skewness and dissection are very similar (Fig. 3.1 D), the altitude standard deviation, however, and the relief of the Buonamico are much higher than that of the other two rivers (Fig. 3.1 C).

Thus altitude standard deviation and altitude skewness prove to be valuable measures of relief and dissection, compatible with field experience, and it should be possible to recognize their influence upon the river-mouth sediments, if they have any at all.

3.1.2 Regional Aspects

It is not the scope of this study, as pointed out, to discuss the geological or geomorphological evolution of Calabria. We only want to quantify the status quo. In the foregoing chapters we presented the morphological variables and we want to discuss their regional distribution in the following section. So this might be the right place to illustrate some specific topics of the Calabrian landscapes with some photographs. South Calabria is characterized by a kind of reversed geomorphology. The upper parts of the most elevated elements, Aspromonte

Fig. 3.2. This is the sharp, razor-like cut of the Zomaro plain, here 950 m; view towards SW from below Canolo Nuovo. Regressive erosion from both the Tyrrhenian and Ionian Sea narrows the plain leading to high relief and steep slopes below 1000 m, whereas the older relief of the Aspromonte above 1000 m is only subdued (Figs. 3.3 and 3.4). Note the many landslides on the lower parts of the slopes (Chap. 15.2.2)

Fig. 3.3. South part of the Zomaro high plain in the Mastrogiovanni area, 1030 m, view towards SW. Wave-cut terrace in the *foreground*, peat bogs vegetated with bracken in the *middleground* (Chap. 4.2), and beech woods in the *background*. In the *rear* the subdued relief of the upper Aspromonte topped by the Montalto (1956 m, *arrow*)

Fig. 3.4. View towards N to the Mt. Scorda (1572 m), the northern limit of the metamorphic Aspromonte massif. Many landslides feed the steep channel of the Vallone Aurea, in the *center*, the main tributary of the upper Buonamico [08]. At *right, at the very rear*, the Zomaro plain. Single examples of pines (*Pinus laricio*)

and Serre, display only a subdued relief, from the summits down to about 1000 m (Figs. 3.3 and 3.4). The Zomaro plain (Fig. 3.2), embedded at about 1000 m between the two massifs, displays almost no relief and it is only below 1000 m that a strong, alpino-type relief develops (Figs. 3.5 and 3.6), forming the "high gradient belt" (Fig. 3.9) which girds the entire area.

3.1.2.1 Altitude Map

The computer-generated altitude map (Fig. 3.7) certainly gives only a small fraction of the visual in-

Fig. 3.5. Steep slopes and landslides of the Potamia, a short but powerful tributary of the Buonamico [08] in the *foreground*. At the crest, *center right*, the ruins of the village Potamia, abandoned after the earthquake of 1783. This is the typical aspect of the slopes of the high gradient belt

formation of the original map. The plot, however, is not intended to supersede maps. It is only one of the nearly infinite possibilities to vary the altitude intervals and to print special combinations for special purposes. The meshes of the point grid, the geomatrix, give an impression of the precision or solution of the data base: to repeat, each point stands for a quadrangle of 250×250 m or 1/16 km^2.

Clearly visible is the stronger morphological accentuation of the Aspromonte Province in the south and its vigorous dissection by very ramified, deep and broad valleys. The selective geological computer map (Fig. 5.11) demonstrates this even more. Here the present-day alluvial sediments of river and coast are printed, this highway system of mobile sediments still covers 16% of the area. Exceptionally extended is the coastal region of the Zomaro Province, dominated by sediments; the same is true for the adjoining shelf section between Bovalino and Gioiosa. The head of the submarine Buonamico-Careri Canyon practically reaches the coast; we mapped this feature to a depth of −900 m.

3.1.2.2 Altitude Versus Distance from River Mouth

Computer and data bank cause quantitative changes in research, especially acceleration. In addition, we notice a qualitative innovation given by the possibility of simply playing with the data, that

Fig. 3.6. Landscape in the Novito basin [13], seen from above Siderno Superiore towards W. The wide valley in the *foreground* consists of low resistant siltstones, the granites of the Mt. Schiavo, in the *center*, are cut by the Novito, creating a classical superimposed valley (Chap. 2.2.5). The scenery is crowned by the Jurassic limestones of the Mt. Mutolo, which lead, with increasing relief, to the sharp cut of the Zomaro plain in the *right background*, above the village of Agnana

is, of creating questions to answer which would have been too expensive and too time-consuming previously, and which, as a consequence, were not asked.

We want to present some graphs where two variables are plotted against each other. These plots, with quite trivial combinations, give surprising insight into the architecture of the mountains. We start with the combination of all altitude values of a basin and their distance from the river mouth. At the beginning, this combination seemed to us especially senseless, but only at the beginning.

The matrix points are simply plotted according to their altitude and distance to the river mouth (Fig. 3.8). Thus, each vertical row represents the altitude values at a segment of a circle drawn around the river mouth with the given radius. Figure 3.8 A explains the method, Fig. 3.8 B shows an example, the Amendolea [02], and Appendix B 01–19 shows the resulting plots of each basin.

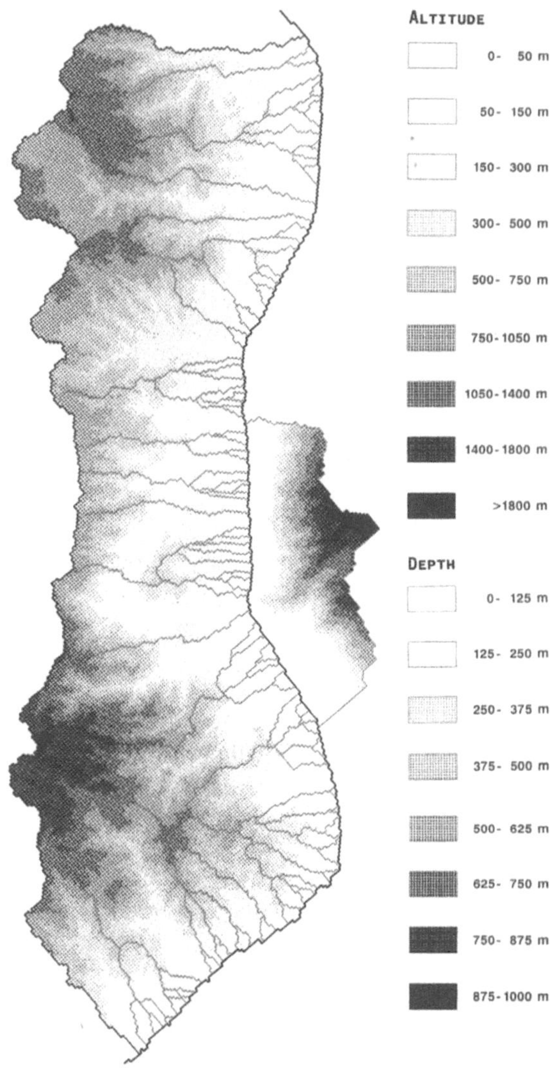

Fig. 3.7. Computer-generated altitude map of the Ionian slope of the Calabrian Massif and depths of the adjoining areas of the shelf and submarine canyon

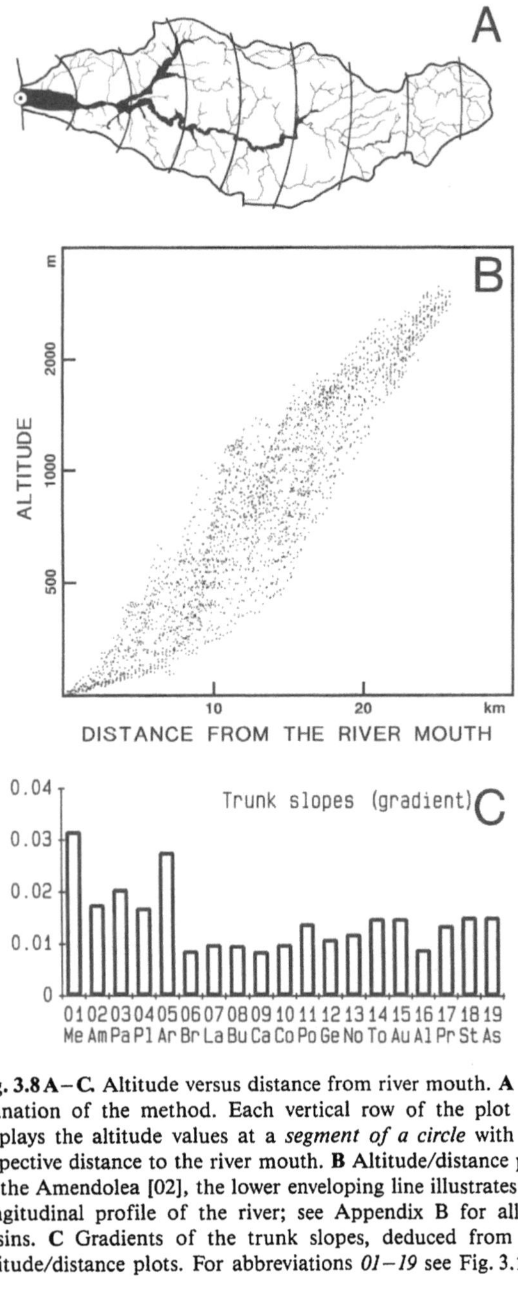

Fig. 3.8 A–C. Altitude versus distance from river mouth. **A** Explanation of the method. Each vertical row of the plot (**B**) displays the altitude values at a *segment of a circle* with the respective distance to the river mouth. **B** Altitude/distance plot of the Amendolea [02], the lower enveloping line illustrates the longitudinal profile of the river; see Appendix B for all 19 basins. **C** Gradients of the trunk slopes, deduced from the altitude/distance plots. For abbreviations *01–19* see Fig. 3.1

Such a segment of a circle leads from the one drainage divide over possible subbasins to the deepest point along this segment, the long profile or thalweg and then back, again over possible subbasins, to the opposite drainage divide. Thus the lowermost point of each vertical point row, assigned to the respective distance to the coast, indicates the altitude of the thalweg, and the uppermost one marks the altitude of the main divide enveloping the basin, here or there, or of a subbasin drainage divide.

The Lower Enveloping Line. The line connecting the lowermost points of the cluster, the lower enveloping line (Fig. 3.8 B), represents directly the thalweg or long profile, a surprising way to filter the linear element of the main river course out of the areal point grid of the geomatrix. However, a possible meandering is not taken into account. Theoretically, there is a small possibility that the lower enveloping line of this graph is not exactly the thalweg but an artefact. This could be the case if a tributary has a somewhat lower gradient than the main channel. The lowermost point at this segment would then be given by the tributary and not by the main channel along which the long profile is usually constructed.

The resolution of the point grid of 250 m is relatively rough, compared to the narrow channels at least in the upper river courses. Thus the deepest point of a profile frequently is not hit and, as a result, the line representing the long profile is often somewhat undulated.

The long profiles of the 19 rivers are either uniform or composed (Appendix B 01–19). The uniform long profiles exhibit the classical concave pattern characterized by a relatively flat main trunk and increasing gradient with increasing altitude (Careri [09]).

The gradients vary noticeably:

	Minimum	Maximum
Lower course:	1:110, Careri [09]	1:30, Melito [01]
Upper course:	1:15, Novito [13]	1:2, Stilaro [18]

Figure 3.8 C shows the gradients of the main trunks; there is a tendency towards higher gradients in the south and central Aspromonte Province, lower gradients in the north Aspromonte and the Zomaro Province and again higher gradients in the Serre Province, correctly supporting the main geomorphological trends. This is meaningful for our provenance problem, the possible participation of morphology shaping the river-mouth sediments. Different gradients of the main trunk of the river courses should control stream velocities and lead possibly to different grain-size properties in the river-mouth sediments (Chap. 10.4.2.1).

The composed long profiles consist of several straight or concave elements. The changing gradients signalize the fall line, changing lithology within the mountains or neotectonics. The rivers joining the plain elements of the Aspromonte or Serre develop retrograde gradients along their upper courses. The long profile of the Allaro [16]

(Appendix B 16) is highly differentiated, as will be pointed out below.

The Upper Enveloping Line. It is not possible to assign the uppermost limits of the cluster to definite morphological elements. They may stem either from the perimeter of the basin or from internal drainage divides which may be higher than the perimeter. Thus the upper enveloping line is an artefact which reflects, however, the homogeneity of the basin.

The Shape of the Plot Altitude/Distance and Its Internal Differentiation. The shape of the point cloud as a whole varies considerably. The Amendolea [02] (Fig. 3.8 B) shows a very regular lanceted pattern corresponding to its rhomboidal basin shape (Fig. 2.6 B); the Careri [09] (Appendix B 09) is characterized by an elongated peak of high altitudes caused by the steep, monadnock-like slopes of the basement which border the wide valleys of the siltstones in the Platí area (Fig. 2.5). The Torbido [14] (Appendix B 14) is truncated by a sharp break caused by the extended main divide at the very end of this pear-shaped basin (Fig. 2.6 C).

The interior of the point clouds frequently shows darker and lighter areas as expression of larger interbasins, Amendolea [02] and Allaro [16] being examples (Appendix B 02 and 16). The cumulative plot of all basins 01 – 19 (Appendix B 20) displays a distinctly separated, appending point cloud which belongs to the Allaro [16]. The altitude distribution of this river is unique compared to the other 18 basins, and the Allaro has to be seen as a distinct outsider: a kind of relictic basin in the high Serre Province is connected with the coast by a long corridor (Fig. 1.3 B).

This is an important tool for our provenance problem. If geomorphology has a visible control on the formation of river-mouth sediments, an exceptional river should create exceptional river-mouth sediments.

3.2 The Slopes (Gradients)

The gradient, the angle of slope, is the first derivative of the altitude. For our investigation the role of the gradients is of great importance. Steep terrain and landslides are so common in Calabria that it is much more difficult to find stable slopes than instable ones. The aspect, i.e. the direction where the slope faces, influences the vegetation, mainly via insulation, and in this way also the slope stability. Our data base makes it possible to determine gradient and aspect for each point of the geomatrix and to combine this with other values: with rock type or altitude, with the distance from the river mouth or with precipitation. The dependency of gradient and aspect of geology, tectonics and rock distribution is surprisingly small, if not absent.

Data Base. Topographic maps 1:25000 of the Italian Istituto Geografico Militare with 25-m contour lines. Geomatrix with a 250-m grid, databank.

Terminology. Gradient: slope angle in a point of the geomatrix. The tangent of the slope angle is the first derivative of altitude. Aspect: direction of the slope in a point of the geomatrix.

Method. Gradient: each point of the geomatrix is surrounded by eight neighbouring points (Fig. 3.12 A). From the reference point the gradient to each of the eight points was calculated (altitude-difference/distance), regardless of whether uphill or downhill, and the highest value was attributed to the reference point in terms of degree. For the statistical treatment and the graphic representation by maps we classified the gradient values into 5° intervals ranging from 0 to 75°. On the computer-generated map (Fig. 3.9), each of the resulting 15 gradient classes bears its gray value as does each point of the geomatrix, the higher the gradient, the darker the gray value.

Aspect: again, each reference point is surrounded by eight matrix points corresponding to the eight directions N, NE, E and so on (Fig. 3.12 A). The computer neglects the points of equal or higher altitude, seeks the highest gradient to the remaining lower points and assigns the resulting slope direction or aspect to the reference point.

3.2.1 Statistics

3.2.1.1 Gradient Frequency Distribution, Non-Rock-Specific

The histograms of the gradient frequency distribution of the single basins are so similar that it may suffice to plot only the mean gradients of the entire area (Fig. 3.10 A).

3.2.1.2 Statistical Parameters

Mean. The mean gradients of the interfluves and basins shift between 12 and 25° (Fig. 3.10 B).

Fig. 3.9. Computer-generated gradient or slope angle map. Each *point* of the geomatrix represents the highest gradient measured against the eight neighbouring points (Fig. 3.12A). *A* Lower "low gradient" or "low energy belt"; *B* "high gradient belt"; *C* upper "low gradient belt"

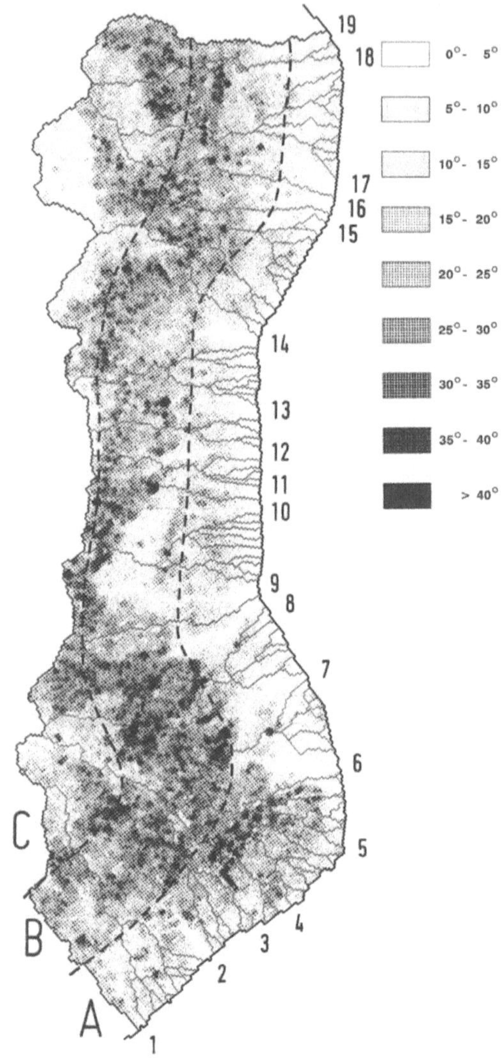

Regionally there is a slight decrease from south to north with a trend of 22 to 16°. This comes mainly from lithology, from the transition of the relatively resistant gneisses and schists of the Aspromonte Province over the Zomaro Province to the less resistant granites of the Serre Province. However, this is valid only for the mean values; in the individual cases the relationships between gradient and rock type and especially erosion rates are much more complicated.

Standard Deviation. The gradient standard deviation of the basins (Fig. 3.10D) varies only slightly, between 7.1 and 10.3°, without a visible trend. This is another hint at the small mutual dependence of gradient and rock type.

Skewness and Land Slide Risk. The overall gradient frequency distribution of the Ionian slope of the Calabrian massif displays a slightly positive skewness with a broad mode of 10 to 20° mean gra-

Fig. 3.10 A–D. Gradients or slope angles. **A** Histogram of mean gradients, basins 01–19. **B** Mean gradients. **C** Standard deviation of the mean rock-specific gradients: *Gra*nite, *Gne*iss, *Sch*ist, *Lim*estone, *Con*glomerate, *Sds*tone (Sandstone), *Si*ltstone, *Arg*illite. **D** Standard deviation of the gradients. Rock-specific mean gradients of the single basins are plotted in Appendix B 01–19. For abbreviations *01–22* see Fig. 3.1

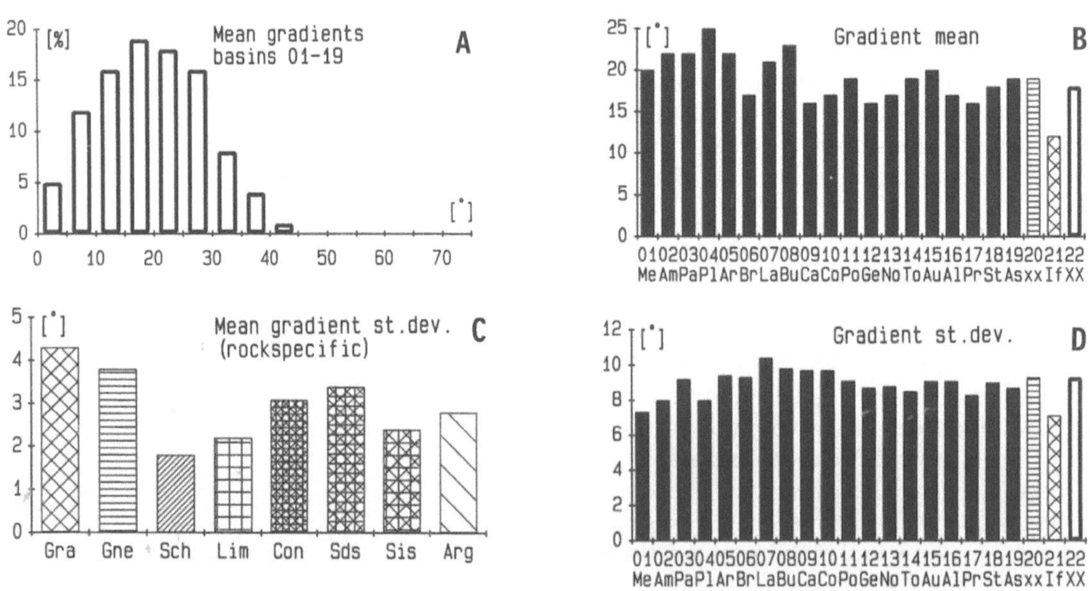

dients (Fig. 3.10A). The slight dominance of lower gradients stems from the coastal areas between the river mouths.

The gradient frequency distributions of the single basins (not plotted) are mainly positively skewed to symmetrical, thus signalizing rather flat terrain. Very positively skewed, among the important rivers, is only Careri [09], whose hinterland is made up by olisthostromes. These only weakly resistant debris flow deposits form wide and ample valleys (Figs. 2.5 and 3.6). The largest basin, Torbido [14], possesses a nearly perfect gradient symmetry; this corresponds to its well-balanced altitude frequency distribution. Amendolea [02], Buonamico [08] and, a little less pronounced, Amusa [15] are the only basins with a negatively skewed gradient frequency distribution. These rivers develop a surplus of steep mean gradients, which holds especially true for the Buonamico; the mode of this basin still lies in the 25–30° regime (Fig. 15.1B). These basins are especially at risk for landslides. The reversed conclusion, that is, basins with positively skewed gradient frequency distributions would be less at risk, is clearly not correct, because local circumstances are not caught by statistics. Thus the area with the highest risk of a megaslide on the Ionian slope, prognosticated by Sorriso-Valvo (1984), lies in the Careri basin [09] above Platì. The Careri, however, exhibits the most positive skewness.

3.2.1.3 Gradient Frequency Distribution, Rock-Specific

The rocks of the Calabrian Massif are grouped into eight lithologic units: granite, gneiss, schist, limestone, conglomerate, sandstone, siltstone and argillite (Chap. 5). Calculating the rock-specific gradients, each reference point (Fig. 3.10C) was assigned to its own rock type, the rock types of the surrounding points not being taken into account. Appendix B 01–20 gives the plots of the single basins and the "average basin".

Let us first consider the mean values of the entire area (Appendix B 20). Granite slopes (20°) are flatter than slopes of metamorphic rocks (24–25°); limestone slopes appear, judged from field experience, too flat (21°); siltstones, which are essentially olisthostromes, and argillites develop the flattest mean gradients – all these statements admittedly would be trivial, if the gradients were determined exclusively by the physical properties of the respective rock types themselves. In this case, however, similar rocks should always develop similar gradients. This is not the case in Calabria. On the contrary, the gradients of the single rock types vary noticeably within the single basins. For example, in nine basins conglomerates have steeper gradients than the sandstones, and in seven basins the contrary is true (Appendix B 01–19).

Table 3.2. Range of mean gradient and standard deviation of the mean gradients of the eight rock types in the 19 basins

Rock	Range (°)	St. dev. (°)	Rock	Range (°)	St. dev. (°)
Granite	17–33	4.3	Conglomerate	13–25	3.1
Gneiss	20–35	3.8	Sandstone	13–24	3.4
Schist	16–27	1.0	Siltstone	11–19	2.4
Limestone	21–28	2.2	Argillite	10–22	2.8

Table 3.2 shows the range of mean gradients of the rock types in the 19 basins and the standard deviations (Fig. 3.10C).

Granite. The granites of the Zomaro Province as a whole are steeper than those of the Serre Province, which fits with the field experience, because the granites of the Serre are widely covered by grus so that it is often difficult to obtain a fresh hand specimen.

Gneiss. The standard deviation of the mean gneiss gradients of $s = 3.8°$ reaches nearly the value of the granites, but it is impossible to realize a similar differentiation of the macroscopically identifiable properties.

Schist. Various metamorphic rocks are hidden under the rock type "schist", so that the high mean gradient of 24° is as astonishing as the low standard deviation, $s = 1.0°$, which represents the minimum of all rock types measured.

Limestone. Limestone outcrops which are large enough to be treated in a statistically correct manner exist only in the basins Novito [13], Precariti [17] and Stilaro [18]. There is no reasonable base for a regional comparison.

Conglomerate and Sandstone. In contrast to the very uniform and concentrated limestones, the deposits of conglomerates and sandstones are widespread over the entire slope of the Calabrian Massif. The standard deviation of the mean gradients, $s = 3.4°$ (sandstone) and $s = 3.1°$ (conglomerate) is not very high; subjective field experience feigns a broader variety. This may be caused by the impression of some spectacular road- or river cuts, but it is not justified by the situation as a whole.

Siltstone and Argillite. Olisthostrome and argillite expectedly display the lowest gradients and also very few standard deviations. Both rocks are extremely homogeneous, compared to all other Calabrian rocks previously described, so that the even low standard deviation could easily by interpreted lithologically. However, seen as a whole, the standard deviations of the mean gradients of the single rock types in the single basins are so high that it is very difficult to harmonize them with the admittedly subjective field experience. We may come closer to an answer if we discuss the regional distribution of the gradients plotted on a map.

3.2.2 Regional Aspects

3.2.2.1 Gradient Map

The computer-generated gradient map (Fig. 3.9) exhibits all gradients of the Ionian slope of the Calabrian Massif in different gray values. At first glance, the distribution of flatter and steeper areas does not make a very meaningful impression. The Aspromonte complex develops the steepest slopes again in the basins of intense relief, in the Amendolea [02] and especially Laverde [07] and Buonamico [08] areas.

Due to the wide olisthostrome areas, the lower parts east of the massif have relatively low gradients, in contrast to its southern slope. The same is true for the Zomaro Province, where argillites occur on a larger scale and for the Serre Province, again because of the argillites outcropping predominantly close to the coast. The granites of the Serre Province do not yield steeper gradients; on the contrary, the higher parts of this massif display extended planations. Even the high Aspromonte has many planations. The high plain of the Zomaro Province (Figs. 3.2 and 3.3), situated between the two massifs, is missing on our maps of the Ionian slope, because its major parts drain westward, on the Tyrrhenian side. As a consequence, the main divide, the central Apenninic divide, is somewhat decentralized and shifted eastward.

Altogether the zones of higher gradients are concentrated on the intermediate parts or altitudes of the two massifs and the internal part of the high plain of the Zomaro Province. The top positions of the massifs and the coastal areas, on the other hand, are flatter. Thus, besides lithology, a further influence upon the gradients of the Calabrian Massif is suggested, namely altitude and distance to the coast.

3.2.2.2 Gradient Versus Distance from River Mouth

A more precise evaluation of the regional distribution of gradients is made possible by the plots gradient/distance of the river mouth of the 19 basins and the "average basin" (Appendix B 01 – 20). The plots display complete or truncated convex bows. The rims are undulated or diffuse, quite in contrast to the point clusters of the altitude-distance plots with their sharp lower limits. This comes simply from the construction of the plots: each vertical point row consists, per definition, of the gradient values at a sector of a circle drawn from the river mouth with the respective radius, see Fig. 3.8 A for the procedure. The position of the points at a given ordinate of the plot has nothing to do with their specific geographic position at the circle segment; they reflect only their gradient values. Thus, the lower and upper undulations of the rims of the point clouds are artefacts without a specific spatial meaning. The convex upper shape of the point cloud is important, indicating the maximum gradients and their completeness or non-truncation. The very regular and curved point clusters of the Amendolea [02] and of the Allaro [16] (Appendix B 02 and 16) belong to first order basins which climb up to the central Apenninic divide and, as a consequence, enjoy the planations round about the altitude of 1000 m. The vertex of the clusters or the maximum gradients are to be found about 14 km from the river mouths, more or less at the center of the basins or slightly shifted towards the interior. Even without knowing the rock distribution of the area, it can be concluded from this that the degree of gradients first of all is a function of their position in the mountains and that the respective rock type plays only a secondary role, unless the different rock types themselves are zoned accordingly. We will meet this problem again when discussing the regional distribution of the erosion rates.

The truncated distributions of the small Aspromonte rivers (Appendix B 03 – 06) stand for second or third order basins which do not reach to the main divide and which consequently have no share in the planations and decreasing gradients. The basins of the Zomaro Province, however, sharply cut, are first order basins that reach the main divide

(Appendix B 09–12). Because the high plains of this province drain mainly towards the west, these distributions end abruptly with the high gradients of the valley heads.

The combined point cluster of all basins 01–19 (Appendix B 20) permit a complex interpretation. It displays a homogeneous segment of a half circle with a somewhat diffuse, but still distinct upper limitation. The maximal distance to the sea is 28 km, the maximal gradient 40°, the extreme values not taken into account. The vertex of this point cloud is somewhat flattened. This means that the gradients of the Ionian slope of the Calabrian Massif can be divided into three zones: zone I, 8 km wide, which follows the cost line, where the maximal gradients increase landward from 0 to 40° (Fig. 3.9); subsequently an intermediate zone II, 9 km wide, with maximal gradients not above 40°, extreme values excepted; and finally an internal or central zone III, 10 km wide, with again decreasing maximal gradients from 40 to 5°. Zone III, however, is developed only in the Aspromonte Province and Serre Province. To understand the Calabrian Massif as a sediment-producing system, it is important to realize that zone II represents a region with increased gradients and slope mobility and that this zone II has different positions in different provinces. In the Zomaro Province zone II lies central; it is internally bordered immediately by the central Apenninic main divide (Fig. 3.9). In the massifs, however, Aspromonte and Serre, zone II has an intermediate position. If and how strongly a rock may become a sediment contributor is determined not only by its physical properties but also by its position in the mountain range.

The three zones can be outlined at the gradient chart. They form elongated, subparallel belts and we call zone I the lower low gradient belt, zone II the high gradient belt and zone III the upper low gradient belt. As will be shown in the next chapters, the erosion rates of the area have to be classified and located in exactly the same way, so that we can also speak of low and high erosion belts, or more

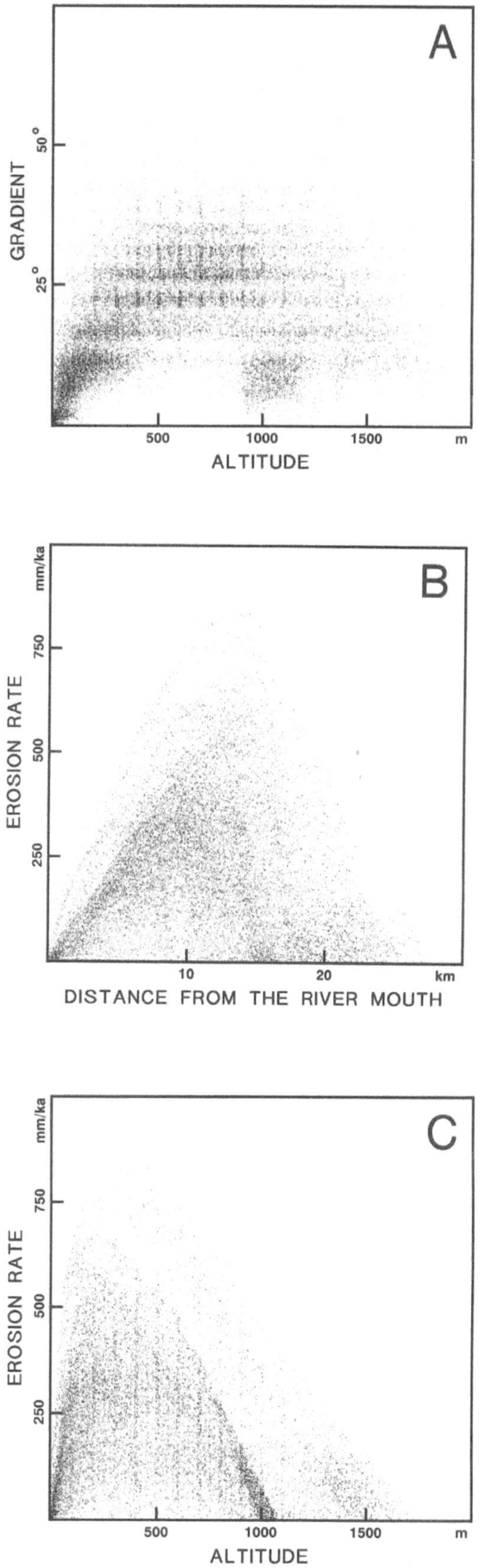

Fig. 3.11 A–C. Overall plots of the basins 01–19; each *point* represents a point of the 250-m geomatrix of the 1832-km^2 study area. See Fig. 3.8 A for the method. **A** Gradient (slope angle) versus altitude; appreciable planations or low gradient areas occur only at 0 to 100 m and at 900 to 1200 m, in other regions gradients usually do not drop below 5°. **B** Erosion rate versus distance from the coast; maximum erosion rates occur only in the "high energy belt" about 8 to 17 km from the river mouth. **C** Erosion rate versus altitude; peak erosion rates do not occur high up in the mountains but at intermediate altitudes of about 500 m

simply of the lower and upper "low energy belts" and the "high energy" belt in between.

The small elongated point cluster at about 17 km from the coast with slope angles up to 70° belong to the Careri [09] (Appendix B 09), this indicates extremely risky steep slopes of granites and schists immediately at the escarpment of the high plain.

Gradient Versus Altitude. This presentation (Fig. 3.11 A) barely differs from the plots gradient/distance treated in the foregoing section. However, the relation gradient/altitude was examined, because it is possible, in this case, to consider the entire area, interfluves included, a relation we were unable to calculate with gradient and distance because of the mostly unknown position of the interfluve outlets to the sea.

The point cluster is not as regularly dotted as the combined plot gradient/distance described above because the interfluves are added, but it demonstrates two distinct gaps at its lower side, one between 100 and 900 m and another one above 1200 m. This proves clearly that appreciable planations only occur at 0 to 100 m and at 900 to 1200 m and that in other regions gradients usually do not drop below 5°.

3.3 The Slope Directions (Aspects)

The second geometrical quality of a slope, besides its angle or gradient, is the direction in which the slope faces, the so-called aspect. Examination of the aspects of our Calabrian research area yielded very surprising geomorphological features but no evidence for a possible control of sediment production, so we will treat this matter only briefly.

Because each point of the geomatrix is surrounded regularly by eight neighbouring points (Fig. 3.12 A), the slope angle directions or aspects belong automatically to one of the eight 45° classes from 0 to 360°. They are plotted in eight computer-generated "single value" maps, N, NE, E, SE and so on (Fig. 3.13). The widely scattered point clusters of all eight maps demonstrate once more the high degree of geomorphological differentiation in Calabria. Furthermore, the maps demonstrate that notwithstanding their frequency, all aspect directions are very regularly spread over the whole area without any dependence on rock type or altitude, as expected from field experience. Thus, aspects or slope angle directions are not further treated here as provenance arguments.

Nonetheless, the eight aspect maps, together with the rose diagram (Fig. 3.13) and the histogram (Fig. 3.12 B) exhibit a surprising symmetry in the aspect distribution. A strong central mode demonstrates the predominance of aspects normal to the general SW-NE strike of the area. That is, 28% of the aspects face southeast and represent elements obviously corresponding to the original virginal surface of the emerging Calabrian Massif and of some subsequent valley sides. Strangely enough, the opposite direction, NW, yields only as little as 4%, although strong subsequent valleys are developed parallel to the general strike. Obviously, these subsequent valleys are non-symmetrical in the sense that they develop large SE- and small NE-facing slopes.

This SE-facing first mode is accompanied by two secondary modes of SW- and NE-facing aspects with 16 and 19%, respectively. These are the valley slopes of the consequent streams, the main rivers of

Fig. 3.12. A Sketch of the point grid of the 250-m geomatrix, orthogonal and diagonal distances from a reference point to the surrounding eight points for the determination of gradients (slope angles) and aspects (slope directions). **B** Histogram of the aspect frequency distribution. The almost perfect symmetry reflects the almost perfect orthogonal arrangement of consequent and subsequent valleys

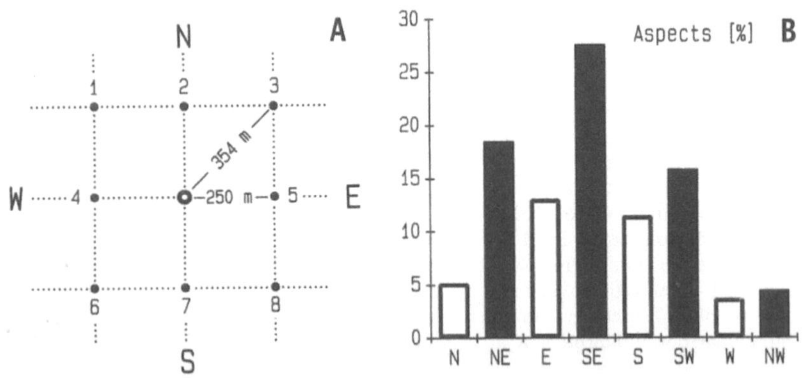

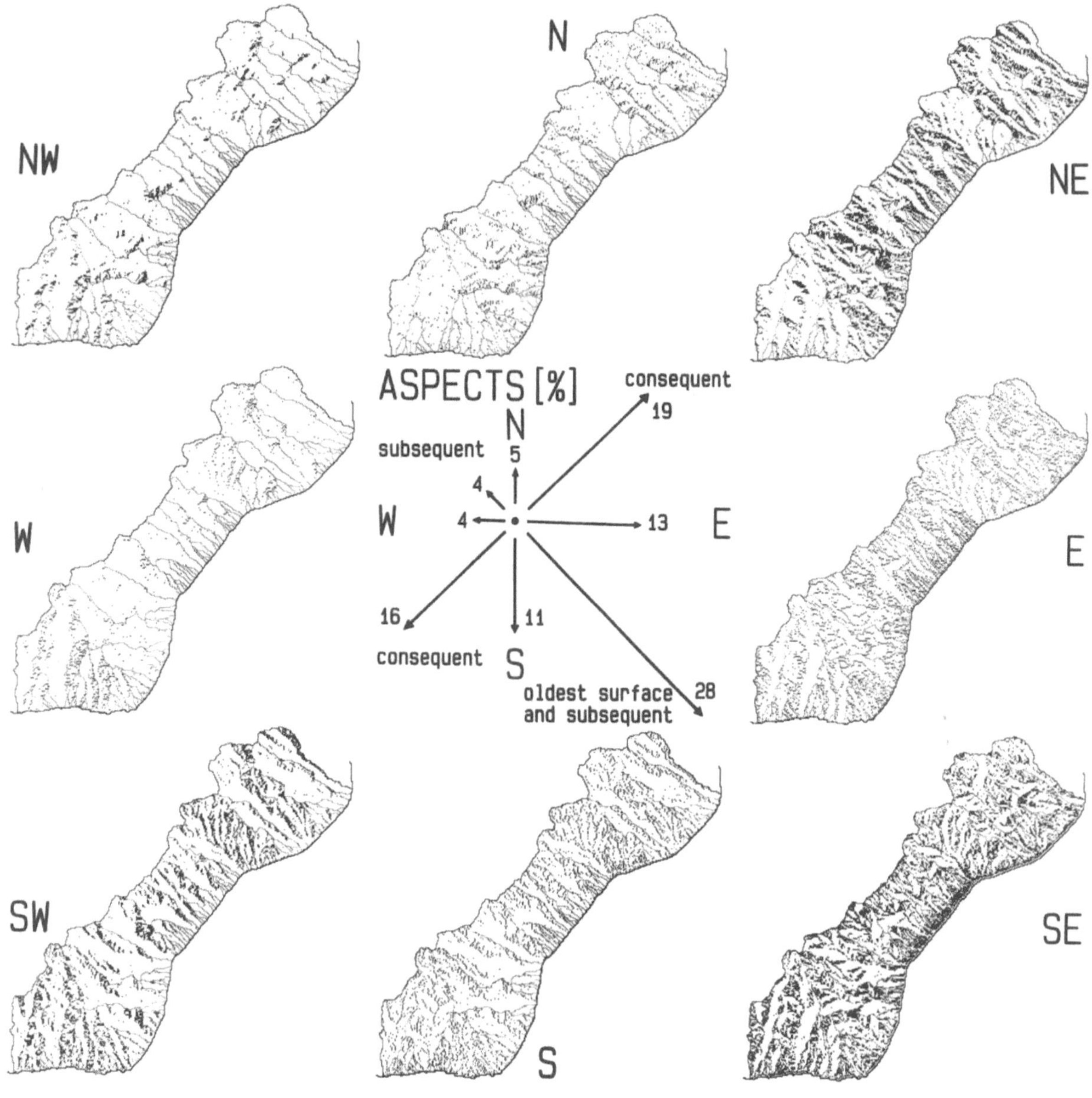

Fig. 3.13. Single value maps of aspects (slope directions) of the octants. All *points* of the geomatrix representing slope directions towards NE are plotted on the graph NE, and so on. Note the striking independence of the aspects from altitude and rock type and the strong symmetry of NE and SW maxima and E and S minima. The rose diagram in the center and the aspect frequency distribution (Fig. 3.12 B) quantify this picture. The primary mode faces SE, this is the oldest surface of the emerging massif. This mode is flanked symmetrically by the secondary modes of the consequent valley slopes. Superimposed are the slopes of the subsequent valley incised perpendicular to the consequent basins and thus subparallel to the primary surface

the Zomaro and Serre Provinces. Because each direction of the graph is made up by thousands of single values from different areas, it is quite easy to over-interpret these findings. However, we treated this basin by basin and found very similar distributions even in the Melito [01] and Amendolea [02], which drain southward.

3.4 Conclusion

It was the goal of this morphometric study to determine relief and dissection quantitatively in order to compare the 19 basins and to detect a possible correlation between basin relief and river-mouth sediment. As will be pointed out in more detail in chapter 10, this expectation was not fulfilled, even if the relief of the 19 basins, the altitude standard deviation, varies between s = 213 m (Aranghia [05]) and s = 506 m (Laverde [07]) and the hypsometric integral, the altitude skewness, goes from sk = −0.14 (Assi [19]) to sk = +1.2 (Careri [09]). Obviously these differences are not large enough to shape the river-mouth sediments differently. Thus it is this type of steep-graded mountain rivers as such which creates this type of river-mouth sediment. Because there are no low gradient rivers in our research area we do not know the critical values of relief and gradient below which a possible influence on the river-mouth sediments may become effective.

The gradients of the Ionian slope of the Calabrian Massif depend on the rock type and on the position in the mountain chain; they are zoned and form an 8-km-wide "low gradient" belt bordering the coast, followed by a "high gradient" belt, 9 km wide and topped by upper "low gradient" areas in the higher Aspromonte and Serre. This concept of zonation would be tautological if the rocks were zoned in the same manner. However, this holds true only for the deposits of argillite and siltstone, which belong mainly to the coastal zones (Figs. 5.8, 5.9). The conglomerates are sometimes zoned, the sandstones even less (Figs. 5.6, 5.7), but the deposits of both rock types also frequently cut the gradient belts. The schists (Fig. 5.4) are zoned only in the Serre Province, while gneiss and granite are by no means zoned and cut the gradient belts everywhere (Figs. 5.2, 5.3). The reconstruction of source area geology by means of its produced sediments is not facilitated by these findings. Unfortunately, it is not only the amount and regional distribution of a rock or its resistency to weathering which determines the amount of its erosional products, but, in addition, also the locus in the mountain chain which defines the efficiency of a source rock. The aspect of the slopes, the slope direction, is by no means dependent on the rock type, either on the position in the mountain range like the gradient values or on altitude. Thus, aspects can definitely be excluded, in Calabria, as factors which influence the formation of river-mouth sediments. Nevertheless, the amazing symmetry of the aspect frequency distribution supports our assumption that the drainage pattern developed randomly, indicating the probability of a more or less homogeneous emergence surface of the rising massif.

Chapter 4

Climate and Vegetation

The interdependence of the factors which share in producing sediments, source area geology, relief and climate, is certainly one of the most interesting aspects of provenance. Plate tectonics create relief, relief models climate, but both tectonics and climate even destroy relief, via weathering, via precipitation and earthquakes, triggering landslides whose debris is transported by floods into nearshore environments. In this chapter we will concentrate on climate and vegetation. Year by year disastrous rainstorms hit the area of the crucial "high gradient" belt not sufficiently protected by vegetation. Every 2 years disastrous floods ravage parts of this or that basin. These are the most striking actuogeologic events in Calabria, but unfortunately also the least predictable ones.

Summary. Calabria has a marine mediterranean climate with a strong, altitude-dependent zonation of temperature, precipitation and consequently vegetation. From the coast to the mountains, mean annual temperatures vary between >16 and <10 °C and mean annual precipitation ranges between <600 and >2000 mm. Rainstorms with precipitations of more than 400 mm a day are the most striking climatic events of the area.

The flora is characterized by an overlap of west-, east- and south-mediterranean plants.

Data Base. The data presented in this chapter, exceptionally, stem from literature. The information about climate and precipitation is extracted from

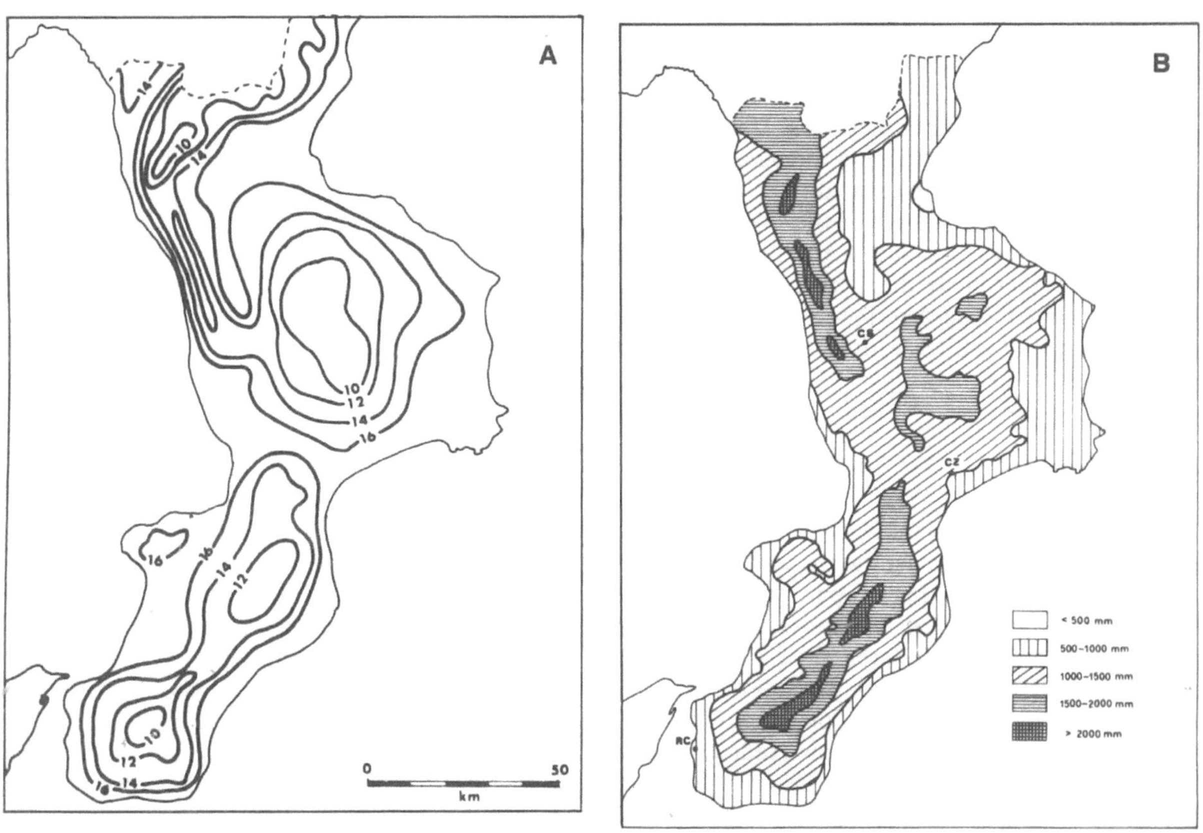

Fig. 4.1. A Mean annual temperatures in Calabria. **B** Mean annual rainfall in Calabria, mean of the period 1921–1970 (Caloiero and Mercuri 1980)

Caloiero and Mercuri (1980) and Versace et al. (1987), the comments on vegetation come from Schneider and Sutter (1982) and from an unpublished manuscript of Drescher-Schneider, who agrees with this condensed version.

Method. For the calculation of the water balance of the area the isohyet map (Caloiero and Mercuri 1980) was digitized and stored in the data bank.

4.1 Climate

4.1.1 Temperature and Precipitation

South Calabria is characterized by a mediterranean climate, with the summers hot and dry and the winters mild. Mean annual temperatures exceed 16 °C in the coastal areas (Fig. 4.1 A) and fall below 10 °C in the Aspromonte.

The mean precipitation accords strictly with the altitude (Fig. 4.1 B). The mean annual precipitation of our research area, the Ionian slope of Aspromonte and Serre, is 1358 mm, compared to the Calabrian mean of 1176 mm and the Italian mean of 970 mm. This mean is calculated over 50 years from 1921 to 1970. Maximum values in the Aspromonte exceed 2000 mm, the coastal areas are relatively dry, below 600 mm in the very south. Berlin, in Germany, for comparison, yields 580 mm.

We digitized the isohyet map from Caloiero and Mercuri 1980; Figure 4.2 shows the resulting computer plot. Weik (1988) measured the chemistry of the ground- and river waters of the 19 basins. The results enable us to calculate the water balance of the area and to estimate the discharge of dissolved load; in Chapter 7 we describe water balance and dissolved load, together with an estimate of the evapotranspiration.

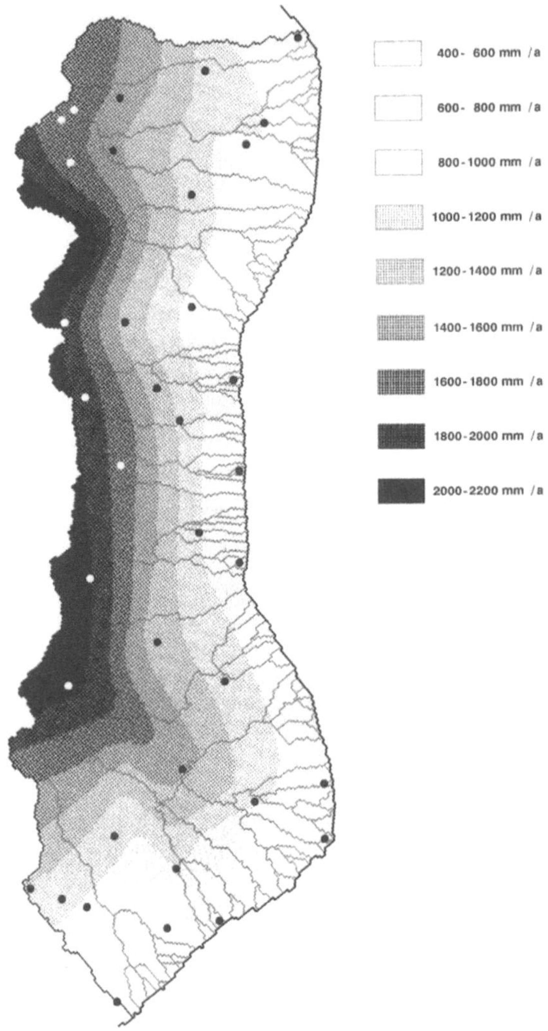

Fig. 4.2. Annual rainfall in Calabria, mean of the period 1921 – 1970; computer-generated isohyet map based on values published by Caloiero and Mercuri (1980), pluviometric stations (●) from Versace et al. (1987)

4.1.2 Rainstorms

For the purpose of this study, at first glance, a regional correlation of precipitation versus gradient or erosion rates would appear appropriate, in order to become aware of possible causal relations to sediment production. However, such a correlation does not exist; the simple comparison of the gradient map (Fig. 3.9) and the precipitation map (Fig. 4.2) makes clear that the "low gradient" belts fit with both the coastal areas of very low and the mountainous top areas with very high mean precipitations. As will be pointed out in Chapter 6, gradients and erosion rates are closely related. Consequently, the same non-relation is valid for the relationship precipitation/erosion and thus mean precipitation proves out to be a useless measure for evaluating erosion in Calabria.

Nevertheless, it is a truism that rain is of striking erosional importance. Anyone, who has ever experienced a Calabrian cloudburst with a precipitation of more than 400 mm a day knows about the erosive force of these events. This is proved by the record given by Caloiero and Mercuri (op. cit.).

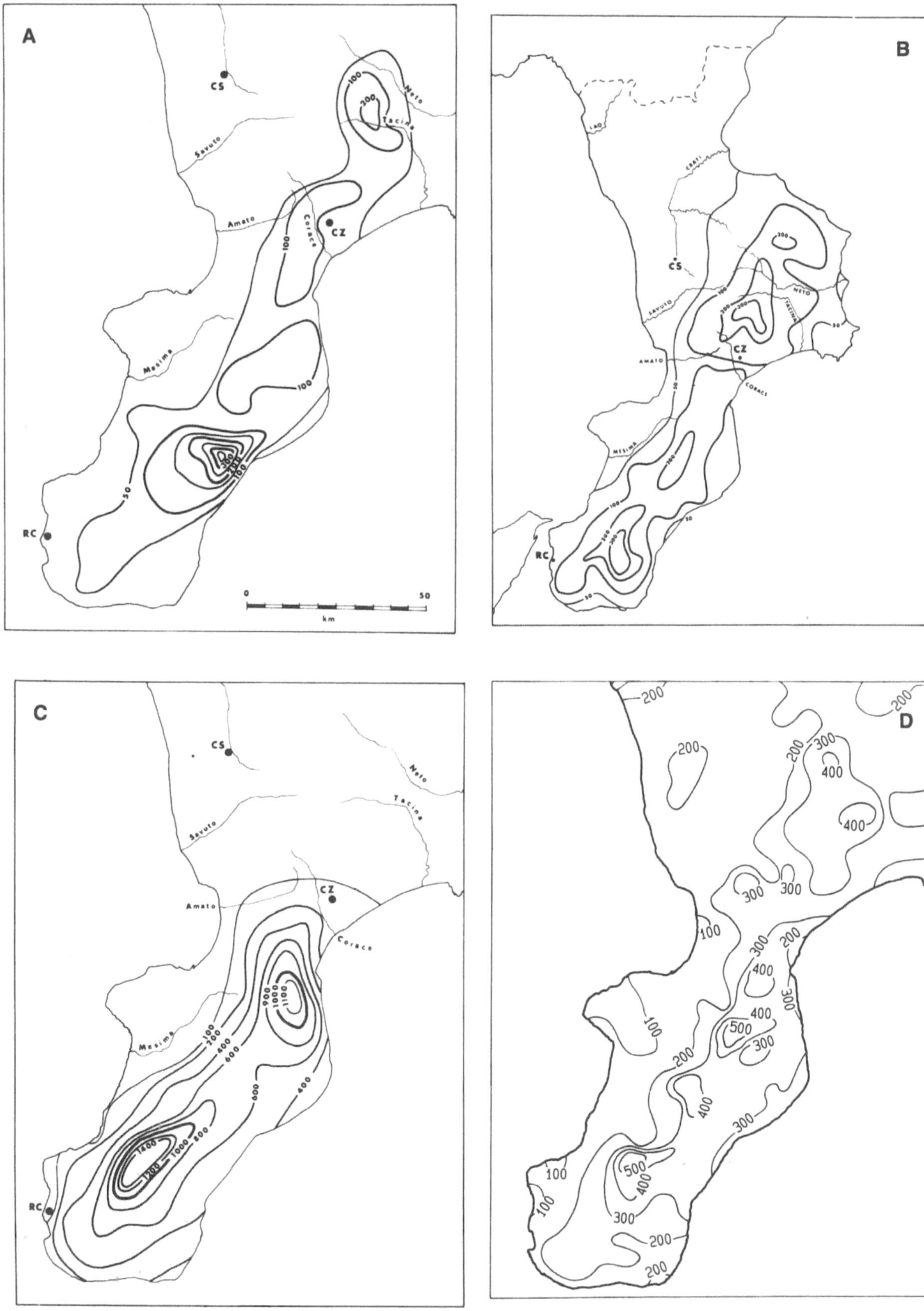
A
CS
Neto
Tacina
Savuto
Amato
Corace
CZ
Mesima
RC
50
100
200
300
0
50
km
B
LAO
CRATI
CS
SAVUTO
NETO
TACINA
AMATO
CZ
CORACE
MESIMA
RC
50
100
200
300
C
CS
Neto
Tacina
Savuto
Amato
CZ
Corace
Mesima
RC
100
200
400
600
800
900
1000
1100
1200
1400
D
100
200
300
400
500

From 1921 to 1970, 31 heavy rainstorms occurred in Calabria, referred to the entire province, and 25 disastrous floods. The isohyet maps of three selected floods are plotted in Fig. 4.3; among them the most disastrous event of the period observed, the 3-day rainstorm from October 16 to 18, 1951. This record of S. Christina d'Aspromonte (Fig. 4.3 C) is 534 mm the first day, 532 mm the second day and 429 mm the third day, adding up to 1495 mm in 3 days, a kind of deluge. Figure 4.3 D illustrates the maximum daily precipitation of the period 1921–1970.

It goes without saying that these rainstorms create disastrous floods, the peak discharge of the Careri [09] for those days being estimated as yielding 1000 m^3/s, that of the Amendolea [02] as 930 m^3/s. Immense damage was registered; in our research area alone there were 64 victims. The huge Ustra landslide in the Buonamico basin [08] was triggered by this event (Fig. 15.6).

4.2 Vegetation

In Calabria the zones of west-, east- and south-mediterranean flora overlap, and thus consists of plants typical of the Iberian peninsula, of the Middle East and of North Africa. In addition, the quasi-insular position of Calabria has been conducive to the formation of endemits, bringing a further increase in plant variety.

The research area, the Ionian slope of the Aspromonte and the Serre, is exposed mainly to the southeast and shows the typical zonation:

1000–ca. 2000 m: *Fagion* (beech woods)
700–ca. 1000 m: *Quercion pubescentis* (deciduous oak woods)
300–ca. 700 m: *Quercion ilicis* (holly oak woods)
0–ca. 300 m: *Oleo ceratonion.*

Man has lived in Calabria for a long time and his influence on vegetation has been considerable, especially in the lower altitudes. In the Greek era the region belonged to Magna Graecia. Along the coast there were several important trading towns with populations vastly more important than those of today's coastal towns. In the Roman era the region was grossly neglected and the number of inhabitants decreased accordingly. It was further decimated around 1000 A.D. by the Saracens, whose repeated attacks drove the population back into the foothills. All these historical changes have deeply marked the vegetation of Calabria.

The former vegetation of the plains and the lower hills, the various associations of *Oleo ceratonion*, have completely disappeared. The plains and the slightly inclined areas are covered with large cultivations of citrus and olive trees. The slopes up to 300 m are used for the cultivation of corn and vegetables or more frequently as pastures. On extensively used pastures *Ampelodesmos mauretanica*, a grass growing to heights of 2 m, may cover more than 80% of the ground.

In certain places the holly-oak woods (*Quercus ilex*) are well preserved, but the surface originally covered by these woods has been strongly reduced: some of it has been transformed into plantations of olive trees, some was cleared or has been used as pasture with shrubs beginning to invade these places again. The undergrowth in the dark woods of *Quercus ilex* is scanty and has been even more reduced by wood pasturing.

The highest evergreen oak woods climb to about 800 m altitude, but between 700 and 1000 m deciduous oak woods are the most common vegetation type. They are characterized by a broad variety of tree species: *Quercus farneto, Quercus cerris, Quercus pubescens, Castanea sativa, Fraxinus ornus, Acer opalus, Acer pseudoplatanus, Celtis australis* and in most places *Alnus cordata.*

Between altitudes of 400 and 1000 m relatively large surfaces are covered only by bracken (*Pteridium*) and grass (Fig. 3.3). These pastures are so highly overgrazed that they are bare in autumn. We are dealing here with the areas of highest landslide risk after rain impact. As soon as these pastures are no longer used, shrubs such as *Sarothamnus scoparius, Calicotome spinosa* and *Erica arborea* begin to spread, and after some time the typical "macchie" develops, covering large areas.

Associations of beech woods are the dominant vegetation on the top levels. Between 1000 and 1800 m beech woods are interspersed with fir trees, *Abies alba.* Higher up, only beech will be found. On the transition between deciduous oak woods and beech woods, frequently plantations of *Pinus laricio* are to be found. These, together with the beech woods, are all that has been left of the old Silva brutica, famous in Roman times and described by Livy, Pliny and Virgil.

Fig. 4.3 A–D. Isohyet maps of the precipitation of selected rainstorms and the daily maximum (mm/day). **A** November 11, 1927. **B** November 21, 1945. **C** October 16, 17 and 18, 1951. **D** Daily maximum precipitation of the period 1921–1970 (Caloiero and Mercuri 1980)

From a study of this vegetation, strongly influenced and partly destroyed by man, the question arises as to what the vegetation looked like before human interference started, whether very different conditions of erosion are to be expected. Pollen analytical studies of two peat bogs (Mastrogiovanni at 1030 m, see Fig. 3.3, and Canolo Nuovo 950 m) show the development of the vegetation during the last 8000 years. The following zones can be distinguished.

Zone A. The area studied was covered by oak forest (*Quercus pubescens, Q. farneto* and *Q. cerris*) mixed with abundant birch, fir and beech. This combination was the result of higher air moisture and perhaps lower temperatures. About 8000–7000 B.C.

Zone B. The forest consisted mainly of holly oaks (*Quercus ilex*) with lower percentage of fir, beech and deciduous oaks. This can be explained with a warmer and drier climate. About 7000–6300 B.C.

Zone C. A combination as in Zone A. About 6300–4500 B.C.

Zone D. Up to about 4500 B.C. the forest composition did not change much except for the introduction of *Olea europaea* by the Greeks and *Castanea sativa* and *Juglans regia* by the Romans. The beginning of the large clearings and the destruction of the woods date back about 1000 B.C.; their possible influence on erosion is discussed in Chapter 15.2.2.1.

4.3 Conclusion

The conclusion to this overview of precipitation and vegetation for our provenance concept is relatively simple. Man's impact on the lowland is admittedly strong as far as the influence on vegetation is concerned, but this does not go hand in hand with increased erosion and sediment production. The lowlands are in the "low gradient" belt, exhibit little relief and receive little rain; excess rainfall is very seldom. The opposite is true of the area of about 400 to 1000 m. Here a kind of positive feedback takes place: steep slopes – we are in the "high gradient" belt – and frequent landslides hinder the development of dense woods, and this, in turn, implies reduced soil and rock protection against sliding. Extended areas of this zone of increased landslide risk are predominantly covered only by bracken, grass and macchie. In addition, mean precipitation climbs up to 1800 mm/a and excess rainfall with >400 mm per day may occur.

Higher up in the mountains, the landslide activity is significantly reduced. In the areas above 1000 m lower gradients, lower relief and an amazing vegetation of beech woods of really mid-European character cause relatively stable conditions, notwithstanding the mean precipitation exceeding 2000 mm/a and the occurrence of excess rainfall.

As a consequence, a causal, unidirectional influence of vegetation on erosion is hard to detect. It goes without saying that vegetation as such protects the underground and this is the reason for the many governmental efforts to establish reforestation programs. Unfortunately, it has been impossible until now to abolish the traditional pasture by goats. By this the positive feedback – or its negative effect – is enhanced, because the pasture by goats concentrates on the steeper slopes not usable for agriculture.

In Calabria, the irregularity of heavy rainstorms in space and time, its "event character", makes the factor climate hard to calculate for provenance and mass balance considerations.

Chapter 5

Rock Units and Their Regional Distribution

The objective of our study is to help answer the question as to in how far, under given geomorphologic and climatic conditions, the rock types of a mountainous region are reflected in the river-mouth sediments or subordinate marine deposits by means of their erosional products.

Therefore, the quantitative analysis of the produced sediments has to be accompanied by the quantitative registration of the geology, i.e. the rock distribution in the source areas. This is best done by digitizing geological maps, whereby, for reasons of economy, the differentiation wanted should be advanced only as far as needed and not as far as possible.

However, such a study of regional rock distribution has to be completed, calculating the respective erosion rates, because the amount of erosional debris delivered from a given rock type into the fluvial system depends, among other causes, not only on its areal extent but also on its erosion rates. We call the geological composition of a source area, weighed by its erosion rates, the effective composition; the procedure is treated in Chapter 6.

In this chapter the mere regional distribution of the rock types is described; Chapter 8 gives the petrographic details.

Summary. We digitized the Carta Geologica Della Calabria (CGDC) 1:25000 using the 250-m geomatrix. Granitic rocks cover 24% of the basins 01–19, metamorphic rocks yield 37%, and 39% are made up by sedimentary rocks. We condensed the 55 lithological rock units of the CGDC to eight source rock types to which the provenance discussion of this book refers: granite (24% area), gneiss (7%), schist (30%), limestone (1%), conglomerate (9%), sandstone (9%), siltstone (12%) and argillite (8%). It goes without saying that source area geology is by far the most important variable controlling the composition of river-mouth sediments.

5.1 Digitization of Geological Maps

5.1.1 Requirements for a Geological Map as Base for Digitization

The comparison of source and sediment has some disadvantages that are in the nature of things and that, as a consequence, are hard to remove. Many important and even mappable properties of the source areas are per se not transmittable to the river-mouth sediments and some important components of the river-mouth sediments are not traceable back to definite regions of the source area.

Barely transmittable elements of source areas are the structures and textures exceeding the dimensions of a rock fragment. They are usually destroyed during erosion and transport and consequently not transmitted. The macrofacies of a fossil sandstone body such as coastal berme, fluviatile channel fill or turbidite is not legible from its erosional products; the same is true for its possible former tectonic position, as part of a nappe or autochthonous or the like. This holds true for its age, the exception of resedimentated fossils not being taken into account. Thus the interpretation of macrofacies, tectonic setting and the age of many source rock formations are mostly beyond the reach of provenance research, again, with seldom exceptions, and to this extent these topics are not demands to be fulfilled by a geological map for source rock registration.

Barely usable elements of the river-mouth sediments means that even clear signals frequently cannot be retraced to specific source areas. The gravel of the fluviatile and coastal sediments of the Aspromonte Province sometimes contains very striking marbles and frequently even more striking pegmatites. These rocks, small exceptions not being taken into account, do not appear on the geologic maps. The pegmatite veins swarm through the metamorphic rocks of the Aspromonte and Zomaro Provinces and the marble occurs as extremely rare, mostly thin intercalations in the same formations. Both rocks are barely mappable and consequently do not appear on the maps. Thus, unfortunately, these easily determinable elements of the fluvial and coastal sediments cannot be retraced to defined source areas. Certainly the quality of the petrographic signal of the gravel as such remains.

These circumstances demonstrate that for the digitization of a geological map certain prerequisites have to be fulfilled in any case, meanwhile other properties, very important for other purposes, are more or less useless. Strong lithological differentiation and high resolution are probably the most important prerequisites. Compared to this, facies interpretation, tectonic setting and datation are less important.

Only against this background is it justifiable to use the geological map of Calabria, the Carta Geologica della Calabria (CGDC). Thirty years old, this voluminous edition does not represent the present-day level of stratigraphic knowledge, and information about structural geology was never intended by the authors. But this diminishes its value only for the interpretation of the geologic development of the Aspromonte and Serre. For the purpose of our actuogeologic analysis of the massif, as a source for the production of terrigenous sediments, this work proves to the a treasure chest, especially because of its high resolution of 1:25 000 and its lithological differentiation. The reader interested in modern analyses of the historical development of Calabria is referred to the authors cited in Chapter 1, position and geological framework.

Data Base. Carta Geologica della Calabria (CGDC) 1:25 000.

Terminology. The formations of the geological map bear only an index with chronostratigraphic and lithologic abbreviations instead of the usual binominal nomenclature. For the purposes of this investigation we reduced this to simple rock names like granite or siltstone.

Method. The information of the geological map was condensed to lithologically meaningful units, the most important rock types. The regional distribution of the combined units was registered by means of the 250-m point grid of the geomatrix and stored in the data bank.

Table 5.1. Abbreviations of the epochs and lithology indexes of the geological map, the Carta Geologica della Calabria (CGDC)

Epochs		*Indices*	
G	Jurassic	cl	conglomerate, conglomeratic
O	Oligocene	ar	sandstone, with sandstones
M	Miocene	s	sand, sandy
P	Pliocene	ss	silt, silty
q	Pleistocene	a	clay or claystone
		ac	argille scagliose
		m	marl
		c	limestone
		g	gypsum
		t	evaporitic limestone

5.1.2 The Information and its Compression for Digitization

The field work for the Carta Geologica Della Calabria, in the following cited as CGDC, was done in 1958–1963 by the Compagnia Aero Ricerche (Burton 1971). The scale of 1:25 000 instead of 1:100 000, unusual for geologic maps, implies an increased linear resolution by a factor of 4. Thus even very small source areas of 10 or 20 km^2 can be registered in detail. In its time this map represented a very modern and differentiated mapping and air photo evaluation. Our area of investigation, the Ionian slope of the Aspromonte and Serre, requires 30 sheets of the CGDC. On these 30 sheets the authors discriminated 55 formations, without, however, the usual binominal terminology of a geographical locality name and a lithological description. The formations are identified only by an index, for example M_{1-2}^{cl-s}, whereby M stands for the epoch, Miocene, cl for clongomerate and s for sandy. The lower index 1–2 stands for additional chronostratigraphic subdivisions (see Table 5.1 for all indices). Furthermore, the legends of the CGDC give additional lithologic data, fossil content and geotechnical hints for each formation.

It was technically impossible for us and not meaningful for our purposes to digitize all these 55 formations. As a first step, we combined all lithologically similar source rocks with similar geotechnical data, but still within the single epochs. For the final evaluation of the single basins we then combined all similar rock types into eight units, now extending over the limits of the epochs. Figure 5.1 gives an even more condensed overview.

At the same time, the much smaller group of the Holocene sediment bodies was also registered.

This strong compression is inevitable for our purposes. It makes no sense to handle too many variables, and many rocks, easy to distinguish in the outcrop, cannot or can only barely be distinguished macroscopically, being small river gravel.

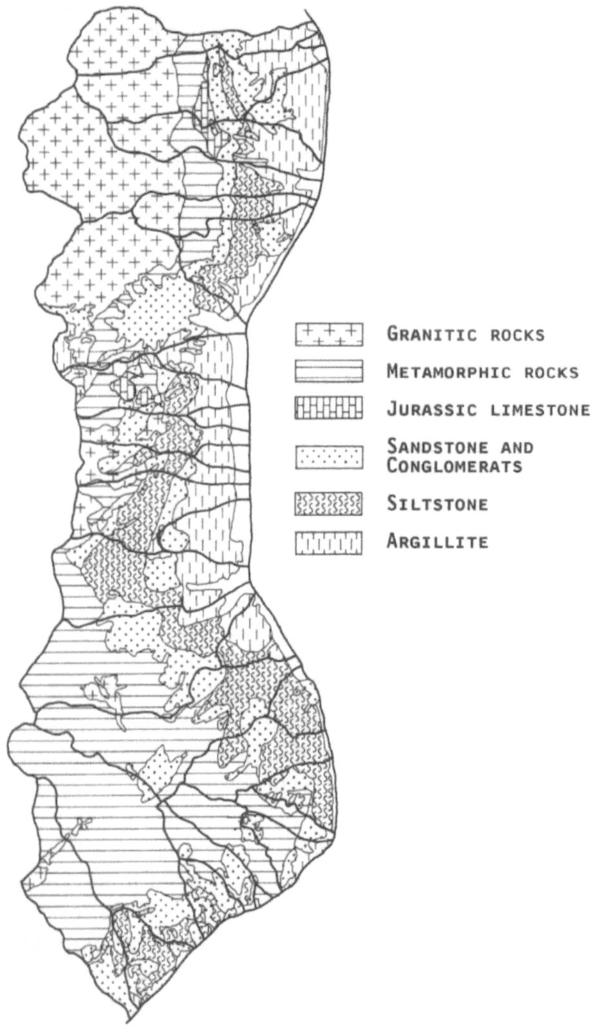

Fig. 5.1. Regional distribution of the most important rock families. For more detailed geological information see Figs. 5.2–5.10

5.2 Regional Distribution of Rock Units

5.2.1 Granitic Rocks

Synonyms of the CGDC

Granites τ
Pegmatites π

All granites of the Stilo and Aspromonte unit and the Cittanova granite (Bonardi et al. 1984, p. 287). On the map, pegmatite veins are outlined only in rare cases.

Petrography. The granites of the Aspromonte Province generally consists of two-mica granites, the Serre Province mainly develops granites between granodiorites and tonalites and quartzmonzonite varieties (see Chap. 8, rocks and soils, for more detailed information).

Regional Distribution. With an area percentage of 24%, granite is the second most common rock after the metamorphic rocks, all area percentages refer to the basins 01–19 only, without including the interfluves. Figure 5.2 shows the regional distribution of the granites on a "single value" map. The most important feature of the granites is the homogeneous, uniform appearance in all provinces, especially the Serre, and their pronounced elevated position. With a mean elevation of 885 m, excluding the interfluves, the granites are the most elevated rocks of the area under investigation (Fig. 5.10A). Nevertheless, the standard deviation of its altitude frequency distribution is relatively high, s = 282 m (Fig. 5.10B). They receive the most precipitation (Fig. 4.2), between 1400 and more than 2000 mm/a, and they carry the most robust vegetation. The histogram of Fig. 5.2 shows the area [km^2] of the granites within the single basins.

In the basins of the Serre Province the granite is always the most common rock type, in the basins of the Zomaro Province it is about equivalent to metamorphic and sedimentary rocks, Novito [13] excepted, and in the Aspromonte Province granites are nearly completely lacking, only the Melito [01] yielding 5%.

The pegmatites hold a contradictory position. Instead of directly forming part of the Serre pluton, the parent rock, the pegmatite veins swarm through almost all metamorphic rocks of the area. Thus the petrographic provenance signal of the pegmatites is not the same as the regional one, the latter being metamorphic.

5.2.2 Metamorphic Rocks

The classification of the metamorphic rocks of the Carta Geologica Della Calabria confronted us with the problem of either putting all of them together and losing information or making a few arbitrary subdivisions identifiable both in the field and among the river-mouth gravels.

Many of the metamorphic rocks come from geosynclinal sediments and are, today, metapelites, metaarenites, metagraywackes or metacalcarenites. They occur individually and uniformly, even over

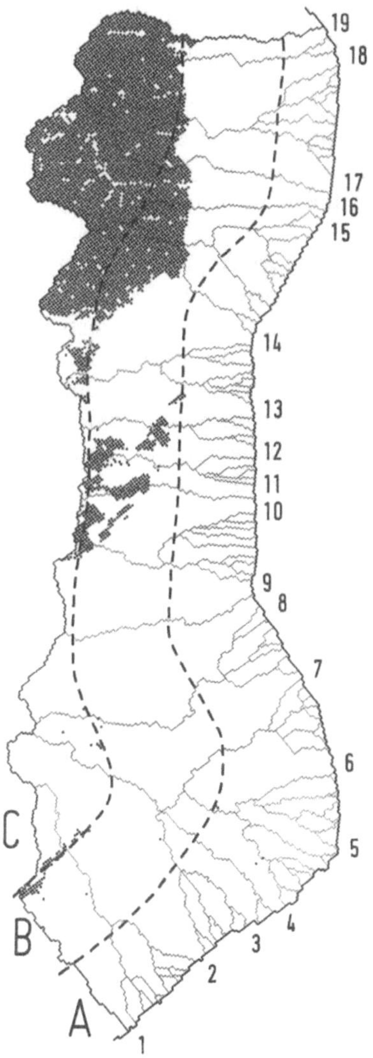

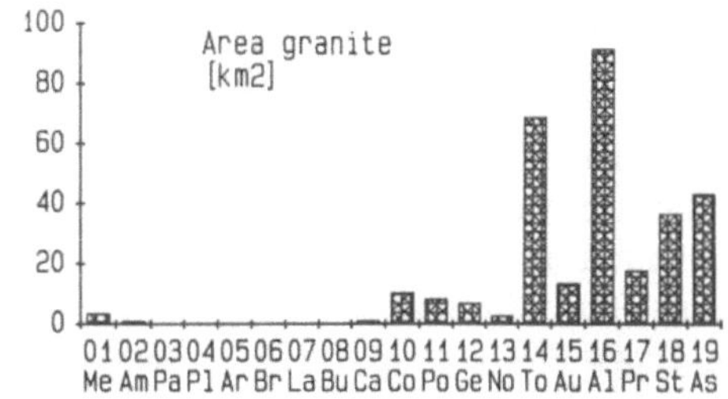

Fig. 5.2. Granite. Computer-generated geological single value map of the Ionian slope of the Calabrian Massif. The Serre pluton is the most uniform rock unit of the area, followed by some scattered exposures in the Zomaro Province. In the Aspromonte granites rarely occur. The histogram gives the area of granite in the 19 basins [km^2]. *01 Me*lito; *02 Am*endolea; *03 Pa*squale; *04 Pa*lizzi; *05 Ar*anghia; *06 Br*uzzano; *07 La*verde; *08 Bu*onamico; *09 Ca*reri; *10 Co*ndoianni; *11 Po*rtigliola; *12 Ge*race; *13 No*vito; *14 To*rbido; *15 A*mu*sa; *16 Al*laro; *17 Pr*ecariti; *18 St*ilaro; *19 As*si. **A** Lower "low energy" belt. **B** "High energy" belt. **C** Upper "low energy" belt (Chap. 6.1.5.1)

larger areas, but frequently they alternate intimately either originally or by thrusting. For example, the augen gneiss often contains portions without macroscopically detectable augen blasts. In the field such a portion would evidently also be mapped as augen gneiss, but the respective river gravel without macroscopically visible augen would never be assigned to an augen gneiss area.

A similar problem exists between schist and gneiss. Grain size and foliation are properties to be judged quite differently at an outcrop of some square meters or when looking at an 16-mm river gravel. But even in the field it is often hard to follow the subdivisions of gneiss and schist made by the authors of the CGDC. On the other hand, if we are dealing with the significant end members of this family, for example the augen gneiss of Roccaforte del Greco in the Amendolea [02] and the phyllites of Stilo in the Stilaro [18], these rocks are easy to distinguish both in the outcrop and among most river gravel.

Therefore we decided to adopt at least the gneiss/schist classification of the map. The main reason for this, however, was the evaluation of the erosion rates (Chap. 6), which demonstrated that the rocks assigned to schists have a significantly (20%) higher mean erosion rate than those assigned to gneiss; 200 mm/ka the schist, 160 mm/ka the gneiss. This demonstrates that the subdivision of the mapping geologists, albeit debatable in detail, makes sense as a whole and should be accepted for our purposes.

Synonyms of the CGDC

Gneiss

Metabasic rocks	β
Amphibolitic rocks	k
Undifferentiated gneiss	g
Migmatite + granite gneiss	gm
Augen gneiss	go
Schist	
Schist and gneiss	sbg
Biotite schists and migmatites	sbm
Biotite schists	sb
Phyllite schists	sf
Quartz-chlorite schists	sm
Marble	cc

All metamorphic rocks of the Stilo and Aspromonte unit (Bonardi et al. 1984, p. 285).

Petrography. All metamorphic rocks are members of the transition between greenschist and amphibolite facies, partially retrograde (see Chap. 8 for more detailed information).

Regional Distribution. Thirty-seven area percent make the metamorphic rocks the most frequent. Figure 5.3 displays the regional distribution of the gneiss according to the definition of the CGDC and Fig. 5.4 shows the schists. The occurrence of metamorphic rocks in the Aspromonte is as consistent and continuous as that of the granitic rocks in the Serre. However, the altitude distribution of the

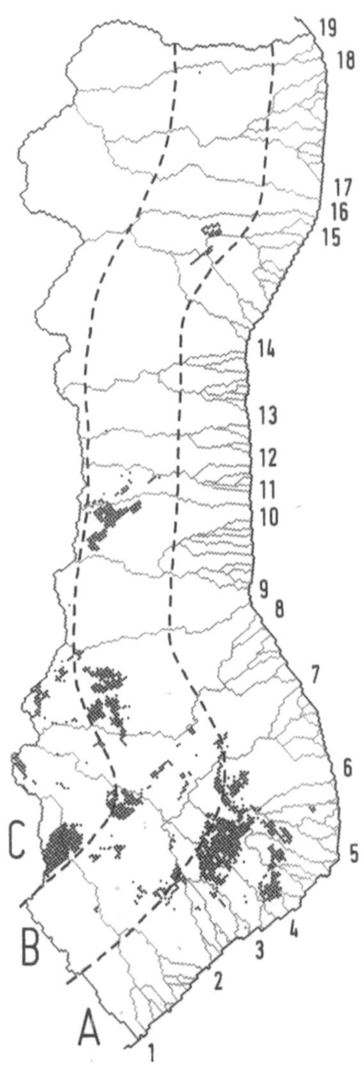

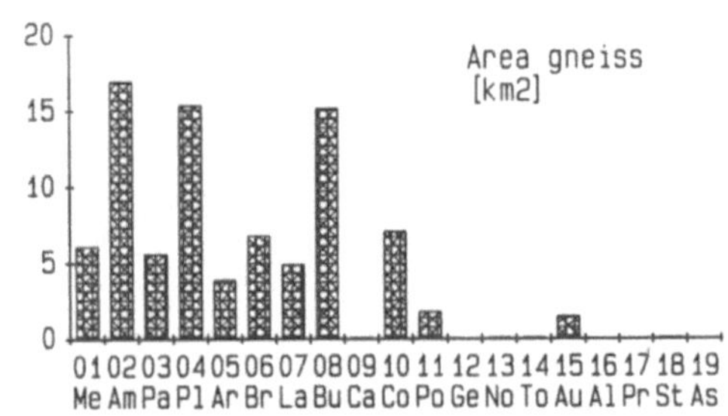

Fig. 5.3. Gneiss. Computer-generated geological single value map of the Ionian slope of the Calabrian Massif. Gneisses concentrate in the Aspromonte, followed by some scattered outcrops in the Zomaro Province, the Serre Province is almost gneiss-free. The histogram gives the area of gneiss in the 19 basins [km^2]. See Fig. 5.2 for abbreviations

metamorphic rocks in the Aspromonte area is much broader, reaching down to the coast, and thus the climatic range to which these rocks are subjected is wider. The altitude of the metamorphic rocks in the Zomaro Province corresponds to that of the granites; in the Serre Province they are concentrated in a narrow belt below the granites, consisting mainly of low resistant phyllites. The phyllites of this area undergo high erosion rates. The mean altitude of the metamorphic rocks, 800 m the gneiss, 854 m the schist, is only a little less than that of granites. The altitude standard deviation of both granite and gneiss is the same (Fig. 5.10B), that of the schists is lower, reflecting more uniform elevations of these less resistant rocks.

All large rivers of the Aspromonte Province are dominated by metamorphic rocks, in the Zomaro Province they balance the granites plus sediments and in the Serre Province the metamorphic rocks drop to about 10% (Figs. 5.3 and 5.4). Thus one may call the Aspromonte a metamorphic province, the Serre a granitic province and the Zomaro Province a mixed area slightly dominated by sedimentary rocks. It goes without saying that this geological composition changing from province to province and thus from basin to basin is an important prerequisite for the reliability of correlations between the river-mouth sediments and their sources.

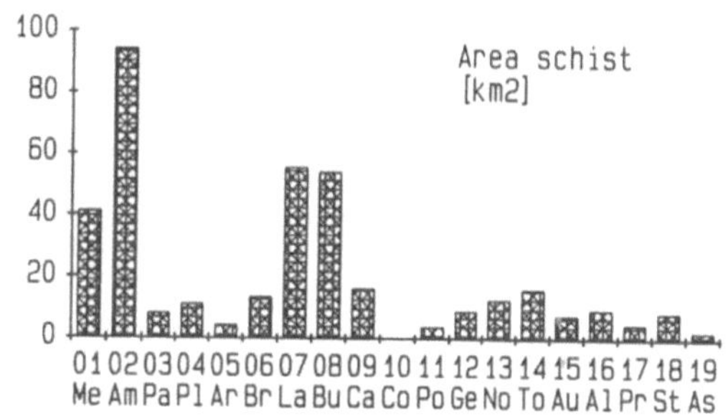

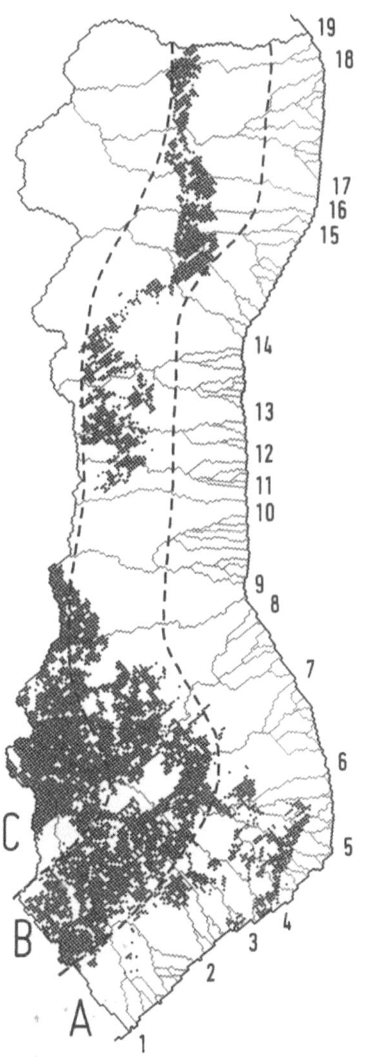

Fig. 5.4. Schist. Computer-generated geological single value map of the Ionian slope of the Calabrian Massif. Schists are almost ubiquitous, concentrating on the Aspromonte Province and north of the Zomaro Province. A narrow belt of schists accompanies the contact of the Serre pluton. The strong decrease of metamorphic rocks from south to north facilitates the discrimination of single basins for provenance calculations. The histogram gives the area of schist in the 19 basins [km^2]. See Fig. 5.2 for abbreviations

5.2.3 Sedimentary Rocks

Sedimentary rocks make up 39% of the basins 01–19. They cover the lower parts of the Aspromonte and Serre Provinces and widen especially in the Zomaro Province. They are nearly exclusively marine siliciclastic sedimentary rocks of the Tertiary and Pleistocene, Jurassic limestones occurring only rarely and yielding less than one area percent.

We grouped the sedimentary rocks into the most important grain-size families, conglomerates and sandstones, siltstones and argillites, plus the above-mentioned limestones. Facies, age and tectonic position were not taken into account. The lithological similarities within these groups, crossing the limits of facies, age and tectonics are so convincing that this decision seems to be justifiable from the perspective of provenance.

5.2.3.1 Jurassic Limestones

Synonyms of the CGDC. G^c of the CGDC. Calcare Portlandiano (Roda 1965), main element of the "Copertura Sedimentaria Mesozoica", the sedimentary Mesozoic cover (Bonardi et al. 1984, p. 285).

Petrography. Solid and compact, mainly massive light limestones, mostly biosparites and dismicrites

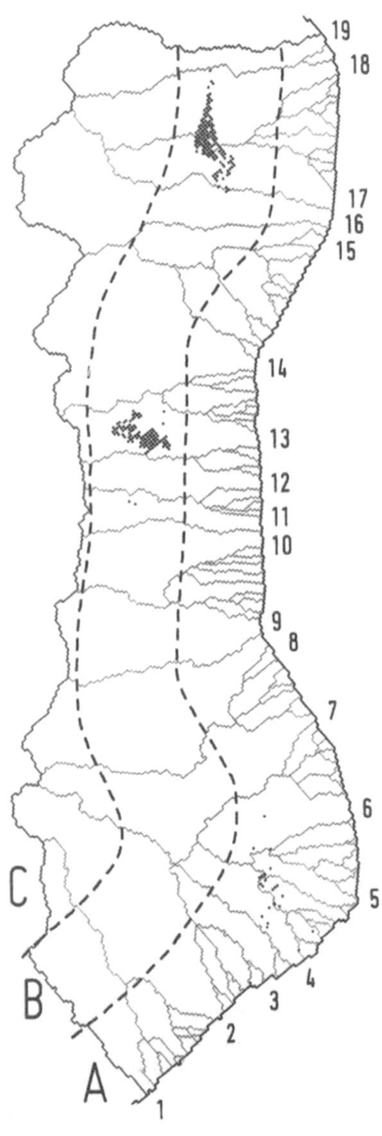

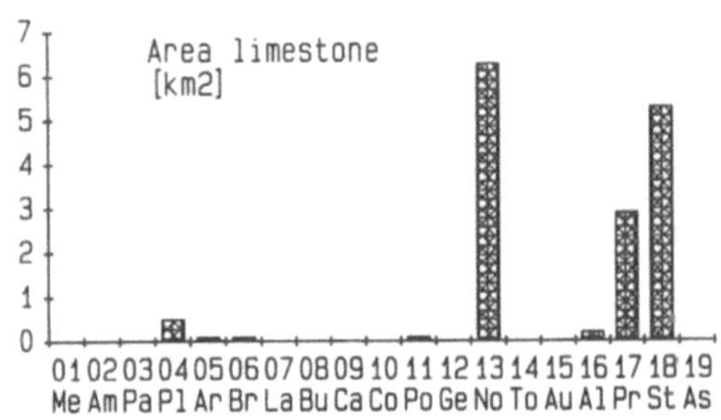

Fig. 5.5. Jurassic limestone. Computer-generated geological single value map of the Ionian slope of the Calabrian Massif. These rocks concentrate on the Novito [13] and the Precariti [17]/Stilaro [18] basins, with very reduced remainders in the Aspromonte Province. The altitude standard deviation of these reef limestones is extraordinarily low. Supposing they all formed in shallow water, today's uniform elevation demonstrates a uniform uplift of the area without major block faulting. The histogram gives the area of Jurassic limestone in the 19 basins [km^2]. See Fig. 5.2 for abbreviations

of a reef and near-reef environment. The fossil content makes it possible to date these limestones as Upper Jurassic, Portlandian (Roda 1965). This very indicative and regionally limited rock produces very easily distinguishable components.

Regional Distribution. Figure 5.5 shows some small, isolated outcrops of the Jurassic limestones at the foot of the south-eastern Aspromonte in the reach of the basins Pasquale [03], Palizzi [05], Aranghia [05] and Bruzzano [06]. Two of the larger and uniform localities lie in the north, one in the Zomaro Province, in the Novito [13], the other one at the margin of the Serre Province, in the basins Precariti [17] and Stilaro [18], coincidentally with the zone of highest erosion rates.

The mean elevation of the limestones is 505 m; the elevations are relatively narrow grouped, so the altitude standard deviation is only $s = 103$ m. This may be seen as a signal for only small tectonic distortions within the area under investigation. The uniformity of the current elevations of the limestones still reflects their uniform position close to sea level during formation.

With a total area of 16.3 km^2 the limestones cover only 0.89% of the area under investigation. However, to provide a small arithmetical example: despite their small source area, the river-mouth sediments of the 19 rivers contain 2.7% limestones in the gravel fraction 25–90 mm. The sandstones, on the contrary, with an area percentage of 9%, yield only 3.4% sandstone gravels. Thus the limestones are eight times more successful in being reproduced in the river-mouth gravels.

5.2.3.2 Conglomerates

Synonyms of the CGDC

Miocene	conglomerates	M^{cl}, M^{cl-ar}, M^{cl-s}
Pliocene	conglomerates	P^{cl}, P^{cl-s}
Pleistocene	conglomerates	q^{cl}, q^{cl-s}.

The conglomeratic portions of the Stilo-Capo d'Orlando Formation and of all younger formations (Bonardi et al. 1984, p. 285).

Petrography. Texture and composition of the conglomerates are very immature, apart from some deposits at the high plain, which are very mature but have no influence upon the river-mouth sediments. The sand portion is always very high. We grouped the sandy conglomerates of the CGDC, (X^{cl-s}), to the conglomerates and we combined the conglomeratic sandstones (X^{s-cl}) with the sandstones. The composition is polymict, very coarse-grained, the components range usually in the dm dimension and sometimes measure more than 1 m^3.

Regional Distribution. The conglomerates yield 9% of the surface distribution of all rocks. Figure 5.6 shows three concentrations of conglomerates, at the south slope of the Aspromonte, in the Zomaro Province, either close to the coast or at the high plain and, finally, the largest and thickest deposits on the lower courses of the Serre rivers Precariti [17], Stilaro [18] and Assi [19].

The regional distribution again makes clear that the conglomerates are always present and that the respective areas, compared over the 19 basins, have the lowest variability. The standard deviation of the amount of conglomerate in the basins 01–19 is $s = 6.2\%$ (sandstone $s = 6.8\%$, argillite $s = 7.4\%$ and siltstone $s = 11.8\%$). As discussed in Chapter 8.1.3.2, the conglomerates produce recycled clasts and are consequently a troublemaker for provenance relationships. The relatively equal amount of conglomerates in each basin, however, makes this influence much easier to calculate.

5.2.3.3 Sandstones

Synonyms of the CGDC

Oligocene	sandstone	O^{ar}
Miocene	sand	M^{s}
Miocene	sandstone	M^{ar}, M^{ar-cl}, M^{ar-ss} M^{ar-a}, M^{s-a}, M^{s-cl}
Pliocene	sand	P^{s}
Pliocene	sandstone	P^{ar-s}, P^{s-cl}
Pleistocene	sand	q^{s-cl}.

The sandstone portions of the Stilo-Capo d'Orlando Formation and of all younger formations (Bonardi et al. 1984, p. 285).

Petrography. The sandstones are very immature, both compositionally and texturally. Thus, distinguishing recycled sand from modern first cycle sand in the river mouths is barely possible. The sandstones are mainly lithic and arcosic or feldspathic arenites.

The diagenetic consolidation differs noticeably from outcrop to outcrop. Thus the amount of sandstone gravels still cemented or already loose sand released to the fluvial transport is hard to calculate. The fact, mentioned above, that the Jurassic limestones are eight times more effective in producing river-mouth gravels, demonstrates that sandstone disintegrates predominantly into sand, in other words, into its initial first cycle sizes.

Regional Distribution. The sandstones yield a 9% coverage similar to the conglomerates, but they are not so uniformly distributed. The main concentration forms the eastern surroundings of the Montalto Massif in the Aspromonte, where the thickness may approach several 100 m. The regional distribution (Fig. 5.7) illustrates the sandstone-rich basins of the north Aspromonte Province and the Zomaro Province. Together with the siltstone, the sandstones are responsible for the abundance of sedimentary rocks, especially of the Zomaro Province and consequently for the lithologic superdivision of the Ionian slope.

5.2.3.4 Siltstones

Synonyms of the CGDC

Oligocene	siltstone	O^{ss}
Miocene	olisthostrome	M^{ac}
Miocene	silt	M^{ss}.

Mainly Argille Varicolori (multicoloured clays), somewhat fewer siltstones of the Stilo-Capo d'Orlando Formation and siltstones of the series

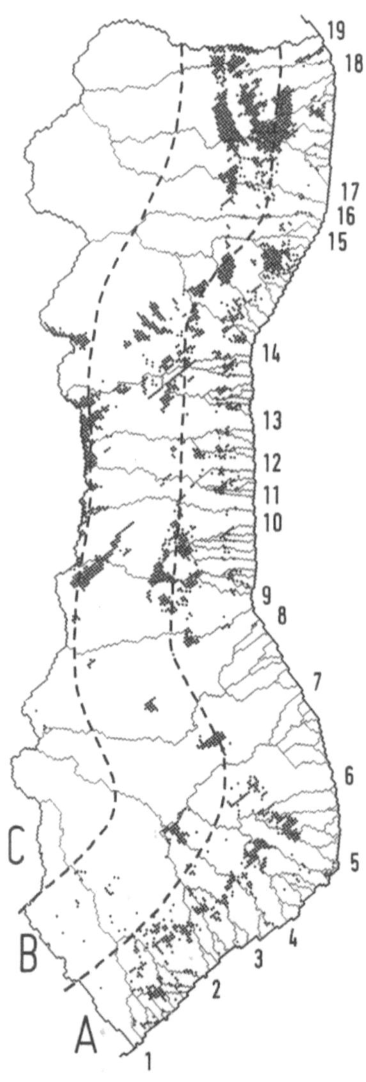

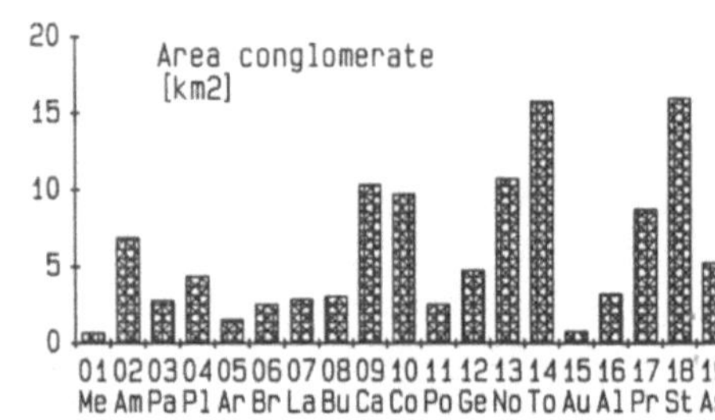

Fig. 5.6. Conglomerate. Computer-generated geological single value map of the Ionian slope of the Calabrian Massif. Conglomerates are widespread, slightly concentrating in the Serre Province. This distribution, together with the sandstone, confirms their "close-by" origin. Larger source areas in the west, however, have to be assumed. The histogram gives the area of conglomerate in the 19 basins [km^2]. See Fig. 5.2 for abbreviations

on top of the Argille Varicolori (Bonardi et al. 1984, p. 285).

The siltstones are, as far as their homogeneity is concerned, the rock of the superlatives in Calabria. Continuity and uniformity of these rocks, elevation, lithology, morphology, development of landslides, vegetation – no other formation is so uniformly developed, so invariable. The only disadvantage of this rock is that the river-mouth sediments contain an average of only about 1% silt and clay, and even the sands of the longshore bars have hardly more fines. This means that roughly 80% of the strong sediment production of this exceptionally uniform source rock unfortunately is not captured in the river-mouth sediments and goes beyond our reach.

Nevertheless, the remaining 20% sand fraction must have some influence upon the development of the river-mouth sediments because of the huge input from the highly eroded siltstones. Apart from this, we have to discuss the possible influence of the siltstones upon the chemical composition of source- and river waters and the petrology of the silts and clays of the river-mouth sediments, even if the amounts are so small.

Petrography. The siltstones are olisthostrome- or debris flow deposits, composed of about 20% sand, 40% silt and 40% clay. The olistholites consist mainly of very mature quartz arenites; micritic limestones occur frequently (see Chap. 8 for more detailed information).

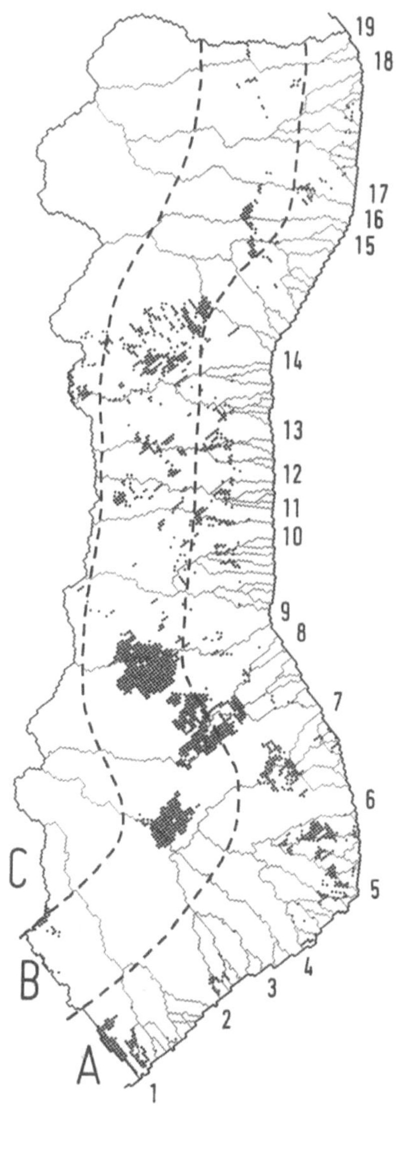

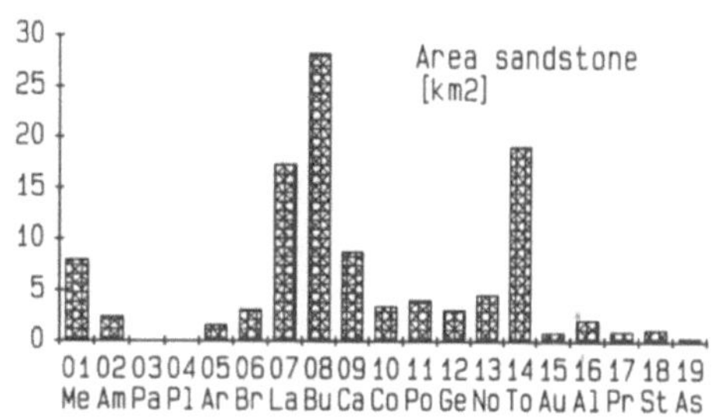

Fig. 5.7. Sandstone. Computer-generated geological single value map of the Ionian slope of the Calabrian Massif. The distribution of sandstones comes close to that of conglomerates (Fig. 5.6); surprisingly, sandstones surround the metamorphic Aspromonte instead of the granitic Serre. The histogram gives the area of sandstone in the 19 basins [km²]. See Fig. 5.2 for abbreviations

Regional Distribution. A percentage of 12% makes the siltstones the most frequently occurring sedimentary rocks of the area. A nearly continuous belt, over 100 km long, leads from the Melito [01] to the Stilaro [18], where it wedges out temporarily at the divide towards the Assi [19] (Fig. 5.8). The maximal thickness is 1000 m, the mean elevation is 221 m. The values of the altitude frequency distribution vary so little that the standard deviation exhibits by far the smallest value of all Calabrian rocks, s = 56 m. In some basins, small and large ones, the siltstones are the area-dominant rocks. The Careri [09] is the most important siltstone river. Due to this the Careri assumes a morphologically exceptional position. The regional distribution demonstrates that the basins of the Zomaro Province display the highest absolute and relative portions of siltstones. The longitudinal, strike-parallel valley of the siltstone is the most striking morphological feature beside the elevations of the Aspromonte and the Serre and the high plain (Figs. 2.5 and 3.6).

5.2.3.5 Argillites

Synonyms of the CGDC

Miocene clay and silt	M^{a-ss}, M^{a-ar}, M^{a}, M^{a-s}
Pliocene clay and silt	P^{a}
Pliocene marls	P^{ms}
Oligocene limestone	O^{c}

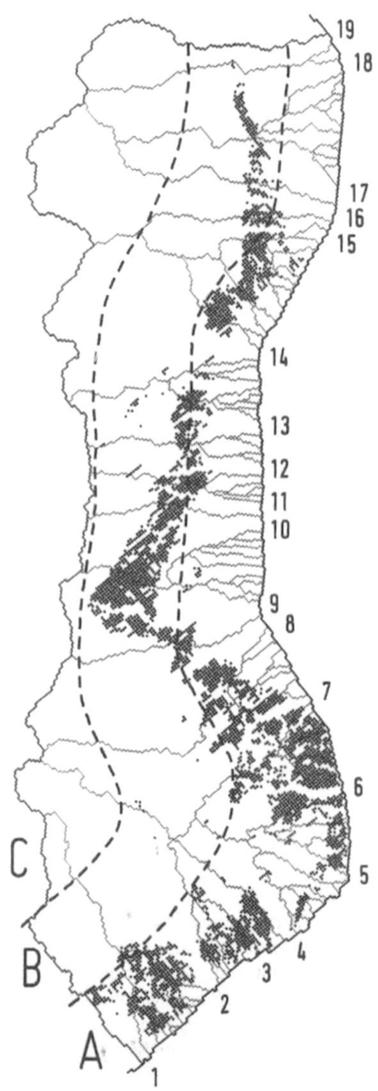

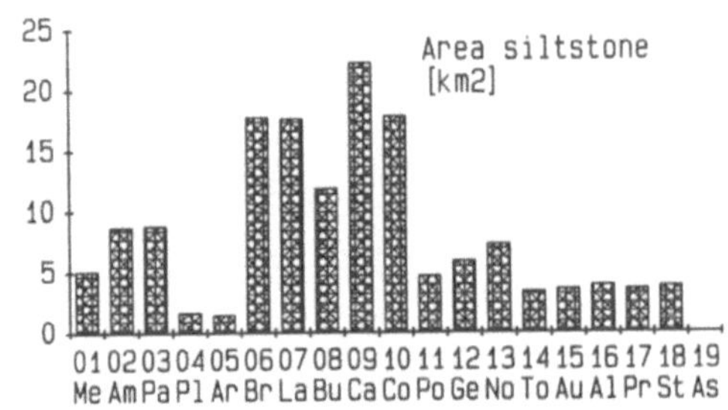

Fig. 5.8. Siltstone. Computer-generated geological single value map of the Ionian slope of the Calabrian Massif. These debris flow deposits, most spectacular and most abundant sediments of the area are >1000 m thick in the Careri [09] area and wedge out completely in the Stilaro [18] basin. The siltstones are barely documented in the river-mouth sediments. The histogram gives the area of siltstone in the 19 basins [km^2]. See Fig. 5.2 for abbreviations

Miocene limestone	M^c, M^{c-ar}
Miocene evaporitic limestone	M^t
Miocene gypsum	M^g
Pliocene limestone	P^c, P^{c-ar}

All claystones and marls on top of the Argille Varicolori (Bonardi et al. 1984, p. 285).

We assigned all the limestones of the CGDC to the argillites, with the exception of the Jurassic reef limestones treated above. This also holds true for the gypsum. This is justified, because we are dealing with very little and very small deposits of a very heterogeneous group of sediments, low resistant and mostly very argillaceous limestones and marls, which hardly ever show up in the river and coastal sediments. They do not merit a grouping on their own, and if assigned to the lithologically completely different Jurassic limestones, they would falsify the appearance of the regional distribution of these much more important rocks. Lithologically, this rock group approaches the argillites, which themselves often develop higher carbonate contents.

Petrography. Not analyzed (see Rio et al. 1989).

Regional Distribution. The argillites hold an area percentage of 8% (Fig. 5.9). They concentrate on the coastal mountains of the Serre and Zomaro Provinces; in the Aspromonte Province they occur less frequently and are limited to the south. An exception to the common vicinity to the coast of these rocks are the basins Novito [13] and especial-

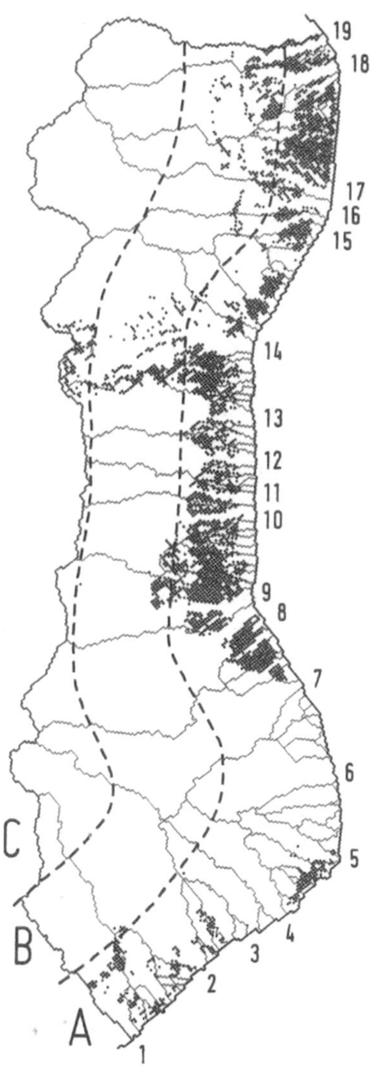

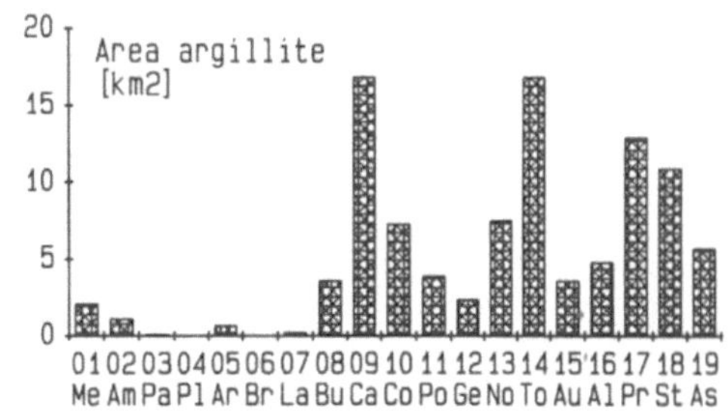

Fig. 5.9. Argillite. Computer-generated geological single value map of the Ionian slope of the Calabrian Massif. Argillites are confined, as a rule, to the coastal areas. The only exception makes the Torbido [14] basin, demonstrating its unique neotectonic position at the south rim of the Serre pluton. Argillites are not documented in the river-mouth sediment. The histogram gives the area of argillite in the 19 basins [km^2]. See Fig. 5.2 for abbreviations

ly Torbido [14]. Here, the argillites continue nearly uninterrupted till the high plain and form an evident transverse element normal to the coast and normal to the common SW-NE strike of the sedimentary cover. The mean elevation is only 200 m (Fig. 5.10A). Thus the argillites are morphologically the lowest element of the area. However, the extension towards the high plain along the Torbido [14] leads to a relatively high altitude standard deviation of the argillites of 121 m (Fig. 5.10B).

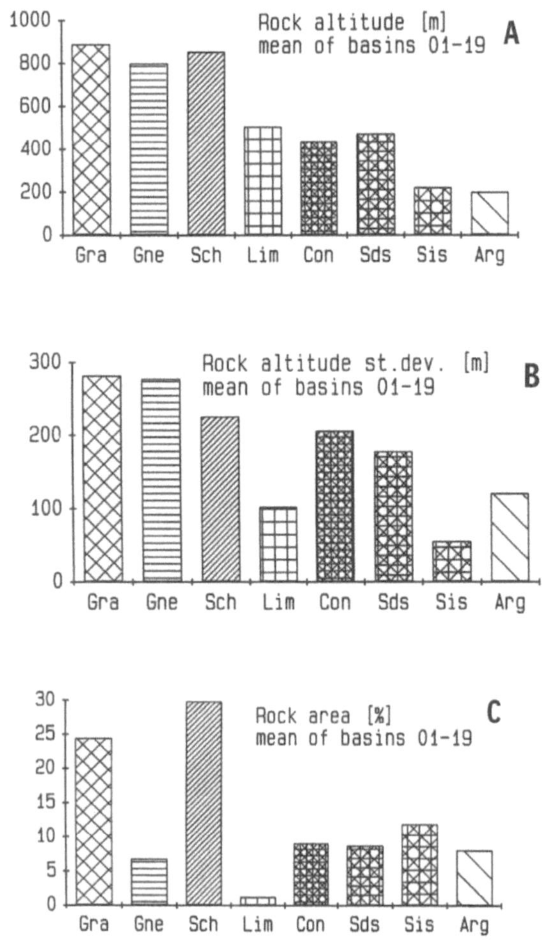

Fig. 5.10A–C. Rock-specific altitudes and areas. **A** Mean altitudes of the single rock types (basins 01–19 only). The position of the crystalline rocks is central and high, the sediments show, as a tendency, decreasing altitude with decreasing age. **B** Altitude standard deviation of the single rock types (basins 01–19). The low value of the neritic limestones makes larger shifts due to block faulting unlikely (cf. Fig. 5.5). **C** Area percentages of the eight rock types. The balanced overall proportion of 24% granite, 37% metamorphic and 39% sedimentary rocks is highly differentiated in the single basins and modified by the erosional budget (Appendix B). Abbreviations: *Gra*nite, *Gne*iss, *Sch*ist, *Lim*estone, *Con*glomerate, *S*an*ds*tone, *Si*lt*s*tone, *Arg*illite

Figure 5.9 shows that in no basin are the argillites enabled to play an important role as source rock. Siltstones and argillites together cover 20% of the area of the 19 basins; however, these are 20% of the source area which do not participate in the formation of the river-mouth sediments.

5.2.3.6 Holocene Deposits

Synonyms of the CGDC

Talus	dt
Landslide	df
Soil creeping	a
Alluvial fan	conoidi
Fixed alluvial deposits	af
Fluvial deposits	ac^1
Eolian sands	d
Alluvial deposits	ac^2

The area percent of the first group, talus, landslide and soil creeping were assigned, if possible, to the appropriate rock type from where they came. The second group, from alluvial fans to fluvial deposits, clearly represents material just transported or being transported. These are transitory deposits ready for reactivation during floods. It is impossible to group these sediment bodies to a source rock or source area.

Regional Distribution. The Holocene sediment bodies cover 16% of the entire area (Fig. 5.11), the fluvial deposits alone adding up to 11%. This demonstrates clearly what a high percentage of the area is used as natural highway or transportation system. For comparison, German traffic roads, both railway and highway, cover about 4% in rural areas and more than 10% in more urban provinces.

Because these sediments do not belong to the source area in the narrow sense of the word, the percent calculation of the eight source rocks described in the foregoing sections was made without the Holocene deposits. They were recalculated to 100% on a Holocene-free basis.

5.3 Conclusion

Our quantitative evaluation of the source area for provenance calculations is a sad compromise between what is desirable and what is attainable. It is much easier to grasp the properties of a sediment body than those of a landscape. The most intrinsic difference is the static character of the "dead" sedi-

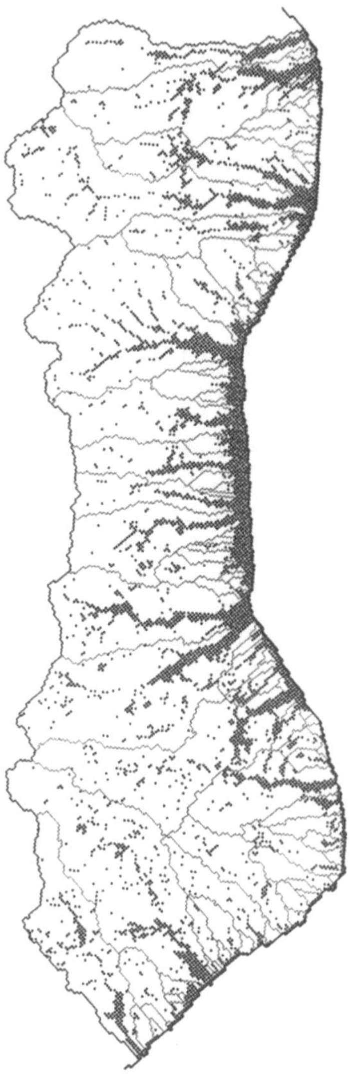

Fig. 5.11. Alluvial deposits. Computer-generated geological single value map of the Ionian slope of the Calabrian Massif. Holocene deposits add up to 16%, the fluvial pathways of transportation alone occupy 11% of the area. The histogram gives the area of alluvial deposits in the 19 basins [km^2]. See Fig. 5.2 for abbreviations

ment and the dynamic character of the "living" landscape, expressed by the much larger complexity of the latter. Thus, to make comparison possible, the variables of the source area have to be reduced. This creates a dilemma, because the dimension and the character of the reduction of "signals" from the source to the sediment is one of the prominent goals of provenance research itself.

The impoverishment of information is enormous: the infinite diversity of nature, already compressed into 55 formations during mapping, was reduced to 38 units for digitization and combined again to a total of eight units for our proper evaluation. On the other hand, despite the small number of only eight rock units, we possess much additional information and we can link them with other variables by means of the data bank. The loss of information by reducing the range of a variable may be balanced by the gain if we pay attention to a stronger internal link of the variables.

The erosion rates, for example, are calculable for each point of the geomatrix. Thus for each rock type a kind of erosional story can be written, as will be done in the next chapter.

Chapter 6

Erosion and Predicted Sediment Composition

Erosion rates determine the amount of weathering debris delivered into the fluvial basins. Rock distribution of the source area and erosion rates are the first among all other factors determining the petrographic and mineralogic composition of the resulting deposits. Thus a quantitative provenance analysis may have only little meaning without a quantitative concept of erosion.

The determination of erosion rates is one of the most difficult tasks of geology, as the problems of continuity or discontinuity of the processes involved and the time span over which an average has to be calculated are seldom really satisfactorily solved. An important stimulus for this provenance study was the fact that the Calabrian conditions enable a quantitative study of the erosional features of this area both long- and short-termed. As a result, for each point of the geomatrix an erosion rate was determinable, and only this enabled us to calculate an erosional budget and to make an attempt to predict the composition of the river-mouth sediments as a reference level for the provenance discussion. In the following sections we first describe the erosion rates, their measurement, regional distribution, statistics and rock-specific aspects, then the erosional budget and finally the concept of the effective source rock composition and of predicting sediment composition.

All these calculations, however, rely on long-term erosion rates, averaged over 1 Ma of uplift until today. Short-term erosion rates, calculated over the last 100, 50 or 20 years are treated in Chapters 15 and 16.

Summary. The mean erosion rate, calculated as long-term (1 Ma) erosional budget instead of sediment yield, is 178 mm/ka for the total area and 206 mm/ka if only the 19 basins are calculated, maximum values being 850 mm/ka. During the reference time of 1 Ma the entire area would have been able to produce a sedimentary deposit of 100×100 km, 55 m thick, with a calculated pore volume of 50%.

Corresponding to the "high gradient belt", a belt of high erosion can be established. Rock-specific erosion rates in the single basins reveal that one and the same rock type yields different erosion rates due to different positions in the mountains.

The effective composition of a basin is the area percentage of its rock types weighed by its erosion rates. This effective composition may also serve as predicted sediment composition, a reference level for the provenance discussion.

Shifting from the pure areal to the effective composition, granitic rocks drop from 24 to 15%, metamorphic rocks go from 37 to 41% and sedimentary rocks from 39 to 44%.

6.1 Erosion Rates

There are three ways to obtain information concerning erosion rates, the determination of basin fill, of transportation rates or of the erosion itself, i.e. the eroded volumes. The classical geologic method investigates the fill of a defined reservoir per unit of time by erosional products and calculates the erosion rates of the respective source area which also has to be definable. Hereby, as a rule, one integrates over longer periods of time, so that maxima or minima or stops of erosion remain unidentifiable. This method is indirect and static; Chapter 15 gives an example; in this case, however, the short-term fill of a reservoir formed behind a huge landslide in the Buonamico [08].

Indirect, too, but dynamic is the second method, the measurement of the transport rates, for example of a river; but this method is not very precise due to the problems of sample techniques (Meade 1969; Walling and Webb 1983, p. 71) and the heterogeneity of runoff conditions. However, under certain circumstances it may be possible to estimate both the erosion rate of the source area and the sedimentation rate of the receiving basin. We have no satisfying information about modern transport rates in Calabria.

The only direct measurement of the erosion rate is the calculation of what has been eroded, of the missing volume in the source area itself or the application of soil loss measurements. These methods, again, are static. To start with the latter, it was not possible to calculate the soil erosion, for example by applying a soil loss equation (Wischmeier, in Toy 1977, p. 54). Doing so, a network of measuring stations would have been established far beyond the economic limits of the project.

In addition to the difficulty of subdividing a mountain belt of sometimes alpino-type relief into areas of equal conditions and supposedly equal erosion rates in order to apply those equations, the main erosive factors of our Calabrian research area are heavy rainstorms, as pointed out in Chapter 4, and landslides, random events for which the soil loss equation is not explicitly designed (Wischmeier, in Toy 1977, p. 54).

However, special conditions made it possible to calculate a long-term average of the eroded and actually missing volume. For this, a reference surface and a time mark is needed. The reference surface should tell us where erosion started and the time mark when this took place. Usually these prerequisites are not fulfilled. The erosion of a young volcano in New Guinea investigated by Ollier and Brown (1971) may be an exception. In Calabria we have the promising condition of being able to apply this direct method. Görler (Ergenzinger et al. 1978, p. 362; Görler and Uchdorf 1980, p. 121) realized the significance of the remnants of the presumably Mid-Pleistocene plain as geometric and time-relevant reference elements for the calculation of the erosion rates. This long-term average is the main base for our calculation of erosion rates. So the reasons for this have to be stated first.

6.1.1 The Mid-Pleistocene Plain (MPP) and Its Use as Upper Enveloping Surface

The idea of Görler (op. cit.) to reconstruct the presumably Mid-Pleistocene plain (MPP) of the Ionian slope of the South Calabrian Massif and to use it as reference plane calculating erosion rates is certainly as elegant as it is risky, as long as the MPP is not exactly dated and defined as such. There are also more prerequisites which have to be fulfilled to avoid miscalculations. However, additional data make us believe today that the concept is fine and the results reliable.

6.1.1.1 Concept

Southern Calabria is famous for its high plains (Figs. 3.2 and 3.3) at altitudes between 850 and 1150 m (Cortese 1895; Gignoux 1911; Lembke 1931). Burton (1971, p. 81) believes that the plain formed during the Mid-Miocene transgression was enlarged during the new Lower Pleistocene transgression. The youngest marine sediments cut by the high plain are Calabrian; however, only in few and very restricted outcrops. Probably immediately thereafter uplift started. Most plains are covered by a thin veneer of gravel or sand, mostly reworked, whose origin, marine or continental, is hard to assign because of the lack of fossils. The morphological and sedimentological similarity of the terraces along the Ionian coasts makes it a reasonable assumption to link these plains with the high plains and to interpret the whole as a kind of pre-erosive emergence surface. Doing this, Görler (op. cit., p. 362) reconstructed this presumably Mid-Pleistocene plain by extra- and intrapolation over the entire Ionian slope of the Calabrian Massif, thus yielding a kind of upper enveloping surface. Then he calculated the eroded volumes from this upper enveloping surface against the lower enveloping surface of the actual morphology by means of a rectangular set of crossing profiles. Today, 10 years later, we possess a noticeably expanded data base. The morphology, the geographic distribution of rock types and the remnants of the MPP are digitized, so that the calculation of the erosion rates could be done by computer and extended to treat numerous special problems.

The following conditions have to be verified to justify the concept: the homogeneous pattern of the surface of the MPP and of the uplift, the dating of the MPP and, if possible, accompanying and independent arguments supporting the concept.

6.1.1.2 Mid-Pleistocene Plain and Uplift

The concept is based on the assumption that there was a continuous, uninterrupted and coeval plain roofing the emerging massif, the higher parts of the Aspromonte and the Serre excepted, and that the entire area raised by tilting, with little uplift close to the coast increasing steadily towards the strongly elevated central area. It is simply impossible either to verify or to falsify the assumption that all elements of this plain are coeval, so this is a working hypothesis. The same is true for a direct proof of the regional pattern of the uplift. However, there is a striking difference to the Tyrrhenian slope of the massif. Neotectonics are very differentiated here and strong block faulting occurs (Ghisetti 1981; Ghisetti and Vezzani 1982). There is no analogy at the Ionian slope, minor faulting excepted. The possible homogeneity of the MPP and of the uplift is furthermore supported by two morphological arguments, the one well known from the beginnings of geological research of the area, the drainage pattern, and the other emerging from our computer evaluation, the distribution of aspects (Chap. 3.3). The drainage pattern is by no means controlled by geology (Lembke 1931, p. 79); different rock bodies are only differently prepared by erosion. The valley of the Novito [13], cutting the isolated granite block of the Mt. Schiavo, for example, is a classical example of an superimposed valley (Fig. 3.6). In addition, all river valleys of the Zomaro Province cut the longitudinal strike valley of the siltstone (Figs. 2.4 and 2.5) without greater distortions of the main divides, thus forming a kind of "consequent" divide (Chap. 2.2.5). The elongated basins (Type A, Fig. 2.6), subparallel in the Serre Province and Zomaro Province, radially arranged in the Aspromonte Province and the similar and regular grouping of the interfluves (Fig. 2.1 B, C) strongly support the probability that the entire area of the mountains up to today's elevation of about 1150 m emerged as a more or less homogeneous element with a randomly developing drainage pattern.

Only the higher parts > 1100 m of the Serre and of the Aspromonte conserve older drainage networks. Certainly a paleorelief existed even in the lower parts, as demonstrated by the Messinian paleovalleys of almost all larger river mouths (Chap. 15.2), but these were filled earlier by Plio-Pleistocene sedimentation.

The independence of geology and geomorphology is even more demonstrated by the distribution of aspects, the slope directions. This would be barely understandable if a paleorelief emerged from the Pleistocene sea or if the uplift were accompanied by stronger block faulting. Figures 3.12B and 3.13 show the striking symmetry of slope orientations. Twenty eight percent, a strong primary mode of aspects faces SE. These SE slopes obviously are the remnants of the original desk-like surface plus some later subsequent elements. Two secondary modes, 16 and 19%, face SW and NE respectively. These belong to the consequent drainage basins; both slope directions are spread over all rock types and elevations, without any significant trend due to geology. The same is true for the "valleys" of the aspect distribution, the S- and E directions with 11 and 13% respectively. The NW-facing slopes, finally, reach 4%, and belong mostly to subsequent valleys portraying the SW-NE strike of most geological units.

Finally there is a very geologic argument demonstrating the absence of larger block faulting at the Ionian slope of Aspromonte and Serre. The Jurassic reef limestones display the lowest altitude standard deviation, after the siltstone, of all rock types (Chap. 5.2.3.1). If we assume this reef sediment to have been formed in very shallow water, the uniformity of today's positions has to be seen as a signal of a relatively uniform uplift.

As a result, we believe that the Ionian slope of the Zomaro Province and of the lower parts < 1100 m of the Aspromonte and the Serre formerly formed a more or less homogeneous element roofed by a more or less continuous surface. The uplift obviously took place mainly by tilting, in contrast to the block faulted Tyrrhenian slope.

6.1.1.3 Dating of the Mid-Pleistocene Plain

As pointed out, the uppermost sediments cut by the high plain are Calabrian, the thin sedimentary cover being free of fossils. Görler and Uchdorf (1980, p. 125) report intercalations of sand and silt whose magnetization is partially inverse, thus pointing towards an age of at least 730000 a. Scarce deposits of tephra (Romagnoli 1969) are not yet evaluated (Guerrera and Veneri 1989). So it was only the general geological framework which led Burton (1971, p. 82) to the decision to assign those plains provisionally to the Quaternary. According to Harland et al. (1982, p. 42), the boundary Calabrian-Emilian is about 0.94 Ma. For simplicity we assume an age of 1 Ma for the MPP as reference plane calculating erosion rates. Because the mean elevation of the high plain is 1000 m, this implies an average uplift rate of 1000 mm/ka.

6.1.1.4 Supporting Data

The assumption of this strong uplift rate of about 1000 mm/ka is supported unexpectedly by two uplift calculations made in the same area but separately and independently by other authors. Aifa et al. (1988) determined uplift rates of 700 – 1000 mm/ka from magnetostratigraphic research in the Trubi Formation at the Tyrrhenian slope, close to the Messina Strait. The Trubi Formation corresponds to main parts of the argillites (Pleistocene clay and silt, p^a, Chap. 5.2.3.5). Dumas et al. (1988) found uplift rates of 850 – 1030 mm/ka applying geomorphic data and amino acid geochronology of uplifted shorelines in the same area, close to Reggio Calabria. This agreement of uplift determinations yielded by three different and independent methods seems to us a reasonable and reliable base for the calculation of the long-term erosion rates of the area.

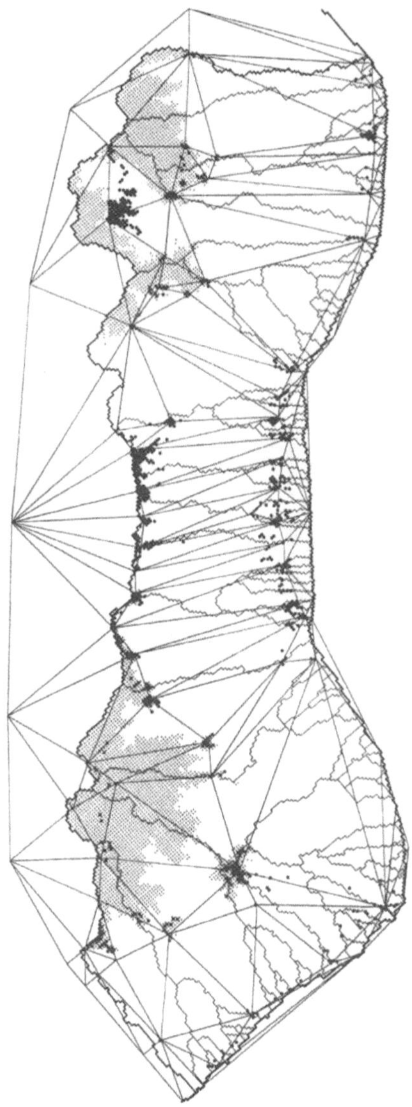

Fig. 6.1. Determination of erosion rates. Computer-generated map of the Mid-Pleistocene planations (*black*) and the altitudes above 1000 m (*grey*). The network of *triangles* is constructed by hand. The position of a corner of a triangle resting on the planation is given by the triple values x, y and z, the coordinates and the altitude. Based on these data the position of the possible entire Mid-Pleistocene planation is interpolated and reconstructed (Fig. 6.3)

Data Base. Topographic map 1:25000 of the Istituto Geografico Militare Italiano with 25-m contour lines, Carta Geologica della Calabria, 1:25000, geomatrix, data bank.

Terminology. We give the erosion rates in mm/ka, 1 ka (kilo-year) is 1000 years. If not explicitly mentioned otherwise, the term erosion rate is always used in the sense of mean erosion rate, for example, of a basin or of a rock type.

Method. The reconstruction of the Mid-Pleistocene plain, the upper enveloping surface: the remnants of the Mid-Pleistocene plain are printed on a computer-generated "single value" map (Fig. 6.1). On this map a network of 153 triangles was constructed by hand, the corners of the triangles resting on the remnants of the plains. Thus, the spatial position of each plane of a triangle is defined by the triple values x, y and z_1 of its corners; x and y are the geographical coordinates, z_1 is the elevation of the present morphology, the lower enveloping surface (Fig. 6.2).

Fig. 6.2. Determination of erosion and erosion rates (schematic). The volume eroded away is defined by the lower enveloping surface, today's morphology (*LES*, z_1) and the reconstructed Mid-Pleistocene plain (*MPP*) forming the upper enveloping surface (*UES*, z_2). Consequently, each point of the geomatrix possesses three values of elevation, z_0, the sea level, z_1 and z_2. The erosion rate in each point is given by $z_1 - z_2$ per 1 Ma

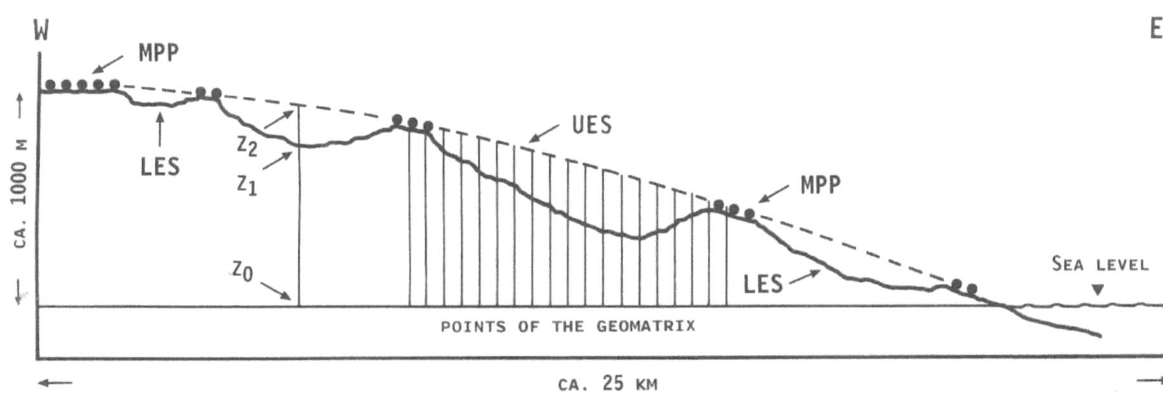

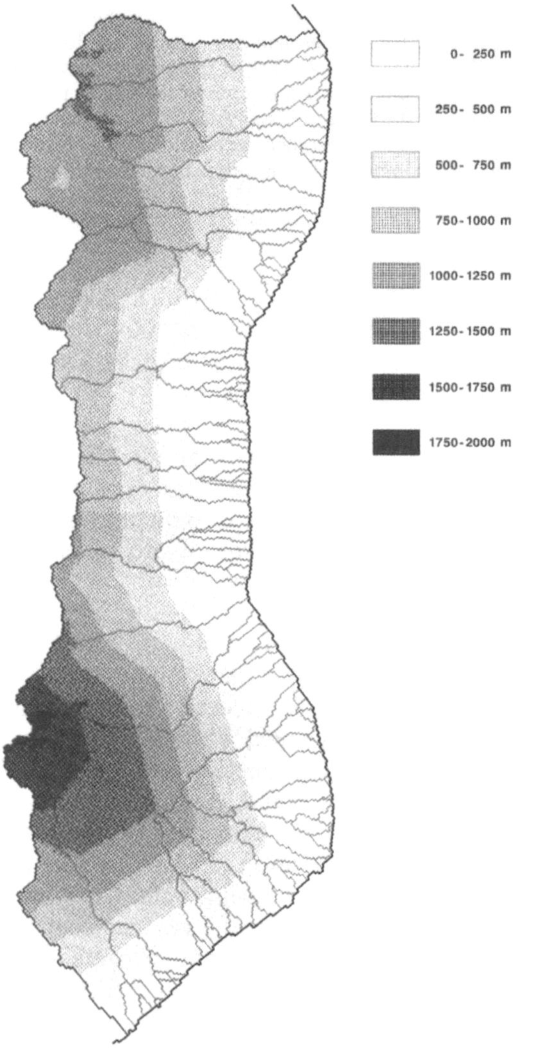

Fig. 6.3. Determination of erosion rates. Computer-generated map of the reconstructed Mid-Pleistocene planation or pre-erosive emergence surface of the Ionian slope of the Calabrian Massif. This surface serves as reference calculating the eroded valley volumes per units of time, the erosion rates (Fig. 6.2)

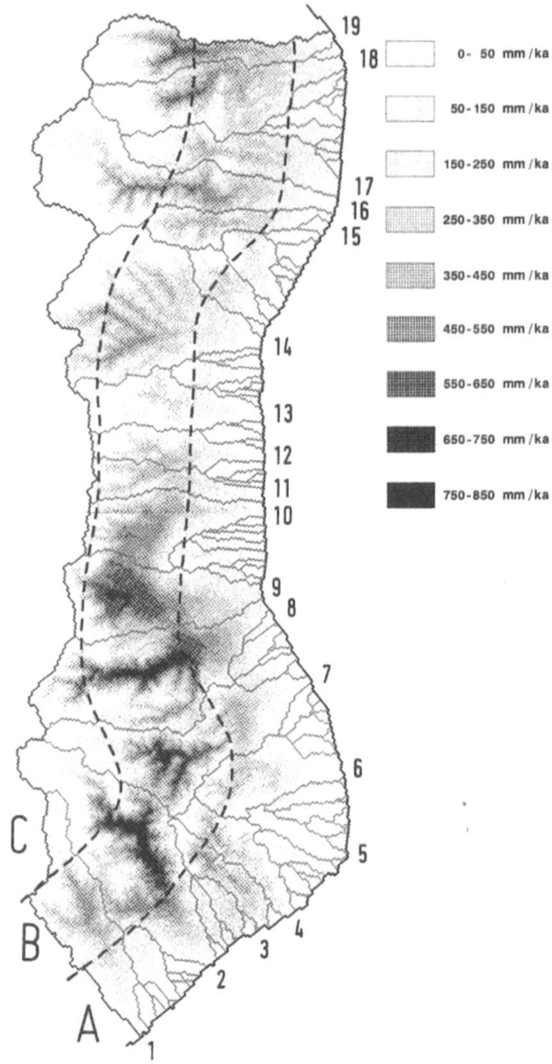

Fig. 6.4. Computer-generated map of the erosion rates. The calculation bases on the eroded volumes defined by the lower enveloping surface of the present-day morphology and the upper enveloping surface of the reconstructed Mid-Pleistocene planation (Fig. 6.2) during 1 Ma. *A* Lower "low erosion" or "low energy" belt. *B* "High erosion" belt. *C* Upper "low erosion" belt (Chap. 6.1.4.1)

For each point of the geomatrix the computer assigned the corresponding triangle and calculated the elevation z_2 of the triangle plane at this point using the x-, y- and z_1-values of the triangle corners. The resulting elevations z_2 represent the reconstructed Mid-Pleistocene plain, the upper enveloping surface (Fig. 6.3). Strangely enough, this reconstructed enveloping surface comes out very regular, considering the relatively rough triangulation from which it is derived. As a result, each point of the geomatrix possesses three elevation values:

z_0, zero point, base- or sea level

z_1, altitude [m] of the actual morphology (lower enveloping surface)

z_2, altitude [m] of the reconstructed Mid-Pleistocene plain (upper enveloping surface, Fig. 6.2).

Elevations of the Aspromonte and the Serre above the MPP are calculated by constructing the upper enveloping surface closely over the actual morphology. This certainly yields relatively low erosion rates; the data base did not stand for more satisfying solutions.

Using the ratio $(z_2 - z_1)/1$ Ma, for each point of the geomatrix, the erosion rate mm/ka can be calculated. We received erosion rates between zero and 850 mm/ka.

6.1.2 Erosion Rate Map

Figure 6.4 shows the computer-generated plot of the erosion rates. Values of high erosion appear dark, lower ones lighter. This map is, with some restrictions, a kind of negative of the computer-generated altitude map (Fig. 3.7). On the altitude map the branching valleys are kept light and contrast with the dark, higher mountainous areas and drainage divides. The erosion rate map works the other way round, the valley floors, deeply eroded and dark, contrast with the lighter, less eroded elevations. It is evident that the strongest erosion takes place neither close to the coast – nobody expected this – nor in the most elevated parts of the mountain chain, as expected, but in between, at about the second quarter of the long profile.

Figure 6.5 shows a computer-generated stereomodel of the research area combining both the

Fig. 6.5 A–C. Computer-generated stereomodel of the Ionian slope of the Calabrian Massif seen from E, from the Ionian Sea. **A** 1 Ma ago, mantled by the presumed emergence surface. **B** 0.5 Ma ago, this theoretical picture was obtained simply by halving the entire erosion, the distances $z_1 - z_2$ of each point of the geomatrix (Fig. 6.2). **C** The present-day morphology. The model implies full uplift 1 Ma and 0.5 Ma ago; it goes without saying that in reality uplift and erosion coincide step by step

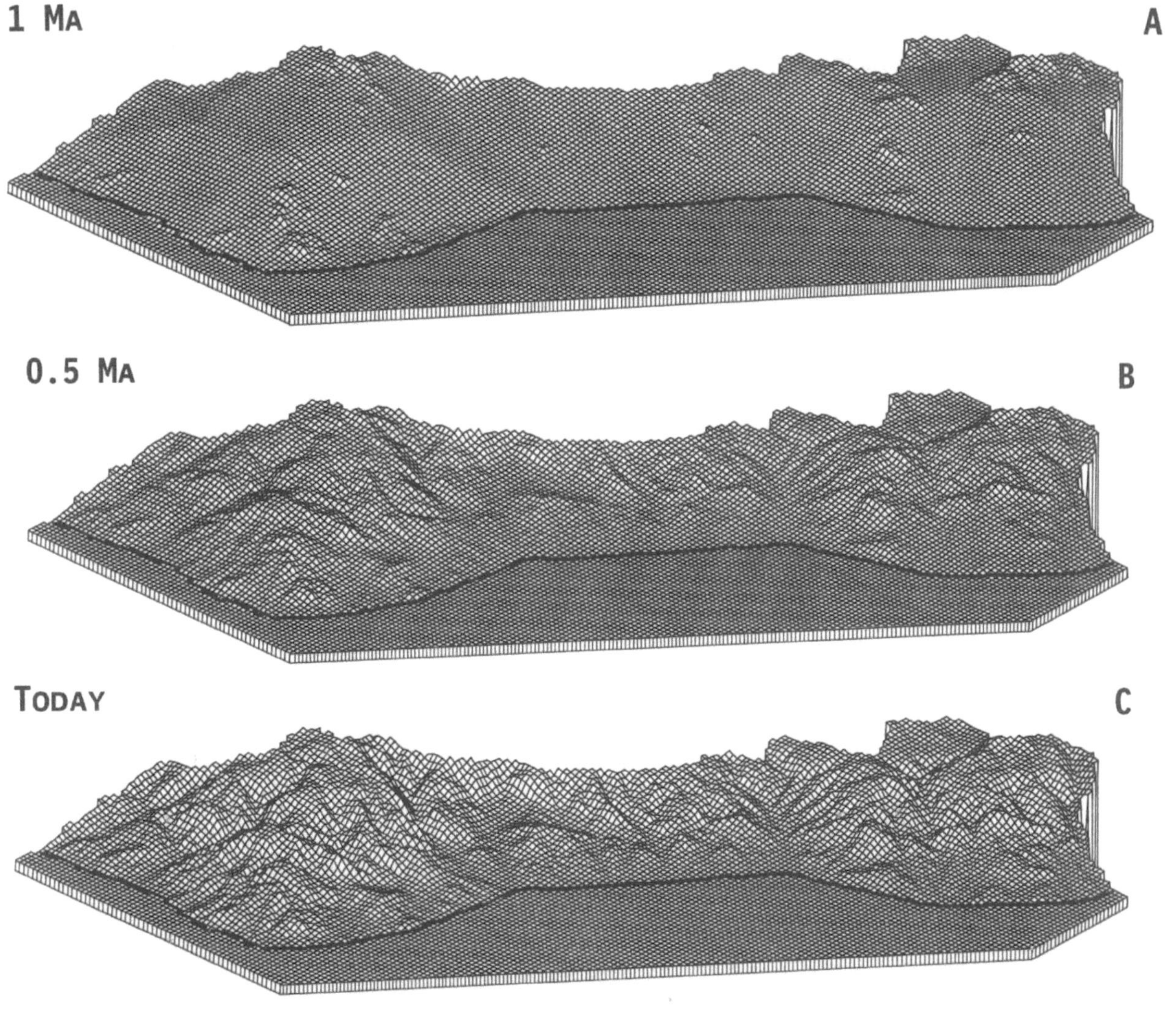

presumed Mid-Pleistocene emergence plain and today's morphology. Neglecting the fact that erosion starts contemporaneously with emersion, we constructed a completely emerged stage, veiled, however, by a non-eroded emergence surface (A). Stage B shows the presumed relief 0.5 Ma ago and stage C illustrates the present-day morphology.

6.1.3 Statistics

6.1.3.1 Erosion Rate Frequency Distribution

The erosion rate frequency distributions of the single basins do not vary noticeably. Thus the overall distribution (Fig. 6.6D) is relatively regular.

About one third of the area is eroded by only <50 mm/ka, a further third by 50–200 and the last one by 200–900 mm/ka. Of the interfluve areas two thirds belong to the lowest erosion class. This relatively high percentage of low erosion rate areas of the Ionian slope comes firstly from the very small uplift of the coastal region; there is still a nearly continuous garland of the remnants of Pleistocene terraces along the coast. A second reason is the presence of the still very elevated areas of the main divides between the basins (Fig. 6.4). A third argument is the existence of extended remnants of the MPP at about 1000 m, which, by

Fig. 6.6 A–F. Statistical parameters of erosion rates [mm/ka] and relations between the variables of the single basins. **A** Mean erosion rates. **B** Standard deviation of erosion rates. **C** Maximum erosion rates. **D** Erosion rate frequency distribution of the entire area (interfluves included). **E** Basin area versus maximum erosion rate. **F** Relief (altitude st. dev.) versus maximum erosion rate. The plots show a predominance of low erosion rates (**D**). Mean standard deviation and maximum of erosion rates differ noticeably but almost jointly from basin to basin. Maximum erosion rates and size of the basins correlate well (r = 0.78*); the Aspromonte rivers, however, are strong eroders, the Serre rivers are weak eroders. The same disproportion holds true for the relation maximum erosion rates and relief, which correlates with r = 0.81*. *01 Me*lito; *02 Am*endolea; *03 Pa*squale; *04 Pa*lizzi; *05 Ar*anghia; *06 Br*uzzano; *07 La*verde; *08 Bu*onamico; *09 Ca*reri; *10 Co*ndoianni; *11 Po*rtigliola; *12 Ge*race; *13 No*vito; *14 To*rbido; *15 Au*musa; *16 Al*laro; *17 Pr*ecariti; *18 St*ilaro; *19 As*si; *20 xx* mean of the basins 01–19; *21 If* mean of the 84 smaller basins and interfluves and *22 XX* total mean (103 basins and interfluves)

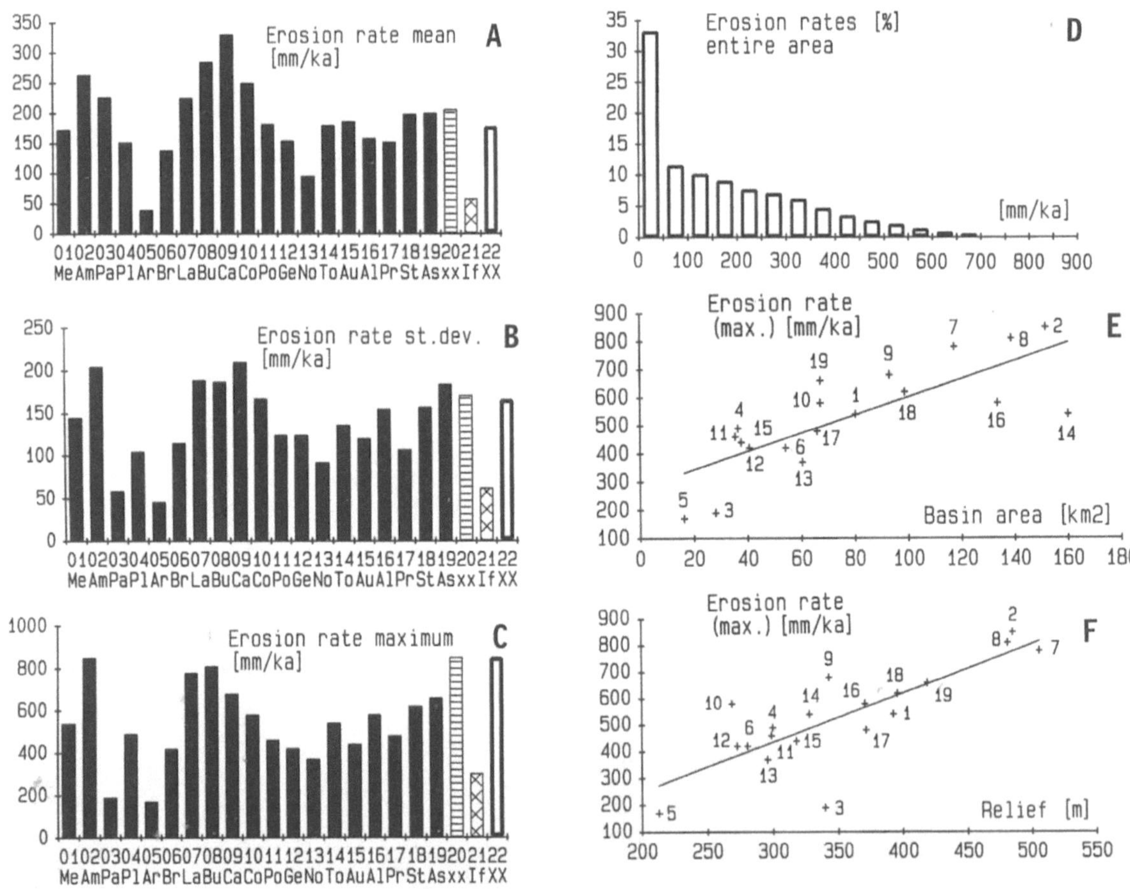

definition, form the upper enveloping surface characterized by zero-erosion (Fig. 6.1).

The erosion data reported here fit what is known worldwide (Young and Saunder 1986; Ahnert 1970; Table 2). These values, however, mainly stem from calculations of the sediment yield (Fig. 6.7), and are short-termed, whereas our data base on a 1-Ma mean. Compared to this, modern erosion is dramatic in Calabria; Chapter 15 is devoted to this topic.

6.1.3.2 Statistical Parameters

Mean. The mean erosion rate of the entire area is only 178 mm/ka (Fig. 6.6A). This is because the interfluves are included, which comprise 19 area percent yielding, however, an erosion mean of only 58 mm/ka. The 19 basins alone achieve 206 mm/ka, top eroder being the Careri [09] with 330 mm/ka; the Aranghia [05] displays minimum erosion, 38 mm/ka.

The three most important rivers of the Aspromonte Province, Amendolea [02], Laverde [07] and Buonamico [08], erode with a relatively high mean of 265, 226, and 285 mm/ka, respectively. The largest river, on the other hand, the Torbido [14], has only 180 mm/ka. Regarding its relief and also its gradient distributions, Torbido has to give way to its counterparts from the south. The Torbido is number 1 by size rank, but only number 11 ranked by mean erosion rates. This is hard to understand, because half of its territory is in the "high gradient belt", the sediment portion is high, mainly argillites, and there are many granites and only few metamorphic rocks. So the conditions of the Torbido come close to those of the powerful Careri, but without its erosional success. The present-day Torbido basin obviously is only a remnant of a much more powerful past. A sign in this direction is the pronounced belt of argillites (Fig. 5.9) reaching up to the main divides, a unique situation in the area, demonstrating long-lasting conditions of local subsidence (Wilkes 1986). Moreover, the paleovalley of the Torbido, more than 100 m deep, is the largest of the area (Chap. 15.2).

Maximum. Maximum erosion rates are found at the loci of the deepest river cuts (Figs. 6.4 and 6.6C). The Aspromonte rivers Amendolea [02], Laverde [07] and Buonamico [08] and the Zomaro river Careri [09] produce the deepest valley incisions with 850, 780, 800 and 680 mm/ka respectively. In the Serre Province only the northern rivers

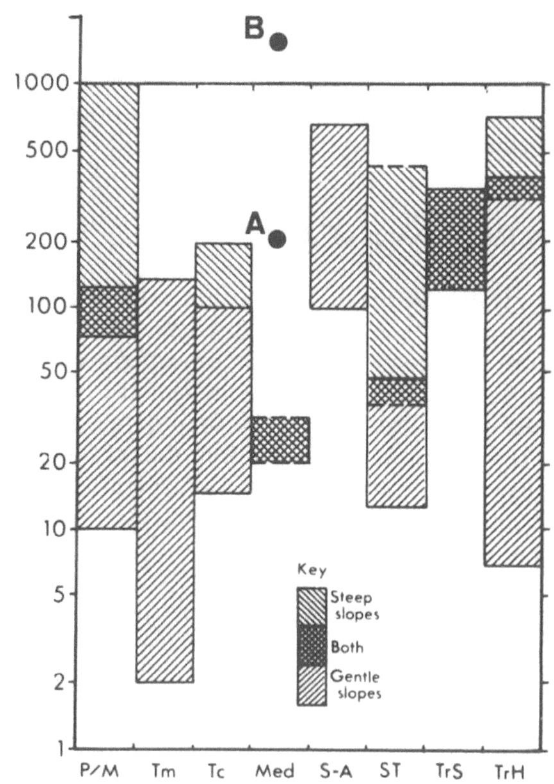

Fig. 6.7. Generalized ranges of observed rates of denudation, grouped according to climate and angle of dominant slopes (Young and Saunders 1986). Scale in Bubnoff units, this is mm per 1000 years or mm/ka; *P/M* polar/montane climate; *Tm* temperate maritime; *Tc* temperate continental; *Med* Mediterranean; *S-A* semiarid; *Arid* arid; *ST* subtropical humid; *TrS* tropical subhumid; *TrH* tropical humid. Calabrian mean erosion rates: *A* 206 mm/ka, basins 01 – 19 only, reference time: 1 Ma; *B* 1527 mm/ka, Buonamico [08] basin only, reference time: 50 years (Chap. 15)

Stilaro [18] and Assi [19] display comparable valley cuts, 620 and 660 mm/ka.

Standard Deviation. All basins contain zones of lowest erosion close to the coast. Thus the standard deviation, the measure of the dispersion of erosion rates, is a measure of how far a basin reaches into higher erosion rates (Fig. 6.6B).

Consequently, high values of erosion rate standard deviation stand for highly eroded basins. Here, too, the big rivers of the Aspromonte Province surpass those of the Serre and especially the Zomaro Provinces.

Skewness. Practically all erosion rate frequency distributions are positively skewed. Hence, areas of high erosion, as a rule, are under-represented. The objective examination by means of the digitized maps forces us again and again to correct the sometimes really spectacular field impression of the erosion of the Calabrian mountain world towards normality or at least towards less dramatic estimates.

6.1.4 Relationships of the Variables

The mean and maximum erosion rates correlate well (r = 0.77*) in the regional comparison of the basins. This connection of the maximum erosion rates with the average ones demonstrates that the former are no extreme values, deep gorges for example with vertical walls, but that the erosion rate frequency distribution and, by this, the morphology is relatively balanced, though with the high gradients and high relief described.

Are the erosion rates influenced by the size of the basins? The mean erosion rate does not correlate very well with the size of the basins (r = 0.59*); the maximum rates, on the contrary, correlate much better (r = 0.78*). The maximum erosion rate or valley incision of the Torbido [14] is "too little", that of the big Aspromonte rivers clearly "too high". However, the Assi [19], the northern boundary of the area, yields the relatively highest incision, compared to its size in the regional comparison (Fig. 6.6E).

Erosion creates and finally destroys relief. So the question arises as to the relationship between the erosion rates and the relief of the basins, that is, the altitude standard deviation. Here, analogously to the dependence on basin size, a weak positive relationship exists between relief and mean erosion rate (r = 0.59*) and a much stronger one with the maximum erosion rate (r = 0.81*, Fig. 6.6F). Here, too, it is relatively easy to interpret the trend and difficult to explain the single case. The larger a basin, here, at the Ionian slope of Aspromonte and Serre, the further it will reach into the "high gradient belt", the zones of more active morphodynamics, of increased erosion rates and valley incision. For the same reason, relief and size of basins correlate well.

6.1.4.1 Erosion Rate Versus Distance

The gradient map (Fig. 3.9) shows a "high gradient belt", a zone of increased slope angles in the interior of the mountains which has to be considered as a locus of increased erosion. Indeed, the erosion rate map (Fig. 6.4) assigns the values of maximum erosion rates to the same strip of terrain. This "high erosion belt" can be derived from the plot's erosion rate versus distance from the river mouth (Appendix B 01–20, Fig. 3.11 B) in the same manner as described with the gradient distribution.

6.1.4.2 Erosion Rate Versus Altitude

Erosion rate and altitude have a clear, but nonlinear correlation (Fig. 3.11 C). Maximal erosion rates increase from the coast till 400 or 500 m elevation and then decrease with a less steep gradient. The combined plot of the entire area is composed of several superimposed elements which can easily be assigned, by visual comparison, to the single basins plotted in the basins ID (Appendix B 01–19).

6.1.4.3 Erosion Rate Versus Precipitation

A dependence of erosion on precipitation is not to be recognized at the base of our stored data. This is nearly absurd, because never and nowhere, if not in Calabria, is it so easy to drastically experience the immediate relationship between rain and erosion. However, those precipitations are cloudbursts or torrential downpours of sometimes more than 400 mm/day which trigger catastrophic landslides.

Nonetheless, this is the place to compare our data with one of the most debated relationships, the connection between sediment yield and mean effective precipitation formulated by Langbein and Schumm (1954) and Schumm (1968) (Fig. 6.8). The philosophy of this theory, the "Langbein-Schumm rule" is the interaction of erosive energy and vegetation density. "Maximum sediment yields occur at an annual effective precipitation of approximately 300 mm (i.e. semi-arid areas). In areas with effective precipitation in excess of 300 mm, vegetation growth is promoted and increases surface protection. In dryer areas, the available energy for erosion and transport is limited" (Walling and Webb, p. 77, in Gregory 1983). However, this data set refers only to the USA and even there the sediment yield of coastal areas of northern California, for example, does not fit into this picture, as pointed out by Walling and Webb (op. cit., p. 78). The Calabrian data go far beyond this relationship, even if we bear in mind that two restrictions have to be regarded: first, we are dealing with the entire erosional budget instead of sediment yield (Chap. 6.2.1), and second, we are comparing precipitation

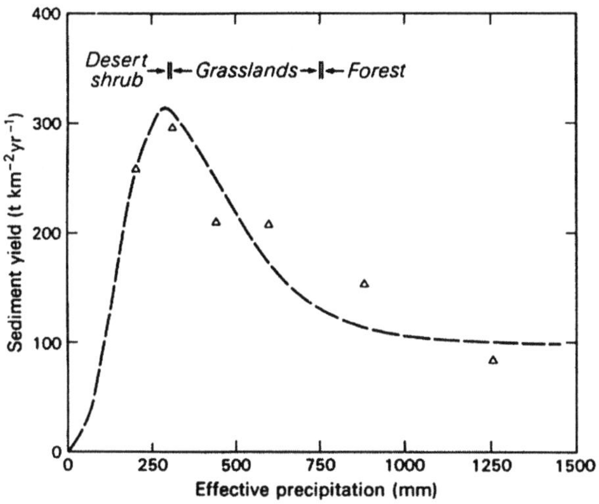

Fig. 6.8. The relationship between sediment yield and annual precipitation proposed by Langbein and Schumm (1954), after Walling and Webb (Fig. 4.3, in Gregory 1983). Calabrian conditions go by far beyond these relations, with a mean effective precipitation of 698 mm, but a total erosional budget, instead of sediment yield, of 550 t/km^2/a

data of a 50 years' mean with an erosion rate bridging 1 Ma.

6.1.5 Rock-Specific Aspects of Erosion

It is certainly a strong simplification if we relate the long-term erosion rates to the different rock types exposed on today's surface. With other words, if at a given point an erosion of $z_2 - z_1 = 300$ m is calculated over 1 Ma (Fig. 6.2), and today's surface is made up of granite, it is probable that this granite was covered by other rocks in former times. Görler and Uchdorf (1980, p. 135) ventured the attempt to reconstruct the pre-erosive geologic section in order to estimate a kind of rock budget of the eroded volumes.

However, the eight rock types into which we compressed and condensed the entire rock suite are almost all very large and thick units, so we believe that the error is tolerable and statistically distributed over all eight rock types. Apart from this, the resulting rock-specific erosion rates follow a very meaningful pattern.

6.1.5.1 Mean Rock-Specific Erosion Rates and the "Energy Belts"

The mean erosion rates of the eight rock types – granite, gneiss, schist, limestone, conglomerate, sandstone, siltstone and argillite – are plotted in the single river ID, Appendix B 01–20. Within the single basins the picture is very differentiated, because the erosion rates are determined by mainly two variables, the resistance to weathering of the single rock types and their location in the mountain chain; the latter seems to be the more important variable.

On the plot of the total area (Appendix B 20) the Jurassic limestones display the lowest mean erosion rate, 103 mm/ka. This fits field experience, the limestones mostly appear as steep-walled blocks (Härtling). But, all in all, the limestones occur so seldom (Fig. 5.5) that they are worthless for a regional estimate. For example, some limestone outcrops of the Aspromonte interfluves refer to only one point of the geomatrix.

Apart from these limestones, granite and argillite develop the lowest erosion rates, 123 mm and 161 mm/ka, respectively. The highest erosion rates suffer schist, siltstone and sandstone with 222, 279 and 236 mm/ka, respectively. These numbers contain the whole philosophy of the Calabrian erosion rates.

Considering only lithology, it is not understandable why the family of the deeply grus-covered granites, the extremely resistant limestones and the argillites with their low resistant badlands should have practically the same erosion rates, 123, 103 and 161 mm/ka, respectively. Just as strange are the erosion rates of sandstone and siltstone, 236 and 279 mm/ka. The sandstones frequently develop small, steep-walled massifs, but the valleys of the siltstones, the olisthostromes, are ample and deeply eroded. This contradiction of high erosion rates – resistant rock or vice versa, easily erodable rocks but only low erosion rates – is only solvable if even the position of these rocks in the mountains is taken into account, its relation to the "high energy belt". The argillites, with some exceptions in the Torbido [14] and Novito [13] area, are situated in the "low energy belt" close to the coast, with precipitations of about 600 mm/a or less, which concentrate during winter. So, despite their high erodibility, little happens to these rocks. The conditions of the granites are more complicated. Rather less than one half of these rocks lies in the upper "low energy belt" of the Serre (Fig. 5.2). The other part, slightly larger, completely occupies the "high energy belt". But here in the north the erosion rates as a whole are considerably lower than in the south, probably caused by a less intense uplift. Thus it becomes plausible that the granites, despite their lithologically low resistance and pronounced grus formation, display erosion rates of only 123 mm/ka.

Even more obvious are the conditions comparing sandstone and siltstone (olisthostrome). We expect that the little-resistant siltstones would be more strongly eroded than the more resistant sandstones. Accordingly, the erosion rates of the Careri [09] are 397 mm/ka for sandstone and 523 mm/ka for siltstone (Appendix B 09). Torbido [14], however, works the other way round, the sandstone is eroded by 238 mm/ka and the siltstone by 167 mm/ka (Appendix B 14). The overall mean (Appendix B 20) shows erosion rates of the two rocks as equal, 206 and 214 mm/ka. Lithologically, no difference is detectable between the single basins. On the contrary, compared to some metamorphic rocks, for example the schists, both rocks are characterized by a high homogeneity over the entire area. The position, however, of these rocks in the mountains is very different and this is the key to solving the contradiction. The sandstone lies at about 70% in the "high energy belt", the siltstone arrives only at 15% (Figs. 5.7 and 5.8). Thus it happens that the sandstones are eroded over-proportionally and the siltstones under-proportionally.

The influence of only the lithology on the erosion rates of the two rocks can be estimated where both sandstone and siltstone outcrops lie at the same "energy level". In the area of the Serre rivers Amusa [15] to Stilaro [18] (Appendix B 15–18) both rocks only appear in the "low energy belt" close to the coast. The erosion rates of sandstone and siltstone in this case behave like 1 : 1.4. In the Careri [09] and Condoianni [10] "high energy" areas, however, the ratio is about 1 : 1.9 (Appendix B 09 and 10). Obviously, a more intensive erosion illustrates the lithological differences more effectively.

The regional comparison of the erosion rates of a rock type through all basins supports the statement that the erosion of the Calabrian rocks depends not only on lithology but also on its position in the mountains. Figure 6.9 A–H shows the regional distribution of the erosion rates of the eight rock types, which differ noticeably between the single basins; the respective standard deviations are plotted in Fig. 6.9 K.

6.1.5.2 Rock-Specific Standard Deviations of Erosion Rates and the "Energy Belts"

The erosion rates of the schists display the highest standard deviation (s = 182 mm/ka) of this rock-specific comparison. The reason for this is, again, their changing situation in the energy levels of the mountains. The same is true for gneiss as well as for sandstone and siltstone. All these rocks show relatively high standard deviations between s = 163 and 173 mm/ka, a signal for their positional heterogeneity, as may be seen on the "single rock" maps (Figs. 5.3, 5.4, 5.7 and 5.8).

Granite and conglomerates exhibit a medium position with standard deviations of s = 142 and 137 mm/ka, respectively, demonstrating in the case of granite that about one half is situated in the "high energy belt" of the Zomaro and Serre Provinces, and the other forms the "low energy belt" of the top parts of the Serre.

The lowest standard deviations of erosion rates show argillite (s = 102 mm/ka) and limestone (s = 98 mm/ka). In the case of the argillites the explanation is simple; the Novito [13] and Torbido [14] areas excepted, the argillites lie nearly exclusively in the "low energy belt" close to the coast. These uniform conditions explain the low standard deviation of the erosion rates. The very low standard deviation of the Jurassic limestone erosion rates is caused by its very homogeneous altitude distribution, even if they occur exclusively in the "high energy belt".

This regional analysis of erosion rates of different rock types demonstrates that even within limited areas such as our Ionian slope, apparently simple properties, in this case the erodibility, may display a noticeable variability understandable only against the background of a much broader context. Isolated research on too limited areas would have yielded misleading results. The better understanding of erosion is a prerequisite for the quantitative provenance analysis, as will be pointed out in the next section.

6.2 Budget Calculations

The quantitative provenance analysis attempts to infer from the portions of certain rock types or minerals in the sediment to the portions of the same rock type in the source area. To write this formulation means to restrain it. It is quite clear that areal proportions of rocks in the source cannot come out directly as quantitative amounts or compositional proportions of the sediments in the receiving basin. Even in the simplest case, if all rock types of an area were lithologically so similar that none were favoured during weathering or transport, such a direct relation could only be expected pro-

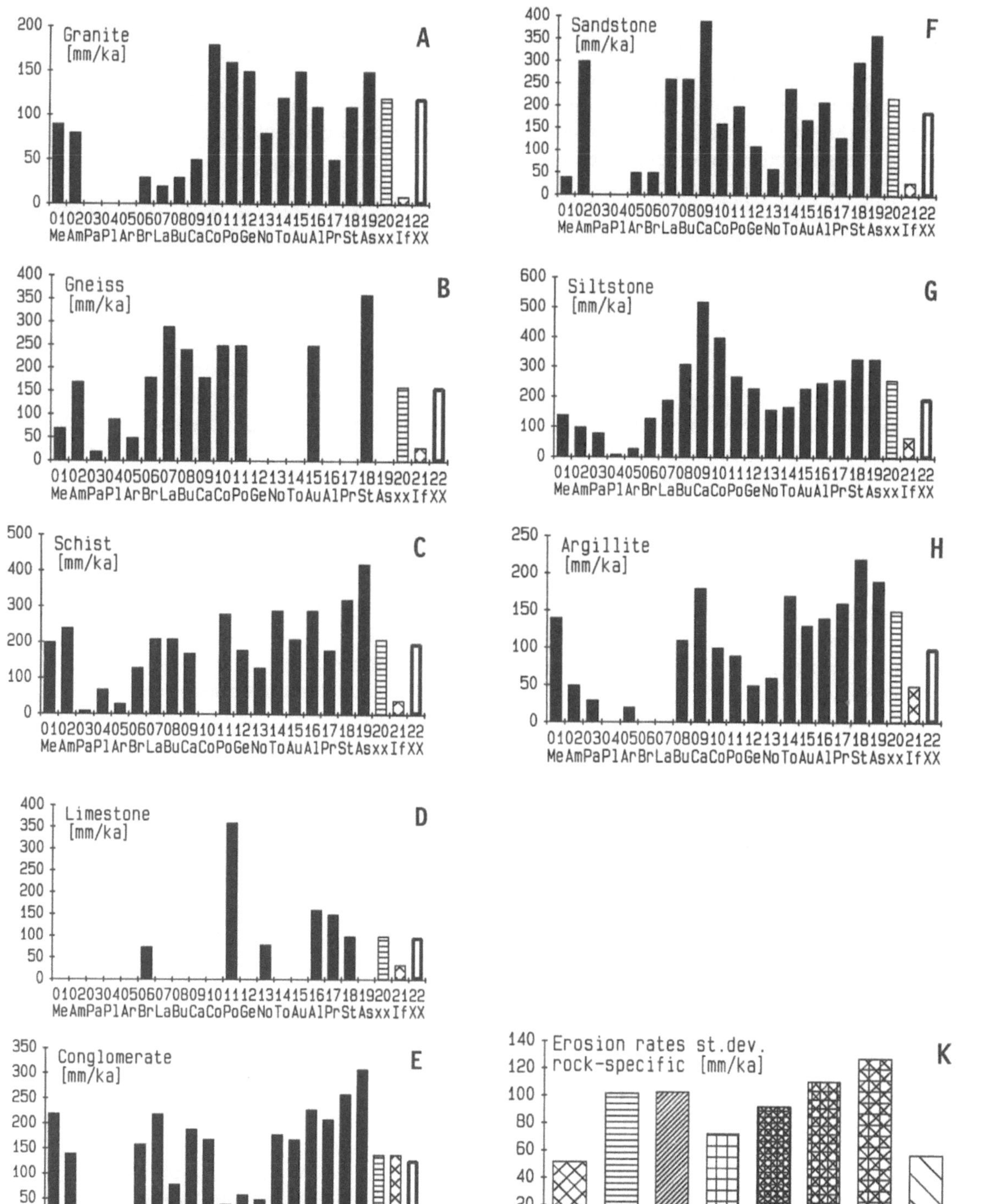

Fig. 6.9. A–H Mean erosion rates of the key rocks in the single basins. Erosion rates are less dependent on lithology than on the "where", the position in relation to the "high energy belt". **K** Standard deviation of rock-specific erosion rates. Granites and argillites with relatively fixed positions outside the "high energy belt" display lowest values. Siltstones, though highly erodible, frequently switch between the belts, thus displaying the highest standard deviation. For abbreviations *01–22* see Fig. 6.6. Abbreviations: *gra*nite, *gne*iss, *sch*ist, *lim*estone, *con*glomerate, *s*and*s*tone, *si*lt*s*tone, *arg*illite

viding similar erosion rates existed over the entire source area, and this is never the case. Thus the minimum prerequisite for a correlation source/sediment is not the simple areal distribution of the rock types in the source area but at least the production of the respective rock debris per unit of time, that is, rock area times erosion rate.

6.2.1 The Erosional Budget

Our knowledge of the rock type and the erosion rate at each point of the geomatrix makes it possible to calculate a rock-specific erosional budget of the area.

At first glance this budget is the sediment yield. The sediment yield is "the amount of material eroded from the land surface by runoff and delivered to a stream basin" (Bates and Jackson 1987). However, this amount is usually determined by measuring the solid and dissolved load of rivers. Therefore, the term sediment yield is definitely product-related. There are two more restrictions: the term sediment yield does not cover the entire erosion of a given area, because each river basin or comparable environment contains transitory deposits whose volume has to be added to the sediment yield to obtain the entire erosion. This is expressed by the sediment delivery ratio D (Toy 1977, p. 15), $D = Y/T$, where Y is the sediment yield at a given point in the basin and T is the total erosion above this point. Consequently, values of D are almost always < 1.

The other restriction is that no conventional method measuring a sediment flux, let us say of a river, is capable of grasping the entire load, either in space, along the river's cross-section, or in time, through a complete event, or with the proportion of bed load, suspended load and dissolved load.

Opposed to these sediment-yield considerations, our calculation is definitely producer-related, based exclusively on area percent and erosion rates without any regard to all modifications by weathering, transport or deposition. Not to confuse these two different meanings, we use the term erosional budget. The erosional budget is the sum of the solid and dissolved matter a basin should produce according to the areal distribution and long-term erosion rates of its rocks.

Data Base. Regional rock type and erosion rate distribution from the data bank.

Terminology. The erosional budget is the volume of erosional products both, solid and dissolved, per unit of time [m^3/a] produced by a rock type, calculated as a long-term average.

Method. The erosional budget EB_R of a rock type of a basin is calculated as

$$EB_R = F_R E_R \ [m^3/a] \ ,$$

F_R is the projected area [km^2] covered by the rock type $_R$ on the geological map, E_R is the mean erosion rate [mm/ka] of this rock type.

In the foregoing section the heterogeneity of the mean erosion rates of the single rock types in the single basins was demonstrated (Appendix B 01–20). Because the portions of the single rock types in the single basins are also very different the erosional budget of the single basins displays a very differentiated picture (Appendix B 01–20). The figures compare the area percentage of a rock (left column) to its erosional budget (right column). We normalized the two variables, because it is not possible to plot the two dimensions, [km^2] and [m^3/a], directly together in a histogram. This normalization, however, leads to the possibility, somewhat confusing at first glance, that the value of an erosional budget may be smaller than that of the respective area; see, for example, the granite of the overall plot (Appendix B 20). However, this only means that in this case the erosion rate of the granite is smaller than the mean erosion rate of all rocks of the entire area. This is, granitic weathering debris is produced under-proportionally or, in other words, the granites share in 20% of the entire area, but only in 15.4% of its erosional budget.

If both columns have the same height, the rock is eroded with the mean erosion rate, if the erosional budget column is higher, the rock produces over-proportionally. The single diagrams of Appendix B 01–20 demonstrate that the single rock types in the single basins undergo very different fates.

It is chiefly the position of the rocks in the mountains which is responsible for these differences. According to Appendix B 20 granites and argillites are eroded under-proportionally. The argillites (Fig. 5.9) belong nearly exclusively to the lower "low energy belt" close to the coast, the granites (Fig. 5.2) half way to the upper "low energy belt", the top positions of the Serre.

The gneisses (Fig. 5.3) produce only an ordinary average, they spread relatively regularly over the entire area, whereas the schists (Fig. 5.4) produce over-proportionally, because they are preferably situated along the "high energy belt". The same holds true for the sandstones (Fig. 5.7). Admittedly, only some 15% of the siltstones are to be found on the "high energy belt" (Fig. 5.8), but the siltstone erosion in that position, the "olisthostrome hole" of Platì in the Careri [09], is so strong that the average siltstone production of our area is over-proportional.

Based on the calculation of our erosional concept, the 19 basins of the Ionian slope of the Calabrian Massif produce an erosional budget of $306\times10^3\ m^3/a$, the interfluves not being taken into account. From this, $50\times103\ m^3$ or 16% are estimated to be lost to the open ocean via dissolved load (Chap. 7). Assuming an average porosity of 50% of the resulting sediment, all grain sizes included, a volume of $512\times10^3\ m^3$ should result.

This makes a regional calculation possible: the proximal part of the Calabrian Rise, the Crotone Spartivento basin, fed more or less directly by our research area, may comprise $100\times100 = 10000\ km^2$. If we spread the estimated sediment volume over this area, a 51-mm layer would result per 1000 years. This calculation is based on the assumption of an average erosion rate of 206 mm/ka over a long-term mean of 1 Ma regarding the 19 basins only without interfluves. Present-day erosion rates are definitely higher, by one magnitude, as will be pointed out in Chapter 15.

Figure 6.10 shows the erosional budget of the combined rock types of the single basins, granite (A), metamorphic (B) and sedimentary rocks (C). Fifteen percent of the possible entire "production" are granite, 41% metamorphic rocks and 44% sedimentary rocks, of which as much as 17% are siltstone. The Allaro [16] is the river producing the highest percentage of granite (65.3%), the Amendolea [02] reaches 89.9% metamorphic rocks and the Careri [09] should produce 88.8% of sedimentary rocks, as forecast solely from source area and erosion rates. These rivers at least should develop distinct and interpretable river-mouth sediments.

In Fig. 6.10D the erosional budget (area weighed by erosion rates) of the 19 basins is plotted against their area (projected km^2 on the geological map). The differences are not overwhelming, as shown by the high correlation coefficient $r = 0.93$*, but the efficiency of the single basins is clearly demonstrated: the important Aspromonte rivers, strong sediment producers, are above the trend, the important rivers of the Zomaro Province and Serre Province work only at or below the trend.

Figure 6.11 shows the qualitative analogy to Fig. 6.10D, the relation area/erosional budget in so far as the shift in source rock geology is demonstrated if one goes from the pure area percentage of granite, metamorphic and sedimentary rocks to the corresponding values of the erosional budget of the respective basin. In Chapter 15 additional calculations are made on the volumes of the modern valleys, of paleovalleys and of the mountainous areas still to be eroded.

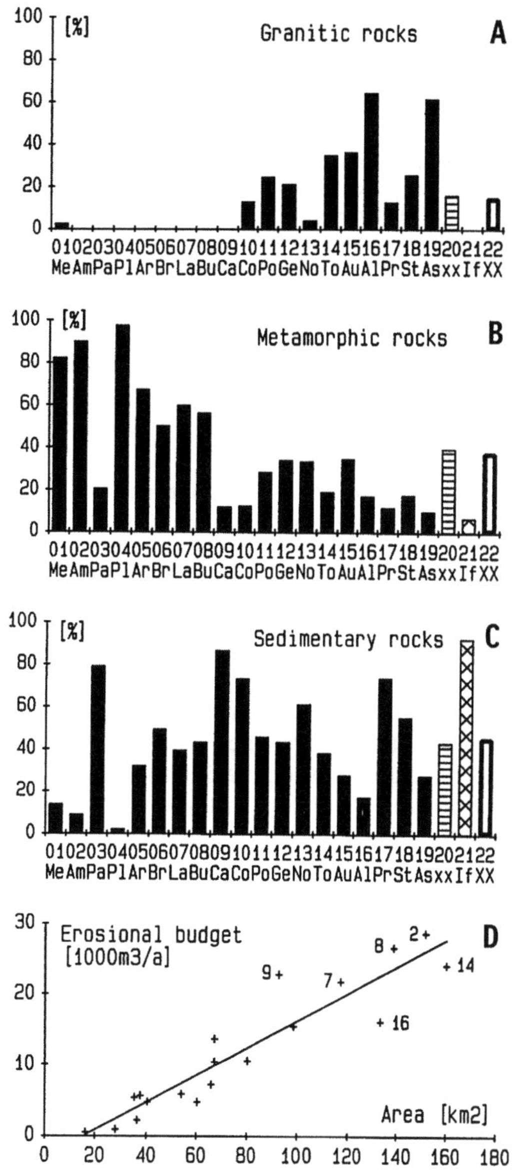

Fig. 6.10 A–D. Erosional budget (area percentage weighed by erosion rates). A Granitic rocks. B Metamorphic rocks. C Sedimentary rocks. D Relation between size and erosional budget of the 19 basins. The Aspromonte rivers Amendolea [02], Laverde [07] and Buonamico [08] and the Zomaro river Careri [09] are relatively strong eroders, whereas the Serre rivers Torbido [14] and Allaro [16] remain below the mean proportion. The correlation coefficient is $r = 0.93$*. For abbreviations *01–22* see Fig. 6.6

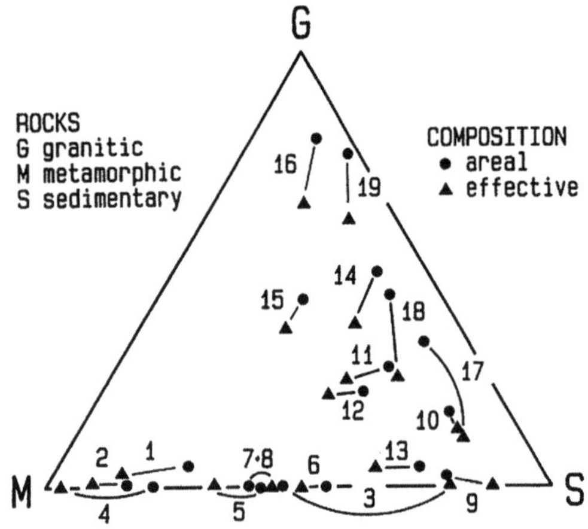

Fig. 6.11. Ternary diagram of granitic (*G*), metamorphic (*M*) and sedimentary rocks (*S*). *Full circles* indicate the simple area percentages of the three rock types as projected on the geological maps. *Triangles* indicate the erosional budget or effective composition, the area percentage weighed by the erosion rates. Most basins move away from granites or sediments because these rocks suffer, as a rule, only low erosion areas. Few rivers are sediment-accentuated regarding the erosional budget, most important the Careri [09] with its extreme siltstone erosion near Platì. For abbreviations *01–19* see Fig. 6.6

6.2.2 Effective Source Area Composition and Predicted Sediment Composition

The erosional budget of single rock types as a producer-related value is not suited directly to give quantitative information about the sediments produced.

The river-mouth sediments, the main reference sediment of this provenance study, are lag sediments and it goes without saying that the amounts and grain-size proportions of these deposits cannot reflect the respective properties of the erosional budget, due to selective weathering, transport and deposition. The qualitative composition of the river-mouth sediment, however, could very well reflect certain rock proportions of the hinterland. Lithologically similar rocks should have similar chances to release particles of similar grain size, shape or specific weight into the river-mouth sediment. But because we do not know the ability of all rock types of a basin to release or not to release a given information into a deposit, we assume all rocks equivalent, equate them and calculate the mean composition of a basin. Hereto the respective properties of the single rock types or soils are combined according to area percentages and erosion rates and recalculated to 100%: we call this the effective composition of a basin. This calculation may refer to rock or soil properties.

The meaning of the term effective composition, however, is twofold. We defined it as the source-area composition, that is the producer-related meaning, but at the same time it may serve as a kind of hypothetical composition of a resulting sediment, in our case the river-mouth sediment. Hypothetical in the sense that we are dealing with a "should-be" composition of the river-mouth sediment, an idealistic composition to be expected if no selection of the information released from the source were to take place due to lithological differentiation, weathering, transport and deposition. This is the product-related aspect and we call this predicted sediment composition. This predicted sediment composition only serves as a reference level for the provenance analysis and we have to discuss why the real, existing river-mouth sediments deviate from the predictions. In this chapter we treated the erosional budget of the single basins compared to the area percentage of their rock types. In Chapters 9, 13 and 14 the attempt is made to predict certain compositional features of the river-mouth sediments as chemical composition, gravel petrography and light and heavy mineral distribution.

6.3 Conclusion

This chapter ends the treatment of the general source area conditions. In the foreground were the morphology with the hypsometry and the geology with the rock distribution. As derived data sets we treated the gradients and the aspects, the erosion rates and the erosional budget. This made it possible to develop the concept of effective source area composition and predicted sediment composition.

Source Area. The area turned out to be highly differentiated; it is impossible to design an "average basin". All 19 basins are distinct individuals. Admittedly, two or more basins may display the same relief or dissection, but their rock distribution or erosion rates differ noticeably, or vice versa. Even if all these properties fit, the shape of the basins differ or one basin would be ten times larger than its neighbour. This is the conceptual chance of our study: if one of these properties exerts a causal influence on the river-mouth sediments, this should become visible on comparing the 19 basins. On the other hand, it should be possible to question certain properties, and simply disqualify them if their assumed relevance cannot be proved.

Erosional Budget. Following our concept calculating the valley volumes against the Mid-Pleistocene plain, the Ionian slope of Aspromonte and Serre, 19 basins plus interfluves, released a solid rock volume of $326 \times 10^9\ m^3$ since uplift started 1 Ma ago. This corresponds to a sediment volume of $546 \times 10^9\ m^3$ assuming a proportion of 16% dissolved load lost to the open ocean (Chap. 7) and an average porosity of 50% of the redeposited solid load. If we extend the calculation presented in Section 6.2.1 to the entire time span of 1 Ma, a deposit 55 m thick would result. However, we have to present additional data until this argument can be resumed in Chapters 15 and 16.

In Chapter 8 we shall describe the special properties of rocks and soils that may be transmitted into the sediment, thus forming provenance signals. This is preceded, however, by the discussion of the dissolved load (Chap. 7), not captured in the river-mouth sediment but an important part of the erosional budget released by the Calabrian Massif.

Chapter 7

Water Budget and Dissolved Load

The erosional story of Aspromonte and Serre makes it possible to estimate the amount and the composition of the erosional products delivered from this system into the Ionian Sea. But what is the proportion of the dissolved load? And what is the dissolved load made of? We have data of the chemical composition of rock-specific groundwaters and of the river-mouth waters. Based on the distribution of precipitation (Chap. 4), a water budget can be established and an average chemical composition of the dissolved load can be estimated. This will be treated in the following chapter.

Summary. The mean effective precipitation in the 50-year period between 1921 and 1970 is 698 mm/a, the mean evaporation rate 48.6%. This results in an average annual runoff of 1278×10^6 m^3. The mean mineralization of the river waters is 322.6 ppm after the summer period and 221.4 ppm in late winter, 250 ppm being a reliable mean. About 47% of this dissolved load is regarded as being recycled or from atmospheric and anthropogenic input. The composition of the Calabrian river water corresponds fairly well to the worldwide mean of river water. Compared to the long-term (1 Ma) mean erosion rates of the 19 basins (206 mm/ka) the proportion of the present-day dissolved load is 16.3%; compared to mean actual effective erosion rates of 1527 mm/ka consequently it is only 2.2%. Today the 19 basins release 319425 metric tons of dissolved load per year into the Ionian Sea, of which 169295 tons (53%) are geogenetic. This corresponds to a solid-rock cube of about 40 m side length.

Data Base. Precipitation data of the isohyet map and mean temperatures from Caloiero and Mercuri (1980) (Chap. 4). Chemical composition of rock-specific groundwaters and of the river-mouth waters of the 19 basins, sampled during both a summer and winter campaign (Weik 1988).

Methods. The precipitation map (Fig. 4.2) was digitized and stored in the data bank. The evapotranspiration was calculated following the formula of Turc.

The field measurement of water chemistry was carried out with the usual field equipment, lab determinations were made by Atomic Absorption Spectometry (AAS) and conventional chemical analysis (Weik 1988). We plot the results as they come from the computer, no round figures.

7.1 Effective Precipitation and Estimated Runoff

The precipitation increases with a steep gradient from the coast line to the mountain chain (Fig. 4.2), from <600 to >2000 mm/a. Ninety-two percent of the annual precipitation falls during the winter period (October to April).

The total absolute annual amount of precipitation results in 2487.8×10^6 m^3/a, which corresponds to a mean precipitation of 1358 mm/a (Chap. 4.1.1).

Evaporation mainly depends on temperature. In Calabria, temperature goes strongly with altitude (Fig. 4.1 A). We calculated the regression altitude/temperature in order to obtain mean temperature values for each point of the geomatrix. From this the real evapotranspiration was calculated using the formula of Turc (cited, e.g. in Hölting 1989, p. 25). The evaporation varies from over 90% at the coast to approximately 25% at the high plain (1000 m) and less than 10% at the Montalto (1956 m). These evaporation rates, based admittedly on very rough assumptions, seem to make sense, considering that in the regions higher than 1000 m much of the annual precipitation is snow, which lasts up to March/April and melts with only little evaporation loss.

The absolute annual effective precipitation (total precipitation minus evaporation) results in 1277.7×10^6 m^3/a, which means an average effective precipitation of 698 mm/a and a mean evaporation rate of 48.6%. The annual effective precipitation falls exclusively in the winter period (October to April), in the summer period (May to September) it is zero.

The precipitation (total and effective) is distributed very differently throughout the single basins as well as among the single rock types, due to the altitude-dependent distributions of both the precipitation itself and the evaporation. This is shown in Fig. 7.1 A for the 19 basins and the interfluves, where the area percentages of the basins are compared with the percentages of the total and the effective precipitation respectively. The larger basins receive disproportionally more water than the smaller ones, because of the higher precipitation and the lower evaporation in the higher parts of these basins. Extremes are the Buonamico [08] and the interfluves. The Buonamico encompasses only 7.6% of the total research area but it gains 12.3% of the total effective precipitation, whereas the interfluves cover 19.1% of the total area and receive only 5.8% of the total effective precipitation. The same is true for the areas of the single rock types (Fig. 7.1 B). It goes without saying that the higher elevated rock types, granitic and metamorphic rocks, receive relatively more precipitation than the lower ones, mainly sedimentary rocks.

In Fig. 7.2 the effective precipitation is plotted in mm/a, thus independently of the size of the area. Considering first the rivers, again Buonamico [08] and the interfluves are the extremes with 1132 and 214 mm/a, respectively. Considering the rock types, granitic areas have the highest effective precipitation with 1275 mm/a and the argillite areas the lowest with 261 mm/a.

7.2 Composition of Rock Waters

In our Calabrian research area some 100 springs are known, each with a water yield of more than 1 l/s. Taking into account only granitic and

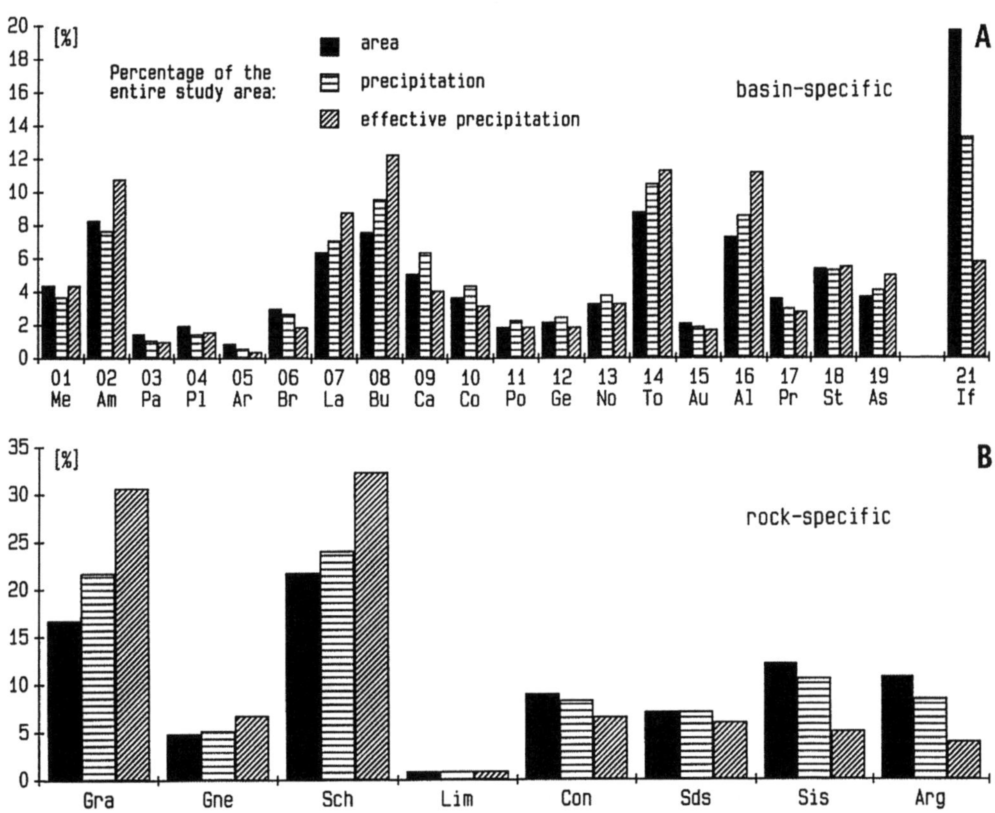

Fig. 7.1. A Drainage area, precipitation and effective precipitation (precipitation minus evaporation) of the single basins, plotted as percentages of the entire area. The high degree of differentiation of these variables between and within the basins is mainly a function of how far these basins reach into the mountain chain. **B** Drainage area, precipitation and effective precipitation (precipitation minus evaporation) of the single rock types, plotted as percentages of the entire area. *01 Me*lito; *02 Am*endolea; *03 Pa*squale; *04 Pal*izzi; *05 Ar*anghia; *06 Br*uzzano; *07 La*verde; *08 Bu*onamico; *09 Ca*reri; *10 Co*ndoianni; *11 Po*rtigliola; *12 Ge*race; *13 No*vito; *14 To*rbido; *15 Amu*sa; *16 Al*laro; *17 Pr*ecariti; *18 St*ilaro; *19 As*si; *21 If* mean of the 84 smaller basins and interfluves. *Gra*nite, *Gne*iss, *Sch*ist, *Lim*estone, *Con*glomerate, *S*and*s*tone, *Si*lt*s*tone, *Arg*illite

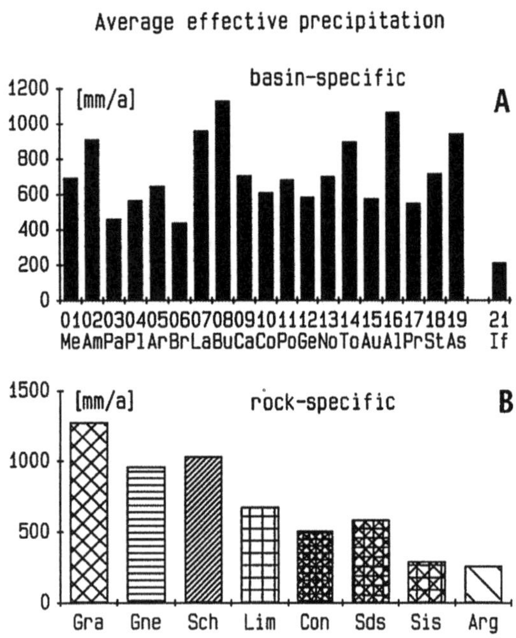

Fig. 7.2A,B. Effective average precipitation [mm/a]. **A** Single basins. **B** Single rock types. Among the rivers, Buonamico [08] and Allaro [16] receive the most average effective precipitation, among the rocks granite and schist, all more than 1000 mm/a. For abbreviations *01–19* and rock types see Fig. 7.1

metamorphic areas, all springs together reach an annual discharge of 1200 l/s, which corresponds to 37.8×10^6 m^3/a (Cassa 1986).

In order to assess the amount of dissolved load coming from the different rock types, 23 spring waters with distinct "mono-rock" drainage areas were sampled and analyzed for the main compounds sodium, potassium, calcium, magnesium and iron as well as bicarbonate, chloride and sulfate. The results from different samples coming from the same rock type vary only little, so that a mean rock-specific composition for each of the eight rock types (Chap. 5) could be calculated. We call this rock-water composition. Mean rock-water compositions are shown in Table 7.1 A. For the conglomerates no rock water analyses were available. As an approximation, the mean sandstone-water composition was used for further calculations.

In Fig. 7.3 the sum of cations is plotted versus the sum of anions. All rock-water analyses lie on the angle bisector, demonstrating the quality of the analyses and the fact that all main components were detected.

The total mineralization [ppm] ranges by ca. factor 5:

granite: 45.6	gneiss: 104	siltstone: 210.4
schist: 48.4	limestone: 113.5	argillite: 224.7
	sandstone: 136.2	

Granite and schist waters show the lowest mineralizations, gneiss, limestone and sandstone waters are medium, and the highest total mineralizations were found in siltstone and argillite waters,

Table 7.1. Average chemical composition of the rock water (spring water of single Calabrian key rocks)[a]

(Number of samples)	Granite (3)		Gneiss (6)		Schist (2)		Limestone (6)		Sandstone (2)		Siltstone (2)		Argillite (2)	
	A	B	A	B	A	B	A	B	A	B	A	B	A	B
Na [mg/l]	7.7	0.0	9.3	1.3	8.7	0.7	9.1	1.1	10.4	2.4	11.8	3.8	10.4	2.4
K [mg/l]	0.9	0.6	1.6	1.3	0.9	0.6	1.5	1.2	1.5	1.2	6.6	6.3	3.4	3.1
Ca [mg/l]	4.4	4.1	13.0	12.7	3.7	3.4	14.7	14.4	18.0	17.7	31.5	31.2	36.3	36.0
Mg [mg/l]	1.3	0.3	5.6	4.6	1.3	0.3	3.3	2.3	5.2	4.2	10.3	9.3	5.5	4.5
Fe [mg/l]	0.1	0.1	0.1	0.1	0.0	0.0	0.4	0.4	0.2	0.2	0.4	0.4	0.0	0.0
Sum of cations [mmol/l]	0.69	0.25	1.56	1.10	0.69	0.24	1.45	1.00	1.82	1.37	3.12	2.66	2.80	2.35
HCO_3 [mg/l]	13.8	0.8	37.0	24.0	12.2	0.0	58.9	45.9	71.8	58.8	29.0	16.0	109.8	96.8
Cl [mg/l]	12.9	0.0	14.8	0.8	16.6	2.6	13.9	0.0	16.1	2.1	13.8	0.0	10.5	0.0
SO_4 [mg/l]	4.5	1.5	22.6	19.6	5.0	2.0	11.7	8.7	13.0	10.0	107.0	104.0	48.8	45.8
Sum of anions [mmol/l]	0.68	0.04	1.49	0.82	0.77	0.11	1.60	0.93	1.90	1.23	3.09	2.43	3.11	2.54
Total mineralization [mg/l]	45.6	7.4	104.0	64.4	48.4	9.6	113.5	74.0	136.2	96.6	210.4	171.0	224.7	188.6

[a] A: Measured composition. B: Geogenetic portion (A minus cycled sea water and atmospheric/anthropogenic influences).

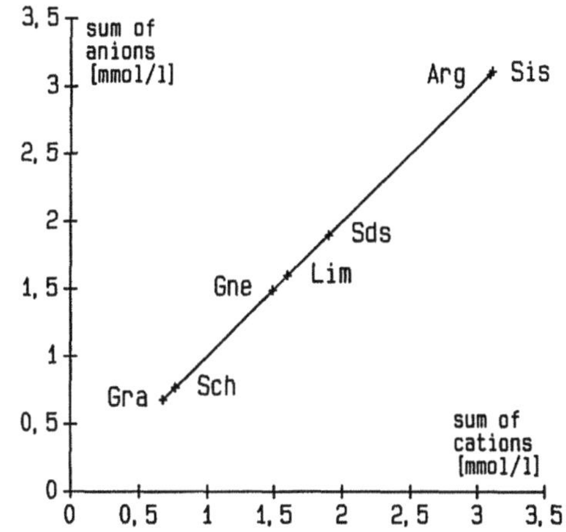

Fig. 7.3. The sum of the cations plotted versus the sum of the anions yields the angle bisector required for the reliability of the analyses and demonstrates the different total mineralizations of the rock waters. For abbreviations of rock types see Fig. 7.1

due to their high content of easily removable evaporitic minerals, gypsum for example.

However, these mineralizations of the rock waters are not entirely a product of rock weathering. Due to the geographical position of southern Calabria in the center of the Mediterranean Sea, with no point more than 20 km from the coast, a good deal must have been cycled through the atmosphere and derives from Mediterranean sea water, via aerosol. This applies in particular to chloride, because there are hardly any primary chloride-bearing minerals in Calabria.

Consequently not only the chloride content but also all other compounds related to sea water composition have to be corrected.

The recycled mean concentrations [ppm] lie in the range of:

chloride:	14.0	calcium:	0.3
sodium:	8.0	potassium:	0.3
sulfate:	2.0	bicarbonate:	0.1
magnesium:	1.0		

These concentrations come very close to the concentrations given by other authors about mean cycled constituents in rain waters from near-shore regions, Bermuda for example (Garrels and Mackenzie 1971, Table 4.7).

Another source of non-geogenetic mineralization of the rock waters is the atmosphere, in particular the anions, bicarbonate and sulfate. The amount of non-geogenetic bicarbonate, which is finally dissolved atmospheric carbon dioxide, can be assessed by looking at its concentrations in those rock waters derived from rocks with a very low content of carbonate minerals. These are granite and schists with approximately 13 ppm bicarbonate. The anthropogenic portion of sulfate in groundwaters compared to the preindustrial era is roughly 40% (Holland 1978, p. 98), which results in 1 ppm in Calabria.

Thus, regarding the total amount of geogenetic dissolved solids, the total mineralizations of the rock waters have to be reduced by about 40 ppm due to both sea-water recycling and atmospheric/ anthropogenic influences, in particular:

chloride:	14.0	cycled sea water
bicarbonate:	13.0	atmospheric carbon dioxide
sodium:	8.0	cycled sea water
sulfate:	3.0	two-thirds cycled sea water and one-third anthropogenic
magnesium:	1.0	cycled sea water
calcium:	0.3	cycled sea water
potassium:	0.3	cycled sea water
	39.6 ppm total	

The corrected, geogenetic dissolved solids of the rock waters are listed in Table 7.1 B.

7.3 Predicting Dissolved Load of River Water

Runoff measurements of the basins were beyond the reach and beyond the scope of this provenance study. However, the low permeability of most rock types, the high relief and the short river courses make it a reasonable calculation to combine the effective precipitation and the basin size by means of the geomatrix and the data bank in order to estimate a water balance of the 19 basins. We follow the same procedure to predict the mean chemical composition of the river water. Although only 10–20% of the precipitation percolates into the rock and reappears in the springs as rock-specific groundwater, the total amount of the effective precipitation was used for the calculations, because the superficially draining water also percolates into the soil and dissolves rock-specific soil material.

With regard to these restrictions, the possible or probable chemical composition of the single river waters was calculated using the area distribution of the rock types, their amount of effective precipitation and the rock-water compositions. The results are shown in Table 7.2 and mainly reflect the geological compositions of the drainage basins. For example, the lowest total mineralizations (54.4 ppm and 55.6 ppm) are predicted for the two rivers with

Table 7.2. Predicted composition of the river waters, the calculation bases on the water yield of each basin, the average chemical composition of the rock waters (Table 7.1 A) and the quantitative rock distribution of the basins

Basin	Na	K	Ca	Mg	Fe	HCO_3	Cl	SO_4	sum
	[mg/l]								
1	9.1	1.3	8.8	2.9	0.07	27.9	15.8	12.4	78.3
2	9.0	1.2	7.0	2.5	0.04	22.4	16.2	10.0	68.3
3	9.6	2.0	12.6	4.3	0.12	31.2	15.7	25.6	101.1
4	9.4	1.5	11.3	4.1	0.10	35.9	15.7	16.4	94.3
5	9.6	1.7	13.3	4.5	0.12	40.6	15.6	20.6	106.0
6	9.7	2.2	13.2	4.5	0.13	31.0	15.6	28.4	104.7
7	9.2	1.4	8.6	2.8	0.07	26.9	16.2	12.7	78.0
8	9.3	1.3	9.5	3.2	0.08	31.4	16.1	12.6	83.4
9	9.9	2.2	15.4	4.4	0.14	41.3	15.4	28.1	116.9
10	9.6	2.4	16.3	5.0	0.18	41.8	14.2	31.0	120.6
11	9.2	1.6	12.2	3.6	0.14	37.8	14.5	17.2	96.1
12	9.3	1.6	11.4	3.4	0.12	35.3	15.2	16.5	92.8
13	9.7	1.8	15.4	4.0	0.16	50.9	15.1	19.5	116.7
14	8.6	1.3	9.8	2.5	0.11	33.0	13.6	10.4	79.1
15	8.4	1.2	7.7	2.2	0.10	22.9	14.0	10.8	67.3
16	8.0	1.0	5.7	1.7	0.10	18.8	13.2	6.0	54.4
17	8.7	1.4	10.5	2.7	0.13	34.7	13.8	12.3	84.2
18	8.5	1.2	8.9	2.4	0.13	29.5	13.7	9.9	74.2
19	8.0	1.0	6.0	1.7	0.10	19.8	13.2	6.0	55.6
Intertfluves	10.4	2.9	22.9	5.7	0.16	63.0	14.3	39.5	158.8
Weighed total	9.0	1.5	10.3	3.0	0.10	31.5	14.8	14.7	84.9

the highest percentages of granitic areas in its drainage basins, Allaro [16] (68.7%) and Assi [19] (64.7%). The highest total mineralization (158.8 ppm) was predicted for the interfluves, whose drainage areas are mainly built up by young sedimentary rocks with high contents of evaporites. The mean predicted total mineralization for the total Calabrian research area is 84.9 ppm, which is equivalent to 108481 metric tons per year. This predicted annual amount of total dissoved material consists of:

11506 t sodium	(10.6%)	130 t iron	(0.1%)
1873 t potassium	(1.7%)	40295 t bicarbonate	(37.1%)
13097 t calcium	(12.1%)	18944 t chloride	(17.5%)
3862 t magnesium	(3.6%)	18776 t sulfate	(17.3%)

Taking into consideration the recycled atmospheric and anthropogenic portion mentioned above, which adds up to about 40 ppm, only 45 ppm or 53% stem from dissolved rock material.

7.4 Measured Dissolved Load of River Water

To determine the differences and similarities between prediction and reality, river waters were sampled at the river mouths and analyzed for the same main compounds as the rock waters. This was done twice, once early in November at the end of the summer period with a very low water discharge, which is exclusively fed by groundwater, and once in February towards the end of the winter period with a high water discharge mainly flowing off superficially. Some rivers could not be sampled in November, because there was no surface water.

The total mineralization of the river waters is higher after the summer period. The mean values for the total research area are 322.6 ppm in November and 221.4 ppm in February, and were calculated by weighing the values of each river due to the effective precipitation in its drainage basin (Table 7.3). Considering the fact that the bulk of the annual water discharge flows off in the winter period, a realistic annual mean value could lie in the range of 250 ppm, three times higher than the predicted value (84.9 ppm). Due to an effective precipitation of $1277.7 \times 10^6\ m^3/a$, this results in 319425 metric tons of dissolved load per year transported to the sea. Assuming a specific weight of $2.65\ g \times cm^{-3}$ (quartz), this corresponds to 120538 m^3. Considering that 47% of this total dissolved load may be recycled or of atmospheric and anthropogenic origin, the numbers must be reduced to 169295 metric tons, which corresponds to 63885 m^3.

Table 7.3. Composition of the river waters, weighed mean of the summer and winter measurement

Basin	Na	K	Ca	Mg	Fe	HCO_3	Cl	SO_4	Total mineralization [mg/l]	
	[%]								Nov.	Feb.
1	5.5	2.1	15.5	4.6	0.4	34.5	8.2	29.3	453.4	193.3
2	5.6	2.8	15.6	4.9	0.6	16.9	7.1	46.7	255.2	183.2
3	5.6	3.1	15.0	4.9	0.2	13.8	6.6	50.9	210.1	210.4
4	4.8	1.1	15.9	4.1	0.0	40.4	6.3	27.4	465.2	430.9
5	8.9	1.2	15.9	4.1	0.0	35.7	14.1	20.4	–	429.7
6									–	–
7	4.6	1.3	17.3	2.5	0.0	52.2	5.8	16.5	289.4	257.2
8	7.9	1.6	16.0	2.5	0.1	47.7	7.4	16.9	296.7	195.2
9	8.6	1.2	17.9	2.9	0.1	54.7	8.5	22.5	505.1	381.7
10	7.1	0.8	15.5	2.6	0.0	46.4	10.1	17.7	593.7	350.8
11	7.3	1.0	18.4	3.0	0.1	55.0	11.0	4.0	–	249.4
12	7.0	0.8	18.4	3.0	0.1	49.5	11.8	11.1	–	406.4
13	7.3	0.9	15.2	3.3	0.0	51.6	9.2	12.8	398.8	322.9
14	5.6	1.4	18.0	2.8	0.3	51.8	9.0	11.5	264.1	188.7
15	7.3	1.3	15.6	3.3	0.8	52.0	9.2	10.6	–	293.2
16	8.3	1.7	12.3	3.7	0.6	51.7	13.1	8.6	192.9	95.6
17	6.5	1.2	13.7	4.7	0.4	59.0	8.7	5.7	–	196.4
18	7.3	1.1	14.8	3.1	0.1	54.4	10.5	8.8	527.9	266.4
19	7.5	1.4	14.1	3.5	0.4	55.9	10.3	7.1	233.6	107.2
Weighed mean	6.6	1.5	15.6	3.3	0.2	45.9	8.9	18.1	322.6	221.4

None of us expected this high amount of dissolved load in Calabria. It is not the amount of dissolved load itself which makes this budget but the enormous precipitation and runoff of this small source area.

7.5 Comparison of Predicted and Measured River Water

Total Mineralizations. The total mineralizations measured are three times higher than the predicted ones. Apart from this, some conformities between the two can be detected (Fig. 7.4A). The rivers with the lowest predicted mineralization, Allaro [16] and Assi [19], also in reality show the lowest ones. The same is true for the rivers with the highest predicted total mineralizations, Careri [09] and Condoianni [10], which also display the highest dissolved loads. All together, a tendency can be detected that rivers with higher predicted mineralizations also show higher measured dissolved loads.

Relative Compositions. The relative percentages of the measured main compounds are very similar

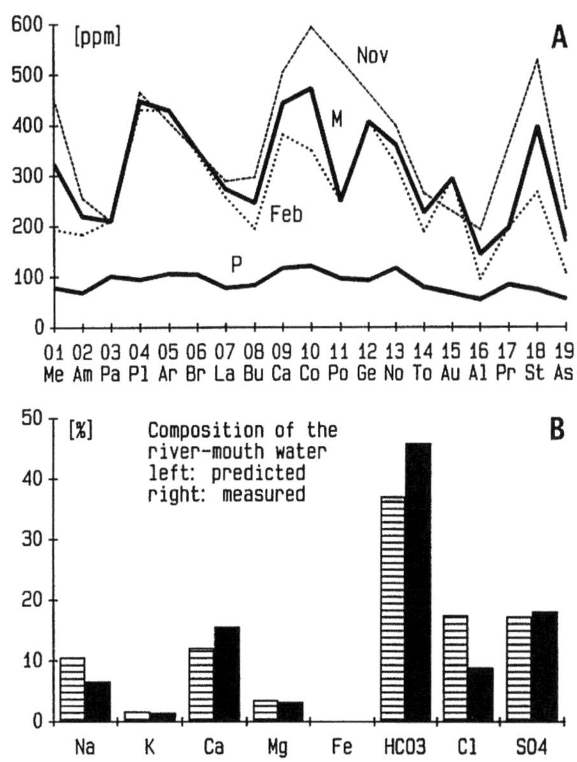

Fig. 7.4. A Total mineralization of the river-mouth waters, measured after winter (*Feb*) and after summer (*Nov*). Mean of the two campaigns (*M*) and predicted mineralization (*P*) based on the concept of effective source area composition. Note the corresponding trends, especially of the mean values measured and the predicted ones. For abbreviations *01–19* see Fig. 7.1. **B** Mean river-mouth water composition, measured and predicted. These values are not corrected, subtracting the cycled and anthropogenic amounts

within the same river during November and February and seem to be more or less river-specific. Therefore, a relative mean composition based on percentages of the dissolved load can be calculated for each river (Table 7.3) and compared with the predicted values. The weighed mean relative composition of the measured total annual dissolved load of our Calabrian research area is:

6.6% sodium	0.2% iron
1.5% potassium	45.9% bicarbonate
15.6% calcium	8.9% chloride
3.3% magnesium	18.1% sulfate

This corresponds with the worldwide mean composition of river water (Fig. 7.5) from Murray (1887) and Livingstone (1963), recalculated from Garrels and Mackenzie (1971, Table 4.3).

The comparison of these values with the predicted relative mean composition of the river waters shows a nice conformity (Fig. 7.4B). Differences are a lower sodium and chloride content and a higher calcium and bicarbonate content in the measured river waters. Because all values are percentages, the higher sodium chloride and the lower calcium bicarbonate content depend on each other. It can be excluded that sodium chloride decreases on the flow path from the spring to the river mouth. Therefore an increasing calcium bicarbonate content on this path must be the cause for the differences. This is explained with an additional solution of carbon dioxide and a resulting higher solubility of calcium bicarbonate in the carbon dioxide and anthropogenic acids containing atmospheric environment.

When examining details, the degree of conformity decreases. Measured and predicted percentages of the eight main compounds are compared separately for the 19 basins. A very heterogeneous

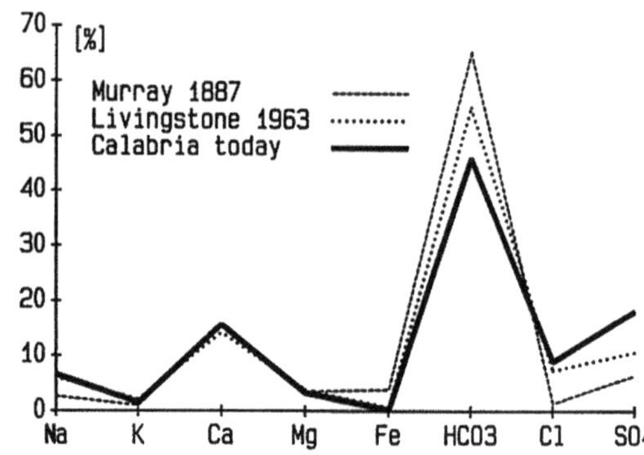

Fig. 7.5. Composition of the dissolved load of river waters: Mean Calabrian river water, worldwide mean after Murray (1887) and after Livingstone (1963). (Garrels and Mackenzie 1971, Table 4.3)

pattern of differences and conformities results and is not discussed here. However, as a concluding result it can be determined that the magnitude of order of measured and predicted values is equal and that the given amounts of dissolved load and its composition give a realistic impression of the actual conditions.

A realistic value of the annual dissolved rock material released into the Ionian Sea must lie between the predicted 21 696 m^3 based on rock water analyses and the 63 885 m^3 measured in river waters (Table 7.4). A realistic compromise seems to be 50 000 m^3, considering that the higher mineralization of the measured river waters is partly of geogenic and partly of atmospheric origin.

7.6 Conclusion

Our data base on water budget and water composition is too small to make reliable statements beyond a general orientation. The data, nonetheless, fit

Table 7.4. Comparison of predicted and measured annual budget of the dissolved load[a]

	Dimension	Calculation based on	Predicted (calculated from measured rock waters)		Measured river waters	
			A	B	A	B
Total mineralization	mg/l	(Measured)	84.9	45.0	250.0	132.5
Total weight	Metric tons (1000 kg)	1277.7×10^6 m^3 annual runoff	108481.0	57495.0	319415.0	169295.0
Total volume	m^3	2.65 g/cm^3 specific weight	40936.0	21696.0	120538.0	63885.0

[a] A: Measured. B: Corrected due to cycled sea water and atmospheric/anthropogenic influences (B = 53% from A).

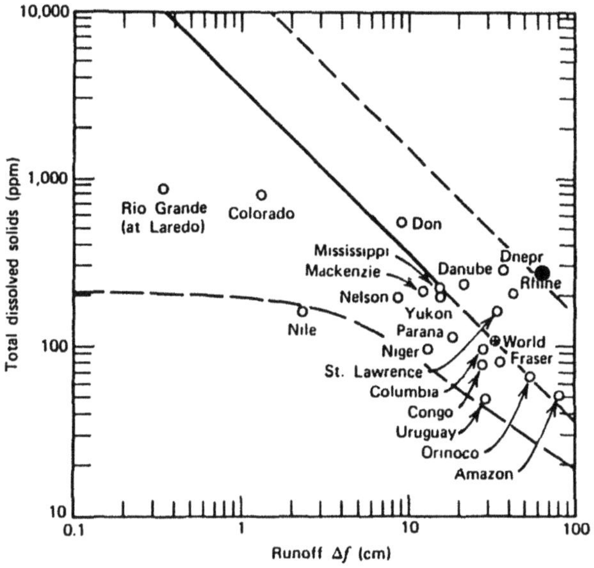

Fig. 7.6. Dissolved load and runoff of major rivers of the world (after Holland 1978, Fig. 4.6). Calabrian conditions are marked *black*, mean dissolved load of 250 ppm by far exceeds the worldwide mean

well what has to be expected if the enormous runoff, unfortunately extremely discontinuous, is taken into account. Holland (1978, Fig. 4.6) plots the total dissolved solids and the runoff of the major rivers of the world (Fig. 7.6). We considered the 19 basins as a whole and inserted this into Fig. 7.6.

The total dissolved load of our Calabrian conditions (250 ppm) is quite normal and lies in the center of the world point cluster. However, this is not true when looking at the runoff. We put the effective annual precipitation (698 mm) equal to the mean annual runoff, surely a maximum estimate. The value lies near the maximum of the world point cluster, exceeded only by the Amazon.

The proportion of solid and dissolved load is hard to calculate and difficult to compare elsewhere because only few data are available. Dissolved load in Calabria is 16.3% if the long-term erosional budget is computed (306×10^3 m^3/a, see Chap. 6.2.1), and only 2.2% if the tremendous present-day erosion rates are calculated (2264×10^3 m^3/a, see Chap. 15.1.5). This is very little compared to most rivers, which vary between 3 and 95% with a mean of 46% (Gregory and Walling 1973, Table 4.4b).

The composition of the Calabrian river-mouth waters corresponds to that of the worldwide mean with less bicarbonate and more chloride and especially sulfate, probably demonstrating increased human activity (Fig. 7.5).

As a concluding result, the absolute quantity of dissolved load produced in Calabria has to be seen as quite normal compared to the main rivers of the world. However, the relative quantity is very low due to the enormous production of clastic material at an active continental margin.

Chapter 8

Composition of Rocks and Soils

This chapter is devoted to the petrographic composition of the source rocks, the grain-size distribution of the soils, the light and heavy mineral composition of the sand fraction of the soils and the mineralogical composition of the silt fraction of the soils. Other features of the source rocks and the soils are more related to other topics of our research and, as a consequence, are treated in other chapters.

The regional distribution of key rocks and related soils forms part of the calculation of the erosional budget, the effective source area composition and the predicted sediment composition. This was discussed in Chapter 6.

The grain-size and grain-shape distribution of the jointed and weathered source rocks and the chemical composition of the sand fractions are features whose development is capable of being traced from the source to the river mouth. Hence, these topics are covered in Part II, from the source to the river-mouth sediment.

The heterogeneity of the source area geology makes it impossible to give a quantitative petrographic picture of the source rocks equivalent to that of the river-mouth sediments. The main trends of source rock petrography, however, are determinable. This and the concept of predicting sediment composition enable us to quantify the differences between source and sediment and to interpret some of them.

Summary. The igneous rocks of the Serre are mostly granodiorites. The majority of the metamorphic rocks belong to the transition from greenschist to amphibolite facies, sometimes retrograde. Among the sediments the conglomerates are a distinct but controllable source of misleading results due to recycled gravels. The sandstones are mainly arkosic. The siltstone or olisthostrome complex contains only 20% sand (40% silt, 40% clay). The soils are dystric and eutric cambisols, orthic luvisols, calcaric regosols; we measured a mean of 71% sand, 19% silt and only 10% clay. Soils of granite and feldspathic sandstone show an arkosic composition (F/Rf and F/L > 1), while the samples from gneiss and schist show a lithic composition (F/Rf and F/L < 1). ZTR values of soil vary greatly; soils from metamorphic rocks contain 12% of ZTR and from granitic or sedimentary ones only 4%.

Data Base. Analyses were made from a selection of 110 sample pairs of solid and fresh source rocks and related soils over the entire area under investigation and of 29 samples of jointed and weathered source rocks. We sampled exclusively typical rocks representative of large areas, based on long-standing field experience (Table 8.1).

Terminology. It was not possible to apply a uniform code of abbreviations for all mineral terms. The practice of hardrock and sandstone petrographers is too different. Even the symbols for rock-forming minerals (Kretz 1983) do not cover the entire topic. Consequently we give our abbreviations at the beginning of each relevant section.

Method. The large samples collected for the grain-size analysis of the jointed and weathered source rocks (Chap. 10.1) were not suitable for petrographic and chemical analysis. Thus, in a different campaign all important rock types or key rocks of the massif were sampled as fresh as possible.

Table 8.1. Number of samples used for rock and soil analyses. Numbers in brackets ( ) refer to original samples which were mixed and combined to bulk samples

	Rock or rock-related soil		
	Granitic	Metamorphic	Sedimentary
Rock			
Petrography (thin section)	10	27	9
Chemistry (RFA)	5	14	9 (15)
Soil			
Grain-size	5	17	12
Mineralogy (light + heavy min.)	6	15	2 + 3 (10)
Chemistry (RFA)	6	16	4 (10)

During the same campaign the soil samples were taken. Here we gathered many single samples over a "clean" area comprising about 100 m^2. Clean means that the sample area was exclusively underlain by the rock and soil type to be sampled and that there was no access by other soil types, for example washed down from higher areas. These subsamples of the usually very thin and patchy soils were mixed in the field, sieved with a 2-mm hand-sieve and split. The grain-size measurement of the soils followed conventional methods (pipette method). The portion of the 63–500-µm sand fraction was sieved wet, one part was prepared for the X-ray fluorescence spectroscopy (RFA), the other for optical mineralogy (light and heavy minerals). The latter samples were subjected to oxalic acid (5%) for removal of iron oxides. Grain mounts for analyses again were made following conventional criteria, utilizing the 63–500-µm fraction for the gross mineral composition and the 63–250-µm fraction for heavy minerals. For mineralogical analysis the silt fraction was separated by conventional Atterberg methods. Strong limitations of accuracy are given by the method of measuring peak area percentages on the X-ray diffractometer plots.

8.1 Source Rocks

Again, it is not within the scope of this study to go into the petrogenesis of the south Calabrian rocks; for details and literature see Bonardi et al. (1984). Predominantly, we want to describe the rocks as they are, not as a finished product of certain processes of the past but as a present-day and active producer or erosional material, both solid and dissolved.

Thin section evaluation of the solid rocks ranks the major minerals by visual estimation, starting from minor constituents; accessory minerals are not quoted. No point counting was made with the exception of the sandstone samples.

Abbreviations:

act	actinolite	hbl	hornblende
alb	albite	kfsp	K-feldspar
alkfsp	alkali-feldspar	mus	muscovite
and	andesine	plg	plagioclase
apt	apatite	qrz	quartz
chl	chlorite	sph	sphene
clzo	clino-zoisite	ser	sericite
cor	cordierite	sil	sillimanite
fsp	feldspar	zo	zoisite
grt	garnet		

The symbols of rock units refer to the Carta Geologica Della Calabria (CGDC, Chap. 5, rock units).

8.1.1 Granitic Rocks

Aspromonte Province

τ Muscovite-biotite-granite
mus + bio + qrz + plg = alkfsp
Bagaladi, Melito [01]

τ Muscovite-biotite-granite
mus + bio + qrz – alkfsp – plg
Platì, Careri [09], Condoianni [10]

τ Muscovite-biotite-granite
mus + bio + alkfsp = plg + qrz
Platì, Careri [09], Condoianni [10]

Zomaro Province

τ Sillimanite-andesine-granite or "gneiss"
sil + and + mus + alkfsp + plg + qrz
Antonimina, Condoianni [10], Portigliola [11]

Serre Province

τ Syenogranite, perhaps "orthogneiss"
bio + qrz = plg + alkfsp
Nardodipace, Precariti [17], Stilaro [18]

τ Biotite-granite or biotite-granodiorite
bio + alb + qrz + plg
Gioiosa, Torbido [14]

τ Biotite-granodiorite to tonalite
bio + alkfsp + qrz + plg
S. Catarina, Assi [19]

τ Biotite-granodiorite
mus + bio + alkfsp + qrz + plg
Ragoná, Allaro [16], Precariti [17]

τ Biotite-granodiorite to granite
mus (sec.) + bio + alkfsp + qrz + plg
Ragoná, Allaro [16], Precariti [17]

τ Biotite-granodiorite
bio + alkfsp = qrz + plg
Ragoná, Allaro [16], Precariti [17]

Large areas of granite are currently deeply weathered and converted into a thick cover of grus. As a consequence, the granites produce much more sand than gravel or boulders. We tried to obtain an idea of how far the grain-size distributions of these sands are determined by the original grain sizes and grain shapes of the granite (Blatt 1967). Ott (1990) developed a stereomodel of the spatial arrangement

Fig. 8.1. Stereomodel of quartz fabric, Serre granite. One plane of a granite cube was ground down in 80-μm steps. Each new surface was feldspar-stained and photographed. The quartz outlines were cut into styropor slices. Mean exaggeration 9× (B. Ott, unpubl.)

of the minerals of several samples of the Serre granites. She ground oriented sample blocks down at one side, step by step in 80-μm intervals. Each new surface was polished, stained and photographed. The outlines of the quartz minerals of each photograph were cut into styropor slices. Finally the whole set was remounted to a stereo model (Fig. 8.1).

This model demonstrates that the quartz as the latest precipitate grows through the remaining interstices, forming a network of ramifications so intricate and polymorph that it is simply impossible, at least in the case of the Serre granite, to speak of "grains" with definable forms and diameters. In addition, the minerals are expected to break not only along the thin connections or weak points due to weathering but even more along the dense net of microfissures present in all samples (Moss 1972a, b). Probably these microfissures are not or not only caused by tectonics, in the broadest sense of the term, but by the strong dilatation of the fabric due to the saussuritization of the plagioclase. Thus, the beginning disaggregation of the granite seems to be established as early as the hydrothermal phase with weathering probably playing a minor role. This is proven, too, by the very little kaolinization of the K-feldspars. This very strong dilatation may be a reason for the deep grus formation of the granites.

8.1.2 Metamorphic Rocks

Aspromonte Province (cf. Figs. 8.2 and 8.3)

go Augengneiss (mylonite)
alb + mus + bio/chl + plg + alkfsp (augen) + qrz
Roccaforte, Amendolea [02]

go Augengneiss (orthogneiss)
alkfsp (microcline-perthite) + mus + bio + qrz + plg + alkfsp
Palizzi, Palizzi [04]

π Metabreccia to muscovite-granite (tectonite?)
mus + alkfsp = plg = qrz + chalcedone
Antenna, Buonamico [08]

sb Garnet-biotite-quartz-schist (metaarkose)
grt + bio + fsp + qrz + plg, kfsp?
S. Lorenzo, Melito [01], Amendolea [02]

sb Amphibolite "amphibole-gneiss"
bio + plg(andes?) + hbl
Bagaladi, Melito [01]

sb Metabasite
sph + fsp + qrz + hbl-act + grt + zo/clzo
Platì, Careri [09]

sb Metaarkose?, fine-grained, feldspar-rich and mica-rich parts changing

Fig. 8.2. "Gneiss" and "pegmatite" of the Lago Costantino area, Buonamico [08]. Strong deformation makes these rocks the main source of gravel supply in the Buonamico basin

Fig. 8.3. Quartz phyllites of the Portella d'Orgaro area E of Samo, Laverde [07]. This extremely foliated rock produces large amounts of fine-grained debris

grt + clzo + mus/ser + bio/chl + qrz + alb
Antenna, Buonamico [08]

sb Garnet-(cordierite)-micaschist with beds of garnet-plagioclase-augengneiss
grt + cor(?) + plg(acid) + qrz + bio/chl + mus
Antenna, Buonamico [08]

sb Garnet-chlorite-sericite/muscovite-micaschist, changing with quartz-micaschist

(with plg-augen) (blastomylonite)
grt + chl + ser/mus + qrz
Platì, Careri [09]

sb Augengneiss
zo + apt + mus/ser + plg + qrz
Platì, Careri [09]

sbm Quartz-porphyry
alkfsp + qrz + granophyritic matrix schlieren of zo + mus + grt
Platì, Careri [09]

sbm Quartz-phyllite with clino-zoisite
bio (Stilpnomelane?) + clzo + mus + qrz + alb(?)
Antenna, Buonamico [08]

sf Sericite-chlorite-phyllite, with metatuff bed
San Lorenzo, Melito [01], Amendolea [02]

sf Quartz-phyllite + limeschist
qrz + ser + chl, qrz + carbonates
Palizzi, Palizzi [04]

sm Metabreccia with parts of metamarl (tectonite?)
grt + clzo + mus + plg + alb + qrz in chalcedone-matrix (sbm?)
Antenna, Buonamico [08]

sm Metaarkose (albitite)
few qrz + alkfsp + much alb
Antenna, Buonamico [08].

Zomaro Province

sf Metagraywacke?, fine grained
plg + chl/mus + qrz + bio
Agnana, Novito [13]

sf Quartz-phyllite ("schist")
chl + ser/mus + qrz
Gerace [12], Novito [13].

Serre Province

sb Sericite-chlorite-phyllite
Nardodipace, Stilaro [18]

sb Garnet-biotite-plagioclase-gneiss (metaarkose)
grt + bio + plg(ol?) + qrz, kfsp?
S Caulonia, Torbido [14], Amusa [15]

sb Garnet-biotite-plagioclase-gneiss (metaarkose)
grt + bio + plg(ol?) + qrz
Gioiosa, Torbido [14]

sb Quartz-phyllite
qrz + chl + ser
Gioiosa, Torbido [14]

sf Chlorite-biotite quartz-schist
chl + bio + qrz = quartzite layers
Stilo, Stilaro [18]

sf Chlorite-phyllite/quartz-phyllite
Nardodipace, Stilaro [18]

sf Metapelite, feldspar-rich
bio + mus + qrz = alb
Stilaro [18]

sf Knotenschiefer to metapelite
cor + mus + bio + qrz
Stilaro [18]

sf Sericite-schist, inhomogeneous (phyllite)
bio + qrz + mus/ser
Allaro [16].

All metamorphic rocks described above belong to the transition from greenschist to amphibolite facies, sometimes retrograde. A primary feature is the strong cataclasis and mylonitization.

The rocks they originate from include a broad variety of primarily eusyncline rocks such as arenites, graywackes, pelites, marls, calcarenites and even some limestones. Sedimentary structures like gradation, slumping or intercalations of small debris flows are sometimes preserved. Granites, eclogites and volcanic rocks such as possible tuff and ignimbrites complete the collection.

It should be strongly emphasized that no ultrabasic or basic rocks of the greenstone suite occur, and no volcanic rocks, basalt or the like.

8.1.3 Sedimentary Rocks

8.1.3.1 Jurassic Limestone

This rock varies from very massive to compact limestones (Fig. 8.4), mainly biosparites and dismicrites of a reef or near-reef environment. The scarce limestone deposits are macroscopically very similar over the entire area, their gravels, regardless of size, are very easy to distinguish from other rock types. Consequently, closer analysis or thin sections was not necessary.

8.1.3.2 Conglomerates

In all basins, the conglomerates are always present, the Novito [13] for example yields an area percentage of 21%, the erosional budget of the Stilaro [18] is 27% conglomerate. The conglomerates are the main confusing factor of this provenance study, because it is almost impossible to distinguish, at the river mouth, between first- and second cycle gravels. There are, however, two ways out of this

Fig. 8.4. Steep walls of Jurassic limestones of the Fossone Novito W of Canolo in the Novito [13] basin. These reef limestones form the steepest relief of the area

Fig. 8.5. Coarse-grained and poorly cemented sandstones and conglomerates of the Stilo Capo d'Orlando Formation in the Stilaro [18] basin immediately below the village of Stilo. The influence of this producer of second cycle clasts is controllable

dilemma. We have basins with no granites outcropping in the source area, Pasquale [03] for example, but the river-mouth gravel of this basin contain 12.1% granite in the 16–80-mm fraction. The only possible source for these granitic gravels is the conglomerates, which should show up in the river-mouth sediments, as calculated by the predicted sediment composition, with 4.4%, or even

somewhat more, if the non-pebble-producing formations of the Pasquale such as the debris flow siltstones are taken into account. If this is done, the high amount of granitic river gravel fits nicely to the source area conglomerates. But even if there are both granitic conglomerate deposits and outcropping granites in the source areas, we can at least estimate the influence of the conglomerates apart from their percentage of the erosional budget, because we possess data about the petrographical composition of some conglomerate deposits itself.

The conglomerates, at first glance, give the impression of being very proximal and representing the Miocene equivalent to the present-day river-mouth, nearshore and canyon-head conditions (Fig. 8.5). All rock types of the gravels of this assemblage are well known in the area and seem to have originated close by. There are, however, exotic rocks, long known to all investigators. These are porphyritic rocks (Ferla and Alaimo 1976), macroscopically very distinct from all other rock types and easily recognized, with percentages reaching 25%. These porphyritic rocks must have originated from important and voluminous source areas – but areas of this kind no longer exist in the present-day neighbourhood. This is especially true for the Aspromonte Province dominated by metamorphic rocks, but even in the Serre Province only tiny veins of porphyry are known. During 20 years of research in the area we detected only one larger, 7-m-thick vein in the upper course of the torrente Ponzo immediately north of our border river Assi [19].

All porphyric gravels of our 19 river-mouth sediments must thus come exclusively from conglomerates, and by this, the influence of this "second-hand" gravel source can, at least, be estimated.

Bonardi et al. (1980, p. 372–380) estimated the composition of the conglomerates along the Stilo-Capo Orlando Formation and came to the following results (Table 8.2). We return to this problem when we discuss the composition of the river gravels in Chapter 13.

8.1.3.3 Sandstones (by Renzo Valloni)

Sandstones play a double game as provenance factor. Their coarse detritus is a quite ordinary gravel source producing sandstone clasts which are easy to distinguish at the river mouth and to relate to the respective source areas. Their production of sand, however, is difficult to detect. The relatively moderate diagenetic cementation stands for an effective disintegration of most parts of the sandstones which should lead to a perceptible contribution directly to the river-mouth sands. The following section is devoted to the petrographic and textural features of the sandstones in order to make distinction possible between first and second cycle sand in the river-mouth sediments. Figure 8.6 shows a typical section of turbiditic sandstones.

Data Base. Calabrian sandstone petrography has been investigated in two areas by Cavazza (1989) and Cavazza and Dahl (1990): the Allaro [16] and Stilaro [18] basins in the Serre (22 samples) and the Amendolea [02] basin in the Aspromonte (10 samples). We analyzed six sandstone samples; they stem from the Amendolea [02] (one sample), from the Buonamico [08] (4 samples) and from the Stilaro [18] (one sample). Results of the two studies are largely coordinated.

Table 8.2. Petrographic composition of some conglomerates of the Stilo-Capo d'Orlando Formation in various basins (after Bonardi et al. 1980)

Basin[a]	[02/1]	[02/2]	[08]	[09]	[10]	[16]	[18]
	[%]						
Granite		20	10	80–90	(+)	50	50–80
Phyllite	80	10			90	20	20
Porphyry	+	20	10	+		+	15
Paragneiss	10	50	70	10–20		10	+
Pegmatite			10	+			
Quartz	10				10		
Jurassic limestone						10	+

[a] Amendolea [02], sample 1 and 2; Buonamico [08]; Careri [09]; Condoianni [10]; Allaro [16]; Stilaro [18].
(+): only traces

Fig. 8.6. Sequence of turbiditic sandstones of the Stilo Capo d'Orlando Formation in the Allaro [16] basin at Ponte S. Giorgio N of Caulonia. The contribution of sand of Tertiary sandstones to the formation of the river-mouth sands is small

Method. Microscope analyses were made by thin section point counting using a mechanical stage; 250 essential grains were counted for each sample following conventional procedures. Grain categories and petrologic parameters are plotted in Tables 8.3 and 8.4.

Terminology. For the computation of point-counting data two main choices are required: (1) the criteria for filing lithic phanerites (Dickinson 1970; Zuffa 1985) and (2) the limit between essential and accessory grains with the former included and the latter excluded from the computation of the detrital mode (Dickinson 1970). In provenance-oriented studies allochemical grains (Folk 1968) are accessory grains, by definition, because of their intrabasinal origin (e.g. fossils).

1. In the point-counting procedure, workers behave differently with respect to the lithic phanerites, as for example a Q+F gneissose grain; they are assigned to the rock fragment pole (McBride 1963), or to the feldspar pole (Folk 1968), or even to the quartz pole (Gilbert 1958) if the microscope crosshair intercepts the quartz crystal of the Q+F grain.

2. Discriminating between essential and accessory grains, the discussion is essentially confined to micas and heavy minerals which are considered either essential (Dott 1964) or, because of their peculiar hydrodynamic behaviour and diagenetic unstability, accessory grains (Gazzi 1966).

In order to assure the comparison of point-counting data a precise definition of individual grain types is needed. This is provided by Table 8.3, which enables the calculation of petrologic parameters according to both the traditional, genetic criteria (Potter 1978) and the late, textural criteria (Dickinson 1970). In the method followed by Potter, focus is on the mineralogy of the grain and all polymineralic grains are considered a rock fragment, while, in the method followed by Dickinson and coworkers, the lithic grains (L) population (Boggs 1968) contains only fine-grained rock fragments, and some critical grains such as mica are excluded from the detrital mode. This method minimizes the well-known variation of mineral composition with grain size (Blatt et al. 1972, p. 302; Ingersoll et al. 1984) and leads to the assignment of sand-sized crystals, which are part of polymineralic grains, to the proper crystal category (Q or F) rather than to the rock fragments (Gilbert 1958). Carbonate grains are also a matter of discussion between "schools" because of their unstability and ubiquitarian provenance. As a consequence it is opportune to specify (Table 8.4) if they are included in (Gazzi 1966) or excluded from (Dickinson 1970) the group of the lithic fragments.

Parameters. A large number of petrologic parameters on mineral composition of arenites can be found in the literature. The most commonly used and a few others here introduced are listed in Tables

Table 8.3. Grain types with codes and ternary grain proportions. Compare with grain proportions of Table 8.4; parameters within brackets (L, Rf) indicate critical components alternatively included in or excluded from the rock fragment pole of ternary diagrams

Lithic grain parameters				Primary parameters		Codification		Components
LmLvLs	QpLvLs	LqLvLs	LcrLvoLse	QFL	QFRf			
							Qu	Single-crystal quartz:
				Q	Q	Qa	– angul.	with roundness from subangular to very angular
				Q	Q	Qr	– round.	with roundness from subrounded to well rounded
							Qc	Composite monomineralic quartz:
				Q	Q	Qg	– granul.	with igneous distinctive mineralogy/texture or <4 crystals grain
	Qp	Lq		Q	Q	Qm	– migran.	with metamorphic distinctive min./texture or >3 crystals grain
	Qp	Lq	se	Q	Rf	Ch	Ch	Chert: microcrystalline and cryptocrystalline quartz
		Lq	cr-vo-se	Q	Rf	LQ	LQ	Quartz crystal (>63 µm) within cryst., volc., sedim. rock fragment
							Fm	Monomineralic feldspar:
				F	F	KF	– KF	potash feldspar
				F	F	PF	– PF	plagioclase
							LF	Feldspar (>63 µm) within cryst., volc., sedim. rock fragment:
			cr-vo-se	F	Rf	LK	– LK	potash feldspar (as above)
			cr-vo-se	F	Rf	LP	– LP	plagioclase (as above)
Lm	Ls	Ls	cr	L	Rf	Lm	Lmet	Fine-grained (<63 µm) metamorphic rock fragment
Lv	Lv	Lv	vo	L	Rf		Lvol	Fine-grained (<63 µm) volcanic rock fragment
Ls	Ls	Ls	se	L	Rf	Ls	Lsed: silicL	Fine-grained (<63 µm) siliciclastic rock fragment
		Ls	se	(L)	Rf	Lc	– : carbL	Marble, dolomite and limestone of any texture
			cr-vo-se		Rf		LM	Mica (>62 µm) within crystalline, volc., sedim. rock fragment
			se		(Rf)	M	Mica	Muscovite, biotite and chlorite
			se		(Rf)	H	H.Min.	Transparent and opaque dense mineral

8.4 and 8.5. Parameters can be grouped into three main categories:

A – Percent frequencies
B – Ratios between primary parameters
C – Ratios within primary parameters

The first category may contain simple grain frequencies (percent Qu+Qc) or grain proportions in a ternary diagram (percent QFL). The second category contains pure numbers, represented by the stability indexes (Q/F+Rf) and provenance indexes (F/Rf), which actually may range from 0.01 to 99.00. The third category is also represented by grain ratios, giving proportions within the quartzose (Qp/Q), feldspathic (P/F) and lithic (carbL/Lsed) components of arenites; all these ratios may take values from 0.00 to 1.00 with the exception of PF/KF (monomineralic plagioclase/monomineralic K-feldspar), which in practice may range from 0.01 to 99.00. Categories B and C are discussed in detail in the soil section (Chap. 8.2.2, Table 8.5).

Ternary diagrams (Okada 1971) are traditionally used to show grain proportions of both the essential constituents (e.g. QFL) and the lithic grains population (e.g. LmLvLs) of the framework.

Framework Essential Constituents. The representations of QFRf (Franzinelli and Potter 1983) and of the micas- and heavies-free QFRf (Van Andel 1958) are based on the traditional criteria where any polymineralic grain is part of the lithics pole (Rf) while the representations of QFL+Lc (Gazzi 1966) and of QFL (Dickinson 1970) are based on the textural criteria (Gilbert 1958), so that only unstable fine-grained lithics enter the lithic pole (L).

Lithic Grains Population. Many different criteria for grouping grain categories have been proposed.

Table 8.4. Terminology of mineral groups and petrologic parameters used for mineral composition. Continued by Table 8.5. Compare with Table 8.3 for codes. Within the category of grain frequencies, monocrystalline quartz (Qu + Qg + LQ), plagioclase (PF + LP) and K-feldspar (KF + LK) are respectively equivalent to Qm, P and K of Dickinson and coworkers. The term non-micaceous is meant for "free of all accessory (LM, M, H) terrigenous grains"

Percent Frequencies

1. Grain frequencies
Qu + Qc = monomineralic quartz (= Q of QFRf)
Qu + Qg + LQ = "monocrystalline" quartz
LQ + LF = lithic phanerites
Fm = monomineralic feldspar (= F of QFRf)
PF + LP = plagioclase (= P of P/F)
KF + LK = K-feldspar
L sed = siliciclastic lithic aphanites and limeclasts
L + Lc = lithic aphanites and limeclasts
Rf = lithic phanerites and aphanites

2. Grain proportions in ternary diagrams
QFRf = total terrigenous framework
QFRf − (M + H) = M- and H-free terrigenous framework
QFL + Lc = non-micaceous terrigenous framework
QFL = non-micaceous non-carbonate terrigenous framework
LcrLvoLse = total rock fragments
LqLvLs = total quartzose and other fine-grained rock fragments
QpLvLs = fine-grained rock fragments
LmLvLs = polymineralic fine-grained rock fragments

In Table 8.4 four criteria are listed. Following an order of increasing detail, the representations are named as follows: total rock fragments (LcrLvoLse; Potter 1978), total quartzose and other fine-grained rock fragments (LqLvLs; Valloni 1985), fine-grained rock fragments (QpLvLs; Graham et al. 1976) and polymineralic fine-grained rock fragments (LmLvLs; Ingersoll 1978).

Textural Components. The sandstone samples collected have a medium grain size and show poor sorting.

Interstitial components are almost exclusively represented by fine detritus (orthomatrix), which is locally recrystallized (epimatrix) and has a quite variable frequency (average 3.2%, st. dev. s = 3.2%). Small quantities of strongly altered labile grains and of replacive calcite are present (average 1.0%, st. dev. s = 1.0%).

Diagenetic processes are mainly related to mechanical compaction which causes deformation of ductile clasts and recrystallization of unstable grains. Cementation is poor, only small amounts of phyllosilicates were found. In addition, Cavazza and Dahl (1990) report little dolomite authigenesis

Table 8.5. Continued from Table 8.4. Petrologic parameters calculated either between or within primary parameters, namely quartzose (Q), feldspathic (F) and lithic (L, Rf) grains. Compare Table 8.3 for grain codes (r.f.: rock fragments)

Ratios between primary parameters

1. Stability index (parameters from ...)

$\frac{Q}{F+Rf}$ (total terrigenous framework)

$\frac{Q}{F+L+Lc}$ (non-micaceous terrigenous framework)

2. Provenance indexes (parameters from ...)

$\frac{Q}{Fm}$; $\frac{F}{Rf}$ (total terrigenous framework)

$\frac{Q}{F}$; $\frac{F}{L+Lc}$ (non-micaceous terrigenous framework)

Ratios within primary parameters (secondary parameters)

1. Quartzose grains

$\frac{Qu\text{-}r}{Qu}$ $\frac{\text{rounded single-crystal quartz}}{\text{total single-crystal quartz}}$

$\frac{Qc}{Qu+Qc}$ $\frac{\text{composite monomineralic quartz}}{\text{total monomineralic quartz}}$

$\frac{Qc\text{-}m}{Lq}$ $\frac{\text{microgranular composite quartz}}{\text{total lithic quartz}}$

$\frac{Qp}{Q}=\frac{Qm+Ch}{Q}$ $\frac{\text{fine-grained lithic quartz}}{\text{total quartzose grains}}$

$\frac{Lq}{Q}=\frac{Qm+Ch+LQ}{Q}$ $\frac{\text{lithic quartzose grains}}{\text{total quartzose grains}}$

2. Feldspathic grains

$\frac{P}{F}$ $\frac{\text{total plagioclase}}{\text{total feldspar}}$

3. Lithic grains

$\frac{Lmet}{L+Lc}$ $\frac{\text{fine-grained metamorphic rock fragments}}{\text{polymineralic fine-grained r.f. and limeclasts}}$

$\frac{Lvol}{L+Lc}$ $\frac{\text{fine-grained volcanic rock fragments}}{\text{polymineralic fine-grained r.f. and limeclasts}}$

$\frac{silL}{Lsed}$ $\frac{\text{fine-grained siliciclastic rock fragments}}{\text{fine-grained siliciclastic r.f. and limeclasts}}$

$\frac{carbL}{Lsed}$ $\frac{\text{limeclasts}}{\text{fine-grained siliciclastic r.f. and limeclasts}}$

and successive abundant calcite cementing some distinct beds. Consequently, the sandstones mostly tend to decay into their former sandy components instead of creating polymineralic sandstone clasts.

Framework grains are of quartzofeldspathic mineral composition; the ratio essential grains/accessory grains is approximately 9/1. Accessory grains are represented by micas and heavy minerals having average frequencies of 8.5% (st. dev.

s = 5.9%) and 0.95% (st. dev. s = 0.8%) respectively.

In general, framework grains are well preserved. Feldspar grains, however, especially plagioclase, show a significant percentage of clay-sericite-altered grains. In the six samples the ratio altered feldspar/total feldspar is very constant and equal to 0.25, indicating that one-quarter of Tertiary sandstone feldspar is altered.

Detrital Mode. The general mineral composition of Tertiary sandstones is shown by the grain proportions of Fig. 8.7, depicting both essential terrigenous grains and rock fragments. Data points indicate a frankly feldspathic composition; quartzose grains average around 50% and fine-grained lithics are mostly below 10%. The data of Cavazza (1989a) are shown by the hexagons of Fig. 8.8 which indicate the Q, F and L mean value ± one standard deviation for the two groups of 10 and 22 samples collected in the Aspromonte and the Serre, respectively. These data largely coincide with ours. The small discrepancy is mostly due to the use of different petrologic parameters (L + Lc versus L).

Analyses clearly separate into two petrofacies centered slightly below and above Q50. These petrofacies properly reflect the metamorphic (Aspromonte) and granitic (Serre) source provinces. As a matter of fact, most rock fragments are represented by granitic and gneissic grains so that their pole averages as much as 31% in the QFRf (M- and H-free) diagram and only 5% in the QFL (Lc included) diagram.

Plots of the lithic grain assemblage show overlap of the two petrofacies. Ternary diagrams of Fig. 8.7 indicate that polymineralic grains are dominated by crystallines (Lcr71 Lvo1 Lse28), which consist mostly of phanerites (Lq70 Lv3 Ls27). Figure 8.8 (from Cavazza 1989b) splits the grain categories of the fine-grained rock fragments (L), which still

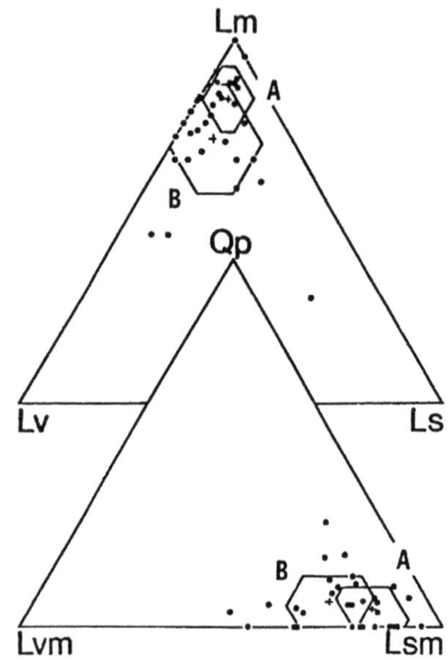

Fig. 8.8. Ternary diagrams showing proportions of the total fine-grained rock fragments (*Qp, Lvm, Lsm*) and of the polymineralic fine-grained rock fragments (*Lm, Lv, Ls*) of the Stilo Capo d'Orlando sandstones (Cavazza 1989b). *A* Aspromonte Province; *B* Serre Province. (+) Average, *polygons* indicate standard deviations for the two groups of samples related to the Aspromonte and Serre Provinces. Plotted values largely overlap and cluster near the poles (*Lm, Lsm*) containing polymineralic metamorphic grains

Fig. 8.7. Framework mineral composition and lithic grain proportions of the Tertiary sandstones of the Stilo Capo d'Orlando Formation. Ternary diagrams represent the essential terrigenous grains of the framework (*Q, F, L + Lc*), the total quartzose and other fine-grained rock fragments (*Lq, Lv, Ls*) and the total rock fragments (*Lcr, Lvo, Lse*) population. *Dots* are samples from the Amendolea [02] and Buonamico [08] basin (Aspromonte Province), and *squares* are from the Stilaro [18] basin (Serre Province). *Triangles* are the averages of the six samples as indicated. *Polygons* are from Cavazza (1989a) and represent standard deviation of Q, F, L values of ten samples from the Amendolea [02] and of 22 samples from the Allaro [16] and Stilaro [18] basins of the Aspromonte and the Serre Provinces, respectively

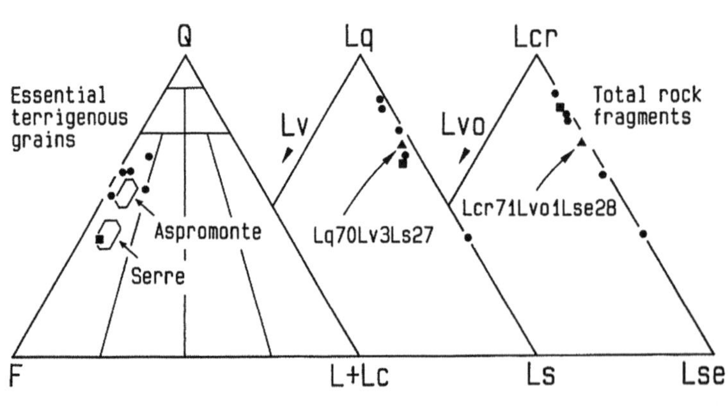

show compositional overlapping for the metamorphic and granitic sources.

To conclude, the small percentage (5%) of fine-grained lithics is almost exclusively represented by metamorphic types (Lmet) because volcanic ones (Lvol) seldom occur and sedimentary types are almost absent. Other grain ratios calculated within primary parameters indicate that among quartzose lithic grains (Lq) the composite microgranular ones (Qc-m of Table 8.3 and Qp of Fig. 8.8) are negligible while among feldspatic (F) grains (altered excluded) K-feldspar slightly exceeds plagioclase (P/F averaging 0.39).

Discussion. The correspondence of the two petrofacies with the rock types outcropping in the underlying basement support the conclusion of a local source, dominated by plutonic and metamorphic rocks in the Serre and Aspromonte respectively. It is not clear, however, where the small amount of volcanic detritus (Cavazza 1989b) comes from. The same is true for the sometimes huge boulders of porphyry in the conglomerates associated with our sandstones (Chap. 8.1.3.2).

Fig. 8.9. Siltstone in the Platì area, Careri [09]. The siltstones are mainly debris flow or olisthostrome deposits which give rise to extended mass movements (Fig. 8.10). Remainders of the olistholites barely show up in the river-mouth sediments, and the erosional products of the siltstone itself are too fine-grained to settle in the river-mouth areas

Nevertheless, the overwhelming part of the Capo d'Orlando Formation seems to have source areas similar to those of our present-day river-mouth sediments. Because the detritus of the Tertiary sandstones is texturally immature there is only little hope of separating the two sources, primary and recycled, in the present-day river-mouth sediments. This problem is discussed in more detail in Chapter 14.

In Fig. 8.13G the average light mineral composition of the sandstones is shown by the frequency of the key grain types; altered feldspar grains are all assigned to plagioclase (PF), probably introducing some error. The average mineralogy of the soils of sandstones is also plotted for comparison (Fig. 8.13F). Although monomineralic quartz is by far the most important constituent of both sandstone rock and soil, processes of soil formation cause some important modifications of the mineral assemblage.

Compared to the outcropping sandstones, soil mineralogy is strongly impoverished with regard to composite quartz (Qc) and polymineralic quartzose grains (Lq); other polymineralic grains (feldspathic and fine-grained metamorphics) are also partly destroyed. Single crystal grains, on the contrary, are much better preserved with unitary quartz exhibiting a strong increase from 32% to 56%. As will be pointed out (Chaps. 8.2.2.3 and 9), this trend does not continue towards the river-mouth sediments where mechanically more stable polymineralic quartzose grains prevail.

Fig. 8.10. Siltstone below Samo, Laverde [07]. Devastated road bends after the rainstorm of January 1973. More than 30% of the study area is endangered by landslides (Fig. 15.1)

8.1.3.4 Siltstones

Siltstones are debris flow or olisthostrome deposits, called in Italy "argille varicolori" or, sometimes, "argille chaotiche", a very appropriate term (Figs. 8.9, 8.10). They are composed of a mixture of a fine-grained matrix and coarse debris or olistholithes. The sand-silt-clay ratio of the matrix is surprisingly constant over the entire outcrop area of 224 km^2 (Fig. 8.11), with a mean of 20% sand, 40% silt and 40% clay. Despite this low sand content, the siltstones have to be regarded as a serious contributor to the sand production of the massif, because their erosion rates are so high.

The light minerals of the siltstones are plotted in Fig. 8.13E, the heavy minerals in Fig. 8.15F. Guerricchio (1986, p. 75) reports two basic types of fine silt and clay mineral composition, one with a strong predominance of illite, the other with kaolinite.

The olistholithes, the other important element of the olisthostromes or debris flows, consist mainly of very fine-grained, well-sorted yellow sandstones, mostly orthoquartzites. Subordinate are micritic limestones, characterized by tiny fissures which are cemented by calcite. These limestones are very rare, but easy to distinguish among the river-mouth gravels.

8.1.3.5 Argillites

Argillites, more or less limited to the coastal areas and of little importance as source rocks, at least for our 19 basins, are very fine-grained, with a mean sand content below 2%. No mineralogical analyses were carried out. Figure 8.11 gives the sand-silt-clay ratio of two sample groups, one from the Aspromonte Province (A) and the other from the Serre Province (B). Paleogeography and paleobathymetry of the argillites were recently treated by Rio et al. (1989).

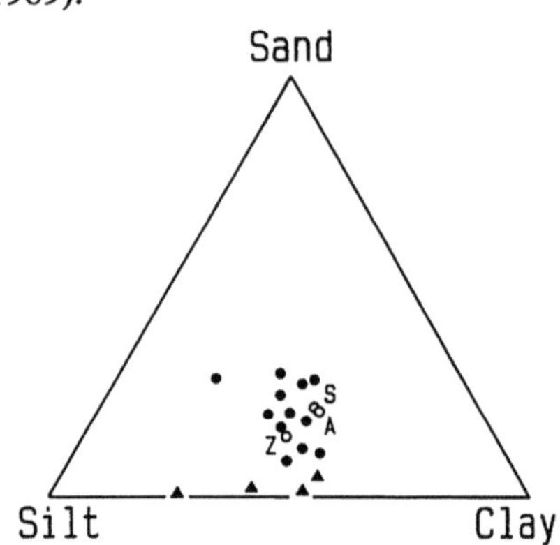

Fig. 8.11. Ternary diagram of sand, silt and clay, grain sizes of siltstone and argillite. *Dots* indicate grain size proportions of the siltstone analyzed by Guerricchio (1986, Fig. 4). *Open circles* indicate our own siltstone values from the Aspromonte Province (*A*), the Zomaro Province (*Z*) and the Serre Province (*S*). *Black triangles* indicate the argillites. Only parts of the sand fraction of the siltstone have a chance to settle in the river-mouth sediment, the finer grain sizes go beyond the river-mouth areas as wash load

8.2 Source Soils

8.2.1 Types and Grain-Size Distribution

The soil cover of the mountainous areas of our area is very thin and patchy, the high plains excepted, which are covered by dense forests, meadows and moorlands as pointed out in Chapter 4, climate and vegetation. These limited areas are of little importance for our study of provenance because they are characterized by nearly zero erosion.

According to the Soil Map of Tavernier (1985), the soils of our study area mainly represent the following types: Dystric Cambisols (Bd-1 cd) in the top areas of Aspromonte and Serre and Orthic Luvisols (Lo-2/4 ab) on the Zomaro plain. In an intermediate region below follow Eutric Cambisols (Be-2/3 cd) and Calcaric Regosols (Rc-3/4 cd). In the coastal areas finally Calcaro Eutric Cambisols (Bec-1/2 bc) prevail or, between Bovalino and Roccella Ionica, Calcaric Regosols (Rc-3/bc).

Grain-size distribution and both mineralogical and chemical composition of the Calabrian soils as a whole are key problems of this study. We will see in Chapter 10, grain-size distributions, that the jointed and weathered source rocks contain only an average of about 4% sand, which does not suffice to feed the rivers and to yield the average sand content of 25% of the river-mouth sediments. Our interest is therefore directed towards the soils as the next most probable source of sand. Indeed, the grain-size distributions of the Calabrian soils are surprisingly sand-rich and uniform, regardless of altitude – again, the high plains excepted – and regardless of the underlying rock type, limestones excepted. However, the chemical composition of the river-mouth sediments in the key fraction 63 – 500 µm does not fit the chemical composition of the respective fraction of the source soils. Thus, the soils obviously play only a minor role in sand formation, as will be shown in Chapter 9. Attrition during fluvial transportation seems to us the major source of sand. For this discussion the study of soil petrography is needed; we restricted the analyses to the most important sand fraction of 63 – 500 µm.

Figure 8.12 displays the narrow point cluster of soil grain-size distributions with an average content of 71% sand, 19% silt and 10% clay.

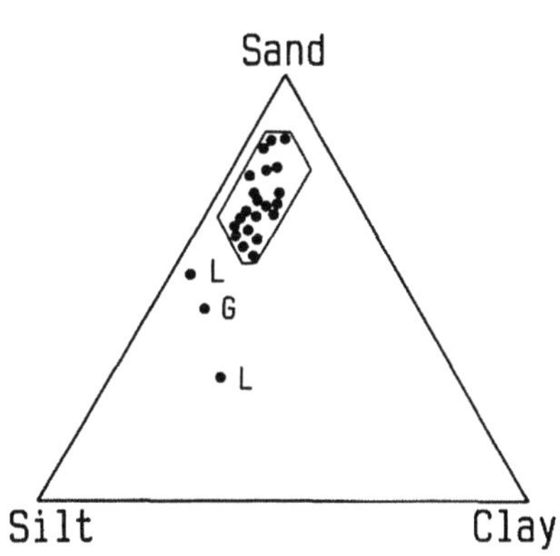

Fig. 8.12. Grain size of Calabrian soils. The soils form a very narrow, uniform cluster, limestone soils (*L*) and only one gneiss soil (*G*) excepted. The average content is 71% sand, 19% silt and 10% clay

8.2.2 Sand Fraction (by Renzo Valloni)

Data Base. See Table 8.1

Method. Light Minerals. The epoxy-indurated 63 – 500-µm-sized fraction was analyzed on thin section using a mechanical stage. Counting proceeded until 200 grains were determined following the procedures described in the sandstone section.

Heavy Minerals. The 63 – 250-µm fraction was analyzed microscopically by strip counting, proceeding until at least 300 transparent non-micaceous grains were determined.

Parameters of Light Minerals. The grain categories used in this study and the petrologic parameters have already been defined in Tables 8.3 and 8.4. Table 8.5 contains additional parameters used in the literature or here introduced for descriptive purposes; they all represent grain ratios calculated between or within primary parameters. Ratios between primary parameters (QFL and QFRf, defined in Table 8.3) have been proposed by Pettijohn (1975) both as stability index (Q/F+Rf) and as provenance index (F/Rf). For rapidly deposited sands the Q/F ratio is also an effective provenance index (Valloni 1985). Ratios within primary parameters are aimed to detail the information hidden in large mineral groups. The grain ratios proposed by Dickinson (1970) are Qp/Q for quartzose and P/F for feldspathic grains. Within the lithic population (L, Rf) are many grain ratios which may show useful results in specific cases; the few proposed in Table 8.5 proved to be significant in this study.

Parameters of Heavy Minerals. The 22 minerals listed in Fig. 8.15 have been counted separately. By summing their percent on total transparent non-micaceous grains, some indexes related to stability or provenance can be calculated (Pettijohn et al. 1987, p. 261–262). In Table 8.7 the most important species are grouped according to their diagenetic stability (Morton 1984), i.e. their resistance to intrastratal solution (Füchtbauer 1974, p. 42).

8.2.2.1 Detrital Modes

Table 8.6 and Figs. 8.13 and 8.14 summarize the mineral composition of individual key rock types of the soils. Provenance indexes (Table 8.5) neatly separate the different types of soil source. In general, the samples from granite and feldspathic sandstone show an arkosic composition (F/Rf and F/L > 1), while the samples from gneiss and schist show a lithic composition (F/Rf and F/L < 1).

The other provenance index, the quartzose/feldspathic grain ratio (Fig. 8.14) also separates the soil data points in two groups; in fact Q/F values are quite similar in both representations and are close to 2 for granite and feldspathic sandstone samples and close to 4 for gneiss and schist samples. Coarse-grained rock fragments of the lithic grains assemblage dominate within sandstone and granite soils, while the fine-grained (Lm) dominate within soils of schist (Fig. 8.13).

In the LqLvLs diagram the four types of soils are widely separated, with the highest content of quartzose lithics (Lq) shown by granitic and the highest content of fine-grained sedimentary and metasedimentary rock fragments (Ls) exhibited by schist samples.

In the other ternary diagrams of Fig. 8.14 (LcrLvoLse) all types of soils plot near the crystallines pole (Lcr) because pole Lse contains only siliciclastic and carbonate lithics (Lsed) plus micas and heavies. The addition of micas in the Lse pole causes the granite soil plot (square) to display the highest Lse content (45%) because the lithic population is relatively rich in micaceous grains.

Table 8.6 summarizes the relevant petrologic data of sand produced from soils of the four rock types. It is worth noticing that, within quartzose grains, the quartzose lithic grains (Lq) assume the highest values on metamorphic substrates, gneiss and schist, with Lq/Q around 0.4, while plagioclase within feldspathic grains takes the highest values on granitic (P/F = 0.44) and the lowest on schistose substrates (P/F = 0.24).

The mineralogic composition of sandstone rock and sandstone soil (Figs. 8.7 and 8.14) is rather similar for both primary (QFL) and secondary parameters (LqLvLs and LcrLvoLse). The overall trend in soil samples is that quartzose grains (Q) increase (+8%), while feldspathic (F) and lithic (L) grains decrease with respect to sandstone rock samples. No trend is shown by some notoriously unstable components, such as volcanic lithic grains and plagioclase, which are not preferentially destroyed. In fact, volcanics of the lithic grains population are slightly higher in the soils than in the rocks, while among feldspar grains plagioclase abundance remains almost unchanged, with P/F = 0.4 for both soils and sandstones. On the contrary, within the group of quartzose and feldspathic grains a drastic drop of the polycrystalline and polymineralic grains abundance is shown by the soil samples (Fig. 8.13).

8.2.2.2 Heavy Mineral Assemblage

The 23 samples detailed in Table 8.1 were supplemented by two limestone samples and one siltstone bulk sample.

Table 8.6. Key grain proportions and grain ratios of the soils of the main rock types (see Tables 8.3, 8.4 and 8.5 for definitions). Ternary grain proportions are plotted in the diagrams of Fig. 8.7[a]

Soil of rock type	Samples	Q	F	Rf	Q	F	L	Lcr	Lvo	Lse	Lq	Lv	Ls	Lq/Q	Qc-m/Lq	P/F
Granite	6	47	29	24	61	37	24	53	2	45	83	5	12	0.13	0.00	0.44
Gneiss	6	33	11	56	52	15	33	81	2	17	47	3	50	0.42	0.18	0.34
Schist	9	19	5	76	27	6	67	95	0	5	14	0	86	0.39	0.27	0.24
Sandstone	2	59	28	13	65	32	3	61	8	31	64	10	26	0.07	0.15	0.41

[a] Lq/Q = lithic quartzose grains/total quartzose grains; Qc-m/Lq = microgranular composite quartz/total lithic quartz; P/F = plagioclase/total feldspar.
Within quartzose grains, the lithic quartzose (Lq) are particularly abundant in gneiss and schist soils (Lq/Q = 0.42–0.39); on total lithic quartzose grains, the microgranular composite quartz (Qc-m) has the highest incidence in schist soils (Qc-m/Lq = 0.27) and equals zero in granite soils. Within feldspars, K-feldspar exceeds plagioclase in all soil types with the highest plagioclase incidence shown by soils of granite and arkosic sandstone (P/F = 0.44–0.41).

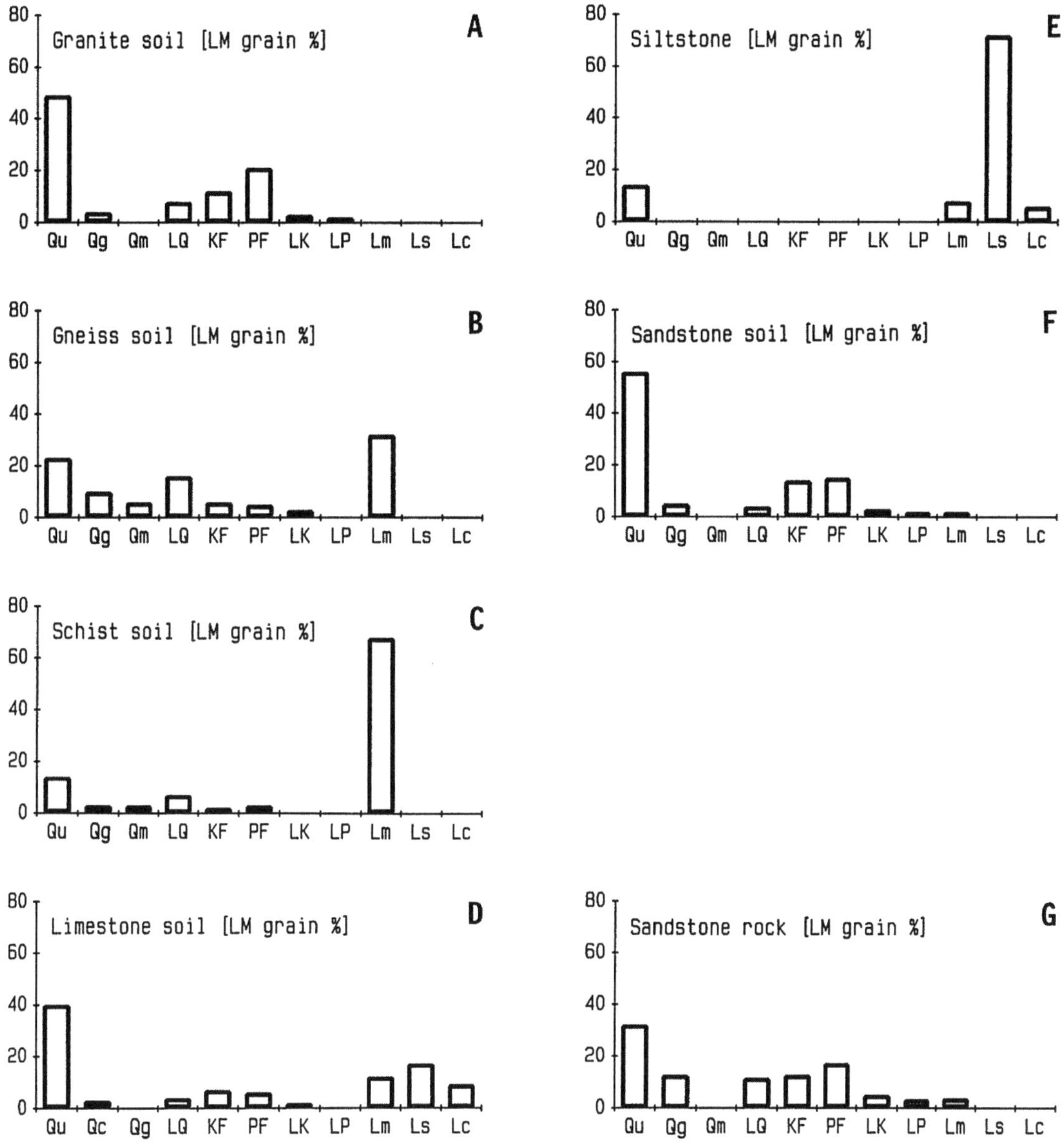

Fig. 8.13 A–G. Histograms of the average light mineral (*LM*) distribution of the soils of granite, gneiss, schist, limestone, siltstone and sandstone. Average of thin section analysis of solid sandstone is also plotted (**G**); soil formation of sandstone causes the selective elimination of polycrystalline and polymineralic grains (*Qg, Lq, LK, LP, Lm*). *Qu* single-crystal quartz; *Qg* composite monomineralic quartz of igneous origin; *Qm* composite monomineralic quartz of metamorphic origin; *LQ* quartz crystal >63 μm within crystalline, volcanic or sedimentary rock fragments; *KF* potash-feldspar; *PF* plagioclase; *LK* and *LP* K-feldspar and plagioclase >63 μm within crystalline, volcanic or sedimentary rock fragments; *Lm* fine-grained (<63 μm) metamorphic rock fragments; *Ls* fine-grained (<63 μm) siliciclastic rock fragments; *Lc* marble (dolomite and limestone of any texture). See Table 8.3 for abbreviations

Mineral percentages counted in each soil type are shown in Fig. 8.15. In addition, the soils were grouped into three units: granitic, metamorphic (gneiss + schist) and sedimentary (limestone + sandstone + siltstone) and the four stability indexes ZTR, GAS, EKS and APA of Nickel (1973) were calculated (Table 8.7). ZTR values of soil (Hubert 1962) vary greatly; ZTR of metamorphic soils (12%) is three times higher than that of granitic or sedimentary soils (4%).

Stable species (GAS) are highest in sedimentary soils (17%) while moderately stable species (EKS) are similar for the three soil types (27–33%). Unstable species (APA) are particularly abundant in granitic soils (62%).

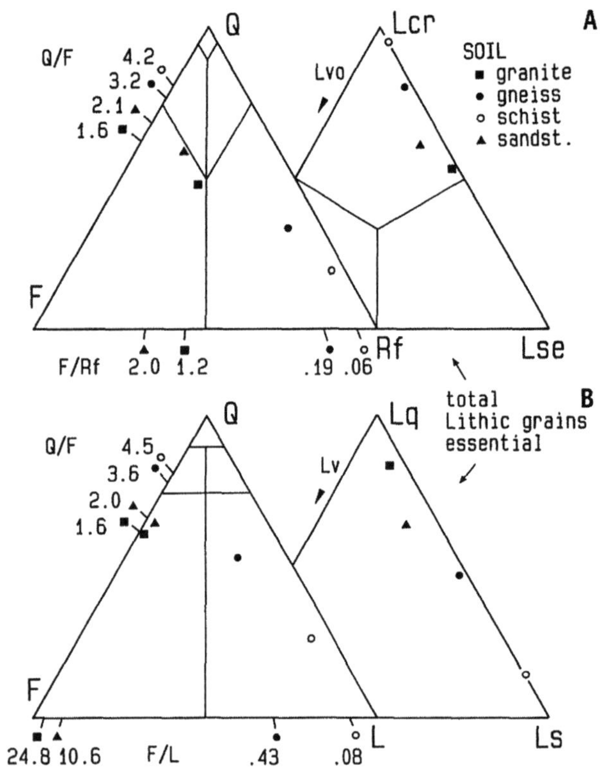

Fig. 8.14 A, B. Framework mineral composition (*QFRf* and *QFL*) and corresponding lithic grain proportions of the soils of the main rock types. Parameters according to the traditional criteria (*above*) and to the textural criteria (*below*). The lithic pole *L* coincides with L+Lc because carbonate lithics are absent; *Lcr* crystalline rock fragment (r.f.); *Lvo* volcanic r.f.; *Lse* sedimentary r.f.; *Lq* lithic quartzose grains; *Lv* volcanic and metavolcanic fine-grained r.f.; *Ls* sedimentary and metasedimentary fine-grained r.f. *Outside the triangles* are values of the Provenance Indexes Q/F (quartzose/feldspathic grains) and F/Rf and F/L (feldspathic/lithic grains). Values of the Q/F ratio cluster around 2 for granite and arkosic sandstone and around 4 for gneiss and schist soils. Soils from granite and arkosic sandstone have feldspathic composition while soils from gneiss and schist have lithic composition. The four soil types plot well apart if parameters LqLvLs are used to represent the lithic grain population

8.2.2.3 Soil Composition

Soils from metamorphic rocks and those from granite and arkosic sandstone differ noticeably (Figs. 8.13, 8.14, 8.15).

Soils from Gneiss and Schist. QFL proportions show a lithic character (L > F), quartzose grains (Q) double passing from schists (27%) to gneisses (52%, Table 8.6). The lithic assemblage is dominated by fine-grained metamorphic rock fragments (Lm, Fig. 8.13). Among grain ratios the high values of lithic quartzose on total quartzose grains are characteristic (Lq/Q is about 0.4, Table 8.6). Tourmaline content is high (5–10%), the same is true for pyriboles; individual values of diopside, augite and hornblende are around 10–15% (Fig. 8.15). The provenance index (Q/F) is around 4.0 (Fig. 8.14) and the stability index (ZTR) is 11.7, a relative maximum (Table 8.7).

The schist soils show high values of fine-grained metamorphic rock fragments (68% Lm grains) and within the transparent non-micaceous heavy-mineral grains the average brookite percentage is >25% (Fig. 8.15). The ratio microgranular composite quartz versus total lithic quartz within the group of quartzose grains is relatively high (Qcm/Lq = 0.27, Table 8.6).

Gneiss soils show characteristically high values of lithic quartzose grains (Lq) represented by the Qm and LQ grain categories which total 22% of essential terrigenous grains (Fig. 8.13 B).

Soils from Granites. Within essential terrigenous grains single-crystal quartz grains (Qu) dominate (49%) and monomineralic feldspar is abundant (33%, Fig. 8.13 A). The P/F ratio is relatively high (0.44, Table 8.6) and altered feldspar plus plagioclase exceeds K-feldspars (PF > KF).

Table 8.7. Order of stability of transparent non-micaceous heavy minerals with proposed codifications (see Table 14.3) and soil-related values [%]. Minerals are grouped to form indexes (Pettijohn et al. 1987, p. 262), related to diagenetic stability, in addition to the ZTR index (Hubert 1962). For the assignment of other stable, moderately stable and unstable minerals to these groups see also Morton (1984) and Füchtbauer (1974)

	Rock-related soil		
	Granitic	Metamorphic	Sedimentary
ZTR[a]	4.2	11.7	3.7
Ultrastable			
Zircone + tourmaline + rutile			
GAS	1.8	7.3	16.8
Stable			
Garnet + apatite + staurolite			
EKS[b]	31.7	32.9	26.9
Moderately stable			
Epidote + zoisite + kyanite + sillimanite + sphene + brookite			
APA[c]	61.6	46.6	35.7%
Unstable			
Andalusite + pyroxene + amphibole			

[a] The ZTR index is low in both granitic and sedimentary soils; soils from granite contain 6% of the stable plus ultrastable species (ZTR + GAS).
[b] The EKS index is similar for all soil types with values constantly around 30%.
[c] Unstable minerals (APA) are prominent in all soil types ranging from 36 to 62%.

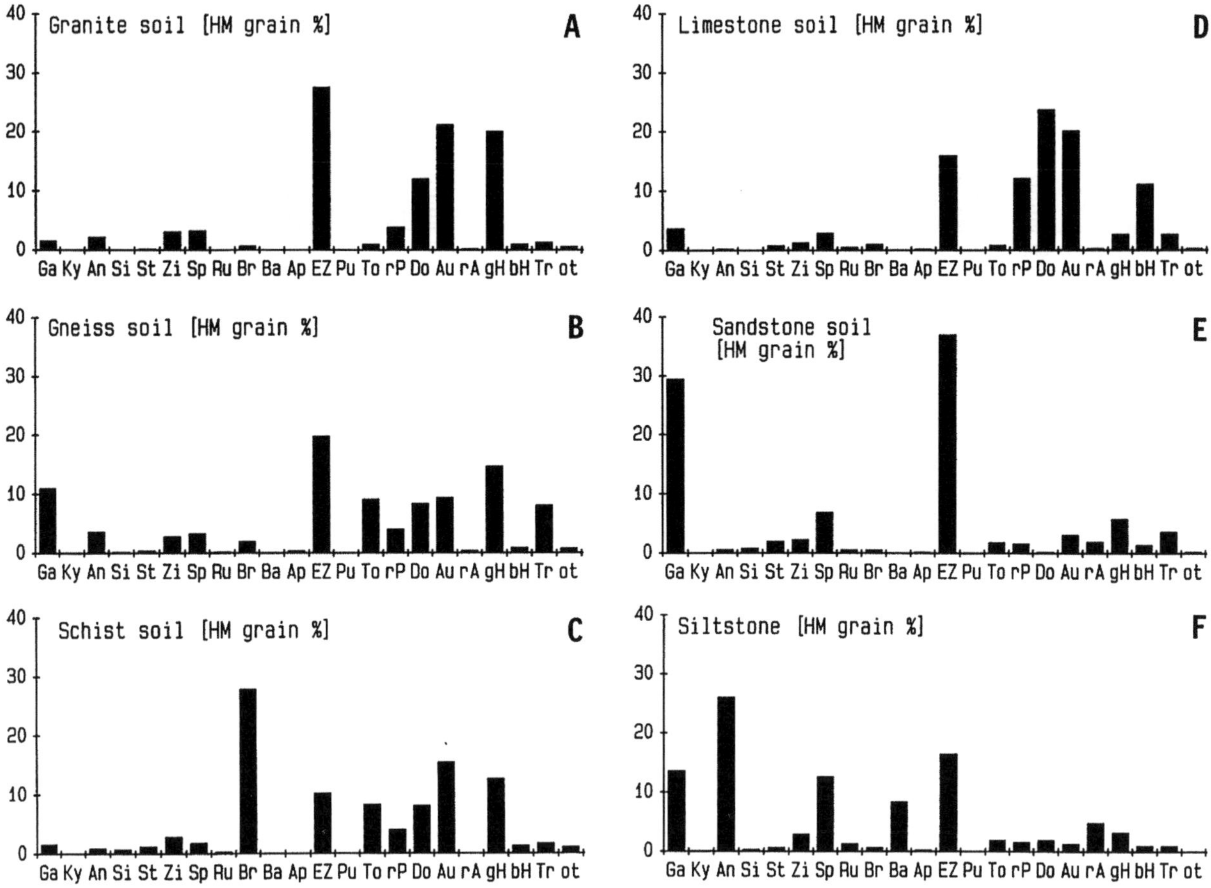

Fig. 8.15 A–F. Histograms of the average heavy mineral distribution of the soils of granite, gneiss, schist, limestone and sandstone. The siltstone values stem from the rock itself, because soil formation is extremely scarce. Soils from sedimentary substrates (limestone, siltstone and sandstone) show various mineral assemblages. The soils of granites are characterized by epidotes and pyroxene + amphibole; those of gneiss by epidotes, tourmaline, pyroxenes and amphiboles, and those of schist by brookite, tourmaline, pyroxene and amphiboles. *Ga*rnet, *Ky*anite, *An*dalusite, *Si*llimanite, *St*aurolite, *Zi*rcon, *Sp*hene, *Ru*tile, *Br*ookite, *Ba*rite, *Ap*atite, *Ep*idotes, *Pu*mpellyite, *To*urmaline, *r*homb. *P*yroxene, *Dio*pside, *Au*gite, *r*homb. *A*mphibole, green *H*ornblende, *b*rown *H*ornblende, *Tr*emolite/actinolite, *ot*hers

Essential rock fragments are dominated by quartzose lithics (Lq 83%, Fig. 8.14). They consist almost entirely of the LQ type, i.e. quartz crystals within phanerites.

The heavy-mineral assemblage (Fig. 8.15) is characterized by the contemporaneous presence of epidote + zoisite, diopside + augite, and green hornblende in comparable amounts (20–40%). Stability indexes show low ultrastables and stables (ZTR 4.2%, GAS 1.8%) and high unstables (APA 61.6%, Table 8.7). Soils of both granite and arkosic sandstone show similar provenance indexes (Fig. 8.14).

Soils from Arkosic Sandstones. Sandstone soils share a high content of single crystal quartz (Qu) and high monomineralic feldspar (Fm = KF + PF) with granite soils (Fig. 8.13). The lithic grain population shows intermediate values with respect to granite and gneiss soils (Fig. 8.14), while the heavy mineral association is characterized by high garnet content (30%) accompanied by epidote + zoisite in comparable amounts (Fig. 8.15).

Soils from Sedimentary Rocks. Both essential terrigenous grains (Fig. 8.13) and heavy mineral assemblage (Fig. 8.15) show distinct mineral associations for the single soil types. Consequently, grand totals for the entire group of sedimentary rocks (Table 8.7) should not be utilized for detailed analyses. The stability indexes of sediment soils surprisingly show a low content of ultrastables (ZTR 3.7%) and similar amounts of the moderately stable grains (EKS 26.9%) in the soils from limestone, sandstone and siltstone (Fig. 8.15, Table

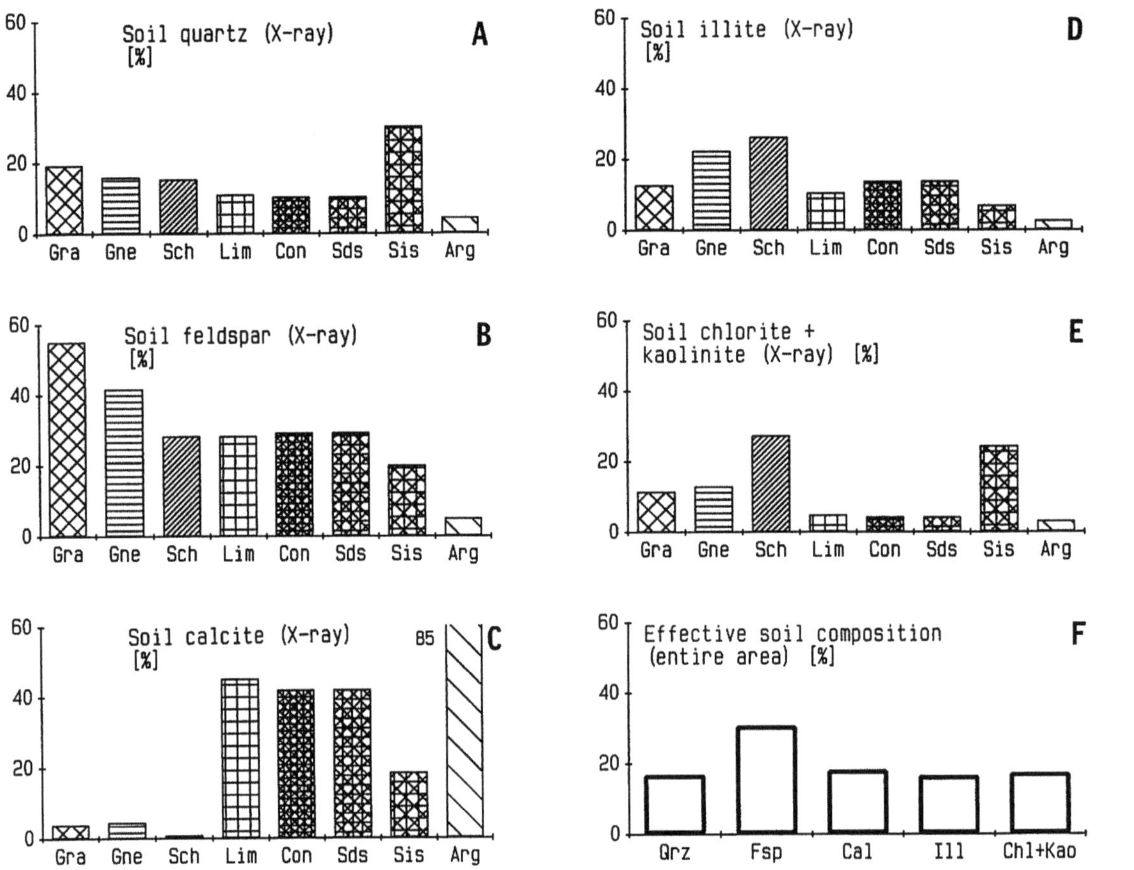

Fig. 8.16. A–E Mineralogical composition (X-ray diffraction) of the silt fraction of Calabrian soils. **F** Effective soil composition of the entire area, silt fraction only. The effective soil composition is the composition of the single rock-related soils weighed by area percentages and erosion rates (Chap. 6.2.2). *Gra*nite, *Gne*iss, *Sch*ist, *Lim*estone, *Con*glomerate, *S*and*s*tone, *Si*lt*s*tone, *Arg*illite

8.7). The essential terrigenous grain categories of limestone soils are mostly siliciclastic because carbL averages only 17%; the heavy mineral assemblage is dominated by pyroxenes (>50%); the siltstone contains much andulusite (25%) and several percent of authigenic (?) barite (Figs. 8.13 and 8.15).

8.2.3 Silt Fraction

Compared to the coarser grain sizes of this study, little effort could be made to examine silt and clay. These analyses have to be seen as a kind of orientation, a byproduct of the sand research and not a self-reliant study. In Chapter 14.2, the composition of the silt fraction of the suspended load and of the river-mouth sediments will be compared with the effective composition of the soils.

Table 8.8 and Fig. 8.16A–E display the mineralogical composition of the silt fraction of the soils of our seven rock types; sandstone and conglomerate are combined. Feldspar dominates in soils of granite and gneiss, the most frequent mineral in the soils of limestone, sandstone and

Table 8.8. Mineralogical composition (X-ray diffraction) of the silt fraction of Calabrian soils (2–63 μm)

	No. of samples	Qrz	Fsp	Cal-cite	Illite	Chlorite + kaolinite
	[%]					
Granite	2	19.3	55.1	3.9	12.7	11.5
Gneiss	6	15.9	41.8	4.5	22.4	13.0
Schist	8	15.3	28.3	0.9	26.3	27.3
Jurassic limestone	3	11.0	28.3	45.1	10.6	4.9
Sandstone/congl.	2	10.7	29.3	42.0	13.8	4.3
Siltstone	3	30.1	20.0	18.6	7.0	24.4
Argillite	2	4.5	5.1	84.6	2.7	3.2
Effective soil composition		16.9	30.7	18.2	16.5	17.3

argillites is calcite; only the siltstones show quartz as the predominant mineral. Illite and chlorite dominate the soils of schist. As pointed out, it is beyond the scope of this study to discuss the origin of this mineral distribution. Our problem is to find a link between the composition of the "original" silts of the source, especially the soils, and those of the river-mouth sediments. For this, we calculated the effective soil composition for each basin and the overall average (Chap. 6.2.2). Surprisingly, on doing so, all minerals, feldspar excepted, show a mean of about 17%; only feldspar climbs up to 31% (Table 8.8, Fig. 8.16F).

The chance of this fine-grained material transmitting a reliable provenance signal into the river-mouth sediment is not good, compared to sand or gravel. During flood discharge the wash load goes beyond the river mouths; because of the low settling velocities of the fines, their mineralogical composition is expected to vary noticeably from deposit to deposit. We will return to this problem in Chapter 14.2.

8.3 Conclusion

Source Rocks. We compressed the 55 mainly lithologic units of the Carta Geologica Della Calabria into 38 units for digitizing the maps (Chap. 5.1.2) and to only eight rock types for provenance analysis: granite, gneiss, schist, limestone, conglomerate, sandstone, siltstone and argillite. From these all granitic and metamorphic rocks are enabled to produce detritus into all fractions of the river-mouth sediments. Compared to this the source sediments are at a disadvantage. Their relatively marginal position in the mountains at low altitudes implies low erosion rates, and their petrographic composition, conglomerates excepted, makes them produce predominantly finer debris, fine sand, silt and clay, which makes part of the wash load barely represented in the river-mouth sediment. Fragments from granites, pegmatites and metamorphic rocks, as a rule, are easy identifiable as such in almost all grain sizes and retraceable to their sources. An exception is the fraction 2–20 mm, mostly too small for macroscopic inspection and too large for microscopical examination.

Admittedly, however, on examining the river-mouth gravel, it is not possible to reconstruct the broad variety especially of the metamorphic source rocks presented in this chapter.

Macroscopically quite similar specimens may turn out to be quite different in thin section. So we were forced to restrict the lithologic classification of the river-mouth gravel to granite, porphyry, pegmatite, gneiss and schist.

Nevertheless, despite these limitations and the absence of basic, ultrabasic or volcanic rocks, the Calabrian rock suite is differentiated enough both lithologically and regionally, from basin to basin, to allow different trends of the source-sediment relations to become evident (Chap. 17).

Of the sediments, only the Jurassic limestones behave like crystalline rocks, producing larger amounts of coarse debris which decrease towards zero, however, if one goes to the sand fraction.

Referring to previous sedimentary cycles, the belt of Tertiary sandstones and conglomerates, outcropping all along the study area, was fed by local or similar sources. This means that the characteristic metamorphic and granitic sources, Aspromonte and Serre, respectively, are barely contaminated by exotic second cycle terranes.

Sandstone grains are texturally and mineralogically immature so that the incidence of second cycle grains, the possible dilution of primary sand in the river sediment, remains difficult to evaluate mineralogically. Nonetheless, the influence of sand recycled from sandstones seems to be only small. Otherwise sandstone-rich basins should produce sand-rich river-mouth sediments. This is definitely not the case.

Sandstone mineral composition is characterized by a quartzose grain (Q) content around 50% (Fig. 8.7) and a lithic population dominated by crystalline phanerites. It is worth noticing that one-quarter of the feldspar grains are strongly altered and that micaceous grains are very abundant, averaging 8.5%. With such a mineralogy, analytical procedure is a key problem: traditional (QFRf) and texture-oriented (QFL) counting of one and the same sandstone sample yields quite different results. The siltstone is destined to transfer no or misleading signals into the river-mouth sediment. From this extended and heavily eroded sediment body 80% passes the river mouth as wash load barely documented in the river-mouth sediment. From the possibly remaining 20% of sand, about 70% is made up by Ls (Fig. 8.13E), and nobody would expect that these sands would come from the same source as the micritic olistolithic limestones. Identifiable detritus of argillites was not found in the river-mouth sediments.

Source Soils. A quite uniform sand content averaging 71% is found for all source terranes (Fig. 8.12). Tables 8.6 and 8.7 and Fig. 8.14 provide baseline data on the mineralogy of the soils of different source rocks in terms of key grain proportions and mineral indexes. The comparison (QFL) of sandstone and sandstone-soil confirms the expected increase of stable quartzose grains and the decrease of the less stable feldspathic and lithic grains due to weathering (Figs. 8.7 and 8.14).

The polycrystalline and polymineralic grain types (Qc, LQ, LK, LP and Lm) are preferentially destroyed during soil formation (Fig. 8.13F, G). Alteration processes are particularly intense on composite monomineralic (Qc) and polymineralic (LQ) quartzose grains which show relatively higher percent reductions. This confirms the conclusions of Harrell and Blatt (1978) on the preferential destruction of polycrystalline quartz grains during chemical processes of soil formation rather than mechanical ones during transport. However, this trend is reversed in river-mouth sediment. In Calabria, sand seems to be formed mainly in the "sedimentary mill" of the rivers via attrition rather than by decomposition during soil formation (Chap. 9).

Part II

From the Source to the River-Mouth Sediment

Part II comprises all topics we are able to trace from the source to the river-mouth sediment and even beyond.

These are the chemical composition of the sands, the overall grain-size distributions, and the sphericity distributions of joint blocks and river gravels. The last chapter of Part II is devoted to some special topics of the lower river courses and especially the beach, to grain-size distributions of the coastal subenvironments and to the development of the sedimentological parameters of a selected, model-like rock type, the "dark gneisses".

At first glance, it may appear mismanaged to start this part with the treatment of the chemical development of the sand. We believe, however, that the interpretation of this chemical development bears a possible solution for the understanding of sand formation in Calabria, at this active plate margin. Consequently, this is essential for all following topics.

Chapter 9

Development of the Chemical Composition of Sand

The past years have brought an increasing interest in chemical composition as a possible indicator of the tectonic setting of sandstones (see Pettijohn et al. 1987, p. 58 for references). Sandstones and modern sands from different realms of plate tectonics like island arcs, both oceanic or continental, or active and passive margins display surprisingly different chemical compositions. Little data are available, however, concerning the differentiation of the chemical composition of a sand during its development from the solid source rocks over transitional stages like soils towards its fluvial and marine successors, combined with the chemistry of the related fountain- and river waters. Our Calabrian data base enables us to draw such a picture, although a rough one. So the goal of our chemical analyses is a double one: we want to answer the question whether the sands of the Calabrian river mouths and of the longshore bar have a bulk chemistry corresponding to that of sandstones of active plate margins, as expected here in Calabria, and, secondly, we want to apply our concept of the effective source area composition, of predicting sediment composition in order to better understand how this chemistry might be created, how it starts from the source rocks and soils and how it is modified during fluvial transportation and deposition.

Summary. The chemical composition of the single types of source rocks and the sand fraction 63–500 μm of the soils, of the effective composition of the source rocks and soils, of the river-mouth sediments and of the longshore bar is compared. Gain and loss and constant-Al_2O_3 calculations reveal that the composition of the river-mouth sands comes closer to the source rocks than to the soils. Attrition during fluvial transport is the most important source of sand at this active plate margin, chemical weathering and soil formation being less effective. The chemical composition of the average Calabrian river-mouth sand fits almost perfectly that of active-margin sandstones; single rivers, however, deviate greatly from this mean and it may be hazardous to draw conclusions from only one or even a limited number of rivers.

Table 9.1. Oxides [%] and most important parameters of chemical composition[a]

	Granite		Gneiss		Schist		X-met		Limestone		Sandstone	
	R	S	R	S	R	S	R	S	R	S	R	S
No. of samples	5	6	6	7	8	9	14	16	1	2	–	2(8)
Column number	1	2	3	4	5	6	7	8	9	10		11
SiO_2	70.64	70.27	64.42	68.21	61.8	64.49	63.11	66.35	1.68	63.68	–	73.68
TiO_2	0.27	0.48	0.93	0.96	0.96	1.16	0.95	1.06	0.05	0.58	–	0.37
Al_2O_3	16.16	16.33	16.15	15.05	18.56	18.25	17.36	16.65	1.07	12.72	–	13.69
Fe_2O_3	2.39	4.07	5.41	7.34	7.87	7.74	6.64	7.54	0.56	4.14	–	2.63
MnO_2	0.03	0.07	0.15	0.11	0.13	0.1	0.14	0.11	0.02	0.12	–	0.06
MgO	1.35	1.59	2.04	2.32	2.54	2.28	2.29	2.3	5.97	2.87	–	0.92
CaO	1.53	1.44	4.54	1.89	2.5	0.99	3.52	1.44	90.36	11.28	–	2.79
Na_2O	4.27	2.74	3.69	2.04	2.04	1.45	2.87	1.75	0.05	1.78	–	2.42
K_2O	3.23	2.86	2.57	2.77	3.36	3.36	2.97	3.07	0.15	2.68	–	3.36
P_2O_5	0.14	0.14	0.18	0.18	0.24	0.16	0.24	0.17	0.09	0.19	–	0.1
Fe_2O_3+MgO	3.74	5.66	7.45	9.16	10.41	10.02	8.93	9.84	6.53	7.01	–	3.55
Al_2O_3/SiO_2	0.23	0.23	0.25	0.22	0.3	0.28	0.28	0.25	0.64	0.2	–	0.19
K_2O/Na_2O	0.76	1.04	0.7	1.36	1.65	2.32	1.03	1.75	3.0	1.51	–	1.4
$Al_2O_3/(Na_2O+CaO)$	2.79	3.91	1.96	3.83	4.09	7.48	2.72	5.22	0.01	0.97	–	2.63

[a] R: Solid rock; S: soil (sand fraction 63–500 μm only); x-met: average of metamorphic rocks; eff. comp.: effective composition or predicted sediment composition; River: average of the 19 river-mouth sediments (63–500 μm); Bar: average of the longshore

Data Base. Single samples were taken from the fresh source rocks, the soils, the river-mouth sands and the Bianco longshore bar. This bar, not yet introduced, is described in Chapter 12.1. Although 23 single submarine samples were analyzed, their value for comparison between the two environments is somewhat limited. The 19 river-mouth sediments are fed by nearly the entire area and consequently represent the average of its broad variety, whereas the longshore bar sampled is only one individual, alimented nearly exclusively by the Laverde [07]. Thus, although the Laverde is a very typical Calabrian river and the chemical composition of its sand is very close to the average of the 19 basins, a serious comparison between river-mouth sediments and their marine successors can only be made between the Laverde and the longshore bar.

Table 9.1 gives the numbers of samples analyzed in each category. The sediments are not completely sampled. The limestones, regionally unimportant, show only one rock and two soil samples. The sandstones are usually weathered and disintegrate to sand, so we sampled only their soils. The soil cover of siltstones and argillites, on the contrary, is so thin, and so hard to distinguish from the underlying solid rocks, that we decided to sample only these soft rocks themselves.

To remain within the economic limits of the project, some bulk samples of argillite, siltstones and a few soils were taken, combinations from several localities. See Chapter 8 for the sampling modalities of the soils, Chapters 12 and 14 for those of the river-mouth sediments and the longshore bar.

The Palizzi [04] was not analyzed geochemically, thus all calculations refer to only 18 basins instead of 19.

Terminology. Total iron was calculated as Fe_2O_3, cited FeO values of other authors were converted to Fe_2O_3. To facilitate the comparison, all data are given on a volatile-free basis, recalculated without ignition loss to 100%.

Method. An automatic X-ray fluorescence spectrometer was used for chemical analysis and an X-ray diffractometer for the mineralogy of silt and clay. From the 18 river-mouth sands simply the average composition was calculated, in the source area the effective rock and soil composition was determined (Chap. 6.2.2).

If only one sample type was available, rock or soil, the missing one was substituted by its counterpart. This procedure has little effect on our calculation, because it concerns only the sediments. Here limestone, siltstone and argillite contribute very little to sand formation, and the conglomerates and sandstones are recycled sediments with soils whose sand composition comes close to that of the rocks themselves.

Siltstone		Argillite		Eff. comp.		River	Bar		Potter	Island Arc		Margin	
R	S	R	S	R	S		1976	1987		Oceanic	Contin.	Active	Passive
3(11)	–	2(4)	–	–	–	19	2	23	34				
12		13		14	15	16	17	18	19	20	21	22	23
61.11	–	44.19	–	64.17	65.69	73.2	74.99	73.25	83.26	58.83	70.69	73.86	81.95
0.93	–	0.56	–	0.71	0.81	0.54	0.4	0.88	0.44	1.06	0.64	0.46	0.49
19.19	–	13.56	–	16.84	16.75	13.7	12.9	12.25	6.68	17.11	14.04	12.89	8.41
8.88	–	5.14	–	5.84	6.21	3.39	3.1	3.88	2.76	8.08	4.82	3.06	3.28
0.14	–	0.05	–	0.1	0.09	0.08	0.06	0.1	0.05	0.15	0.1	0.1	0.05
0.26	–	2.72	–	1.66	1.61	1.31	1.46	1.63	0.88	3.65	1.97	1.23	1.39
3.9	–	28.13	–	5.08	3.84	2.04	1.94	2.03	3.45	5.83	2.68	2.48	1.89
0.37	–	2.97	–	2.34	1.8	2.71	2.82	2.86	1.24	4.10	3.12	2.77	1.07
2.55	–	2.5	–	3.08	3.05	2.95	2.27	2.24	1.25	1.6	1.89	2.9	1.71
0.19	–	0.16	–	0.18	0.15	0.11	0.09	0.17	0.06	0.26	0.16	0.09	0.12
9.14	–	7.86	–	7.5	7.82	4.7	4.57	5.51	2.75	11.73	6.79	4.63	2.89
0.31	–	0.31	–	0.26	0.25	0.19	0.17	0.17	0.08	0.29	0.16	0.18	0.1
6.89	–	0.84	–	1.32	1.7	1.09	1.09	0.78	1.01	0.39	0.61	0.99	1.6
4.49	–	0.44	–	2.27	2.97	2.88	2.88	2.51	1.42	1.72	2.42	2.56	4.5

bar off the Bianco coast: 1976 and 1987 campaign; Potter: average of large modern rivers (Potter 1978, Table 6); Island arc and margin: average chemical composition of sandstones of various plate tectonic settings (Bhatia 1983, Table 10).

9.1 The Oxides

It should be emphasized again that, the fresh source rocks excluded, only the 63 – 500-μm fraction of all sands was analyzed. These medium- to fine-sand fractions not only seem to be the most common in the geological record (Blatt 1970, p. 260), but their composition also seems to be relatively constant compared to the neighbouring fractions. The coarser ones are characterized by increasing portions of rock fragments, the finer ones, silt, by increasing heavy minerals and decreasing quartz (Potter 1978, Fig. 3). The relation between grain size and the composition of the sand fraction is discussed in more detail in Chapter 12. The restriction of our analysis on this 63 – 500-μm window means that we cannot perceive trends that depend on larger amounts of rock fragments, typical for coarser fractions. The same is true for the finer fractions, silt and clay, which play an important role in soil formation. Finally all phyllosilicates, important source-rock forming minerals, form part mainly of the suspended or wash load, thus being under-represented in the river-mouth sediments.

These limitations have to be considered when we compare our data with ancient sandstones, which frequently contain larger amounts of fines via matrix and rock fragments. An additional problem in comparing our modern data with ancient sandstones is the deficiency of cement and the lack of any diagenetic differentiation.

Table 9.1 shows the percentages of the oxides usually measured, calculated for the Calabrian environments and set against values of literature. These values are hard to compare directly, so we follow Garrels and Mackenzie 1971 (p. 163) in plotting gain and loss diagrams and constant oxide calculations in order to estimate the relative and absolute variations of chemical composition going from the source to the sediment.

9.1.1 Gain and Loss

In Fig. 9.1 A the mean weight percentages of the oxides of the fresh granites, gneisses and schists are divided by the percentages of their weathered residues in the respective soils, the results are multiplied by 100 and plotted on a linear scale.

SiO_2, Al_2O_3, MgO and K_2O differ only slightly, the same being true for the CaO of granites. CaO of gneisses and schists, on the other hand, displays the greatest relative loss. This is because the CaO content of these sometimes calcareous metasediments (Chap. 8.1.2) comes mainly from calcite and not from the silicates. Seen as a whole, Calabrian soils are very immature because relative chemical deviations from the fresh source rocks are so small. Only little K_2O is dissolved and carried off, and only a little more Na_2O, which may be due to the greater solubility of plagioclase compared to K-feldspar.

The mean chemical composition of the Calabrian river water corresponds perfectly to this picture (Fig. 7.4 B): Ca^{2+} is the most frequent cation, followed by Na^+ and K^+.

It goes without saying that it is not possible to compare the chemical composition of the single source rocks directly with that of the river-mouth sediments. The concept of the effective source rock composition, however, makes this possible. If all rocks, or soils, were able in like manner to deliver our reference fraction of sand of 63 – 500 μm into the fluvial system, then the composition of the river-mouth sands of this fraction, regardless of the absolute amounts produced, should correspond to the effective source rock- or soil composition, that is, the rock area percentages weighed by the respective erosion rates (Chap. 6.2.2). This consideration takes no sorting effects into account such as selective transportation and deposition of heavy minerals, for example, or phyllosilicates.

Figure 9.1 B shows the relative gains and losses of the couples rock/soil, soil/sediment and rock/sediment. In this plot the entire rock suite participates: granite, gneiss, schist, limestone, conglomerate, sandstone, siltstone and argillite. Again, the immaturity of the soils is reflected by only small relative deviations from the rocks. CaO and Na_2O expectedly differ the most, K_2O remaining relatively constant.

If we compare the relations soil/sediment and rock/sediment, we realize that these are either almost equal (SiO_2, Al_2O_3, MgO and K_2O) or the values of the rock/sediment ratio are closer to 100 than those of the soil/sediment ratio (TiO_2, Fe_2O_3 and Na_2O). In other words, the chemical composition of the river-mouth sediments is closer to the source *rocks* than to the source *soils*. The only exception is CaO, where the relative loss of the relation soil/sediment is exceeded by the relation rock/sediment. The reason for this is that the effect of solution during soil formation is surmounted by solution plus mechanical destruction during fluvial transportation. The relative gain of Na_2O of the relation rock/sediment is in fact the low Na_2O content of source rocks such as limestones and especially argillites, rocks which contribute only lit-

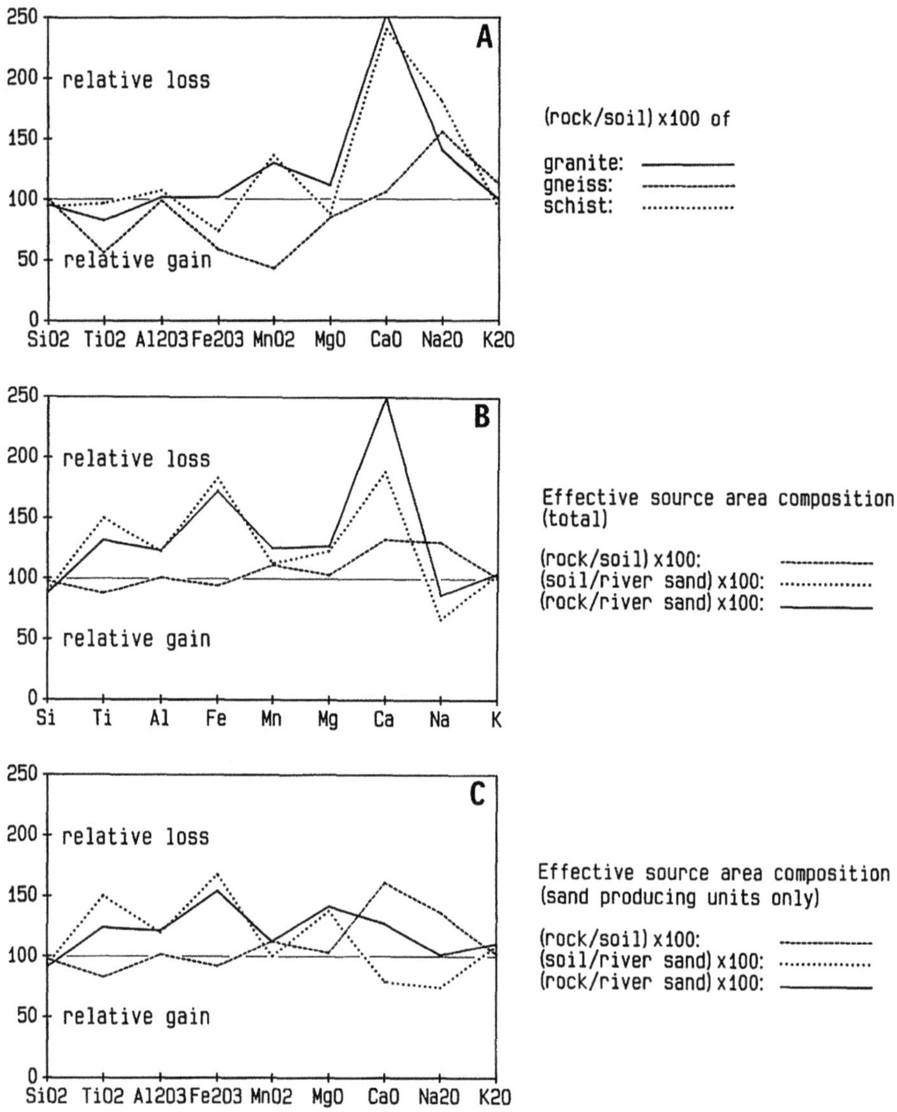

Fig. 9.1 A–C. Gain and loss calculations comparing the chemical composition of source rocks, source soils and river-mouth sediments (63–500-μm fraction only). **A** Comparison between rock and related soils of the most important first cycle rocks, granite, gneiss and schist. **B** Comparison between the effective composition (source area percentages weighed by erosion rates) of all source rocks and source soils and the river-mouth sands. **C** Same as **B**, but considering only sand-producing source units

tle to the 63–500-μm fraction of the river-mouth sediment. To give a more realistic picture of the source, Fig. 9.1 C illustrates this calculation without limestones, argillites and siltstones, based on sand-producing units only. The result, however, is not very different. The relative loss of CaO of the relation rock/sediment is smaller than that of rock/soil, indicating less solution, and almost the same is true for Na_2O, which is reduced in the soils but remains relatively constant in the river-mouth sediment if compared with the rocks. We try to give the explanation in the conclusion (Chap. 9.4).

9.1.2 Constant Oxide Calculation

We follow a suggestion by Maynard (pers. commun.) and recalculate both the composition of the effective soil and the sediment (Table 9.2 A, columns 2 and 5) assimilating their Al_2O_3 values to that of the average effective rock (column 1). Subtracting this modified soil composition (column 3) from the rocks (column 1), the gains and the losses become visible (column 4): SiO_2, TiO_2 and Fe_2O_3

Table 9.2. Constant-Al_2O_3 calculation [%] comparing rocks, soils and river-mouth sediment. **A** Total area; **B** Reduced area, sand-producing units only, Jurassic limestone, siltstone and argillite excluded[a]

Column	1 Rock	2 Soil	3 Soil (Al)	4 Soil-Rock (Al)	5 RivS	6 RivS (Al)	7 RivS-Soil (Al) (Al)	8 RivS-Rock (Al)
A *Total area*								
SiO_2	64.17	65.69	66.04	1.87	73.2	89.98	23.94	25.81
TiO_2	0.71	0.81	0.81	0.10	0.54	0.66	−0.15	−0.05
Al_2O_3	16.84	16.75	16.84	0.00	13.7	16.84	0.00	0.00
Fe_2O_3	5.84	6.21	6.24	0.4	3.39	4.17	−2.07	−1.67
MnO_2	0.1	0.09	0.09	−0.01	0.08	0.10	0.01	0.00
MgO	1.66	1.61	1.62	−0.04	1.31	1.61	−0.01	−0.05
CaO	5.08	3.84	3.86	−1.22	2.04	2.51	−1.35	−2.57
Na_2O	2.34	1.8	1.81	−0.53	2.71	3.33	1.52	0.99
K_2O	3.08	3.05	3.07	−0.01	2.95	3.63	0.56	0.55
CIA	69.91	73.09			64.11			
B *Sand-producing units only*								
SiO_2	66.78	68.26	69.13	2.35	72.2	87.59	18.46	20.81
TiO_2	0.67	0.81	0.82	0.15	0.54	0.66	−0.16	−0.01
Al_2O_3	16.62	16.41	16.62	0.00	13.70	16.62	0.00	0.00
Fe_2O_3	5.22	5.68	5.75	0.53	3.39	4.11	−1.64	−1.11
MnO_2	0.09	0.08	0.08	−0.01	0.08	0.1	0.02	0.01
MgO	1.85	1.8	1.82	−0.03	1.31	1.59	−0.23	−0.26
CaO	2.59	1.61	1.63	−0.96	2.04	2.47	0.84	−0.12
Na_2O	2.74	2.01	2.04	−0.70	2.71	3.29	1.25	0.55
K_2O	3.26	3.20	3.24	−0.02	2.95	3.58	0.34	0.32
CIA	67.36	74.8			64.11			

[a] Abbreviations: Rock: Effective source area rock composition; soil: effective source area soil composition; soil (Al): soil values recalculated with Al_2O_3 put equal to the rock value; RivS: river-mouth sediment; CIA: chemical index of alteration. In most cases the composition of the river-mouth sediment comes closer to the rocks than to the soils, especially if regarding only the sand-producing units (B). The CIA values demonstrate this clearly.

are enriched and MgO, CaO, Na_2O and K_2O decrease; Al_2O_3 is zero by definition. This is to be expected on going from rock to soil, and should continue on going from soil to sediment because soils are thought to represent the main source of sand. However, the opposite is the case (column 7), Na_2O and K_2O are enriched, the former more than the latter (see Fig. 9.1 B), while TiO_2 and Fe_2O_3 are depleted. Thus the sediment is closer to the rock than to the soil. Performing the last step of comparing sediment and rock (column 8), we realize that again the sediment is enriched in SiO_2, Na_2O and K_2O. The enrichment of quartz is expected, but not that of Na_2O and K_2O. An enrichment in feldspar can be excluded because the calculation bases on Al_2O_3 held constant.

The difference in Na_2O is larger, so we are dealing mainly with a sodium problem. The Na_2O measurement of the 19 river-mouth sands is reliable, Na_2O values display the lowest standard deviation and correspond to those of the longshore bar (Table 9.1 A, column 17 and 18). The calculation of the effective rock and soil composition, however, includes limestone and siltstone, rocks and/or soils characterized by extremely low Na_2O contents. As pointed out, both rocks evidently contribute only little to the river-mouth's 63 – 500-μm fraction of sand. Consequently, we recalculated the effective rock and soil composition based on sand-producing units only, that is, without limestone, siltstone and argillite (Table 9.2 B), as done in the case of the gain and loss calculation. Executing the same operations, the problematical enrichment of Na_2O and K_2O is much reduced. As a result, the calculation of gains and losses points towards a closer geochemical relationship between sediment and rock instead of sediment and soil. The same is true for the geochemical parameters.

9.2 The Parameters

Table 9.1 shows the percentages of the oxides and the most common parameters. For better comparison with other authors we follow Bhatia (1983), who emphasizes the plots of the following parame-

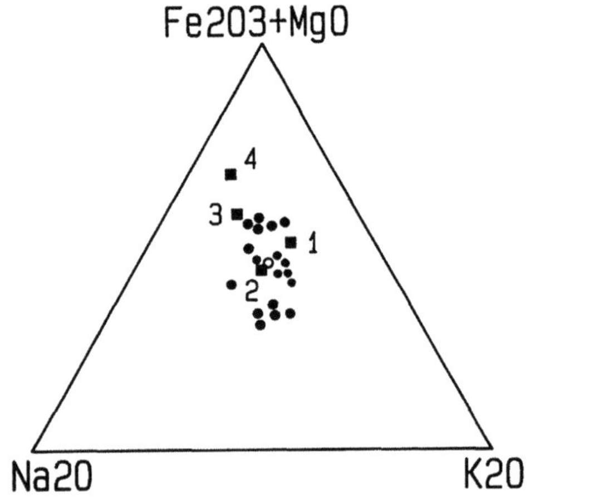

Fig. 9.2. Chemical composition of the river-mouth sand (63 – 500 μm). Ternary diagram of the system $Fe_2O_3 + MgO/Na_2O/K_2O$ %. Eighteen basins (●), without Palizzi [04]; average of the 18 basins (○). Average chemical composition of sandstones (■) of various tectonic settings (Bhatia 1983, Table 10); *1* passive margin; *2* active margin; *3* continental island arc; *4* oceanic island arc. Note the broad scatter of the 18 Calabrian river-mouth sediments and the good agreement of their average chemical composition (○) with that of active margin sandstones

ters as most discriminating: $Fe_2O_3 + MgO$ % versus TiO_2 %, Al_2O_3/SiO_2, K_2O/Na_2O, and $Al_2O_3/(CaO + Na_2O)$. The chemical index of alteration (CIA) introduced by Nesbitt and Young (1984) is also plotted (Table 9.2).

We agree with Potter (1978, p. 443) that the key variables $Fe_2O_3 + MgO$, Na_2O and K_2O, when plotted on a ternary diagram, fail to show very distinct groupings based upon our data. This type of diagram was used by Blatt et al. (1972, Fig. 9.2 and 1980, Fig. 20.2) and, among others, resumed by Maynard et al. (1982, p. 556), who renamed the data in terms of modern plate tectonics. The AFM diagram ($A = Na_2O + K_2O$, F = total iron as FeO and $M = MgO$), introduced into sedimentology by Robinson and Leake (1975) and applied on Californian arkoses and gneisses (Van de Kamp et al. 1975), groups our data even worse. The only difference between the two ternary diagrams is that the one combines Fe_2O_3 and MgO versus Na_2O and K_2O separately (Fig. 9.2) and the other combines Na_2O and K_2O versus Fe_2O_3 and MgO separately.

Probably the best distinction is made by plotting Al_2O_3/SiO_2 versus K_2O/Na_2O, as done by Maynard et al. (1982, p. 557), discriminating turbidite sandstones and shales, and by Pettijohn et al. (1987, p. 57), classifying the whole sandstone family. The latter use the reciprocal ratios.

Figure 9.3 compares these parameters in the different environments. In addition, all plots display, for reference, the mean sandstone composition of the four most important plate tectonic settings,

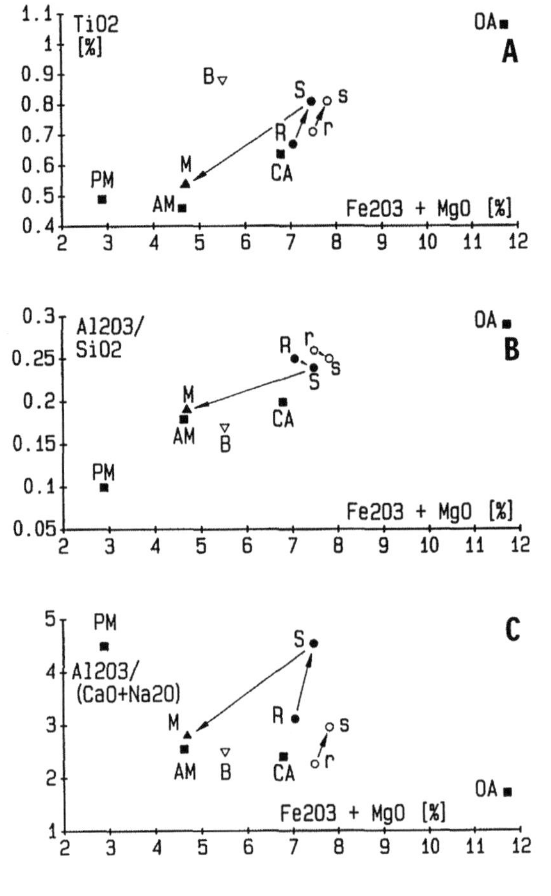

Fig. 9.3 A – D. Chemical composition of the solid source rocks, of the sand fractions 63 – 500 μm of soils, of the river-mouth sediment and of the longshore bar, compared with the average sandstone composition of various tectonic settings (Bhatia 1983, Table 10). Note that the composition of river-mouth sediments is always closer to the source rocks than to the soils. Abbreviations: sandstone composition (Bhatia 1983, Table 10) of: *PM* passive margins; *AM* active margins; *CA* continental island arcs; *OA* oceanic island arcs; *r* effective source rock composition, total area; *s* effective source soil composition, total area; *R* effective source rock composition, reduced area, sand producing units only; *S* effective source soil composition, reduced area; *M* river-mouth sediment; *B* longshore bar off the Bianco Beach

oceanic island arc, continental island arc, active continental margin and passive margin (Bhatia 1983, Table 10).

9.2.1 TiO_2 % Versus Fe_2O_3+MgO % (Fig. 9.3A)

Single Rocks and Soils. Fe_2O_3+MgO % are found in detrital iron oxides or ferromagnesian minerals, such as biotite and hornblende. During diagenesis, the iron generally is redistributed only locally, so that the total content remains more or less constant after burial (Blatt et al. 1980, p. 377; Argast and Donnelly 1987). The mobility of Ti is even low, so also the TiO_2% is regarded as a discriminating factor (Bhatia 1983, p. 619; Basu and Molinaroli 1989).

Both solid granite and granite soils display the lowest TiO_2 contents; gneiss and especially schists, on the other hand, show maximal values, and there is an increase from solid rock to soil by about 0.15%, gneiss excepted. Among the sediments only the siltstones belong to this high TiO_2-group.

We do not know where the TiO_2 comes from. It is not possible to relate the Ti to the sphene or rutile of the heavy mineral fractions of the gneiss-, schist- and siltstone soils, as the concentrations are too low (Table 9.3).

Table 9.3. Heavy minerals (total), sphene and rutile of schists and siltstones

Sample	Heavy minerals [%]	Sphene [%]	Rutile [%]
Schist (soil)	1.4	1.8	0.3
Siltstone	0.5	12.7	1.3

A first step to identifying the possible source of TiO_2 is to check the correlation between the erosional budget of TiO_2-rich source rocks, gneiss, schist and siltstone and the TiO_2 content of the river-mouth sediments through the 18 basins. The percentage of TiO_2 of the river-mouth sediments correlates with the erosional budget of schist with $r = 0.65$ (Fig. 9.4); with schist plus gneiss the yield was $r = 0.60$ and with schist plus gneiss plus siltstone we obtained $r = 0.64$. So the TiO_2 content of the river-mouth sediment seems to depend mainly on the schists. This is because the erosional budget of gneiss is only 6%, and, more important, although the erosional budget of the siltstone is 17%, this has to be reduced by 60% to only 6.8% because the average sand content of these debris flow sediments is only 40%, and this 40% sand is the only signal transmittable to the river-mouth sediments, if there is any at all.

Suppose the good correlation ($r = 0.65$) between the percentage of schists of the source areas and the percentage of TiO_2 of the single river-mouth sediments is causal – the broad scatter of the values (Fig. 9.4) demonstrates that there are low-schist and high-TiO_2 basins and vice versa, so we are dealing with a trend only. The neighbouring

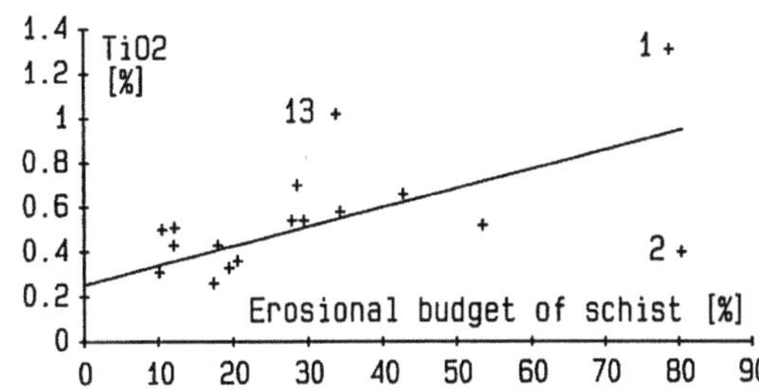

Fig. 9.4. TiO_2 content of the river-mouth sediments (fraction 63–500 μm) versus the erosional budget of schist of 17 basins: *1* Melito; *2* Ammendolea; *13* Novito

rivers Melito [01] and Amendolea [02], for example, display similar percentages of schist – 78.7 and 80.4% – but differ noticeably in TiO_2, 1.31 and 0.40%. This has to be seen once more as a strong limitation to the significance of "one-river-only" examinations.

The Fe_2O_3+MgO % values are spread over a much broader range, 3.5 to 10.5%, again with low contents of the granites and high ones of the metamorphic rocks and siltstones and their soils. This is mainly due to the larger variety of Fe_2O_3 compared to MgO (Table 9.1). From rock to soil, MgO increases only slightly in granites and gneisses and decreases in schists. Fe_2O_3, on the other hand, increases noticeably in granites and gneisses and decreases only slightly or remains constant in schists. The increasing Fe_2O_3 may be interpreted by formation of minerals of the former limonite group, especially hematite and goethite, during weathering.

The high TiO_2- and Fe_2O_3+MgO values of the siltstones obviously depend on the high content of ophiolithic and metamorphic rock fragments of the sand fraction of these debris flow sediments, reported by Guerricchio (1986, p. 75).

TiO_2- and Fe_2O_3+MgO correlate with $r = 0.92$ if the effective rock composition of the 18 basins is calculated.

The Environments. TiO_2- and Fe_2O_3+MgO increase from rock to soil (Fig. 9.3A) as expected and demonstrated by the gain-and-loss calculation but they are strongly depleted towards the river-mouth sediments. So the TiO_2- and the Fe_2O_3-bearing minerals are no longer visible through our 63–500-μm window. We suppose this deficit to be caused by the phyllosilicates, abundant in Calabrian rocks and soils but not represented in the river-mouth sediment, because they make part of the wash load and settle beyond the river mouths elsewhere in the marine realm.

We have no explanation for the high average TiO_2 content of the Bianco longshore bar, especially because the feeder river, the Laverde [07], has no exaggegrated TiO_2 values.

The standard deviation of the TiO_2 values of the 23 longshore bar subsamples is very high, so local concentrations may exist.

It is satisfying to realize the nearly perfect fit of the ratio TiO_2 % to Fe_2O_3+MgO % of our modern sands to that of the worldwide mean of ancient sandstones of active plate margins.

It is less satisfying, however, to realize that the trend from rock to soil is reversed from the soil to the river-mouth sediment and that both TiO_2 % and Fe_2O_3+MgO % drop even below the values of effective rock composition calculated with both the total and the reduced rock suite.

9.2.2 Al_2O_3/SiO_2 Versus Fe_2O_3+MgO % (Fig. 9.3B)

The ratio Al_2O_3/SiO_2 gives an indication of the quartz enrichment in sandstones (Bhatia 1983, p. 619). In the following, the content of Fe_2O_3+MgO % is not further mentioned, because it fits with that of the diagram described above.

Single Rocks and Soils. Maximum values of single rocks, that is, little SiO_2 or much Al_2O_3, respectively, are yielded by schists, siltstones and argillites. Limestones show an extreme ratio, the absolute content of SiO_2 and Al_2O_3 being very low, of course, but the ratio of the two, 0.64, is the highest of all rocks. From these four rocks, only the schists, at first glance, have a chance of transmitting their high Al_2O_3/SiO_2 ratios into the river-mouth sediments, because of their high erosional budget of 34%. Siltstones, on the other hand, contribute only 6.8% sand, as pointed out in the foregoing section. The sand content of argillites is also minimal, as is their importance as contributors to the river-mouth sediments. Even limestones are very rare, and the limestone content of the river sediments depletes drastically with decreasing grain size. The lowest ratios show granite as solid rock and sandstone as soil. The relative maturity of the sandstone soils is understandable, because the sandstones run at least one cycle and represent an arkose with a tendency towards a subarkose. As a whole, the Al_2O_3/SiO_2 ratios of both single rocks and soils span only a small range between 0.19 and 0.31, regardless of the limestone, and this feature will be maintained until the shelf is reached.

The Environments. To conclude, Fig. 9.3B shows the averages of the Al_2O_3/SiO_2 ratio, which move from the effective rock composition (0.27) over the soils (0.26) to the river-mouth sediments (0.19). The values of the reduced rock suite show only little deviation from the overall mean. There is no trend reversal. The Al_2O_3/SiO_2 ratio turns out to be the most stable parameter of chemical composition. Again, the Al_2O_3/SiO_2 ratio of the river-mouth sand (0.19) fits almost perfectly that of the average sandstone of active margins (0.18).

The sand of the longshore bar fits its feeder sand of the Laverde [07]. Thus there is no jump in compositional maturity between the fluvial and marine environment. This is in distinct contrast to the gravel situation (Chap. 12.3).

9.2.3 K_2O/Na_2O Versus Fe_2O_3+MgO % (Fig. 9.3C)

The ratio K_2O/Na_2O is a measure of the K-feldspar and mica versus plagioclase content (Bhatia 1983, p. 619).

Single Rocks and Soils. Table 9.1 demonstrates the differences between the single source rocks and their soils. Limestone excepted, which is unimportant for our concept, all rock types show a slight increase of the K_2O/Na_2O ratio from rock to soil. This should be caused by a loss of sodium corresponding to the lower stability of plagioclase compared to K-feldspar. The mica content of the rocks, however, should be mainly preserved in the soils.

The Environments. The average K_2O/Na_2O ratio of the river-mouth sediments corresponds almost perfectly to the worldwide mean of active margin sandstones. The K_2O/Na_2O ratio, however, increases from rock to soil and decreases from the source area to the river-mouth sediment, returning practically to the standard of the most important source rocks. So we are confronted again with a reversal of trend, as in the case of TiO_2 % and Fe_2O_3+MgO %. In terms of feldspar petrography, the increase of the K_2O/Na_2O ratio from rock to soil is understandable because it corresponds to the fact that K-feldspar is the most stable feldspar under weathering conditions, followed by albite, and, least stable, by anorthite. The reversed trend, however, a decrease of the K_2O/Na_2O ratio, is hard to understand.

These differences between the K_2O/Na_2O ratio of the river-mouth sediment (1.09) and the overall effective rock composition (1.32) are much smaller, calculating the reduced rock suite (1.19).

The K_2O/Na_2O ratio of the river-mouth sediments of the Laverde [07] corresponds to that of the sands of the longshore bar.

9.2.4 $Al_2O_3/(CaO+Na_2O)$ Versus Fe_2O_3+MgO% (Fig. 9.3D); CIA

The ratio $Al_2O_3/(CaO+Na_2O)$ is a ratio of the most immobile to the most mobile elements (Bhatia 1983, p. 619). It seems to be somewhat problematic to use this ratio, because there are only little changes to be expected by our limitation on the fine- and medium-sand fraction. Total samples or analyses including silt and clay fractions should yield more sensitive results. Nesbitt and Young (1984) developed a chemical index of alteration (CIA) using the molecular proportions:

$$CIA = [Al_2O_3/(Al_2O_3+CaO^*+Na_2O+K_2O)] \times 100 \ ,$$

where CaO* is the amount of CaO in silicates only. This index also expresses mainly the degree of alteration of feldspars to clay minerals, which are almost all excluded, by definition, from our sand analyses 63–500 μm.

Single Rocks and Soils. Among the single rocks and soils (Table 9.1), only argillite and limestone clearly exhibit values of $Al_2O_3/(CaO+Na_2O)$ near or below one. The soil sand of schists, on the other hand, has the lowest content of soluble elements.

The Environments. The plot of the average values of the $Al_2O_3/(CaO+Na_2O)$ ratio, again shows the good fit of our modern fluvial sediments with the ancient sandstones of active plate margins, and, well known in the meantime, a further reversal of trend during the development of the sand: the content of soluble minerals decreases from rock to soil, that is, the ratio increases, but from the soils to the river mouth this ratio decreases. This holds even more true when the reduced rock suite is calculated disregarding limestone, siltstone and argillite.

The $Al_2O_3/(CaO+Na_2O)$ ratio of the longshore bar comes very close to that of the Laverde [07] river-mouth sediment. Thus this first real marine successor of the fluvial sediments demonstrates no change in compositional maturity.

Our application of the CIA is limited because we did not determine the CaO of silicates. However, dealing with the reduced rock suite, disregarding limestone, siltstone and argillite, one may argue that most CaO is from silicates. This limitation provided the CIA of the effective source rock composition is 67.4, the effective source soil yields 74.8 and the river-mouth sediment has 64.1. Here again, the geochemical composition of the river-mouth sand resembles the source rocks and not the source soils.

9.2.5 Al_2O_3/SiO_2 Versus K_2O/Na_2O (Fig. 9.5)

This diagram (Fig. 9.5A) gives no really new information, because the ratios Al_2O_3/SiO_2 and K_2O/Na_2O were already plotted in Fig. 9.3B and

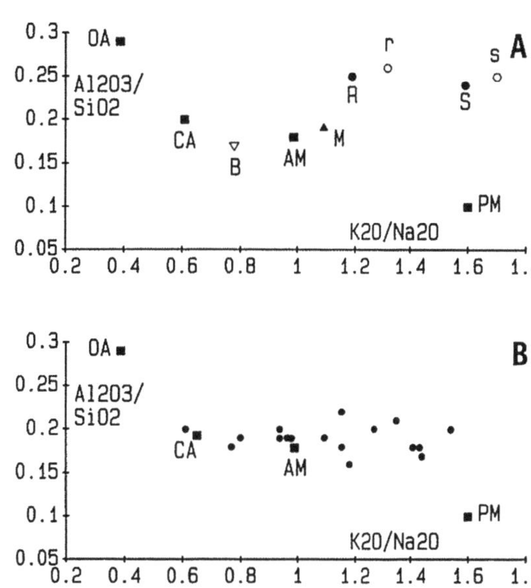

Fig. 9.5. A Al_2O_3/SiO_2 versus K_2O/Na_2O (see Fig. 9.3 for abbreviations). Again, the mean river-mouth sediment is closer to the source rocks than to the source soils and it comes closest to the average sandstone composition of active margins. **B** Same as **A**, sandstones of the various plate tectonic realms; instead of the mean river-mouth sediment, however, the 18 single river-mouth sediments are plotted to demonstrate the extreme variability of these rivers and, by this, the reduced reliability of "one river only" results

C. However, the combination of these two probably most important ratios summarizes the trend reversal going from the rocks to the soils and the sediments, be they fluviatile or marine. The changes from rock to soil are normal, but the sediments resemble the source rocks and not the source soils.

9.2.6 Trend and Single Measurement

Terminating this section on chemical development, we wish to underline a fact which is becoming more and more one of the probably most important findings of this study: the strong reliability of trends and the impressive unreliability of single measurements, even of only one or two complete river basins.

Figure 9.5B shows the ratios Al_2O_3/SiO_2 and K_2O/Na_2O of the average sandstones of the four plate tectonic realms, the corresponding ratios of 18 river-mouth sediments and their mean, which coincides with the Laverde [07]. This average Calabrian river-mouth sediment fits almost exactly to the sandstones of the active plate margins. The single basins, however, differ widely, forming relatively distinct clusters portraying the Aspromonte, the Zomaro and the Serre Provinces, that is, the mainly metamorphic, sedimentary and granitic realms.

Only four of these 18 rivers come close to the proper value of this active plate margin; single and isolated analyses of all other basins would have yielded misleading results, headed by the Amendolea [02] being classified as a "continental island arc" river.

9.3 Comparing Small and Big Rivers and the Problem of Na_2O

The comparison of our chemical data from small rivers and ancient sandstones of different tectonic settings revealed an almost perfect fit to the conditions of active plate margins (Figs. 9.3 and 9.5). This brings up the question as to the nature of the differences between these sediments and those of present-day big rivers, which predominantly drain passive margins. The mean chemical composition of these rivers (Potter 1978, Table 6) is recalculated (Table 9.1, column 19) and compared with our small rivers by means of a gain and loss calculation presented in the foregoing sections (Fig. 9.6). It

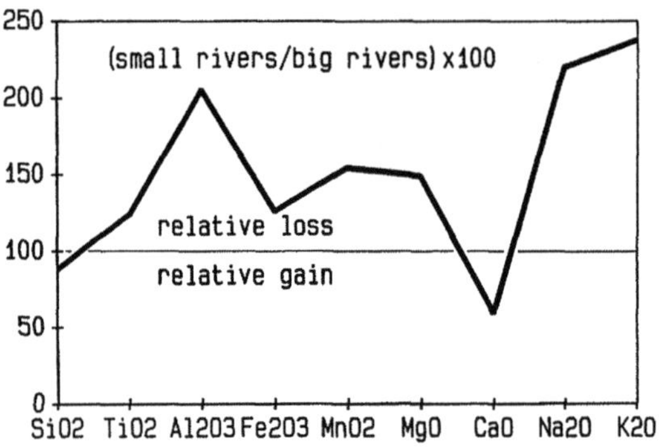

Fig. 9.6. Comparison of the chemical composition of the river-mouth sediments of big modern rivers (Potter 1978, Table 6) and of the 18 small Calabrian rivers (column 16 and 19 of Table 9.1), executed as gain and loss calculation (small rivers/big river × 100). Calabrian river sediments, as expected, prove to be very immature

goes without saying that this method in this case is purely descriptive, and that furthermore Potter's data represent a worldwide average whereas our data stem from a very restricted area without ophiolithic and volcanic rocks.

Being much more mature, the big-river sands display the expected relative gain of SiO_2 but, strangely, CaO is also high.

Comparing our Calabrian data with ancient sandstones, such an increase in CaO would be understandable due to the effects of cementation, but, more mature as they are, the big-river sands behave in an opposite manner. We have no explanation for this. The effective composition of the Calabrian source area shows 5.43% CaO for the rocks and 3.1% for the soils; percentages which never show up in the river-mouth sediments despite low chemical weathering and only short transportation. Bhatia's (1983) ancient sandstones (Table 9.1, column 22), on the other hand, show the expected higher CaO values of active margins.

At the loss side, K_2O of the big-river sands shows the largest relative difference to our small rivers, followed by Na_2O, Al_2O_3, MgO and Fe_2O_3. The relatively high contents of K_2O and Na_2O of Calabrian river sands mainly reflect the high content of both sodium and potassic feldspar. The ratio K_2O/Na_2O (Fig. 9.5) is high at passive margin sands and low at active margins. This corresponds to the Calabrian ratio of 1.09, that is, a balance between the two feldspars.

This leads to the problem of Na_2O summarized by Pettijohn et al. (1987, p. 174), to the question whether the high Na_2O content of graywackes is mainly a post-depositional feature or not. The

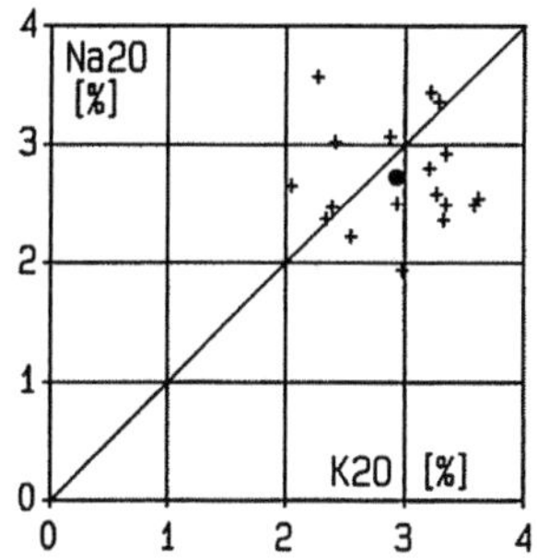

Fig. 9.7. Na_2O and K_2O content of Calabrian river-mouth sediments in the fraction 63–500 µm. *Dot* indicates the average. This high average of 2.71% Na_2O indicates an original, non-diagenetic origin of the high Na_2O-content typical for graywackes

Calabrian river sands reveal an average content of Na_2O of 2.71%, ranging between 1.94% Novito [13] and 3.57% Amendolea [02], st. dev. s = 0.44%. This fits the Na_2O content of graywackes reported by Pettijohn et al. (op. cit., Fig. 5.22) and has to be seen as an argument in favour of a primary, detrital origin (Fig. 9.7).

The Na_2O content of the 18 river-mouth sediments (Fig. 9.8) correlates with the content of plagioclase of the 63–500-µm fraction of these sands, r = 0.59*. There is no corresponding correlation between potassic feldspar and the K_2O content of the river-mouth sediments.

9.4 Conclusion

The development of the chemical composition from rock to soil of both selected single source rocks and soils and effective rocks and soils is quite normal. As far as this development can be expressed by the comparison of bulk samples of the solid rocks with the 63–500-µm fraction of soils, the following statements can be made: both gain and loss and constant oxide calculations reveal some alteration, small but in the sense of normal weathering. Among the parameters, only Al_2O_3/SiO_2 decreases, all other parameters or ratios, Fe_2O_3+MgO %, K_2O/Na_2O, $Al_2O_3/(CaO+Na_2O)$ and the CIA increase. It makes no difference whether the entire spectrum of rocks is calculated, limestone, siltstone and argillite included, or if only the sand-producing units are plotted. All this is consistent with what has to be expected during chemical weathering.

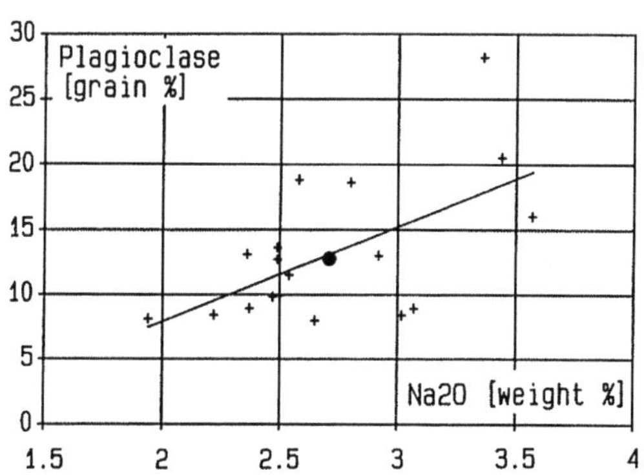

Fig. 9.8. The plagioclase content of the river-mouth sediments of the fraction 125–500 µm correlates with r = 0.59* with the Na_2O percentage of the samples. *Dot* indicates the average

Going from the soils to the river-mouth sediments, however, all ratios, Al_2O_3/SiO_2 excluded, discontinue this trend; they not only return to the effective rock composition but they frequently go beyond this and sometimes exceed the values of single rock types (Table 9.1). Thus, in a certain way, the river-mouth sediments are fresher or less weathered than the source rocks; one is tempted to say, more original than the original. Of course, this is supported by the results of the gain and loss and constant oxide calculations.

It turns out that the Calabrian soils fail as a serious producer of sand. Even if they contain 71% sand, the soil cover is so thin and patchy that the soils cannot play a great part in sand formation. Otherwise the sands of the soils, chemically altered by weathering, should have a visible influence upon the composition of the river-mouth sediment, and this is not the case. But also the other possible hinterland source of sand, the jointed and weathered source rocks, has to be excluded, because they can only offer a mean sand content of 3.8%.

Consequently, the only important mechanism of sand formation that remains in order to fit the river-mouth sediments quantitatively and qualitatively is attrition, the grinding and milling machine of fluvial transport during bankfull stages. This process obviously works like a rejuvenation treatment. During flood transport, weathered rocks are more quickly destroyed than fresh ones, and if there are remaining nuclei of the clasts, they are more resistant than the cover and the corners which are chipped off. The attrition of these nuclei is reduced and the resulting fines become fresher. The same is true for all minerals. Less stable minerals are more easily ground to fines than resistant ones. Thus, a selection takes place in the sense that the less stable particles move with higher probability into finer fractions than the more resistant ones, and consequently they have an advantage in transportation and leave the fluvial basin earlier than the relatively coarser and fresher material. The reference fraction of 63 – 500 μm is constantly refilled by the fresh-ground and milled material from coarser fractions and it is left by less stable, finer fractions. Instead of a solubility rank, Mohs' hardness scale or an appropriate measure of mechanical resistance should be applied in order to establish the history of selection during fluvial attrition. This is in contrast to Rogers et al. (1963), who believe that large grains are rarely abraded down to the fine sand or silt range.

It thus happens that the river-mouth sediment, as a fluviatile relict sediment, contains the relatively late and relatively fresh elements of a discharge. The chemical composition of these river sands corresponds to that of the freshest source rocks.

In this context only the phyllosilicates play a special role. These sheet minerals are part of the very suspended wash load, and settle predominantly only in the marine realm, together with silt and clay. Consequently, they are under-represented in the river-mouth sediment and do not show up in the geochemical record. This may be the main reason, apart from the rough calculations of effective source rock and soil composition, for the inconsistencies in our comparison rock, soil and sediment, with special emphasis on Na_2O and K_2O.

This scenery may be seen as a fluvial equivalent to the active reworking of detritus in the shallow marine environment described by Mack (1978).

To conclude, the experience of the striking predominance of mechanical over chemical impact in the Calabrian environments is supported by the increase of polymineralic quartzose grains going from the soils to the river-mouth sediment. As noted in Chapter 8.2.2.3, the content of polymineralic quartzose grains decreases from sandstones to sandstone soils because, according to Harrell and Blatt (1978), polycrystalline quartz is more resistant to mechanical destruction than to chemical weathering, compared to monocrystalline quartz. In other words, as the soils cannot account for the surplus of polycrystalline quartz in the river-mouth sediment, this has to be seen as one more sign of attrition.

The residence time of the river-mouth sediments is also too short to allow chemical weathering to proceed noticeably. We make the rough calculation from climatic data (Chap. 4) and from the occurrence of landslides (Chap. 15.1.2) that the recurrence interval of heavy rainstorms may be about 25 years for a given basin. We believe from this that large parts of the river bed's surface are reworked within this time interval. Speaking in terms of Johnsson and Stallard (1989), the physiographic control on sediment composition in Calabria is characterized by short transportation paths and short residence times which lead to remarkably immature sediments.

So it becomes understandable that it would be much easier in Calabria to obtain a good hand specimen of rocks among the river gravels than from the outcrop.

Delivering mainly fresh and little altered sand, active plate margins may become more suitable objects for provenance research than passive ones. This seems to be justified even from a quantitative point of view, because also small rivers are known to be serious producers of sediment to the oceans.

Chapter 10

Development of Grain-Size Distributions

Almost every fluvial flow regime is able to move sand. The broad range of gravel and boulder sizes, however, with its open-end scale at the coarse side, needs a broad spectrum of current velocities to be entrained and moved. Thus the medium "gravel" is more suited to illustrate differing competence than the medium "sand" and thus to demonstrate the possible downstream or longshore development of textural maturity. This is the main reason, without regard to the fact that Calabrian rivers are gravel rivers, to devote our special interest in this study to the grain-size distributions of gravels.

Summary. Calabrian river-mouth sediments are both bimodal or unimodal. The gravel mode of the bimodal distributions, mostly Rosin-distributed, is interpreted as a first-order element, a primary feature inherited from the jointed and weathered source rocks, whereas the sand mode can be seen as a second-order element acquired during fluvial attrition. The shape of the gravel mode varies significantly from the jointed and weathered source rocks over the river mouths to the shoreface, the latter exhibiting lognormal grain-size distributions.

If the percentage of the fraction 1–20 mm exceeds about 35%, the distributions became unimodal. Thus this fraction is crucial for modality and is called, consequently, a third-order element.

All single factors assumed to control modality alone, such as relief, basin shape, source rock petrography, the petrographical composition of the river sediment itself or postdepositional refill of interstices, fail to exert a significant or mono-causal influence on modality. In addition, no hydrodynamic conditions are known which specifically select, as a rule and within the grain-size range of gravel and sand, this fraction of 1–20 mm. Thus, other possibilities being excluded, only availability remains, but not in the sense of source area-related sediment production. It is assumed that different floods drive the "sedimentary mill" of these mountain rivers differently and that the formation of the fraction 1–20 mm is sensitive to these differences.

Data Base. 133 samples (see Table 10.1 for sample numbers of the single environments). The only purpose of these samples was to serve for the grain-size analysis of the jointed and weathered source rock materials.

Method. The mean sample size is 271 kg, the maximum grain size 256 mm. The samples were taken as channel samples (Fig. 10.1) by means of a crowbar (handspike) or motor hammer. The fraction >16 mm was sieved in the field by means of a transportable sieving and splitting machine in 1/3 phi steps (Ibbeken 1974), the fraction <16 mm was split and taken to the lab. The results of the sieve analyses are stored in the data bank.

The coastal samples were taken with the help of a shovel excavator; see the method section of Chapter 12 for the procedure.

Following the concept of this book, we describe the development of the grain-size distributions from the source to the sediment and not conversely. Our Calabrian data base enables this to be done through five distinct environments arranged in sequence:

A Jointed and weathered source rocks (in situ),
B Transition (regolith, talus debris, alluvial cone),
C River course (downstream of the fall line),
D River mouth,
E Beach (Shoreface).

We start with a discussion of the gravel mode, followed by the treatment of the sand mode and the

Table 10.1. Grain-size parameters of the gravel modes of the single environments from the source to the beach

Environment[a]	No. of samples	Mean [mm]	St. dev. [mm]	Skewness	Kurtosis	RG ratio[b]
A	25	64.1	1.26	1.17	5.1	4.0
B	22	36.4	1.39	0.74	3.6	3.6
C	17	41.0	1.56	0.87	3.1	5.0
D	35	24.6	1.51	0.47	2.4	8.0
E	34	21.3	1.31	0.22	2.7	0.56
	133					

[a] A: Jointed and weathered source rocks, in situ; B: transition, regolith, talus debris, alluvial cone; C: river course, downstream of the fall line; D: river mouth; E: shoreface; in the environments C–D only bimodal distributions were evaluated.
[b] RG ratio: Rosin-Gauss ratio, proportion of the number of Rosin- and Gauss-distributed samples of an environment.

Fig. 10.1. Channel sampling of Buonamico [08] river sediments. View upstream from below San Luca towards W, towards the fall line where the river widens. The light, oblique element in the background is the Costantino slide (Chap. 15.1.2.2)

1–20-mm fraction. We conclude with a section dedicated to the problem of uni- and bimodality.

10.1 The Gravel Mode (First-Order Element)

The granulometric interpretation is illustrated by the mean histograms of the five environments under examination (Fig. 10.2). All grain-size parameters, however, are derived from the complete data set of the respective sample groups and not from the plotted mean histogram. Table 10.1 summarizes the mean granulometric parameters of the gravel modes.

There is a certain inconsistency in the data set, because the samples from the environments A to D are spread over the entire area under consideration, nearly 2000 km^2, whereas the shoreface data, environment E, stem only from repeated sampling of the 9-km shore strip of the Bianco beach between the Laverde [07] and the Buonamico [08] (Fig. 12.1). Field experience from diving at almost most Calabrian coasts tells us, however, that the Bianco beach is evidently representative of the Ionian beaches of the area under investigation.

The sequence of the five gravel modes A–E (Fig. 10.2) is characterized by decreasing maximum grain sizes and the fining modal class. The most striking feature is the peakedness of the gravel modes and their positive skewness in the cases A–D. Only the shoreface gravels seem to be symmetrically distributed. However, if we apply the classical grain-size parameters, mean, sorting, skewness and kurtosis (Table 10.1), to describe and interpret this sequence, the results yielded are not very satisfying.

The decrease of the skewness and kurtosis values is linked with decreasing mean grain size. The correlation coefficient mean versus skewness is $r = 0.95$, and mean versus kurtosis correlates with $r = 0.97$. This relation, however, is partly trivial and partly an artefact. Because we consider only the gravel mode, the distributions are fixed by definition at the fine end, the 2-mm size limit. Attrition and selective fluvial transport make maximum grain sizes and mode shift towards the fines. The maximum grain sizes decrease only slightly by two size classes, whereas the mode moves strongly by eight size classes. This process is linked with the production of fines, but all new grain sizes <2 mm go beyond this artificial limit and are no longer part of the gravel mode. Consequently, the distributions flatten and skewness and kurtosis decrease with decreasing mean. This development from the mountains to the sea is linked to a decreasing variability of the samples as a whole.

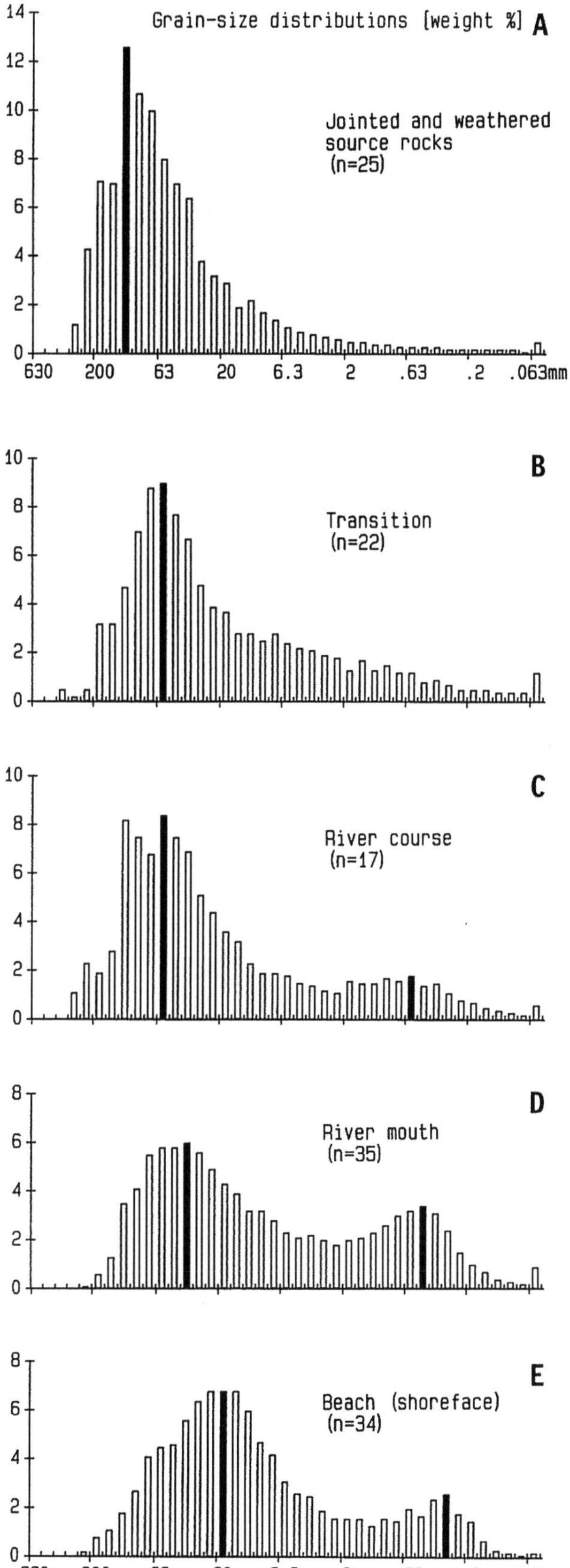

Fig. 10.2A–E. Average histograms of the grain-size distributions of the single environments from the source to the beach. **A** Jointed and weathered source rocks (in situ). **B** Transition (regolith, talus debris, alluvial cone). **C** River course (downstream the fall line). **D** River mouth. **E** Beach (shoreface). *n* Number of samples. The river samples **C** and **D** refer to bimodal distributions only. The modal class is marked *black*

Figure 10.3 shows the frequency distribution of the skewness values – gravel mode only – of the single environments. There is a strong tendency towards higher symmetries, as pointed out, but the broad scatter of 12 classes of the jointed and weathered source rocks narrows step by step towards the river mouth and shore face down to only four or five classes. We know this reduction from other variables of the source areas like petrography and geochemistry which are strongly filtered out and concentrated if we follow the development of sediment from the source areas to the river mouth and shelf deposits.

The value of this picture of an increasing symmetry of the gravel mode is curtailed by its connection with the decreasing mean and its descriptive character. A measure not depending on grain size and founded on a geological concept would be much more satisfying. We believe that the application of the Rosin-Gauss concept yields a better, geologically more meaningful understanding of the provenance and development of gravel distributions. The log-hyperbolic distribution, another fit technique independent of grain size, will be discussed in Chapter 10.1.3.

10.1.1 The Gauss-Rosin Concept

Classical grain-size parameters are derived on the assumption that the data set is normally distributed; this was also the concept of the first modern grain-size study of the Calabrian fiumaras (Viparelli 1964). However, if we apply a test of normality on the Calabrian grain-size data (Ibbeken 1983; Schleyer 1987), it turns out that the majority of the distributions do not follow the log-normal or Gauss distribution, but the Rosin law.

Rosin distributions are asymmetrical, positively skewed and typical for breaker and technical mill products (Rosin et al. 1933; Krumbein and Tisdel 1940; Kittleman 1964). Because, on the other hand, mature and well-adapted sediments tend to display symmetrical, log-normal distributions, the Rosin distribution of the jointed and weathered source rocks may be seen as a well-defined and fixed pro-

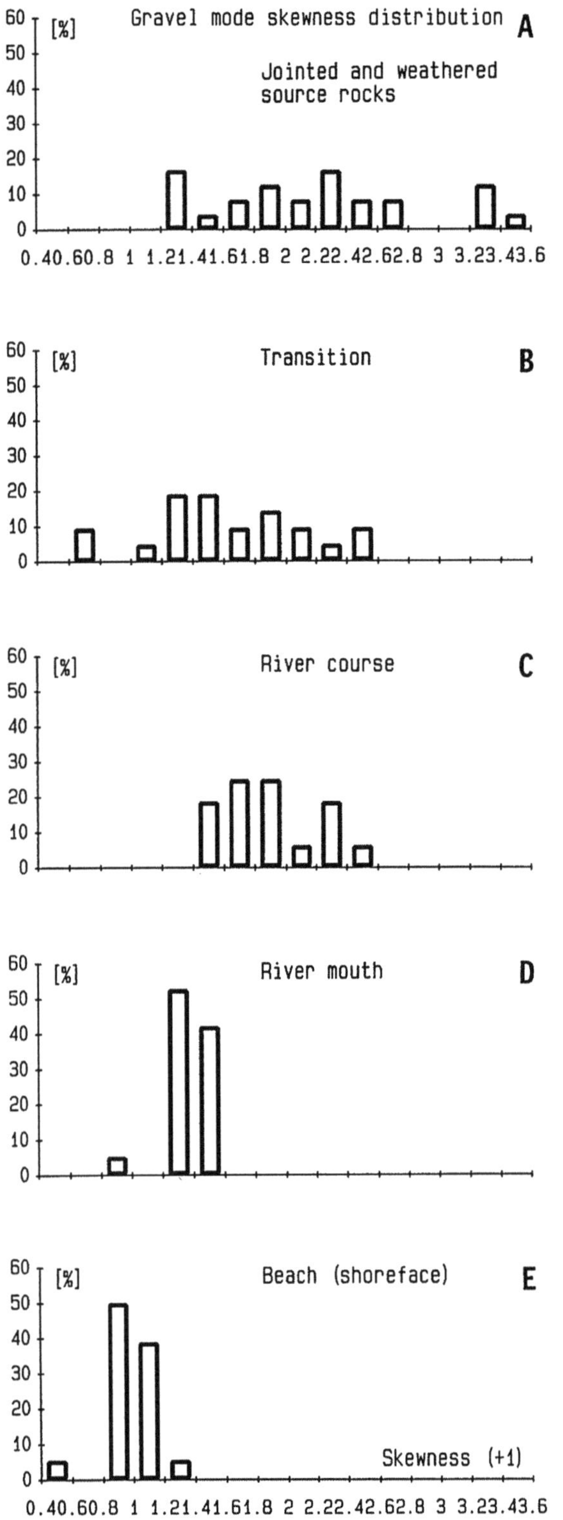

Fig. 10.3 A–E. Histograms of the skewness distribution of the gravel modes. **A** Jointed and weathered source rocks (in situ). **B** Transition (regolith, talus debris, alluvial cone). **C** River course (downstream the fall line). **D** River mouth. **E** Beach (shoreface). From the outcrop to the beach, symmetry increases and the variability of the skewness values decreases

venance property, to which the development of later and younger distributions can be referred. Or, in other words, the Rosin distribution of a terrigenous deposit is an inherited property and the log-normality an acquired one.

Consequently, in addition to the statistical treatment, the quantitative fit of the gravel mode grain-size distribution versus both the corresponding ideal Rosin or Gauss distribution (Fig. 10.4) was calculated and plotted (Schleyer 1987). We call a distribution Rosin-distributed when it yields a higher value for the Rosin fit and vice versa. It is not possible to characterize a sample group by its mean Rosin or Gauss fit, because higher Rosin fits are usually correlated with higher Gauss fits. But in each environment there are both Rosin- and Gauss-distributed samples, and it is sedimentologically more meaningful to note whether one type prevails over the other or not. Thus, the "RG-ratio" stands for the quotient of the number of Rosin- to Gauss-distributed samples of an environment.

10.1.2 The Sequence of the Five Environments

The jointed and weathered source rocks (Fig. 10.2A) are predominantly Rosin-distributed, as shown by a RG-ratio of 4. This is true of all rock types able to form clasts, be it sandstone or granite, reef limestone or gneiss (Ibbeken 1983, p. 1219). We do not know the reason, but field evidence is so striking that we stopped further sampling on this topic.

The genetically very broad transition group (Fig. 10.2B) comprises all stages between the fixed source rock and the fluvial transport such as regolith, talus debris, alluvial cone and the like. Nonetheless, it shows a RG-ratio of 3.6. The Rosin tendency of the gravel modes of this only little transported material is interpreted as a primary feature inherited from the jointed and weathered source rocks.

The sediments of the short river courses (Fig. 10.2C) are mostly bimodal, sometimes unimodal. In this section we refer to the gravel modes of only the bimodal distributions, because the fine tail of the gravel mode of unimodal distributions is hidden under additional fine gravels and difficult to reconstruct (Ibbeken 1983, Fig. 8).

The gravel modes of the bimodal distributions display an even stronger Rosin tendency, with an RG-ratio of 5. This may be attributed to the argument that both formation of lag sediment during increasing flow regime and deposition during

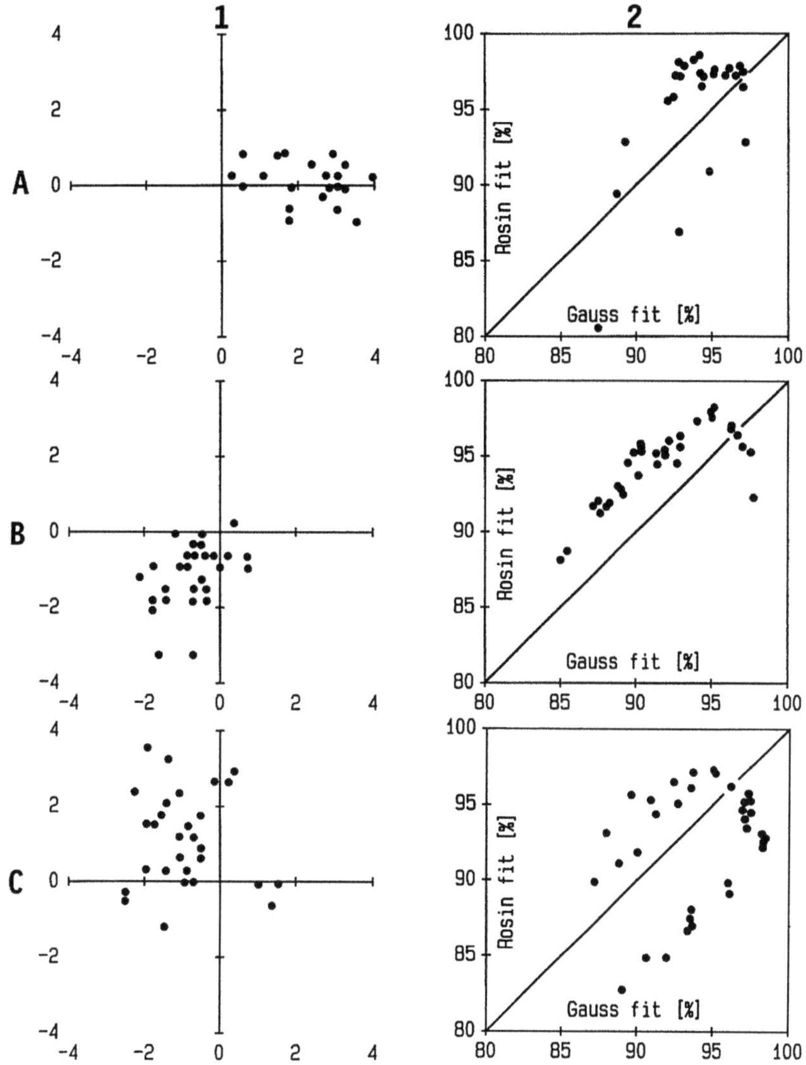

Fig. 10.4 A–C. Discriminant analysis and Rosin-Gauss fit of the gravel modes of selected environments; the grain sizes are taken as the only variables. **A** Jointed and weathered source rocks. **B** River mouth. **C** Beach (shoreface). Column *1* Discriminant analysis, the ordinates are the canonical discriminant functions 1 and 2, respectively; column *2* Rosin-Gauss fit

decreasing flow regime favour positively skewed distributions (McLaren 1981; McLaren and Bowles 1985). Thus the Rosin tendency of the river gravels is interpreted as primarily inherited from the source rocks and secondarily strengthened by fluvial transportation, an interpretation confirmed recently by Shih and Komar (1990).

This is even more true for the river-mouth sediments (Fig. 10.2 D), which are mostly bimodal. In this context, again, we consider only the gravel modes of bimodal distributions. Compared to the transition and river-course environment, the modal class of the river-mouth sediments fined up by 2/3 phi steps, compared to the jointed and weathered source rocks by 5/3 phi steps, corresponding to a decreased volume by factor 4 and 32 respectively. This means that even steep-graded mountain rivers of only 20 km length are not able to transmit larger amounts of very coarse grain sizes into the marine realm.

The Rosin tendency appears still stronger, with an RG-ratio of 8, supporting the hypothetical considerations of McLaren and Bowles (1985). Obviously, our short mountain rivers are not only unable to bring about a normalization of the gravel modes, but they have the opposite effect and intensify their Rosin tendency. This is mainly due to the decreasing competency of the rivers and the increasing content of fines. The basis of the gravel mode narrows at the coarse side, but the coarse flank does not steepen because the mode shifts towards the fines and flattens as a whole. At the same time, the granule and sand content increases, flattening the fine flank.

There is a marked change in the last environment investigated, the beach with its strip of the shoreface break (Fig. 10.2E). As the zone of most active transportation, the shoreface strip is the most important element of the beach. Only here does a clear Gauss tendency appear, which is attributed to three processes independent of each other: the basal width of the gravel mode remains preserved, the modal class migrates towards the fines and many granules become lost.

The basal width of the gravel mode remains preserved, because there is practically no reduction of the coarse extreme values along the shoreface strip, notwithstanding the high energetic environment. The competency of the longshore transport does not decrease noticeably if, during storm conditions, extreme grain sizes are transported longshore, a fundamental contrast to the fluvial environment.

The modal class, however, migrates towards the fines even though by only one phi step. Transport wear and selection by wave action reduce the finer grain sizes of the shoreface strip much more effectively than the larger ones. This shift of the modal class means a volumetric reduction of its particles by factor 8, adding up the modal fining over the entire range of environments to factor 64. Even though the "valley" of the shoreface gravel also migrates towards the fines by one phi step, there is a concomitant decrease in mean sand content towards 21%, accompanied by a loss of granules, and consequently the fine flank of the gravel mode steepens: the result is a Gauss tendency or normalization expressed by an RG-ratio of this environment of 0.65.

The driving force of this normalization of the gravel mode is the fining modal class, combined with nearly constant coarse extreme values and reduced granule and sand content, all typical of the shoreface environment.

The primary provenance signal of the Rosin-distributed river-mouth gravel modes, secondarily strengthened during fluvial transportation, is effaced by the longshore transport and replaced by normalization. A Rosin-Gauss transformation takes place. Information about longer rivers is needed to decide whether later stages of fluvial transportation may lead to similar results.

We conclude the discussion of the development of the gravel modes with a short statistical treatment. To remain within meaningful limits we compare only the three most important environments, the jointed and weathered source rocks as the origin and the river-mouth deposits as the final stage of fluvial transportation and finally the beach (shoreface) as one of the first steps of the marine development.

The data set of the jointed and weathered source rocks (25 samples), of the river-mouth gravel (35 samples) and of the beach or shoreface gravel (34 samples) was subjected to a Rosin-Gauss goodness-of-fit test, a discriminant analysis and a hierarchical cluster analysis, in each case with the grain sizes as the only variables.

The Rosin-Gauss Fit (Fig. 10.4, column 2). In this diagram the Rosin fit values of the single samples are simply plotted against their Gauss fit values. Higher Rosin fit values clearly correlate with higher Gauss fit values. The values of the jointed and weathered source rocks form a relatively narrow cluster with high Rosin fits, RG-ratio 3.6 (Fig. 10.4A2). The river-mouth gravels display a less homogeneous group, the RG-ratio, however, is 8 (Fig. 10.4B2). The shoreface group is even more scattered, but the RG-ratio is 0.65 (Fig. 10.4C2). Although the point clusters display clear trends, there is too much overlap for a graphically satisfying presentation.

Discriminant Analysis (Fig. 10.4, column 1). Discriminant analysis yields much better results as far as the graphical presentation is concerned. The three environments form very distinct clusters with only little overlap, demonstrating that the gravel modes are in fact characterized by significantly different shapes.

Cluster Analysis (Fig. 10.5). Surprisingly, the samples of the three environments are also clustered into three subgroups, corresponding to the three environments with a 79, 64 and 79% fit respectively. Seven shoreface samples were "misplaced" by the cluster analysis as fluvial. However, these samples were taken immediately off the Laverde delta (Figs. 12.1B, 16.8), and obviously still carry the fingerprint of their fluvial past.

As a result, there is a significant differentiation of the gravel modes of bimodal distributions from the jointed and weathered source rocks over fluvial stages to the coastal environment.

In 1983 R. Ehrlich wrote an editorial entitled "Size analysis wears no clothes ...". Ehrlich may be right as far as sand is concerned; we believe, however, that our gravel modes wear clothes, real clothes.

10.1.3 The Log-Hyperbolic Distribution

The log-hyperbolic distribution was first introduced by Barndorff-Nielsen (1977) in sedimentology. Its name is derived from the fact that the curve of its frequency distribution is a hyperbola when applied to the logarithm of both size and frequency. The approximation of a log-hyperbolic distribution to natural grain-size distribution seems to gain better results than log-normal or Rosin distributions, especially with regard to the tails of the distributions (Bagnold and Barndorff-Nielsen 1980; Barndorff-Nielsen et al. 1982; Durst and Macagno 1986). However, the concept is not much used by sedimentologists, probably because of the very complex computations for determining the four parameters α, β, δ and μ. Nevertheless, some recent papers and discussions, especially from the Aarhus group in Denmark, show the importance of this concept (Wyrwoll and Smyth 1985; Christiansen and Hartmann 1988a; Wyrwoll and Smyth 1988; Schleyer 1987, 1988a; Hartmann 1988). Therefore, the log-hyperbolic concept should be applied to the gravel modes of our five mean distributions and be compared with the Rosin-Gauss concept.

The four parameters of the best-fitting log-hyperbolic distributions were calculated using the program LHPDF (Log-Hyperbolic Probability Density Function) from Bhatia and Durst (1988). The attempt to apply the program SAHARA (Statistical Analysis Hyperbolic AccuRAcy, Jensen 1988; Christiansen and Hartmann 1988b) for this purpose failed, because the calculation procedure based on iterations did not come to an end.

Figure 10.6 shows the five average gravel modes from the five environments, source rock (A), transition (B), river course (C), river mouth (D) and shoreface (E) as histograms together with the curves of best fitting log-hyperbolic distribution (solid line) as well as Rosin distribution (dotted line) and log-normal or Gauss distribution (hatched line). The better fit of the log-hyperbolic distribution is evident, especially at the tails of the

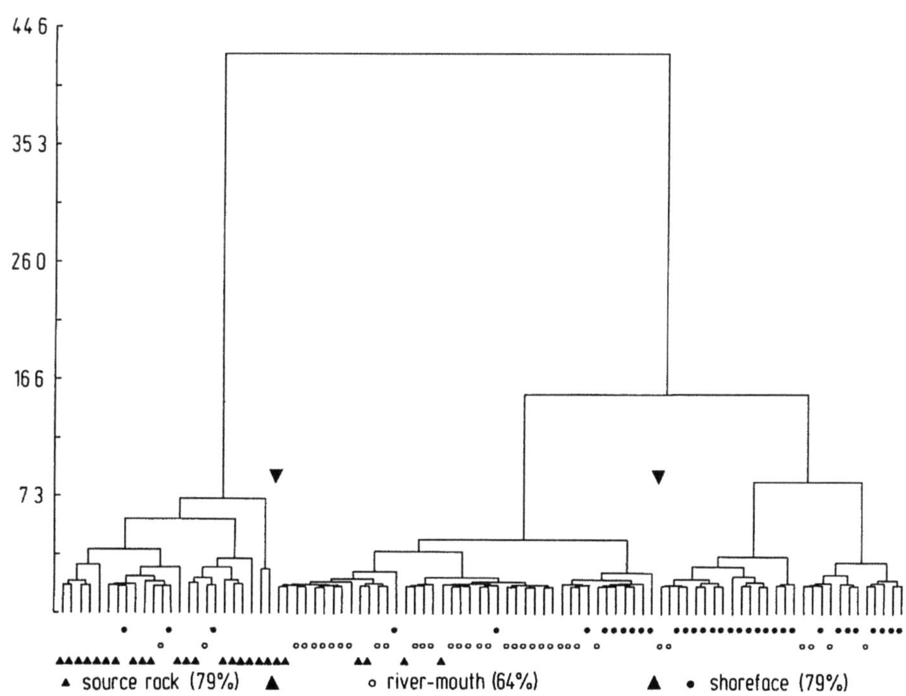

Fig. 10.5. Cluster analysis of the gravel modes of selected environments. Jointed and weathered source rocks, river mouth and beach (shoreface); the grain sizes are taken as the only variables. The source rocks are placed with a 79% fit as is the shoreface gravel. Nine shoreface samples were "misplaced" by the cluster analysis as fluvial. However, seven of these samples were taken immediately off the Laverde delta, they obviously still carry the fingerprint of their fluvial past

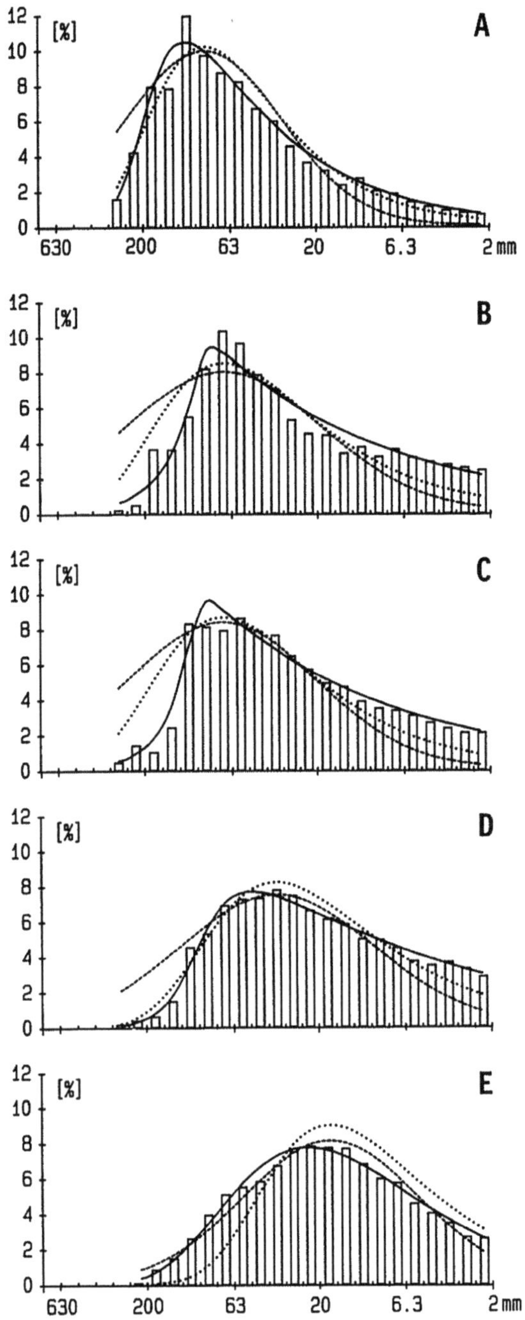

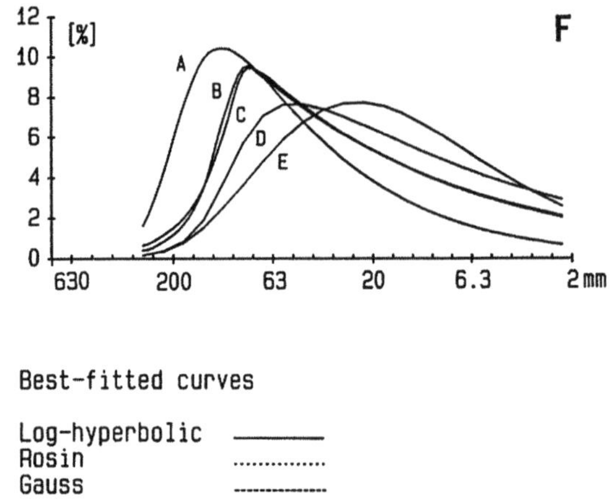

Fig. 10.6 A–F. Histograms of the average grain-size distributions. **A** Jointed and weathered source rocks (in situ). **B** Transition (regolith, talus debris, alluvial cone). **C** River course (downstream of the fall line). **D** River mouth. **E** Beach (shoreface). **F** Log-hyperbolic curves of the five environments **A–E**

distributions. Figure 10.6F shows the amazing homogeneous sequence of the five hyperbolic curves through the five environments.

The parameters of both the Gauss-Rosin and the log-hyperbolic techniques may be plotted as scattergrams (Fig. 10.7). Ibbeken (1983) introduced the Gauss-Rosin plot (Fig. 10.4, column 2), where the Gauss fit of a distribution is simply plotted against its Rosin fit. In Fig. 10.7 only the average of each environment is plotted.

Barndorff-Nielsen et al. (1985) and Barndorff-Nielsen and Christiansen (1986) introduced the log-hyperbolic shape triangle, which in principle is also a scattergram of the two parameters χ and ζ which are calculated from α, β and δ (Fig. 10.7B).

The comparison of the two plots shows nearly identical patterns. For comparison, mean and skewness of the five gravel modes are also plotted (Fig. 10.7C), as proposed and discussed in many papers by Friedman (1961, 1967, 1979). At first

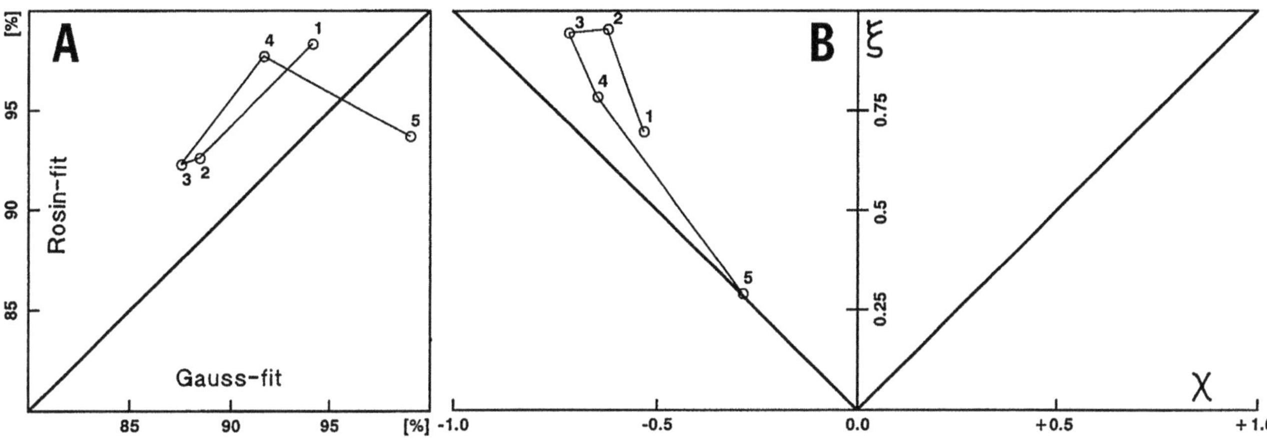

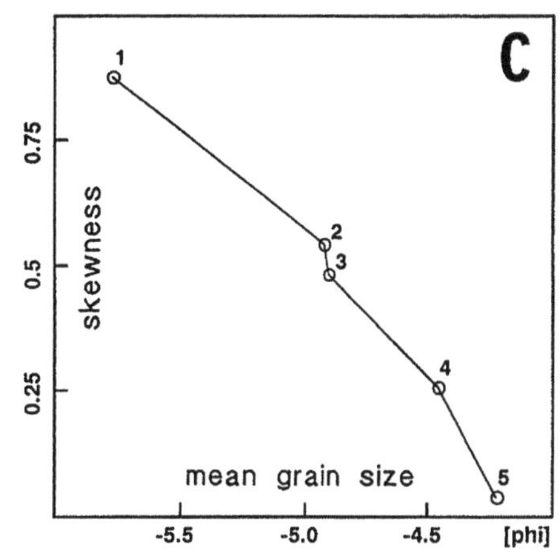

Fig. 10.7 A–C. Comparison of grain-size parameters describing the development of the gravel modes through the five environments. **A** Rosin-Gauss plot (see Fig. 10.4, column 2, for the entire point cluster). **B** Log-hyperbolic shape triangle. **C** Skewness versus mean grain size. *1* Jointed and weathered source rocks (in situ); *2* transition (regolith, talus debris, alluvial cone); *3* river course (downstream of the fall line); *4* river mouth; *5* beach (shoreface). The straight and regular pattern of plot **C** is not very meaningful sedimentologically, skewness and mean are not independent variables. Rosin-Gauss plot and log-hyperbolic shape triangle are surprisingly similar

glance this graph seems to be the best, illustrating the sedimentological development of gravel modes from the weathered and jointed source rock to the shoreface, because the five mean distributions lie almost perfectly on a straight line. Unfortunately, this apparently good fit is of only little meaning sedimentologically, because the two variables are dependent on each other, as pointed out in Chapter 10.1. In both the Rosin-Gauss diagram and the log-hyperbolic shape triangle no absolute grain-size values but distribution shapes are compared, and this is the essential difference to other efforts to find a correlation between grain-size parameters and environment.

To come to a conclusion, the advantages and the disadvantages of the Rosin-Gauss concept have to be compared with those of the log-hyperbolic concept. The main advantages of the log-hyperbolic distribution are the better fit to observed data and the fact that only one distribution type is needed to describe multiple distribution shapes. Disadvantages are the very complicated calculations of the four parameters, which can only be done by computer programs. These calculations are based on maximizing the log-likelihood function by iteration, consequently they are very time-consuming, up to 20 min for one distribution on an IBM-AT, and they often gain no results. The loss of the log-hyperbolic concept is the gain of the Rosin-Gauss concept and vice versa. Both Gauss and Rosin fit are easily and rapidly calculated and fit values result for any type of grain-size distribution. The lower quality of fit is made up by the possibility of a meaningful interpretation instead of the pure description: the Rosin distribution is interpreted as a source-rock-specific, inherited feature of immaturity, which can be explained by the hypothesis that the sum of processes which disintegrate a rock body is similar to that of a crusher. Normality, on the other hand, may be interpreted as a property acquired during fluvial attrition and especially during fluvial transportation. Unfortunately, the log-hyperbolic concept fails to contribute such a simple explanation and sedimentological interpretation. Wyrwoll and Smyth (1988, p. 162): "Using the distribution purely for its fit of some sediment, grain-size distribution provides limited gain and retraces long-worn and sterile paths". We believe that the best method is to apply both the mathematically more satisfying log-hyperbolic concept and the sedimentologically more convincing Gauss-Rosin concept.

10.2 The Sand Mode (Second-Order Element)

The sand budget of both bimodal and unimodal river-mouth sediments is exactly balanced – the average sand content is 24.95% (bimodal sediments) and 25.19% (unimodal sediments); 60% of the bimodal and 61.9% of unimodal river-mouth sediments have sand contents under 25%. In single cases the sand content varies considerably, between about 10 and 50%. Thus, compared to the always present gravel mode, the Calabrian sand modes are a relatively unstable feature.

However, starting a discussion of how sand is being formed at this active plate margin, it has to be demonstrated that the Calabrian river-mouth sands are mainly first cycle deposits independent of the erosional budget of sandstone in the hinterland. This is definitely the case. The sand content of the river-mouth sediments correlates – or does not correlate – with the erosional budget of sandstone of the 19 basins with $r = -0.12$. Thus the sandstone deposits, mainly of the Stilo Capo d'Orlando Formation, exert no significant influence on the formation of the river-mouth sediment.

We compare the sand modes of the single environments (Fig. 10.2) analogously to the gravel treatment. The sand content increases continuously from the jointed and weathered source rocks (A), over the transitional stage (B) and the river course (C) to the river mouth (D). The shoreface area (E) is left out, because it represents only a small, least important domain of sand of the Calabrian beaches. For calculations of the sand budget the entire beach profile has to be considered. This will be done in Chapter 12.

In Table 10.2 the sand contents are listed, the mean, minimum and maximum values of the environments and the standard deviation of the samples.

It is impossible to calculate a budget of sand creation and sand transfer from environment to environment because fluvial transportation is an open system where only lag sediments are preserved and most of the fines are washed into the sea (see Chap. 12 for general calculations). However, it follows from Fig. 10.2 and Table 10.2 that only negligible amounts of sand come from the original jointed and weathered source rocks. A mean of 4.6% is simply too little to create the average 25% of the Calabrian river-mouth sediments plus the amount not recorded in these deposits. Even the role of the transitional stage with a mean of only 13.6% is not much more promising. Nevertheless, for many years we believed in these deposits and especially in the soils as the main source of sand, because the mean sand content of the soils of nearly all rock types, the limestones excluded, is 71% (Fig. 8.12). But we learned from geochemistry that the composition of the soil sands by no means becomes evident in the river-mouth sands, and the same is true for the weathered debris of the transitional stages. The geochemical composition of the river-mouth sands, on the contrary, corresponds to that of the very fresh, unweathered source rocks. As a consequence, all prefluvial stages of sand formation or accumulation in Calabria, mainly jointed and weathered source rock, regolith and soil, fail to act as a distinguishable source of sand for the river-mouth sediments.

The only solution remaining is that attrition during fluvial transportation is the most important source of sand in Calabria. This stands in contrast to the generally valid opinions (Pettijohn et al. 1987, p. 253), which favour disintegration as a main source of sand. Abrasion of pebbles, it is said, does not produce sand, only silt. This is certainly true if one considers the gravel load of relatively mature rivers, the particles of which are already selected and rounded. However, the conversion of land slide- or slope wasting masses into such pebbles by

Table 10.2. The sand content of the single environments from the source to the river-mouth sediment: mean, standard deviation of the mean, minimum and maximum sand content

	Mean [%]	St. dev.	Minimum	Maximum
A[a]	4.55	3.32	0.2	11.7
B	13.61	9.83	0.2	29.7
C_u	17.83	4.33	12.6	25.0
C_b	16.48	5.2	6.8	28.0
D_u	25.19	6.16	17.2	33.4
D_b	24.58	8.41	12.4	54.8

[a] A: Jointed and weathered source rocks (in situ); B: transition (regolith, talus debris, alluvial cone); C: river course (downstream of the fall line); D: river mouth. The indices refer to unimodal distributions (u) and bimodal distributions (b).

steeply graded mountain rivers creates enormous quantities of fines. A simple joint block, be it a cube or a cuboid, loses 47% of volume if the corners are chipped off and rounded until the sphere or ellipsoid inscribed is reached. Moreover, a "conventional" gravel never represents exactly the inscribed sphere or ellipsoid or the like of a former joint block but, if anything at all, only the small most resistant nucleus of a much larger body. Thus each river clast stands at least for the same volume, if not much more, of material lost into finer fractions. The modelling of such a fractional process (Kolmogoroff 1941) is beyond the scope of this provenance study.

The grain-size parameters listed in Table 10.3 show no clear trend that runs with the sequence of environments. This may be caused by the fact that sand, easily activated by even weak currents, reacts in a less differentiated manner on the broad spectrum of the hydrodynamic conditions of these mountain rivers than the gravel mode. In contrast to the gravel modes, however, all sand modes of both unimodal and bimodal river-mouth sediments (57 samples) are log-normally distributed, with only one exception. The sand modes of the river-course sediments are also log-normally distributed. The sand modes of the earlier stages, transition and jointed and weathered source rocks, are so weakly developed that we did not apply the fit procedure.

We interpret the sand modes of the Calabrian river-mouth sediments as a secondary property, predominantly acquired during fluvial transportation by fluvial attrition. This means that there is no chance to use the granulometric characteristics of the sand modes as a provenance signal. The continuous production of new sand during bankfull stages, its continuous reworking and selection during transportation, makes it impossible to speak of defined starting conditions that may be transmitted as an inherited property to later stages as in the case of the coarse gravel fractions.

Table 10.3. The grain-size parameters of the sand modes of the single environments from the source to the river mouth: mean grain-size, standard deviation of the mean and skewness

	Mean [mm]	St. dev. [mm]	Skewness
A[a]	0.52	1.4	0.49
B	0.56	1.3	0.66
C_u	0.64	1.1	0.69
C_b	0.56	1.2	0.49
D_u	0.62	1.1	0.51
D_b	0.54	1.1	0.29

[a] A: Jointed and weathered source rocks (in situ); B: transition (regolith, talus debris, alluvial cone); C: river course (downstream of the fall line); D: river mouth. The indices refer to unimodal distributions (u) and bimodal distributions (b).

One contradiction, however, remains that we are not able to reconcile. As pointed out in Section 10.1.1, the Rosin distribution of the gravel fraction of the jointed and weathered source rocks may be interpreted as an expression of their mechanical destruction (Ibbeken 1983). If the sand, on the other hand, is also produced by mechanical impact, by attrition, why are the sand modes almost exclusively log-normally distributed instead of being Rosin-distributed? We have no answer for this. We can only speculate that selective transport and selective deposition during and after bankfull stages shapes the expected original Rosin distributions of the crushed sand into log-normal ones. We return to this problem in the concluding section of this chapter.

10.3 The 1–20-mm Fraction (Third-Order Element)

Before we discuss the possible relations between modality and other variables we have to briefly determine Calabrian modality as such and to look for possible internal links of its different features. Figure 10.8A shows the average of 52 bimodal distributions. The gravel modes (A1) shift from 12.5–128 mm, the sand modes (A2) vary between 0.135 and 1.6 mm, but the position of the "valley" in between (A3), much more stable, ranges only between 1.25 and 4 mm. The "kink" of the 1.6–2-mm class stems from sample splitting at 2 mm.

Figure 10.8B displays the average of 22 unimodal distributions, the much less peaked modes vary between 8 and 80 mm.

As pointed out, the sand content of both bi- and unimodal distributions is well balanced (Fig. 10.8) ranging from about 10 to 50% with an average of 25%, regardless of bi- or unimodal distributions. Thus it is not the amount of sand which is responsible for modality. This is demonstrated by Fig. 10.8C. Here the difference between the two types of distribution is calculated simply by subtracting the mean unimodal from the mean bimodal distribution. It turns out that the fraction 1–20 mm is the essential one which makes the difference, in the sense of the word, between bi- and unimodal distributions. It is called, consequently, a third-order element (Ibbeken 1983, p. 1230). Figure

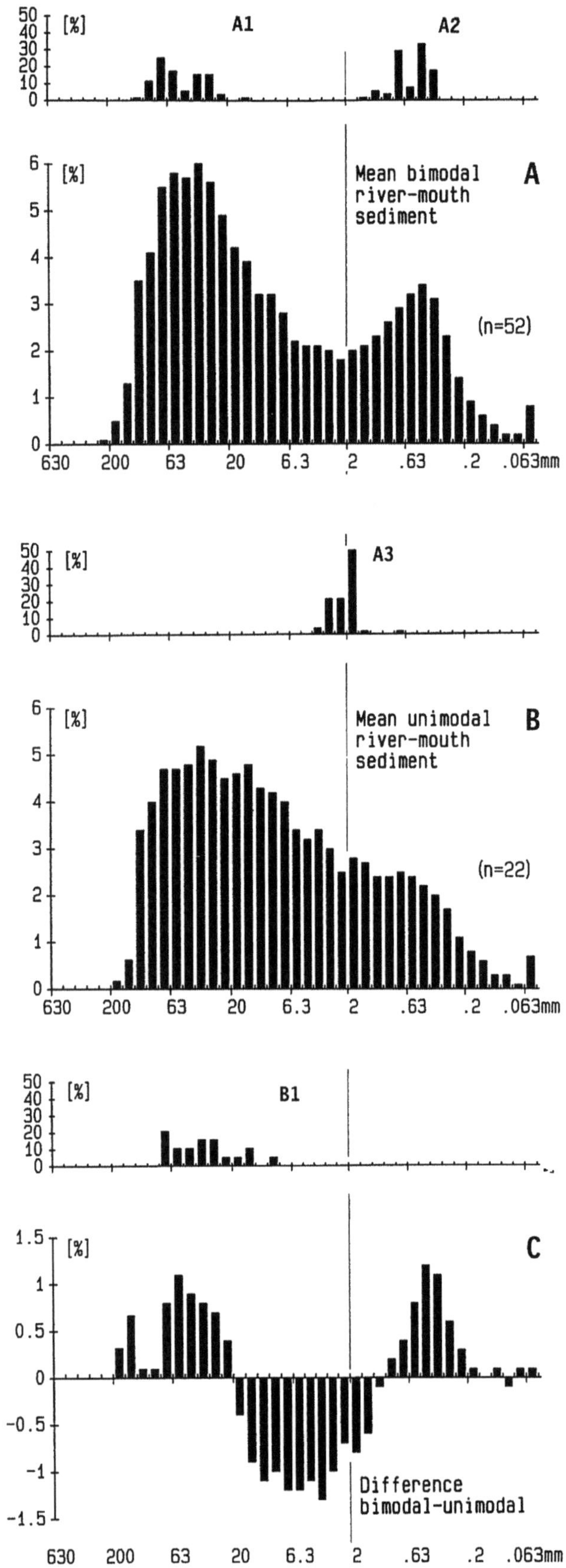

Fig. 10.8 A–C. Features of grain-size distributions in Calabria. **A** Average of 52 bimodal distributions. *A1* Distribution of the gravel modes. *A2* Distribution of the sand modes. *A3* Distribution of the "valleys" (lowest percentage) in between. **B** Average of 22 unimodal distributions. *B1* Distribution of the gravel modes. **C** Average bimodal distribution (**A**) minus average unimodal distribution (**B**). The presence or absence of the fraction 1–20 mm decides on uni- or bimodality. There is also no correlation between the position of the modes and the "valleys" or between the respective percentages; all these features vary independently from each other

10.10A shows that most unimodal rivers contain more than 35% of this fraction, the Bruzzano [06] excepted. Ibbeken (op. cit., Fig. 8C) demonstrated that it is possible to split the average Calabrian unimodal distribution into a Rosin-distribued gravel mode (first order) and a log-normal distributed sand mode (second order) and a remainder which comprises the fraction 1.6–12.5 mm, corresponding almost perfectly to the 1–20-mm interval which makes the difference between the average bimodal and unimodal distribution. We believe that this third-order element, crucial for modality, is more dependent on availability than on hydrodynamics and we shall take up this problem again in the conclusion section after having presented our source area data.

As far as the internal variables of the Calabrian grain-size distributions are concerned, we made correlations between all these to detect any relation, but they yielded no correlation coefficient with more than r = 0.35 despite the high number of 51 samples. We will follow up this argument when we discuss the possible link of modality to external variables as source area properties and others in the next sections.

10.4 Unimodal and Bimodal Distributions

The problem of bimodality or, more generally, of deficiencies of certain grain-size classes within grain-size distributions of terrigenous sediments, expressed by "kinks" of the cumulative frequency distribution, finds much interest among sedimentologists. We refer exclusively to the deficiencies of sandy gravel deposits somewhere in the range of 1 to 20 mm. Shea (1974) gives an excellent overview of the problem. The author states that most contributions to the deficiency complex are not accompanied by statistically rigorous techniques and that many theories are simply contradictory. Shea presents a histogram of an overall weighed average of 11212 analyses, which shows no deficiencies. However, many of our modern Calabrian deposits evidently display both bimodal and unimodal distributions and our knowledge of the source area conditions should lead us to a better understanding of the problem. This is of special interest, because some authors believe bimodality to be a provenance factor; but this can only be partly true. As pointed out in the foregoing sections, only the shape of the gravel mode has to be seen as an inherited element, whereas the sand mode represents an acquired one, created mainly during fluvial transportation. However, bimodal or unimodal, in Calabria this depends mainly on whether the amount of the fraction 1–20 mm exceeds 35% or not.

In the following we want to support this with the results of an extended test sampling in the river-mouth areas, with the description of bimodality itself and with the correlation of predominantly unimodal river-mouth sediments with quantitative geomorphological and geological source area properties.

10.4.1 Test Sampling

It was the objective of the test-sampling campaigns to find out if modality depends on the following circumstances:

- sampling technique (operator bias)
- position within the river bed (close to banks or midstream)
- downstream development (distance from the river mouth)
- repeated sampling over several years.

We took randomly chosen samples at certain distances from bank to bank and midstream at the walls of gravel bar cuts. The mean size of the channel samples is 271 kg (maximum grain size is about 120 mm), thus guaranteeing that we are dealing with real bulk samples cutting several sedimentation units. To demonstrate this, nearly all samples were taken as twin or double samples, that is, after having taken and treated a channel sample we continued in the same sample channel to manage the second or twin sample (Fig. 10.9).

Taking channel samples of the walls of river cuts, it is not possible to realize and to predict if a unimodal or bimodal distribution occurs (Fig. 10.1). Consequently, an operator error can be excluded.

The Amendolea [02] is one of the largest rivers of the area. The source area is dominated by metamorphic rocks, and conglomerates feed masses of recycled granitic gravels into the river-mouth sediments. The twin samples with distances of several 100 m, taken within some days under constant conditions, show surprisingly uniform sand modes, less stable gravel modes and an obviously stable bimodality (Fig. 10.9 A). From this, one is tempted to speak of evidently bimodal river-mouth sediments of this river. Two other samples taken in other years (Appendix B 02) disturb this picture, because one of these samples displays only a very weak sand mode close to unimodality.

The Buonamico [08], one of the largest and most characteristic Calabrian Fiumaras with a mostly metamorphic source area, was chosen to investigate a possible downstream development of modality. The three twin samples from upstream, 10 km from the coast (Fig. 10.9 B), display nearly unimodal distributions with relatively stable gravel modes. The midstream section, 3.6 km from the coast, shows a bimodal and a nearly unimodal distribution close to the banks as well as the same ambiguous situation in a triple sample sequence in between (Fig. 10.9 C).

The river-mouth sediments yield the same picture (Fig. 10.9 D). Thus, in this campaign, the Buonamico appears as a producer of very slightly bimodal sediments; but only during this campaign. Additional sampling in other years (Appendix B 08) in the river-mouth area show a strictly unimodal and, 1 year later, a strictly bimodal Buonamico.

The Torbido [14] is the largest river under investigation, the source area is granite- and sediment-dominated. A river of this type is commonly expected to be a powerful producer of sand. But

Fig. 10.9 A–F. Sampling test, position (●) of the randomly taken twin samples and the respective grain-size distributions. **A** Amendolea [02] river-mouth area. Relatively stable sand modes contrast with relatively variable gravel modes. **B** Buonamico [08] upstream, 10.6 km from the coast. Sand modes are only weakly developed or missing, gravel modes are relatively stable. **C** Buonamico [08] midstream, 3.6 km from the coast. As in the upstream section sand modes are only weakly developed or missing, gravel modes are relatively stable. **D** Buonamico [08] river-mouth area. As in the up- and midstream section (**B** and **C**) sand modes are only weakly developed or missing, gravel modes, however, are somewhat less stable. **E** Torbido [14] river-mouth area. South bank samples are nearly unimodal, midstream and north bank samples are not. **F** Amusa [15] river-mouth area. This small river displays the most homogeneous grain-size frequency distributions during this campaign. The basic granulometric structure of unimodality is accompanied by many small variations

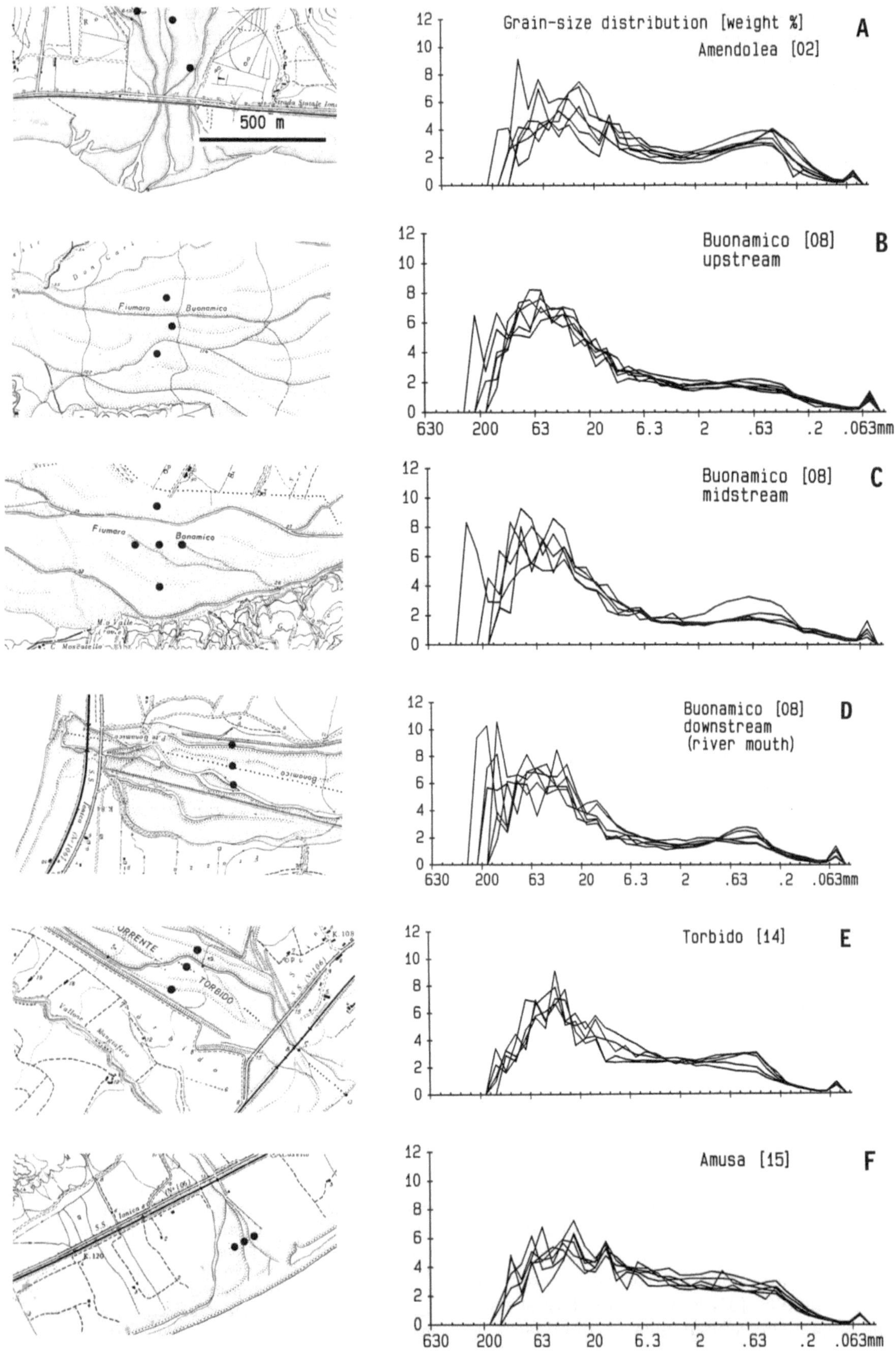
500 m
Grain-size distribution [weight %]
Amendolea [02]
A
Fiumara Buonamico
Buonamico [08]
upstream
B
Buonamico [08]
midstream
C
Buonamico [08]
downstream
(river mouth)
D
TORBIDO
Torbido [14]
E
Amusa [15]
F
12
10
8
6
4
2
0
630 200 63 20 6.3 2 .63 .2 .063mm

again, the sand modes are very modest or missing (Fig. 10.9E). The modality of these river-mouth sediments depends on where these bulk samples are taken. Additional sampling in different years makes this even more evident (Appendix B 14). Here, we obtained a weakly bimodal sample, as during the test sampling, and a year later a strongly bimodal one.

The Amusa [15] is a very small river with a predominantly granitic source area, followed by metamorphic rocks. This river displays a strong unimodal tendency over all samples (Fig. 10.9F), one sample of the SW bank excepted. The same is true for the additional samples (Appendix B 15).

The results of the sampling-test campaign demonstrate that none of the criteria listed at the beginning of this section displays a significant and mono-causal influence on bi- or unimodality. Both types of distribution may occur more or less irregularly under all circumstances. This becomes even more striking if we compare the overall grain-size distributions of the river-mouth sediments of the 1972 and 1973 campaigns, before and after the disastrous January flooding (Appendix B 01–20). Strong differences in modality from year to year signalize that nearly all rivers are able to produce both types of distribution, uni- and bimodal. From this it becomes evident that modality is not a given characteristic of a given river in space and time, and it may appear useless to look at possible links between modality and source-area conditions or other properties which have to be assumed as constant in space and time. However, the phenomenon of modality exists as such, and, as pointed out in the introductory section, our understanding should be improved at least by narrowing possible solutions at the expense of falsified ones.

10.4.2 Modality and the Source Area Variables

Looking for explanations for bi- or unimodal river sediments, we will ignore the possible long-term changes in modality of a given river and trust the results of the 2-year campaigns plotted in Appendix B 01–19. Here the river-mouth sediments of seven rivers display at least a certain unimodal tendency and an acceptable similarity between the two campaigns:

Melito	[01]	Novito	[13]
Palizzi	[04]	Amusa	[15]
Aranghia	[05]	Precariti	[17].
Bruzzano	[06]		

The Palizzi, however, was only sampled once. Suppose these seven rivers produce predominantly unimodal sediments and the other 12 rivers favour bimodal ones, we are now able, in the case of Calabria, to check most arguments in favour of or against bimodality cited in the literature. These are morphology, petrography and primary grain-size distributions of the source areas, the petrography of the river-mouth sediments itself, packing, post-depositional refill of interstices and, interpretable only indirectly, runoff- and hydrodynamic conditions.

10.4.2.1 Source Area Morphology and Modality

If runoff and hydrodynamic conditions characteristic of high-gradient mountain rivers were responsible for the uni- or bimodality of fluvial sediments, and if these runoff and hydrodynamic conditions were relatable to morphological features of the source area such as relief, dissection, basin shape or the gradient of the trunk slope, then these features, in turn, should correlate with the modality of the river-mouth sediments. Strahler (1964, p. 4–45), for example, expects low but extended peak flows from elongated basins and sharp peaks from rotund basins. High relief rivers should create river-mouth sediments different from low relief rivers. Or, current velocities of a rivers with a steep trunk slope should be higher than those of rivers with lower gradients. However, the Calabrian data do not reveal the slightest relationship between the morphology of the source area and the modality of river-mouth sediments. Figure 10.10B shows the relief of the 19 basins in terms of the altitude standard deviation, Figure 10.10C presents the hypsometric integral or dissection in terms of the altitude skewness and Figure 10.10D displays the basin shapes (Chap. 2.2.5) according to Horton's formula: on all plots the rivers with predominantly unimodal river-mouth sediments are marked. There are certain tendencies, but they are never exclusive to one modality alone. For example, all unimodal river-mouth sediments belong to elongated basins characterized by low Horton ratios. But Laverde [07], Portigliola [11] and Assi [19], with partly even lower Horton ratios, produce predominantly bimodal river-mouth sediments. Figure 10.10E shows the trunk slopes of the 19 rivers, the gradients of the main channel of highest order: these gradients vary noticeably between about 0.03 and 0.01, this is, by factor 3, but unimodality is combined with both lowest and highest gradients.

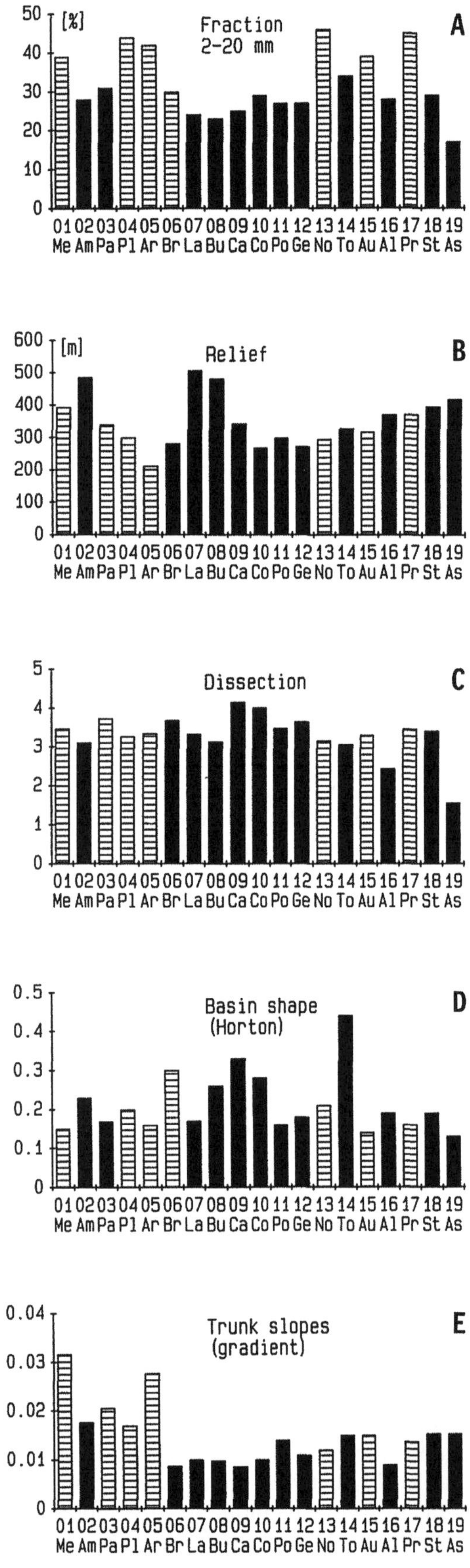

Fig. 10.10 A–E. Uni- and bimodality of the 19 rivers versus sediment and source area variables. River-mouth sediments with predominantly unimodal distribution are *hatched*, bimodal ones are marked *black*. **A** Percentage of the fraction 2–20 mm. Rivers with more than about 35% of this fraction are unimodal, the Bruzzano [06] excepted. So this fraction is crucial for modality. **B** Relief (altitude standard deviation) does not correlate with modality. **C** Dissection (altitude skewness) does not correlate with modality. **D** Basin shape R_f (Horton 1932) does not correlate with modality. **E** Slope of the trunk (river segment of the highest order) does not correlate with modality. *01 Me*lito; *02 Am*endolea; *03 Pa*squale; *04 Pal*izzi; *05 Ar*anghia; *06 Br*uzzano; *07 La*verde; *08 Bu*onamico; *09 Ca*reri; *10 Co*ndoianni; *11 Po*rtigliola; *12 Ge*race; *13 No*vito; *14 To*rbido; *15 A*musa; *16 Al*laro; *17 Pr*ecariti; *18 St*ilaro; *19 As*si

10.4.2.2 Source Area Petrography and Modality

Source rock petrography is probably the most popular argument made responsible for bimodality. The missing grain sizes – mostly granules – are said to be less frequently produced by certain rock types. There is no support for this in Calabria. Figure 13.5 A, B shows the area percentage and the erosional budget of the 19 basins in terms of granitic, sedimentary and metamorphic rocks. The two plots demonstrate some combinations of unimodality and granite-poor source areas, but not exclusively. Five "bimodal rivers" also fall into this category. We repeated the test using various combinations of metamorphic and sedimentary rocks – without results; no preferred correlation occurred. We finally plotted a combination of granite, sandstone and conglomerate of the source areas versus the modality in order to find out if these predominantly sand-producing units correlate with the modality of the river-mouth sediments – again without success. In Calabria the modality of river-mouth sediments is definitely not controlled by source rock petrography, as supposed and proposed in other regions by Hough (1942), Yatsu (1955) or Rogers et al. (1963).

10.4.2.3 Source Area Grain-Size Distributions and Modality

Wolcott (1988) proposed the size distribution of the input material as being largely responsible for the size distribution of channel deposits. It is hard to explain how such a hydrodynamically delicate structure like bimodality is maintained during fluvial transport over several kilometers (Chap. 10.4.2.5) – in the case of Calabria it has been known for years (Ibbeken 1983, Fig. 4, Table 1) that most jointed and weathered source rocks are extremely unimodal and most river-mouth sediments are strikingly bimodal. Figure 10.2 displays the development of modality through the five environments. The biomodality of river-mouth sediments in Calabria is thus definitely not controlled directly by the granulometric properties of the source rocks.

10.4.2.4 Petrography of River-Mouth Sediments and Modality

Very little attention has been paid to the petrography of the sediments itself as a possible reason for modality (Davies et al. 1978). However, in Calabria this question also leads into a blind alley. Figure 13.5 illustrates the petrographic composition of the 19 river-mouth gravels of the fractions 16–80 mm and 1–20 mm respectively. There is definitely no correlation between the character of modality and any petrographic composition of the two gravel fractions. This is especially important in the case of the range of 1–20 mm, the third-order element, because this fraction makes the main part of the mysterious "gap" which defines modality. The same is true for the petrographic composition of the sand mode. Figure 14.5 shows the light mineral indices of all 19 river-mouth sediments. The content of rock fragments or the quartz/feldspar ratio may account for the breakdown of certain fractions and may be responsible, in this way, for modality – but the graphs do not reveal the slightest correlation between these or other compositional parameters and modality. So even the petrographic composition of the river-mouth sediments itself can definitely be excluded as a possible factor controlling modality – in Calabria.

10.4.2.5 Packing and Refill of Interstices and Modality

In 1935 an important statement concerning our topic was made by Fraser, barely cited thereafter (Plumley 1948, p. 542; Pettijohn 1957, p. 45). Fraser (1935, p. 987) says: "... it is difficult to understand how a river capable of carrying all sizes up to large boulders can deposit them at the same time. At any instant, the largest particle that may be moved varies at the sixth power of the current velocity". Fraser calculates that the velocity of a stream capable of carrying 250-mm pebbles has to decrease 60% before 1 mm sand could be deposited. "Hence it seems improbable that large pebbles and fine sand are necessarily deposited simultaneously", and "the periods during which river gravels may be moved, sized and sorted are usually brief and are followed by much longer quiescent periods during which infiltration and arrangement of the finer sizes may be accomplished" (op. cit., p. 984). Fraser calculates the effects of various gravel grain sizes enabling finer fractions to enter the framework after deposition and defines a "critical ratio of entrance" (op. cit., p. 919). Plumley (1948, p. 544) agrees with Fraser and found, by theoretical considerations, ratios between 22 and 32% for the sand content of gravels. Plumley found this confirmed by his samples and states that the weight ratios of the infilling fractions increase with decreasing mean size of the sediment.

As pointed out in Section 10.3, we checked this and all other features of the grain-size distributions and found no correlation: the grain-size classes of both gravel and sand mode, the "valley" in between, the sand content and finally the percentage of the third-order elemént, the fraction 1–20-mm, all vary independently among themselves.

Only the mean sand content of 25% fits nicely to the sand contents claimed for filling interstices, but this, unfortunately, is irrelevant for uni- or bimodality.

10.4.2.6 Hydrodynamic Conditions and Modality

Many authors believe in selective transportation due to different hydrodynamic behaviour of different grain sizes as a main factor controlling modality. These arguments are largely refuted by Shea (1974, p. 1000). The most-cited authors in this context are Sundborg (1956) and especially Russel (1968), who argues "that the fractions 1–6 mm are more readily entrained and more rapidly transported than grains of larger or smaller size. Currents, mainly of wave origin, concentrate these grains in beaches, where they are present in more-than-ordinary abundance" (op. cit., p. 31). The first problem which arises accepting the idea of Russel is that the Calabria grain-size gap is much larger than from 1–6 mm. As pointed out by Shea (1974, p. 1000), no hydrodynamic mechanism is known which favours the entrainment and transportation of grain sizes between 2 and 20 mm. We are not able to check the hydrodynamic qualities of this argument, but we do have the bulk grain-size distribution of the beach to check the consequences claimed by Russel. This sample (Fig. 10.11) represents the average of 32 excavator samples, 7 tons as

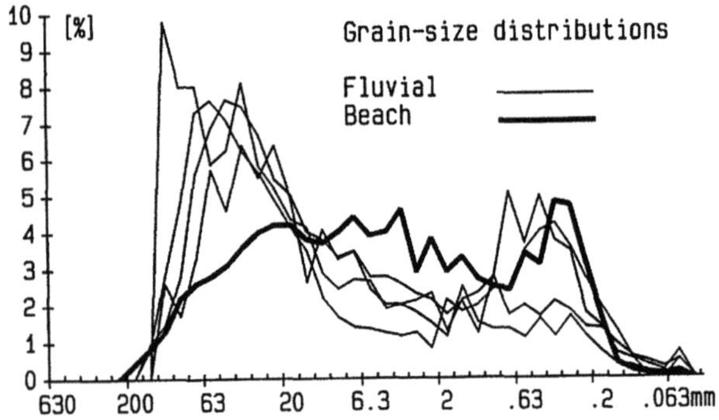

Fig. 10.11. Comparison of fluvial and coastal grain-size distributions, Laverde [07] and Bianco beach. The Laverde is the main feeder of the Bianco beach. The four Laverde samples stem from the 1970, 1972, 1973 and 1983 campaign (*light lines*). The *heavy line* refers to the Strand Profil, a continuously sampled beach-normal profile 70 m long from 2 m above to 4 m below sea level, comprising a total of 7 t. The profile tries to bridge the entire beach in motion (Chap. 12.2.1)

a whole, of a 70-m profile of the entire beach in motion, normal to the coast from the backshore 2 m above sea level to the offshore strip 4 m below sea level (Chap. 12.2.1). From the Laverde river mouth [07] which feeds this beach we have four samples, from 1970, 1972, 1973 and 1983. The single fluvial distributions vary between nearly unimodal and strongly bimodal. The coastal distribution admittedly fills the fluvial gap (Fig. 10.11) if referred to the bimodal distributions. However, the coastal sand mode surmounts the gravel mode slightly, and there is also a deficiency, somewhat shifted towards the fines. Thus from the merely descriptive point of view Russel is not wrong, but the question remains whether the fill of the gap in the coastal realm stems from increased fluvial input or simply from attrition of the coarse mode during longshore transportation. We will come back to this problem in Chapter 12.2.1.

As a result, none of the reasons assumed to be responsible for uni- or bimodality can be accepted in Calabria. Our own interpretation focusses attention on attrition and selective transport or deposition, as already proposed by Shea (1974, p. 1001).

10.5 Conclusion

We return to the synoptic comparison of the river-mouth sediment's grain-size distributions of the 1972 and 1973 campaign (Appendix B 01 – 20). The basic or first order feature present in all samples is the positively skewed gravel mode. We interpret this Rosin-distributed gravel mode as an inherited, primary source rock element (Ibbeken 1983, p. 1227), which is maintained during fluvial attrition and selection until the river-mouth areas are reached. It goes without saying that during this development the gravel mode is modified, shifting towards the fines and flattening. Thus the gravel mode carries both a compositional and a textural provenance signal.

The other important or second-order feature is the sand mode of the bimodal distributions, usually log-normally or Gauss distributed. This mode may be very pronounced (Careri [09] and Condoianni [10]) or nearly absent (Melito [01], Palizzi [04] and Precariti [17]). Our geochemical data urge us to assume that the overwhelming portion of fluvial sand does not come from the jointed and weathered source rocks or soils, or from disintegration or decomposition, but from attrition mainly during the floods of the high-gradient mountain rivers. This is the only way to create sand whose geochemical composition is fresher and more original than that of the debris of slopes and soils. We do not know why the sand modes of the bimodal distributions almost exclusively display log-normal instead of Rosin distributions as expected from the process of sand formation by attrition. Log-normality is usually seen as a transport-specific feature. Obviously the movement of sand, much more frequent than that of gravels, is sufficient to realize a Rosin-Gauss transformation. Thus the sand mode carries only a compositional or petrographic provenance signal, but not a textural one.

Bimodality, in Calabria, may be simply explained as the result of mixing between these different modes, gravel and sand. Ibbeken (1983, Fig. 10) mixed an ideal, coarse-grained Rosin distribution and an ideal fine-grained log-normal distribution in accordance with the grain sizes and parameters of the average Calabrian river-mouth sediment, 70 and 30%, and yielded a bimodal distribution which corresponds nearly perfectly to that of the mean bimodal river-mouth sediment in Calabria. We believe from this that the bimodal Calabrian river deposits originate from gravel modes with a long fine tail, superimposed by a sand mode.

This picture is seriously disturbed by the presence of unimodal distributions. These are characterized by amounts of more than about 35% of the fraction 1 – 20 mm, the third-order element. The crucial question is what controls the appearance or disappearance of this grain-size interval. We falsified and excluded all source area variables and all variables of the sediment itself, as petrographic composition or grain-size parameters, so the only variables not directly tested which remain are hydrodynamics and availability. We believe that hydrodynamics can also be excluded. As pointed out in Section 10.4.2.5, there is no reason why the fraction 1 – 20 mm should be selected somehow by hydrodynamics. The complex process of attrition during bankfull stages of mountainous rivers has not been solved yet. Consequently, the following can only be a matter of belief: the process of crushing during floods differs, in space and time, between bed load, transitional stages and suspended load. The suspended load, at any given time, is to a great extent protected against further crushing, the bed load is not. Thus floods of varying intensity may produce varying grain sizes, and the third-order element, the fraction 1 – 20 mm, may be especially susceptible to these differences, being destroyed or not, so that bi- or unimodal lag sediments result. We are forced to speculate on this topic, availability via attrition, in order to explain the confusing variability of modality in space and time: all other variables cited can definitely be ruled out, in Calabria.

Chapter 11

Development of Gravel Sphericity

We investigated Calabrian gravel sphericities both intensively and extensively and, as a result, obtained uniform and meaningful results, especially the dependence of sphericity on grain size and on petrography and the independence from environment. Thus the value of sphericity and shape as a provenance signal remains very limited, or better, it does not exist. If, in the geological record, a fluvial gravel should be preserved, its fluvial origin may be determined much easier by other features of its macrofacies than gravel sphericity. The same may be true for most coastal deposits.

Summary. The axes of more than 40000 pebbles were measured. Sphericity is strongly dependent on lithology and increases with grain size. It increases from the jointed and weathered source rocks towards the river-mouth sediments and decreases along the beach, seen as a whole, that is, the sphericity distribution of the coast returns exactly to its starting pattern or joint block origin. So sphericity fails to be a useful provenance signal.

Data Base. The field sieving work was continuously accompanied by measurements of the axes of the clasts of selected rock types and size fractions. The final sum, measured over many years and by many operators, adds up to 42211 clasts measured. The data were stored in the data bank.

Terminology. *Sphericity.* According to Sneed and Folk (1958, p. 118), sphericity psi_r (maximum projection sphericity)

$$psi_r = \sqrt[3]{S^2/LI} \; ,$$

whereby L is the long, I the intermediate and S the short axis of the enveloping cuboid.

Shape. This is a descriptive term characterizing different proportions of the axes, for example compact, platy and elongate (Fig. 11.2).

Rock Types. Rock types of the routine measurements are listed in Table 11.1.

Method. Clast shapes were measured by means of a primitive measuring box (Ibbeken and Denzer 1988). The device measures the three axes of the enveloping cuboid.

Table 11.1. Mean sphericity and standard deviation of the gravel populations of all environments, of the most important rock types and grain-size classes, number of samples

Environment	Rock type	Grain size [mm]	Sphericity		No. of samples
			Mean	St. dev.	
All	All	All	0.660	0.228	41211
Sphericity versus grain size					
All	All	16–31.5	0.638	0.231	20012
All	All	31.5–63	0.680	0.226	9240
All	All	>63	0.700	0.222	1739
Sphericity versus rock type					
All	Limest.	All	0.722	0.218	2426
All	Granite	All	0.710	0.221	3222
All	Sandst.	All	0.681	0.225	1511
All	Marble	All	0.652	0.231	1504
All	Gneiss	All	0.619	0.234	18221
All	Schist	All	0.493	0.245	1240
Sphericity versus environment					
Source	All	All	0.637	0.232	1630
Transition	All	All	0.696	0.223	702
Riv. course	All	All	0.696	0.222	4974
Riv. mouth	All	All	0.702	0.222	6326
Shoreface	All	All	0.663	0.228	18011
Backshore	All	All	0.609	0.235	9568
Sphericity versus main environments					
Source + Transition	All	All	0.655	0.230	2332
Riv. course + Riv. mouth	All	All	0.699	0.222	11300
Backshore + Shoreface	All	All	0.644	0.230	27579
Sphericity of pebbles of only one rock type (gneiss) versus environment					
Source	Gneiss	All	0.562	0.240	416
Riv. course	Gneiss	All	0.600	0.237	753
Riv. mouth	Gneiss	All	0.655	0.229	1074
Shoreface	Gneiss	All	0.634	0.232	10996
Backshore	Gneiss	All	0.586	0.238	4942

11.1 Overall Trends

The overall distribution of the entire data set (Fig. 11.1 A) displays a very well-balanced sphericity frequency distribution with a mean sphericity of $psi_r = 0.66$, minimum is $psi_r = 0.275$, maximum is $psi_r = 0.975$. The curve is slightly negatively skewed, indicating a slight excess of low sphericity values which stems from schist and gneiss. Table 11.1 summarizes the data set listing environments, rock types, grain-size intervals, mean and standard deviation of sphericities and number of measurements. Standard deviations are relatively high, about 0.228 psi_r, but very similar comparing the single groupings of Table 11.1.

11.2 Sphericity Versus Grain Size

We measured grain sizes from 16–80 mm in selected intervals of 1/3 phi. Most rock types are not fine-grained enough to yield homogeneous particles in size classes <16 mm. Sphericity increases with increasing grain size (Fig. 11.1 B). The sphericity distribution of the 16–31.5-mm fraction is broad and flat whereas the coarser fractions, 31.5–63 mm and >63 mm, yield narrower, more peaked distributions; the mean sphericity climbs from $psi_r = 0.638$ over 0.68 to 0.7. Consequently, the comparison of shape measurements has to be done within defined size classes only.

11.3 Sphericity Versus Rock Type

Here the most striking differences occur (Fig. 11.1 C). Granite and limestone clasts display a very similar sphericity distribution with the highest mean sphericities $psi_r = 0.722$ and 0.71 respectively, followed by sandstone (0.682), marble (0.652) and gneiss (0.619), a sequence which reflects exactly the decreasing homogeneity or increasing textural and structural anisotropy of these rocks, due to layering, foliation and cleavage. The very platy phyllites form an almost isolated sphericity distribution characterized by a mean of only $psi_r = 0.493$. The relevance of this feature for transportation and, finally, provenance is evident. Clasts of different shapes are differently entrained by flowing water (Komar and Li 1988), but also sliding, rolling and saltation during further transport depend partly on shape and decide on advanced transportation or not. Moreover, it depends

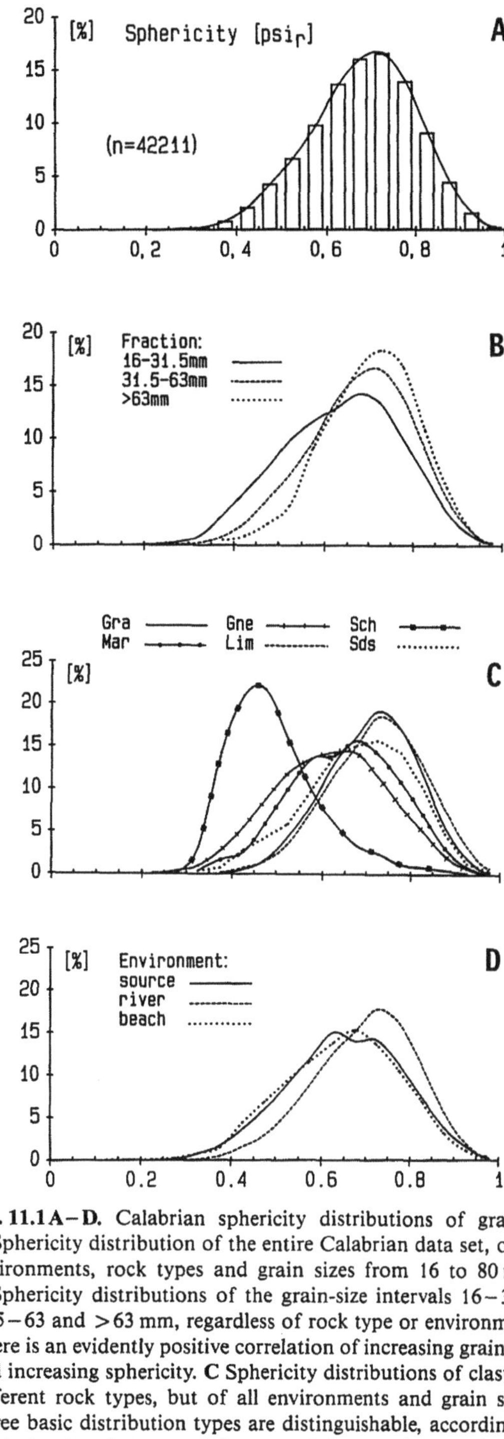

Fig. 11.1 A–D. Calabrian sphericity distributions of gravels. **A** Sphericity distribution of the entire Calabrian data set, of all environments, rock types and grain sizes from 16 to 80 mm. **B** Sphericity distributions of the grain-size intervals 16–31.5, 31.5–63 and >63 mm, regardless of rock type or environment. There is an evidently positive correlation of increasing grain size and increasing sphericity. **C** Sphericity distributions of clasts of different rock types, but of all environments and grain sizes. Three basic distribution types are distinguishable, according to the textural and structural features of these rocks: *gra*nite and Jurassic *lim*estone with relatively isotropic textures display highest mean sphericities, lowest values present the *sch*ist clasts characterized by strong cleavage. *S*and*s*tone, *mar*ble and *gnei*ss form clasts of intermediate sphericities. **D** Sphericity distributions of clasts of all grain sizes and rock types in the main environments source, river and beach. There is a striking trend reversal, because the end members, source and beach, present identical distributions whereas the river system obviously favours higher sphericities

on the lapse of time of a flood event if, at the river mouth, such a possibly advanced population of clasts becomes relatively enriched or impoverished. Add to this that this process is superimposed by different attrition of different rock types. Consequently, it is hard to decide which of these variables is the most responsible for the composition of the river-mouth sediment and, by this, for a provenance calculation. Figure 13.5 shows the source area composition of each of the 19 basins, the effective composition which equals the predicted sediment composition and the composition of the river-mouth gravel of both the 4–16-mm and the 16–80-mm fraction, all this in terms of granitic, metamorphic and sedimentary rocks. Going from the source to the river-mouth gravel, most trends move away from the sedimentary units of the source area; this is a strong tendency towards metamorphic and granitic rocks, the latter being evidently favoured. The miscarriage of sedimentary clasts being represented in the river-mouth sediments is evidently caused by attrition. The preferential treatment of granites, compared to metamorphic rocks, may be at least influenced by sphericity, this is, the more compact shapes and higher sphericities of granitic clasts. However, we are not able to quantify and to prove this.

11.4 Sphericity Versus Environment

The sphericity development from the source over the rivers to the beach undergoes a striking reversal (Fig. 11.1 D). The sphericity distributions of the end members, source and beach, are nearly identical with mean values $psi_r = 0.655$ and 0.644, whereas the fluvial system in between displays a more peaked sphericity distribution shifted by about 0.05 psi_r-step towards higher sphericities, regardless of if the river courses or the river mouths are concerned. The mean sphericity is $psi_r = 0.699$. However, the coastal distribution reflects only the average, disregarding the strong zonation of this area (Chap. 12.4.2.2). It is this similarity of sphericities which presents further proof for the well-known fact that sphericity cannot be used as an environmental signal.

11.5 Conclusion

In Calabria, sphericity and shape of gravels are not provenance signals. The overall sphericity distributions of various grain sizes, rock types or en-

Fig. 11.2. Shape triangle (Sneed and Folk 1958). The clast shapes corresponding to the mean sphericities approximately cover the *hatched area* on the ternary diagram. *S* Source; *R* river mouth; *B* beach

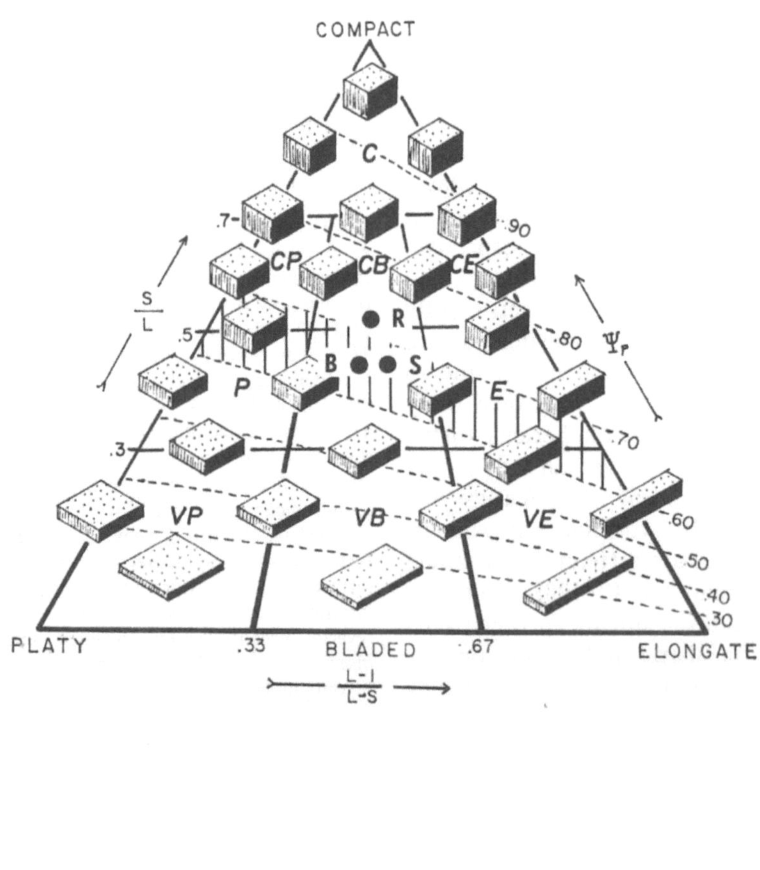

vironments demonstrate that very large samples are needed to obtain reliable results and that these results fail to make sphericities of single samples an interpretable provenance factor.

This may be illustrated by Fig. 11.2, the ternary shape diagram: this classical diagram, introduced by Sneed and Folk (1958, Fig. 6) shows the girdle of most common sphericities and clast shapes in Calabria. The average clast from the source (S), the river mouth (R) and the coast (C) is also marked. The neighbourhood of these values is conclusive proof of the uselessness of shape or sphericity as a provenance factor. Techniques such as the oblate-prolate index introduced by Dobkins and Folk (1970) in order to refine shape differentiation fail in the case of Calabrian pebbles, because the center of the ternary diagram is barely spread by this procedure.

Chapter 12

Special Topics of Sediment Development of Lower River Course and Beach

River-mouth sediments definitely represent the final stage of the fluvial development and the initial stage of many marine sediments. The main interest of our study is focussed on the relation source/river-mouth sediment. The many ways of further marine differentiation are more difficult to grasp. In terms of provenance, the end product "river-mouth sediment" is well definable, but beach sediments, for example, are not. Discontinuous input by floods and longshore transportation create a complex steady state in time and space and make it impossible to define a reference point "beach" comparable to the reference point "river mouth". Or, in other words, a fluvial basin is a kind of converging system, the information "basin" is compressed towards the river mouth. Compared to this, the marine realm has to be seen as a diverging system, the information "river-mouth sediment" is differently spread over different pathways, thus being much more mutilated. We want to follow one of these pathways, the coast, in order to quantify some developments away from the river-mouth sediment as we did towards the river-mouth sediment in the foregoing chapters.

Summary. *Grain size.* We sampled the entire spectrum of grain sizes. Grain-size distributions vary only little longshore but strongly normal to the beach. This is proven by 13 large samples taken in the foreshore/shoreface strip with 800 m distances and by 32 samples from a beach-normal profile (Strand Profil) from 2 m above sea level down to 4 m below, more or less continuously sampled. We were unable to find "the" grain-size distribution typical of beaches. The overall grain-size distribution of the entire beach in motion comes very close to the mean river-mouth sediment, being shifted, however, towards the fines by one phi step. The gravel modes of the foreshore/shoreface strip are predominantly Gauss-distributed, in contrast to the chiefly Rosin-distributed river-mouth gravels. This may be seen as the only grain-size characteristic predominantly developed on the beach – in Calabria.

Rock Type. Not to get lost in details, we concentrate on one rock type, "dark gneiss", and the fraction 20–31.5 mm only, when we describe the petrographic differentiation and the development of sphericity and roundness. The amount of "dark gneiss", a fine-grained, low resistant rock type, drops from about 55 to 45% during final fluvial transportation (7.5 km) and from about 45 to 20% during longshore transportation (9 km).

Sphericity. Mean fluvial sphericity decreases from $psi_r = 0.61$ to 0.57. Along the beach, mean sphericity drops in the direction of net transportation, strongly along the backshore strip, less evidently along the shoreface and offshore strip (distance as above). The Strand Profil displays an amazing regular, unidirectional trend of increasing sphericity from land to sea. Sphericity is no provenance signal.

Roundness. This property, measured as equivalent of modified Krumbein pebble-images via Fourier analysis increases from $r_{FO} = 0.48$ to 0.57 during fluvial transport and from $r_{FO} = 0.6$ to 0.8 during longshore transport (distances as above). Roundness is not a sensitive provenance signal, end members of very bad and very good rounding not being taken into account.

12.1 Laverde, Buonamico and the Bianco Beach

The Bianco Beach (Figs. 2.3 and 12.1) is limited by the Laverde [07] in the south and the Buonamico [08] in the north. We sampled the lower courses of these rivers, 7.5 km of the Laverde trunk and 6 km of the Buonamico, see Appendix B 07 and 08 for the characteristics of these rivers.

The Bianco beach, 9 km long, is quite unique (Fig. 12.1); it has to be seen as a well-definable natural laboratory: the sediment supply, gravel and sand, comes nearly exclusively from the Laverde [07] (Chap. 16.4.1) and its delta (Fig. 16.8). The net longshore transportation is northward, Capo Bruzzano south of the Laverde serves as a barrier preventing the access of additional material from the south. There is evidently only little input from the Buonamico river-mouth area in the north because the Buonamico [08] has no delta and delivers its bedload, during floods, almost completely and immediately into the head of its submarine canyon.

The Bianco Beach is paralleled by a longshore trough, maximum depth 10 m, and a single longshore bar. Figure 12.1 shows the state in August 1977. Immediately south of Bianco, 2.5 km N of the Laverde delta, a beach lab was run for several years, the so-called Strand Profil. This profile bridges the entire beach segment evidently in mo-

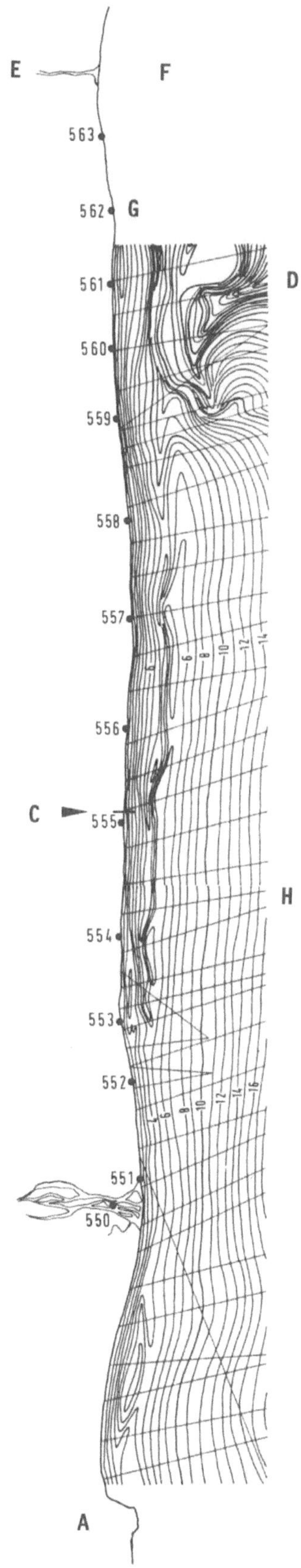

tion even during storms; the transition to the more distant offshore areas is admittedly uncertain. The profile is 72 m long, running from 2 m above sea level down to water depths of 4 m, the coastline is exactly halfway (Fig. 12.2).

The influence of tides is negligible, we measured a semi-diural tide of about 25 cm. Swell comes mainly from the SE, with significant wave heights up to 2.5 m, the Bianco beach is a wave-dominated high energy environment (Heward 1981; Davis and Hayes 1984).

We are thus confronted with a distinct fluvial point source and a distinct net transportation northward, necessary prerequisites to quantify coastal differentiation.

Data Base. Thirteen large samples from longshore (shoreface only), 1983 campaign, 10 triple samples (backshore, shoreface, offshore), 1988 campaign and 32 samples from the Strand Profil, the beach-normal profile, sieved in 1/3 phi steps, 1988 campaign; data stored in the data bank. Percentages of the dark gneisses (20–31.5-mm fraction only).

Terminology

Beach. We follow the conventional terminology (Bates and Jackson 1987) and call the backshore the zone above the reach of high tide or normal wave activity, usually dry. Foreshore is the zone of uprush or swash and backwash and/or within the tidal range. Shoreface is the strip following seaward of the foreshore, usually characterized by a distinct angle of the beach profile, continuously water-covered, the zone beneath the breaking wave and consequently the locus of highest sediment movement. Offshore finally is the continuation seaward; its outer limit is hard to draw. We stopped sampling at the end of the gravel carpet, this may be within 5 to 35 m off the shoreface strip (Fig. 12.2A).

Fig. 12.1. Bianco Beach and longshore bar, see Fig. 1.3 for the geographic setting. The distance *B–E* is 8 km, north on top. *A* Capo Bruzzano serves as a barrier preventing the access of exotic debris to the Bianco Beach from the south; *B* The Laverde and its delta (cf. Fig. 16.8). Both coast and longshore bar are fed by the delta. The net longshore transportation is directed northward. The longshore bar moves into the tributary of the Buonamico-Careri Canyon (*D*), the coastal debris moves further north and is lost, in part, in the main canyon head (*F*). *C* Position of the "Strand Profil"; *D* Canyon tributary; *E* Buonamico [08] river mouth; *F* Head of the Buonamico-Careri Canyon; *G* Excavator samples (Spannagel 1985; Köster 1986). Echosounding was done in July 1976 with a Fahrentholz shallow water echosounder by Ibbeken and Hofmann (Hofmann 1980). *H* Main echosounding tracks

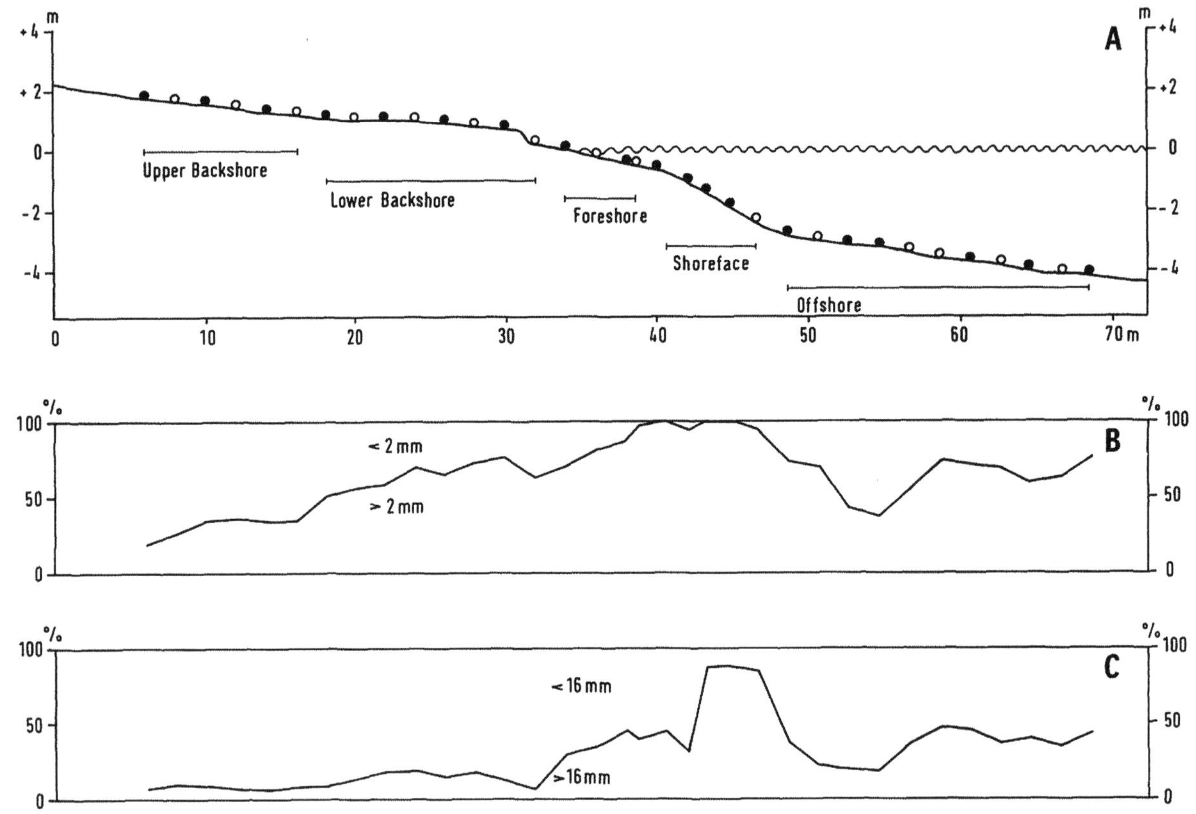

Fig. 12.2 A–C. The beach-normal profile (Strand Profil), subenvironments, gravel and sand content. See Fig. 12.1 for the position of the profile. **A** Morphology, terminology of subenvironments and sample position. *Open* and *full circles* indicate samples evaluated for grain-size distribution (Fig. 12.3), *full circles* indicate samples evaluated for the elimination of dark gneiss and sphericity. **B** Beach-normal development of the proportion sand/gravel. **C** Beach-normal development of the proportion <16 and >16 mm

Rock Type. The "dark gneiss" is a fine-grained, mica-rich and strongly foliated gneiss.

Sphericity and Shape. See terminology of Chapter 11.

Roundness. Slightly modified Krumbein (1941) roundness values.

Method

Sampling. All land samples, shoreface included, were taken by a heavy shovel excavator yielding a mean sample weight of 300 kg. The longshore samples (Fig. 12.1) were taken with about 800 m distances, the beach-normal Strand Profil (Fig. 12.2 A) was sampled continuously. The samples are defined by the movement of the excavator shovel, a half-circle 2 m in diameter and 1 m deep. The offshore samples were dug by divers operating from a pontoon. Maximum sample depth is 50 cm, so these samples do not correspond exactly to the more proximal excavator samples. The samples of the beach-normal profile add up to 7 t. For the determination of grain-size distributions each sample was evaluated, for that of petrography and sphericity only each second one.

Clast shapes were measured by means of a primitive measuring box (Ibbeken and Denzer 1988). The device measures the three axes of the enveloping cuboid. Roundness was determined evaluating photo images of the clasts via Fourier analysis (Diepenbroek et al. 1991).

12.2 Longshore and Beach-Normal Development of Grain Sizes

12.2.1 The Beach-Normal Profile

It goes without saying that it is impossible to capture the entire beach in motion and to compress it into a single, definite grain-size picture. Even if such a snapshot were possible, no conclusion could be drawn on the overall mass balance, because the single segments of the Bianco beach, back- and foreshore, shoreface and offshore move with different velocities. However, we believe that we come close to this picture because our sample profile comprises almost continuously the entire moving beach, from the scarcely vegetated backshore to the offshore limit of the gravel carpet. Both longshore trough and longshore bar (Fig. 12.1), however, are excluded from these considerations, as they are not coastal, but shelf features.

Figure 12.2A shows the morphology and the sample sites of the beach-normal profile. The gravel content increases steadily from the back- to the shoreface, from some 10 to 100%, decreases seawards rapidly to about 50% and ends abruptly. The coarse gravels >16 mm follow this only roughly, with a distinct peak in the shoreface area (Fig. 12.2B, C).

Figure 12.3 shows the grain-size frequency distributions of the beach-normal profile. Each beach strip is characterized by very similar distributions, even in the very large and deep samples which ignore any bedding structure.

Upper Backshore (Fig. 12.3A). The six samples are very uniformly distributed; this is the uppermost beach close to the scarce vegetational cover, sand-dominated (the mode is 315 μm), thus strongly negatively skewed, with some shingle from earlier storm events.

Lower Backshore (Fig. 12.3B). The eight distributions of the lower backshore are nearly symmetrical, with a 2-mm mode. This may be the locus and the kind of distribution that Russel (1968) assumed to be typical for beach sediments when he developed his model explaining the bimodality of fluvial sediments.

Foreshore (Fig. 12.3C). The four samples of this narrow strip dominated by the breaking waves display an amazing similarity with the mean Laverde river-mouth sediment (Figs. 10.11, 12.4), however, shifted entirely towards the fines by almost 2 phi steps.

Shoreface (Fig. 12.3D). This is evidently the high energy strip of the beach, characterized almost exclusively by gravel and only little sand. The mean gravel mode is at about 32 mm. The variability of the single gravel modes is high, but all distributions

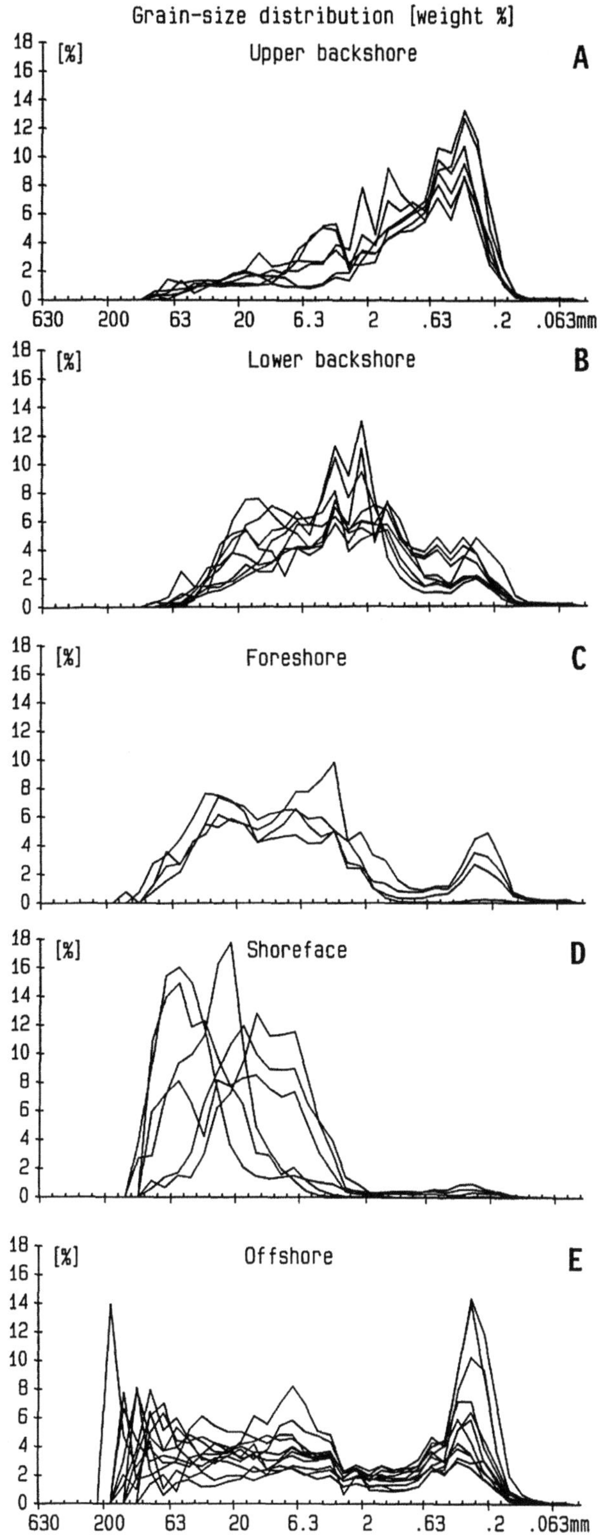

Fig. 12.3A–E. Grain-size distributions of the beach-normal profile (Strand Profil). See Fig. 12.1 for the position and Fig. 12.2 for terminology. Grain-size distributions vary only little within the single subenvironments but strongly between: **A** Upper backshore. **B** Lower backshore. **C** Foreshore. **D** Shoreface. **E** Offshore

share in an abrupt cut towards the fines somewhere in the 2–4-mm range.

Offshore (Fig. 12.3 E). This family of ten samples also shows very similar trends. Peaked gravel modes disappeared and made room for a kind of broad plateau of gravel classes, a 1–2-mm valley is suggested and followed by some very peaked sand modes. These are the increasing fine sands from the shelf and the longshore bar which tend to veil the gravel carpet following a storm event.

If we accept the Bianco beach as an example of a wave-dominated sandy gravel beach with a unidirectional net transportation, then we have to take note of the fact that these beaches have no single and distinctive grain-size-type distribution but a broad variety of both positively and negatively skewed and also symmetrical distributions which cover a size range from cobbles to fine sand. Many of them are identical with fluvial grain-size distributions. This means that mere grain-size distributions of these environments cannot serve as a provenance signal.

Figure 12.4 shows the overall average grain-size distribution of the Strand Profil. This overall distribution may be seen as a first approach to describe the entire sediment strip of a sandy gravel beach in motion. Certainly we are dealing with a very unbalanced cut, because each strip or subenvironment of the profile must move with different velocities longshore, as evidenced by the different trends of changing petrographic composition and sphericity. However, the similarity of the two distributions surprises. The fluvial distribution has to be shifted by only one phi step to obtain maximum congruence (Fig. 12.4 B). Certainly, not only the position but also the volumes of the modes differ noticeably between the two environments, as demonstrated by subtracting the coastal histogram from the shifted fluvial one (Fig. 12.4 C), but the bimodality is preserved, and this despite the many sedimentological differences between the two environments. We have no explanation for this.

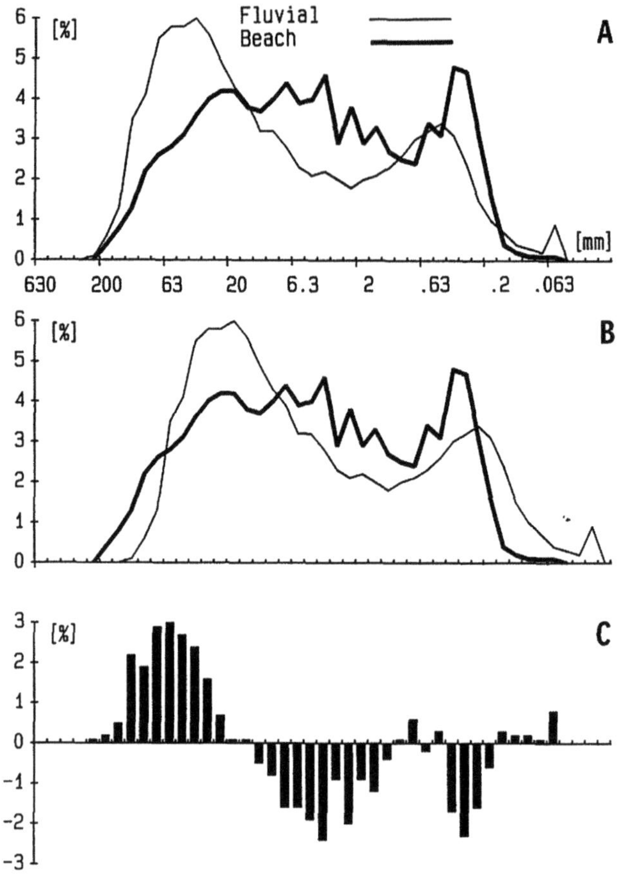

Fig. 12.4 A–C. Comparing the shape of overall grain-size distributions from the fluvial and coastal environment. **A** Mean of the 1972 and 1973 campaign of the 19 river-mouth sediments; mean of the Strand Profil, the continuously sampled beach-normal profile 70 m long from 2 m above to 4 m below sea level. Intervals are 1/3 phi steps. **B** Same as **A**; the fluvial distribution, however, is shifted by 1 phi step towards the fines to obtain maximum congruence and comparability of the distribution shape. **C** Difference between the distribution shape "river" minus the distribution shape "coast" (from **A**). The coast also displays a gravel mode, however, finer and less peaked, which occupies parts of the "2-mm valley" of the river-mouth sediment and which is topped by a somewhat finer and more peaked sand mode

12.2.2 The Longshore Section

The samples presented in this section stem exclusively from the shoreface/foreshore strip (Fig. 12.2 A). It was beyond our possibilities to sample the other strips adequately.

The longshore variations of the grain-size frequency distributions along this foreshore/shoreface strip are very limited. Figure 12.5 shows the ensemble plot of the 13 frequency distributions, Fig. 12.6 the longshore development of mode and mean of both the gravel and sand modes. The grain-size class of the sand mode varies the most (Spannagel 1985; Köster 1986), followed by the variation of the gravel mode, which decreases leaving the Laverde delta. All other variables, however, remain more or less constant along this longshore section, regardless of if we look at the gravel mode or the sand mode. The most striking contrast to the river-mouth sediments is the symmetry of the gravel

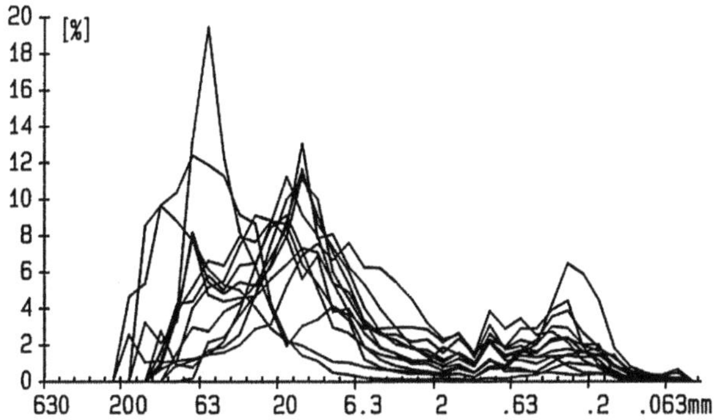

Fig. 12.5. The longshore profile. Ensemble plot of the grain-size distributions of 13 samples of the foreshore and shoreface strip, see Fig. 12.1 for sample positions (Spannagel 1985; Köster 1986). Even within given subenvironments the distributions vary only little longshore

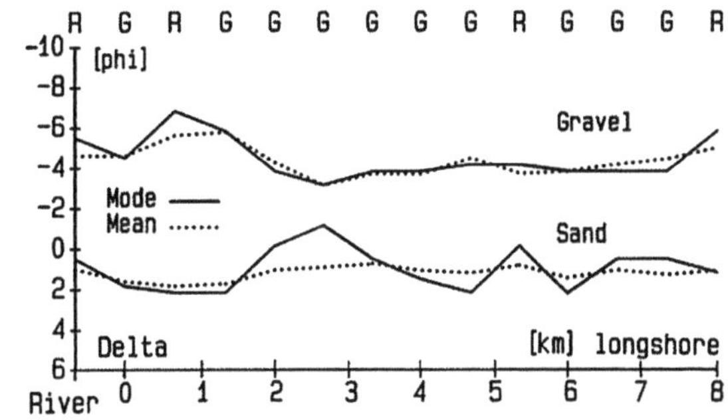

Fig. 12.6. The longshore profile. Development of mode and mean of both the gravel and sand modes of the 13 samples, see Fig. 12.1 for sample positions (Spannagel 1985; Köster 1986). Sample "River" is fluvial, from the Laverde [07] river mouth. *G* Gauss-distributed, symmetrical gravel mode; *R* Rosin-distributed, positively skewed gravel mode. The gravel mode decreases beyond the delta, the slight increase of the last (8 km) sample signalizes the Buonamico [08] river mouth. Also the sand mean changes only little, the sand modes are less stable. Consequently, the prevailing Gauss distribution of the gravel modes may be seen as the only provenance signal of this area, indicative of the beach environment

mode: the fluvial gravels are positively skewed and Rosin-distributed the shoreface gravels being predominantly symmetrically and Gauss-distributed. If this low variability longshore were accompanied by the same low variability normal to the beach, a distinct grain-size distribution "beach" could serve as an useful provenance signal. The extremely high variability normal to the coast, unfortunately, made this impossible.

12.3 The Reduction of Dark Gneiss

The low variability of most grain-size parameters in the direction of net longshore transportation is a fact, but it would be wrong to link this with the assumption that there are no changes of the petrographic composition. Figure 12.7 shows the percentages of dark gneisses in the single fluvial and coastal subenvironments.

The elimination of dark gneisses starts already during fluvial transportation. The short distances of only 5 or 7 km suffice to make the dark gneisses drop by almost 10%.

Longshore we are confronted with a strong decrease from about 45 to about 20% in the direction of net transportation, most pronounced along the offshore strip, followed by the shoreface and the backshore strip (Fig. 12.7). In Calabria, the fluvial environment is characterized by the elimination of sedimentary clasts. In the coastal environment this part is taken by the dark gneisses, the second low resistant rock type which vanished first during longshore transport.

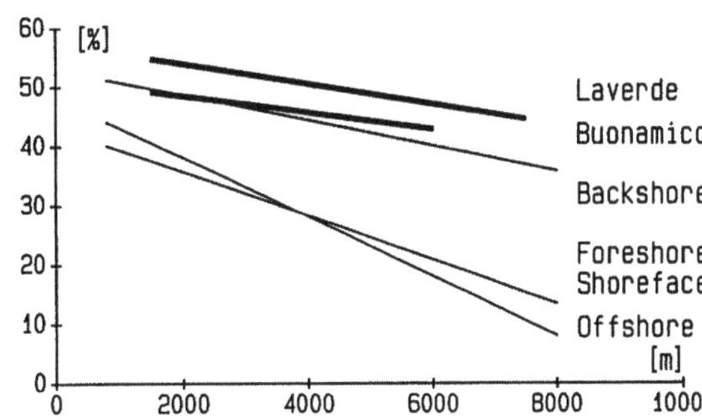

Fig. 12.7. The elimination of dark gneiss within the fraction 20–31.5 mm during fluvial and coastal transport, trend lines. *Laverde* [07]: five samples with 1500-m distances; *Buonamico* [08]: four samples with 1500-m distances; *backshore, shoreface* and *offshore*: ten samples each with 800-m distances. Long residence times of gravel in the backshore environment make the low gradient understandable. We have as yet no answer as to why the offshore gradient is slightly higher than the trend within the more strongly moved shoreface (Bartholomä in prep.)

12.4 Downstream and Longshore Development of Sphericity

12.4.1 The Rivers Laverde and Buonamico

The fluvial sphericity development of the dark gneisses corresponds to that of the overall mean (Fig. 11.1 D), that is, only slight or very slight changes occur (Laverde [07] and Buonamico [08]). These differences are too small to serve for a reliable interpretation.

12.4.2 The Bianco Beach

The occurrence of low sphericity gravels and shingles along beaches is certainly one of the most common experiences, even of non-geoscientists. The zonation of gravel shapes on beaches attracted the interest of sedimentologists, of whom we cite only Bluck (1967), who probably published the still most comprehensive study on this topic.

12.4.2.1 The Beach-Normal Profile

The Strand Profil displays amazingly homogeneous trends (Fig. 12.8): sphericity increases from land to sea. Each landward sample has a lower average sphericity than its neighbour seaward. Mean sphericity ranges from $psi_r = 0.4$ to 0.8; and larger grain sizes show higher sphericities and higher correlation coefficients (distance) than lower ones.

However, this well-balanced profile develops only during fair weather or low energy conditions. During high energy or storm conditions the entire gravel suite is assembled in the offshore strip close to the longshore trough. Only when the storm passes on does the perfect sphericity profile develop again within a day or even less.

This means that the Bianco beach offers nearly the entire spectrum of sphericities, but its distribution or zonation is not fixed but changes continuously. Bluck (1967) states that beaches do not produce low sphericities or flat pebbles but select and arrange them. This can only be proved by comparing the sphericity of the relevant grain sizes of both the fluvial input and the entire coastal environment (Bartholomä in prep.).

12.4.2.2 The Longshore Section

We learned from the beach-normal profile that sphericities increase seawards, or platy gravels concentrate towards the backshore. Obviously the higher pivotability of gravels with higher sphericities enables them to return more efficiently into the wave engine during swash and backwash than those with lower sphericities, which are advanced in landward transportation. The resulting zonation of different sphericities cannot be ignored when investigating the longshore development of sphericity.

The trends of changing sphericities of the ten triple samples are plotted in Fig. 12.9.

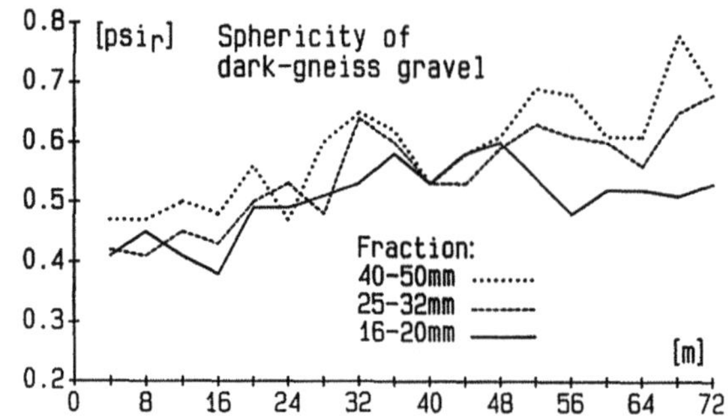

Fig. 12.8. Beach-normal development of sphericity (psi_r) of dark gneiss gravel along the 70 m Strand Profil, see Fig. 12.1 for position, backshore *left*, offshore *right*. Larger clasts fit the trend better than smaller ones. Surprisingly, the trend of increasing sphericity of the coarser grain sizes is not interrupted in the shoreface but continues seaward with the same ascent (Bartholomä in prep.)

Backshore. The selection favouring low sphericities during longshore transportation is most pronounced along the backshore strip. The correlation coefficient sphericity versus distance along the coast is r = 0.86*.

Shoreface and Offshore. Based on the results of the beach-normal profile, we expected a more striking longshore differentiation of the sphericity development between shoreface and offshore. However, the distances covered by the beach-normal profile are much larger than those between our sample strips, only 10–20 m, due to changed weather conditions. Thus the sphericities of the two subenvironments

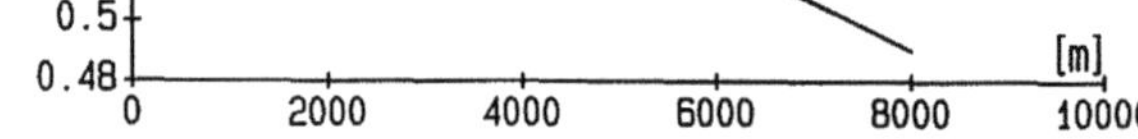

Fig. 12.9. Trends of the sphericity (psi_r) development during fluvial and coastal transportation; dark gneiss of the fraction 20–31.5 mm. *Laverde* [07]: five samples with 1500-m distances; *Buonamico* [08]: four samples with 1500-m distances; *backshore, shoreface* and *offshore*: ten samples each with 800-m distances. The sphericity spectrum of the river's bedload is split into two families entering the marine realm: a high sphericity or compact family which moves along the shoreface and the offshore strip characterized by a low gradient of decreasing sphericity, and a low sphericity family, very platy, which moves on the backshore strip, characterized by a high gradient of decreasing sphericity (Bartholomä in prep.)

come close; they are much higher than along the backshore, of course, but the decrease during longshore transportation is less pronounced and more irregular.

12.5 Downstream and Longshore Development of Roundness

The roundness data of gravel presented in this study are obtained applying Fourier analysis, our procedure avoids misleading results due to the elongation of clasts (Diepenbroek et al. 1991). Thus our roundness data are not exactly compatible with those from other authors. This limitation is still more valid for values obtained by visual estimation of pebble images (Krumbein 1941). We learned from various tests that operator error makes visual estimation close to useless.

Roundness is evidently no provenance signal per se, easy to generalize and to transmit for comparison with similar or other environments. In the case of our provenance study, however, it is a useful tool to test the consistency with other data, the elimination of dark gneisses and the decreasing sphericities.

12.5.1 The Rivers Laverde and Buonamico

The roundness development of gravel of the Laverde [07] and the Buonamico [08] clasts is surprisingly similar, from $r_{FO} = 4.85$ to $r_{fo} = 5.74$ (Laverde) and from $r_{fo} = 5.3$ to $r_{fo} = 5.7$ (Buonamico) (Fig. 12.10). Certainly, possible final stages of fluvial rounding are not reached by these values; further fluvial transport would certainly increase roundness, but Calabrian rivers are simply too short for this.

12.5.2 The Longshore Section

The longshore development of roundness starts, at the Laverde delta, more or less where the Laverde ended, without a jump, and the trend of increasing roundness continues longshore, directed northward (Fig. 12.10). There are no striking trends normal to the beach, sometimes the landside clasts are more rounded, sometimes those of the seaside. The longshore trend is evident, roundness increases from about 6.0 to 7.5, but the last three sample stations, close to the Buonamico [08] river mouth display slightly decreasing roundness, the backshore excepted. We interpret this as the influence of less rounded river input from the Buonamico that had drifted southward. Sphericity does not follow this slight trend reversal, but the longshore depletion of clasts of dark gneisses does so. The last triple sample of the longshore section before the Buonamico [08] is reached displays slightly increased percentages of dark gneisses, again, a signal of south drift from this river mouth.

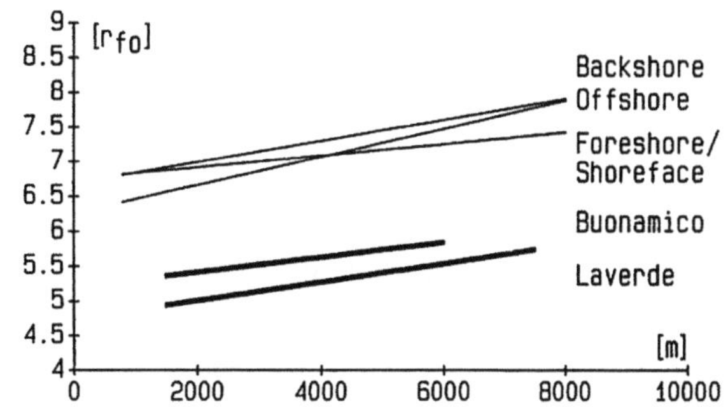

Fig. 12.10. Trends of the roundness (r_{fo}) development during fluvial and coastal transportation; dark gneiss of the fraction 20–31.5 mm. *Laverde* [07]: five samples with 1500-m distances; *Buonamico* [08]: four samples with 1500-m distances; *backshore, shoreface* and *offshore*: ten samples each with 800-m distances. Gradients of increasing roundness do not differ between the fluvial and the coastal realm. Roundness values of the final fluvial and the initial coastal stage, in reality, fit better than indicated by the trend lines (Bartholomä in prep.)

12.6 Conclusion

At least two questions arise from these features of longshore development, the sedimentological interpretation and the relevance to provenance. It is not possible for us, in the context of this study, to study sedimentology. To give only a brief comment, there are two main sedimentological questions. First are the reduction of dark gneiss or other types of rocks or minerals and the reduction of sphericity mainly caused by selective transport or by attrition and, second, which strip moves the fastest along the coast, backshore, foreshore or offshore? These problems are linked, but to fully understand this, the fluvial input and the coastal sequences have to be studied quantitatively fraction by fraction (Bartholomä in prep.).

And provenance? The overall sphericity distribution (Chap. 11) taught us that sphericity is no provenance signal. The study of a selected, sensitive rock type like the dark gneiss leads to similar results: short fluvial transportation yields slightly increased mean sphericities, but only if compared to the jointed and weathered source rocks, not within the fluvial system itself. Sphericities of the beach-normal and longshore profiles, on the other hand, are distinctively differentiated, due to selective transportation.

However, this differentiation is very sensitive to the temporary energy level due to changing weather conditions; concentration and redistribution alternate frequently. Consequently, sphericity cannot successfully be used as provenance factor in Calabria.

The same is true for the roundness. Early stages of fluvial roundness and the late stage of the coastal development are easy to discriminate, but mature fluvial roundness and the immature coastal one are more or less identical in Calabria.

Mixed Provenance. As pointed out in the introductory section of this chapter, the main interest of this provenance study is directed towards the relation source area/river-mouth sediment. However, the model-like character of the Bianco Beach offered the almost unique possibility to follow directly one of the most important marine pathways for further modification of gravel and sand, the longshore transportation. Probably the next important marine pathway of terrigenous sediments, the direct river input submarine canyons, is also realized in our study area by the Buonamico-Careri canyon. The unique function of the system Laverde, Buonamico and Bianco Beach makes it possible at least to estimate the meaning of provenance mixing and to learn that sediments of quite different maturity may stem from exactly the same point source.

The head of the Buonamico-Careri canyon is fed directly, during floods, by the river's bed load, which consist of about 60% dark gneiss. The longshore transportation, however, contributes a more or less continuous supply of detritus with only about 20% dark gneiss. Even if roundness and sphericity differ to some degree, it would be difficult to reconstruct these conditions from a gravelly turbidite released from this environment and resedimentated on a submarine fan.

Part III

River-Mouth Sediment and Provenance

Part III presents the river-mouth sediments and tries to decipher provenance. Chapter 13 is devoted to the gravel fractions, Chapter 14 to the treatment especially of sand and, to a smaller degree, of silt and clay.

Chapter 13

River-Mouth Gravel

In 1787 Goethe wrote that "there is no readier way of forming an idea of a mountainous district, than by examining the nature of the stones which are washed down by the streams". This is certainly correct, so we initially describe the petrographic suite of the river-mouth gravels. But it does not suffice to examine only the various rock types; the statistical treatment requires information about the homogeneity of the petrographic composition of the gravel body over the entire river-mouth area and the significance of the samples. In addition, we need information on the relationship between grain size and petrographic composition, because a certain petrographic signal from the source area linked to certain grain sizes may be distorted due to selective fluvial transport. We follow up these questions before we discuss the provenance complex of gravels.

Summary. One sample of about 270 kg suffices to represent the petrographic composition of the river-mouth sediment of all fractions < 80 mm. The "petrographic mixing" of the rivers is so perfect that even after larger floods there is no change in composition.

Petrographic composition is closely connected with grain size, thus an uncritical comparison of different fractions must yield misleading results.

The amount of granitic and metamorphic debris in the river-mouth sediments (16–80-mm fraction) correlates nicely with that of the effective rock composition of the 19 feeder basins or, in other words, the predicted sediment composition (granite: r = 0.76*, metamorphic rocks: r = 0.71*); but the sedimentary rocks do not (r = 0.45). Only the Jurassic limestones climb to r = 0.87*. The correlation source/sediment basing on the 2–16-mm fraction is much worse.

13.1 The Petrographic Composition

13.1.1 The Petrographic Suite

Data Base. Seven large samples (mean 271 kg) for statistical tests, 18 samples from the 1972 campaign and 19 from 1973.

Method. Portion > 16 mm: Determination of the clasts fraction by fraction (1/3 phi steps), 1972 and 1973 campaign. Portion 4–16 mm (1973 campaign only): determination of every second fraction. All determinations were made exclusively by macroscopical examination of the wet gravels. The smaller the clast, the more difficult the correct determination. Eighty grains were counted as a minimum, maximum 300. Evaluation in grain number percent.

13.1.1.1 Rocks

Granite. Usually easy to distinguish. In size classes < 20 mm granites may be mixed up with coarse-grained and not visibly foliated parts of gneisses. The petrographic variety of the recycled granites from the conglomerates is much higher than that of the Serre pluton. Maximum grain sizes of 50 cm are not seldom.

Porphyry. Very easy to identify. Light phenocrysts in a dark groundmass. All porphyry clasts are recycled. At least we do not know of any outcrop of porphyry within the limits of our area. Maximum grain sizes do not exceed 20 cm. Even 4-mm grains are usually easy to identify.

Gneiss and schist. This is quantitatively the most important and qualitatively the most problematic group of clasts. It is no problem to distinguish the "clean" end members of this group, well-foliated dark or light gneisses and phyllites with well-developed schistosity. However, the many transitional variations in between are difficult to assign. And even if this were possible, as pointed out (Chap. 5.1.2), it would be impossible to correlate this with definite source areas, at least in most cases, because many gneiss areas contain small intercalations of schists and vice versa. We therefore followed the geological map of Calabria, plotting gneiss and schist separately, especially in the hope of obtaining more differentiated information. But unfortunately it is not practicable to do this also with the clasts of gneiss and schist. Thus, for the provenance correlation, we have to combine the clasts of both schist and gneiss to metamorphic rocks which are compared to the combined schists and gneisses of the source area.

Maximum grain sizes of 50 cm are possible. Phyllites are very easy to identify even in grain sizes below 4 mm; in the case of the light gneisses this is impossible.

Marble. These rocks are often striated, bluish gray and white, usually very folded and thus very easy to identify.

Intercalations of marble in the metamorphic rocks are seldom and irregular, and never occur on the maps. Thus these clasts carry only a qualitative provenance signal, which is useless for regional correlations. Maximum grain sizes of 50 cm occur.

Pegmatite. Larger clasts of pegmatite show very little feldspar, but large micas, frequently masses of tourmaline up to several centimeters thick, and quartz. It is more or less impossible to distinguish macroscopically very small clasts of pegmatite from coarse-grained, light granites. Thus, we determined all relevant clasts as granite or pegmatite in the fraction >16 mm, calculated the ratio granite/pegmatite and split all relevant clasts of the fraction <16 mm according to this ratio into granite and pegmatite. Only in this way can one special problem of the area be solved. The pegmatites occur nearly exclusively as veins which swarm through the metamorphic provinces. Thus the pegmatites are a regional signal for the metamorphic rocks, mainly of the Aspromonte, not for the granite family of the Serre, to which the rocks petrographically belong (Chap. 5.2.1). A provenance comparison of the 16–2-mm fraction of the river-mouth sediments with the relevant clasts counted only as granites would suggest much recycling in the Aspromonte, because the granite clasts of this area stem mainly from conglomerates. The pegmatites of this area, however, are a powerful source of first cycle debris, and may not be neglected. Maximum grain sizes of 50 cm of pegmatites are not seldom.

Jurassic Limestone. There is no difficulty in identifying quickly and precisely the clasts of this light, neritic limestone, unique in the area; 30-cm boulders occur. In finer size classes a distinction from marbles is possible by the finer grain and slightly yellow colour.

Olistolithic Limestone. The micritic limestone is very easy to distinguish, because of its characteristic slightly greenish colour and the many thin sparry veins. This limestone stems from the olistolithes of the debris flow, the siltstone. Maximum grain size seldom exceeds 15 cm. Even small particles are easy to identify by the micritic finest grain.

Sandstone. It is no problem to identify the sandstone clasts quickly and precisely, in all fractions. Well-cemented clasts frequently contain conglomerates. All this was classified as sandstone. Many sandstones of the source area bear conglomeratic intercalations and vice versa, and the content of sandstone clasts in the river-mouth areas seldom exceeds 1%. Thus, a subdivision of sandstone and conglomeratic clasts yields no more information about provenance. Large boulders of more than 50 cm occur. Sandstone clasts are easy to distinguish in all grain sizes.

13.1.1.2 Minerals

Quartz, K-feldspar and Plagioclase. In the fractions <5 mm the minerals were identified by staining. If it was impossible to assign a rock fragment to a rock type, this particle was assigned to the largest mineral of which it was composed. Dealing with single minerals we are leaving the proper domain of gravel or conglomerate and we enter the world of sand and sandstone even in grain sizes >2 mm. Vein quartz particles up to 10 cm occur, single feldspar grains with more than 4 mm are seldom.

Anthropogenic Clasts. A very special property of modern deposits is the content of anthropogenic debris, of all kinds of products of human intelligence like glass, brick, plastic, metal and so on. In 1972/73, when we conducted this petrographic analysis, Calabrian economics and standard of living were still so low that the rivers had no problem in suppressing this human influx. Today, unfortunately almost all river-mouth areas are misused as waste disposal sites and our work seems, to us, barely repeatable.

13.1.2 Sample Reliability and Dependence of Petrographic Composition on Grain Size

13.1.2.1 Sample Reliability

The Significance Test. The labour-intensive technique of field sieving allowed only one large sample per river mouth and campaign. Therefore, it had to be tested to determine if such a single sample is representative of the petrographic composition of the entire river mouth area or not. The answer is yes. This is supported by two arguments independent of each other:

- randomly taken samples along three test sections,
- the correlation of the results of repeated measurements of the 19 rivers.

Randomly Taken Samples. To obtain an idea about the homogeneity of the petrographic composition over the river-mouth area and its possible development in the lower course of the river, we sampled three cross-sections of the Buonamico [08] (Fig. 13.1). The first is 10.6 km from the coast, at the fall line, where the river leaves the crystalline basement and the valley widens. The third is at the river-mouth area, but still within the fluvial regime, and the second one is in between, 3.6 km from the coast. The distance of the samples along the profiles is between 60 and 150 m (see Fig. 13.1 for sample positions). The samples are dominated by metamorphic rocks, all other components are accessories (granite, pegmatite, marble, limestone,

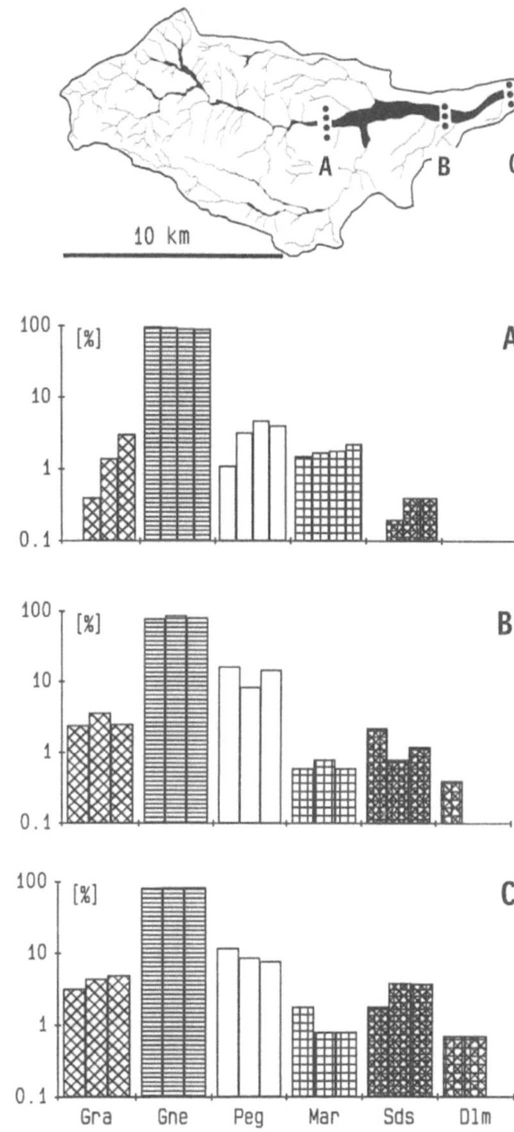

Fig. 13.1A–C. Significance test. Petrographic composition of river gravels (Buonamico [08]) at various locations. Mean of the fractions 12.5–50 mm (sieved in 1/3 phi intervals). The variability of the variables is significantly smaller within the test sections than between them. Profile **A** Four samples in 90-m distances. Profile **B** Three samples in 150-m distances. Profile **C** Three samples in 60-m distances, all from bank to bank. *Gra*nite, *Gne*iss, *Peg*matite, *Mar*ble, *S*and*s*tone and *D*ebris-flow *l*i*m*estone; log scale

Table 13.1. Average petrographic composition of the Buonamico [08] river gravels in sample sections from bank to bank. From each sample, the fractions 12.5–16, 16–20, 20–25, 25–31.5, 31.5–40 and 40–50 mm were evaluated and combined to the average of the sample. These mean samples were then combined to the average of the section[a]

	No. of samples	Grains counted	gra	gne	peg	mar	sds	lim
			[%]					
Upstream	4	3300	1.2	93.5	3.3	1.8	0.3	–
99% c.l.[b]			0.56	1.28	0.93	0.69	0.28	–
st. dev.			1.31	3.1	1.56	0.29	0.28	–
Midstream	3	1906	2.8	81.9	13.1	0.7	1.4	0.2
99% c.l.			1.13	2.69	2.31	0.57	0.81	0.31
st. dev.			3.0	4.18	4.18	0.12	0.72	0.21
River mouth	3	1949	4.2	82.3	9.3	1.1	3.1	0.5
99% c.l.			1.36	2.59	1.97	0.71	1.17	0.48
st. dev.			0.87	0.62	2.1	0.58	1.13	0.4

[a] *gra*nite, *gne*iss, *peg*matite, *mar*ble, *s*an*ds*tone, *lim*estone.
[b] c.l.: confidence limit.

Table 13.2. Reliability of petrographic sampling and comparison of predicted sediment composition and measured composition of the river-mouth sediments

Fraction	Correlation coefficients[a]		
	A 1972/73 16–80 mm	Predicted sed. composition	
		B 16–80 mm	C 4–16 mm
Granite	0.98*	0.77*	0.51*
Gneiss/schist	0.97*	0.64*	0.23
Pegmatite	0.97*	0.49*	0.6
Metamorphic rocks	0.98*	0.71*	0.56*
Jurassic limestone	0.99*	0.87*	0.87*
Sandstone	0.95*	0.06	0.18
Olistholitic limestone	0.59*	0.47*	0.59*
(Quartz)	0.85*	–	–
Sedimentary rocks	0.94*	0.45	0.52*

[a] Correlation coefficients of the 19 basins: A: Petrographic composition of the gravel fraction 16–80 mm, evaluation of the 1972 campaign versus the 1973 campaign; B: predicted sediment composition versus measured river-mouth gravels of the fraction 16–80 mm; C: predicted sediment composition versus the river-mouth gravels of the fraction 4–16 mm.
* Significant correlation (95% level).

Table 13.1). Between the two upper profiles, limited source areas of conglomerate, sand and siltstone still contribute to the Buonamico. For this reason the respective components increase downstream. Granite and pegmatite gravels stem from the conglomerates, the limestones from olistolithes of the debris flow (siltstone). Despite this, the homogeneity evidently improves downstream.

The upstream- and the river-mouth profile are more stable in petrographic composition than the midstream section, because the latter has to digest a new influx of recycled clasts. Where this influx is added to larger amounts of still existing rock types such as granite, pegmatite and metamorphic rocks, the standard deviation of the river-mouth profile is not greatly increased. In the case of the rare sandstone and olistolithic limestone this contribution is more striking.

An analysis of variance of the variables granite, gneiss, pegmatite, marble and sandstone shows that the variability is significantly smaller within a given test section than between the test sections. The F-value of all these variables exceeds the corresponding table value on a level of significance of about 1%.

Correlation of Repeated Measurements. The nearly perfect petrographical homogeneity of the river-mouth area is not only valid at a given time. After the extensive flooding at New Year 1972/73 the petrographic composition still remained unchanged.

Table 13.2 A shows the correlation coefficients of the petrographic composition before and after this event. The two compositions are nearly identical. So we may expect that the mixing engine "river" delivers more or less the same petrographic spectrum over longer periods.

We did not expect this homogeneity because of the event character, the discontinuous activity of the Calabrian fiumaras:

- most important contributors are occassional landslides, mostly with a limited lithological spectrum;
- powerful floods, the only really "counting" phases of transportation, occur only seldom;
- even during such a flood the full width of the braided stream is only seldom activated.

These more or less isolated events would supply good evidence in support of different petrographic compositions at different parts and different times of the river bed. However, the opposite is the case. We know that during floods the river bed may be reworked to a depth of more than 1 m. Obviously the sum of the repeated events, especially the reworking of the river bed, finally yields the petrographical homogeneity. Dott jr. (1982) formulated the question: "Episodic sedimentation – how normal is average ..." – we are tempted to conclude, from our findings, that the sum of Calabrian discontinuities seems to produce a surprisingly stable average, admittedly in a microscale, compared to historical geology.

13.1.2.2 Petrographic Composition and Grain Size

The conventional Calabrian river-mouth sediment comprises a grain-size range from about 50 cm to clay minerals; these are seven orders of magnitude. The compositional and textural maturity is extremely low. Thus, to speak of the petrographic composition of single fractions which are more than 10 1/3 phi steps apart from each other means speaking of quite different sediments. This also holds true for the gravel fraction. Figures 13.2 and 13.3 show the petrographic composition of four important rivers, determined fraction per fraction in 1/3 phi steps: Amendolea [02] and Buonamico [08] of the Aspromonte Province, Torbido [14] and Assi [19] of the Serre Province. Figure 13.4 demonstrates the somewhat smaller range of the Novito [13] of the Zomaro Province.

The petrographic composition of the gravel fractions of these five rivers displays a strong and variable dependence on grain size. Unfortunately, there is no uniform trend, the 16-mm limit excepted, and no fixed grain-size class whose petrographic composition can be regarded as representative for the entire gravel population. The existence of such a reference fraction would be very time-saving. Thus, for the regional comparison between the single rivers and their source areas, only the average of the entire gravel fraction 16–80 mm was calculated.

The most important grain-size limit is that at about 16 mm. In a certain sense this is the very beginning of the formation of sand (Figs. 13.2, 13.3). Here, quartz starts to become an increasingly important constituent of the fractions, followed a little later, at about 4 mm, by K-feldspar and especially plagioclase, all at the expense of the rock fragments. It may thus be meaningful and justified to make the following grouping:

- fractions 16–80 mm: rock fragments and vein quartz
- fractions 4–16 mm: overlap of decreasing rock fragments and increasing single minerals
- fractions <2 mm: classical sand.

The fractions 16–80 mm. The subdivision fraction by fraction makes different compositional trends visible. The predominant components are metamorphic rocks and granite plus pegmatite. However, in the case of the Amendolea and Buonamico (Fig. 13.2A, B) the proportions of these two rock types remain more or less constant going from the 100-mm to the 16-mm fractions,

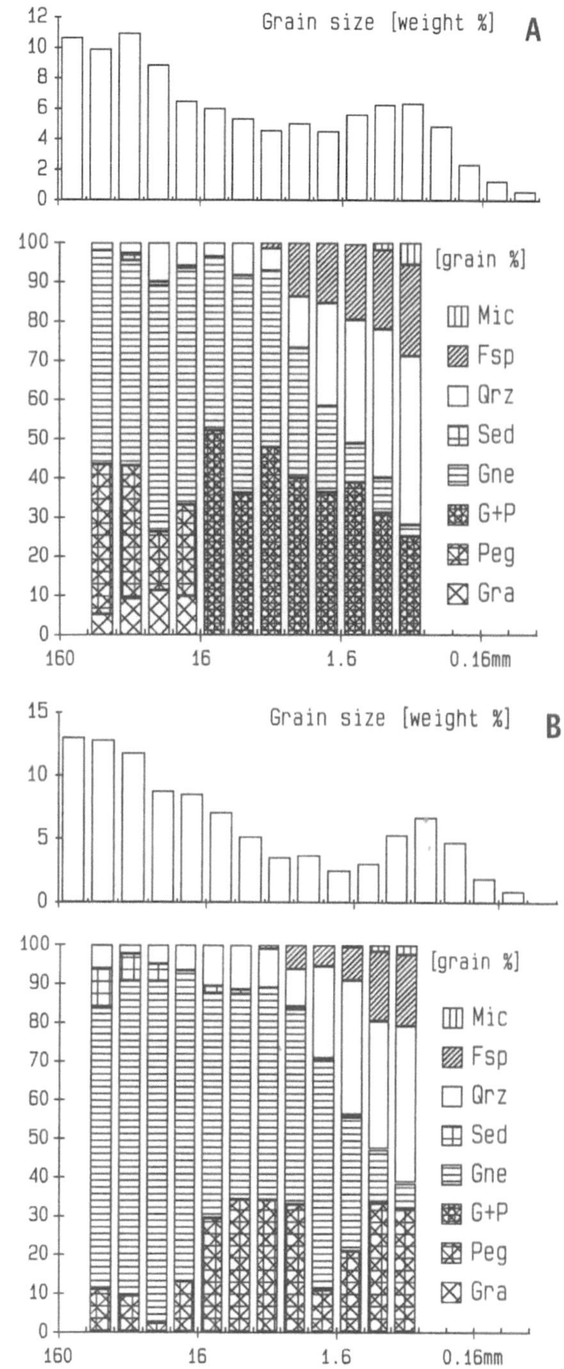

Fig. 13.2 A, B. Petrographic composition and grain size. Histogram of the grain-size distribution and petrographic composition calculated to 100% of each fraction. **A** Amendolea [02]. **B** Buonamico [08], both Aspromonte Province, mostly metamorphic rocks. Amounts of metamorphic rocks, pegmatites and granites remain almost constant through coarse fractions, single minerals increase starting with about 16 mm. Almost all granitic gravels are second-cycle elements from conglomerates. This preselection is the reason for the non-increase of granite towards coarser grain sizes. *Gra*nite, *Peg*matite, *G+P* granite+pegmatite, *Gne*iss, *Sed*iments, *oSe* other sediments, *Lim*estone, *Qrz* quartz, *Fsp* feldspar, *Mic*a

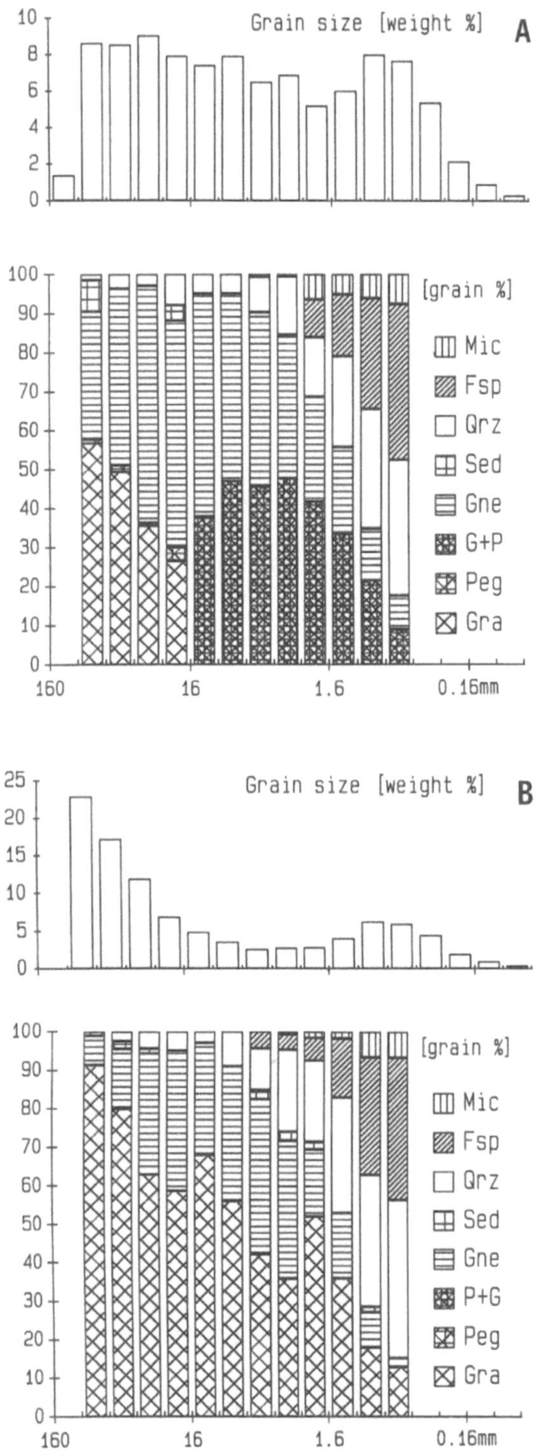

Fig. 13.3 A, B. Petrographic composition and grain size. Histogram of the grain-size distribution and petrographic composition calculated to 100% of each fraction. **A** Torbido [14]. **B** Assi [19], both Serre Province, mostly granitic rocks. Amounts of granites decrease through coarse fractions, single minerals increase starting with about 16 mm. Granitic gravels are mixed from both first- and second-cycle origin, granite content increases towards coarser grain sizes. For abbreviations see Fig. 13.2

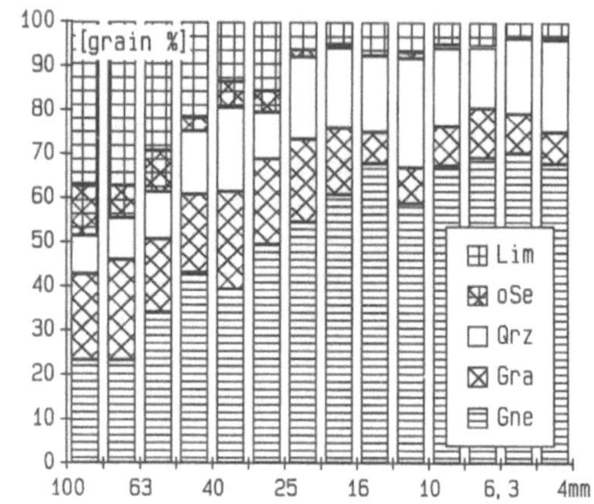

Fig. 13.4. Petrographic composition and grain size. Novito [13] from the Zomaro Province, mostly sedimentary rocks. Content of limestone strongly decreases from 37% (fraction 80–100 mm) to 3.2% (fraction 4–5 mm) (Ibbeken 1983, Fig. 4). For abbreviations see Fig. 13.2

whereas Torbido and Assi (Fig. 13.3 A, B) display a distinct increase in the metamorphic rocks over the same range. This reflects the ability of the source rocks to release large clasts or not. The granite clasts of the Amendolea and Buonamico are completely recycled, and the majority of the pegmatite clasts stem from highly jointed pegmatite veins (Fig. 8.2). Consequently, the resulting clasts are preselected to the disadvantage of the coarser grain sizes, especially of granite. The metamorphic rocks of the area release a priori smaller clasts.

Torbido and Assi, on the other hand, directly drain the granite pluton of the Serre, but the granites tend to deliver both larger clasts or grus, and only little in between. Thus granite is the dominant element of the coarser fractions, and the metamorphic rocks increase towards the fines.

The same argumentation is valid for the third example, the sediment-dominated Zomaro Province. The limestones of the Novito [13] basin also tend to release coarser grain sizes instead of fines. Figure 13.4 shows that these compositions shifting from fraction to fraction make it impossible to find a representative fraction which corresponds to the overall mean. In the case of the Novito the limestone percentage of the river-mouth sediment ranges between 37% (fraction 80–100 mm) and 3.2% (4–5 mm). In other words, the limestones of the Novito basin are unable to represent themselves adequately, even in the fine gravel fraction.

The fractions 4–16 mm. The petrographic picture of these fractions is even more differentiated. In addition, it is difficult to determine this interval, most particles being too small to be identified macroscopically and too large to be caught in thin sections. Thus, for the purpose of our work further subdivisions are not useful. All four type samples display a strong increase of single minerals of quartz, K-feldspar and mica towards the finer fractions until the grain size domain of sand is reached.

As a result, all source-sediment correlations or provenance calculations of this examination are based on both the >16-mm and the 16–4-mm interval. These are, in terms of the Wentworth scale, small cobbles to coarse pebbles and medium pebbles to very fine pebbles.

13.1.3 Reading Provenance from the Gravels

The examination, here in Calabria, of only one individual river basin, even a very detailed examination, would probably yield a convincing picture of the relationship between the areal distribution of the source rocks, their "weighing" by the erosion rates and the nature of the gravel products in the river-mouth areas. The argumentation would be thorough, with good reasons for the pros and cons, and finally the author would lead us to a more or less stabilized interpretation, combined with proposals for the application in other areas. In other words, because of the limited data base – one river only –, the extent of the result would be limited, but its reliability would appear relatively high. The synopsis of our 19 basins (Fig. 13.5), however, creates uncertainty, revealing a broad variety of contradictory developments. Seen as a whole, there are only some few trends which seem to be generally valid.

We present the gravel data in three sets of diagrams; one shows the basins and their sediments in the ternary diagrams granitic-/metamorphic-/sedimentary rocks, Appendix B 01–19 shows the single cases, Fig. 13.5 gives the conclusion; one compares the gravel petrography of all basins (Fig. 13.6) and one correlates the predicted sedi-

Fig. 13.5A–D. From the source to the sediment; the shift in composition, reduced to (*G*) granitic, (*M*) metamorphic and (*S*) sedimentary rocks if one goes from the area percentage of the source (**A**) to its erosional budget or effective composition (**B**), which is consistent with the predicted sediment composition, and finally to the river-mouth sediment itself, whose composition is given in the gravel fraction 16–80 mm (**C**) and 4–16 mm (**D**). Basins which produce mainly unimodal river-mouth sediments and their gravels are marked by a *triangle*, *dots* indicate bimodality. Single basin developments are plotted in Appendix B 01–19. Table 13.3 shows the corresponding numerical values (average only)

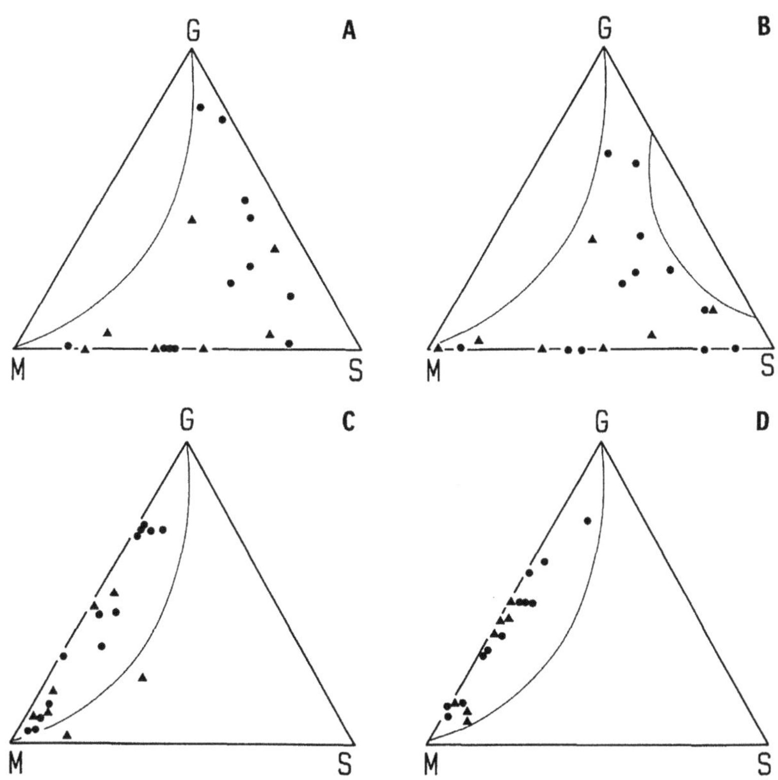

ment composition (effective source-area composition) and the measured gravel composition of the 19 basins (Fig. 13.7). In Table 13.3 the percentages of the three rock families are listed: source area, effective source area, river-mouth gravel 16–80-mm and 4–16-mm fraction. The ratio granitic/metamorphic rocks remains surprisingly constant going from the source to the river-mouth sediments, whereas the ratio granitic+metamorphic/sedimentary rocks increases dramatically.

13.1.3.1 Trends Within the Basins

Areal Distribution (Fig. 13.5A, Table 13.3A). The ternary diagram granitic/metamorphic/sedimentary rocks of the source basins shows a kind of "miscibility gap" between granitic and metamorphic rocks. This means that there are no drainage areas composed of only granites and gneiss plus schist and that all basins contain at least 10% sediment. The most "metamorphic" river (85%) is the Amendolea [02], the most "sedimentary" (77%) is the Careri [09] and the most "granitic river" (79%) is the Allaro [16] (single river plots of Appendix B).

Effective Source Area Composition (Fig. 13.5B, Table 13.3B). This is the areal distribution, weighed by the erosion rates (Chap. 6.2.2). The miscibility gap remains as it is at the foregoing plot, but most rivers move away from the granite and sediment corner. Consequently, a second gap at the right side of the triangle opens. If all rock types were able to produce the same grain-size distribution of gravel debris, the composition of the river-mouth gravels should correspond to this plot. Just the opposite is the case.

The Petrographic Composition of the River-Mouth Gravels (Fig. 13.5C, D, Table 13.3C, D). The most striking feature of the petrographic composition of the river-mouth gravels is that they occupy exactly the field of both the "miscibility gap" of the drainage areas and the effective source rock composition: the misfit is perfect.

The following distinct trends or developments are ascertained.

1. Most sedimentary clasts vanish by fluvial attrition and are barely recorded in the gravel fractions. This is the explanation for the strong shift of the composition of almost all basins away from the sediment corner toward granitic and metamorphic compositions, if one goes from the effective source rock composition to the river-mouth sediment. This tendency increases with decreasing grain size, from the fraction 16–80 mm to the fraction 4–16 mm (Fig. 13.5C, D).

2. Nine rivers start with an effective source rock composition of granite which is zero or a little more, but their river-mouth sediments contain granite clasts. Thus a kind of granite creation seems to exist in between. This, in fact, is the case, via recycling from conglomerates and conglomeratic sandstones (Appendix B 01–09).

3. In 12 of the 19 basins the relation between granitic and metamorphic clasts shifts towards the granitic components if one moves towards the finer gravel fraction. This could be a selection at the ex-

Table 13.3. From the source to the sediment, shift of the petrographic composition in terms of granitic, sedimentary and metamorphic rocks and calculated as average of the 19 basins

	Source		River-mouth sediment fraction	
	Area extension	Effective source rock composition	16–80 mm	4–16 mm
Granitic rocks	24%	15%	33%	41%
Metamorphic rocks	37%	41%	62%	57%
Sedimentary rocks	39%	44%	5%	2%
Ratio gran/met [a]	0.6	0.4	0.5	0.7
Ratio gran+met/sed [a]	1.6	1.3	19	49

[a] Both granitic and metamorphic rocks are enriched at the expense of sediments, this effect increases towards the finer fractions, as demonstrated by the ratio granitic+metamorphic/sedimentary rocks. The ratio granitic/metamorphic rocks, however, remains almost constant.

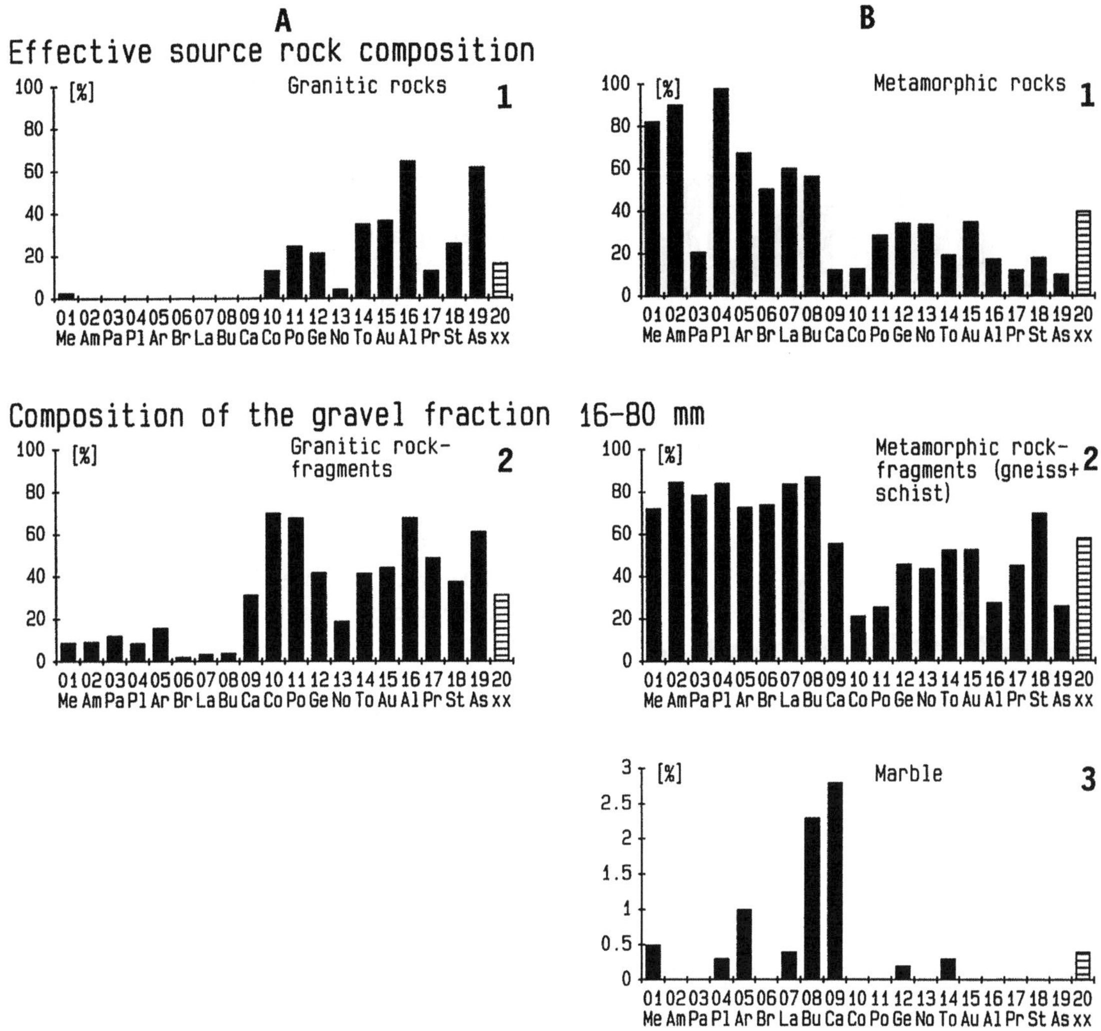

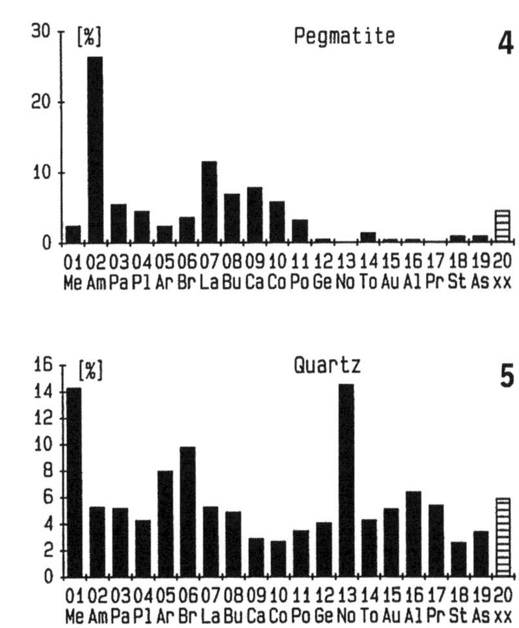

Fig. 13.6 A–C. Effective source area composition and petrographic composition (average of the fractions 16–80 mm) of the 19 basins, campaign 1973. **A** Granitic rocks. **B** Metamorphic rocks. **C** Sedimentary rocks. The occurrence of pegmatite veins is limited on metamorphic rocks. This is the only reason for the geographic assignment of pegmatites to metamorphic rocks. Scales differ from plot to plot. *01 Me*lito; *02 Am*endolea; *03 Pa*squale; *04 Pa*l*izzi; *05 Ar*anghia; *06 Br*uzzano; *07 La*verde; *08 Bu*onamico; *09 Ca*reri; *10 Co*ndoianni; *11 Po*rtigliola; *12 Ge*race; *13 No*vito; *14 To*rbido; *15 A*m*u*sa; *16 Al*laro; *17 Pr*ecariti; *18 St*ilaro; *19 As*si; *20 xx* mean of the basins 01–19

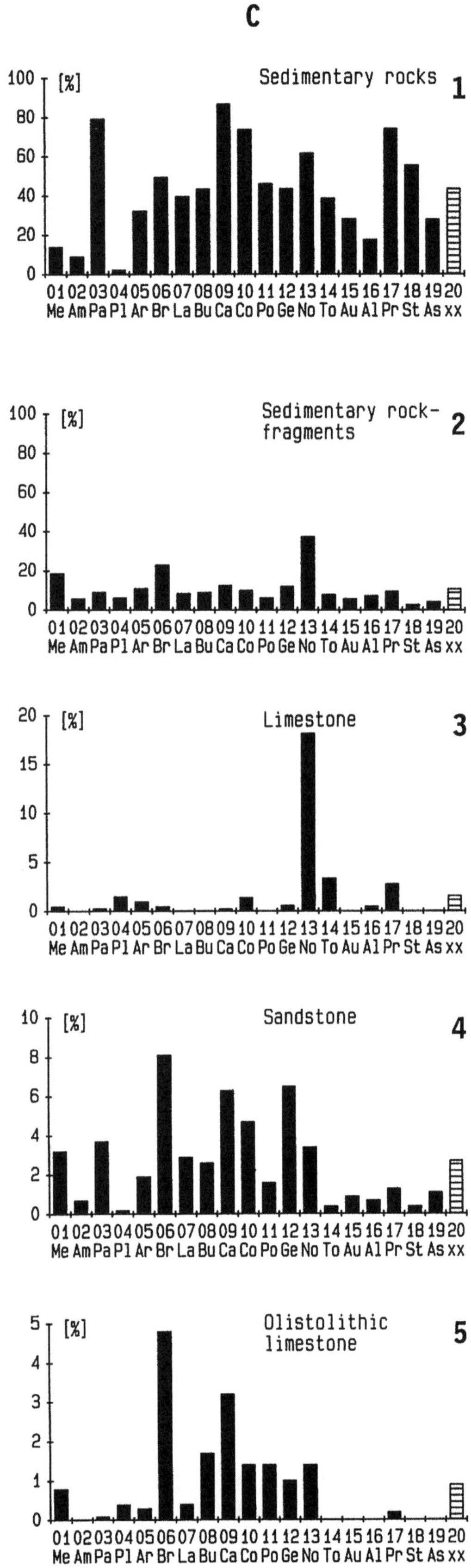

pense of the somewhat less resistant metamorphic rocks. However, seven basins prefer the opposite direction, we have no explanation for this.

13.1.3.2 Comparison of the 19 Basins

The petrographic composition of the mean gravel fraction 16–80 mm of the 19 river-mouth sediments correlates better with the effective source rock composition of the source areas (Table 13.2B) than the fraction 4–16 mm (Table 13.2C), the sediments excepted. Thus, the coarse fraction is a more reliable guide and we concentrate on this fraction in comparing the 19 rivers.

Figure 13.6 compares the petrographic composition between the 19 river-mouth sediments (percentage of single rock types of the mean fraction 16–80 mm). The effective source rock composition, the reference basis of the source areas, is also plotted.

Granitic Rocks. The granite content of the river gravel of the 19 rivers correlates fairly well with that of the source areas, r = 0.77* (Fig. 13.6, 1A, 2A). Because the basins 02–08 of the Aspromonte Province are practically free of granite, their granite clasts in the river-mouth sediments must be recycled from conglomerates. Suppose that the granite content of these conglomerates varies only slightly among the eight rivers, then the amount of granites in the river-mouth sediment should correlate with the amount of conglomerates in terms of the effective source rock composition of these rivers. This is the case, the correlation coefficient being r = 0.60*. Because the mean content of granite clasts in the river-mouth sediments of these eight granite-free basins is 11%, we corrected the granite file of the 19 river-mouth sediments by cancelling the recycled granites of the granite-free rivers and by reducing the granites of the other granite-bearing basins by 10%. Doing so, the resulting correlation coefficient – granite of the river-mouth sediment versus granite of the effective source rock composition – improves slightly from 0.77* to r = 0.80*. At first glance, the granite debris of the Calabrian rivers seems to originate close by and to form first cycle clasts. The Aspromonte conglomerates, however, with many granitic components, but without granitic source areas, indicate that probably a good deal of all granites are recycled, 10% being a reliable estimate. Based on field experience only, the petrographic spectrum of the granites from conglomerates appears broader than that of the granites of the Serre Province. Thus the granite clasts of the Aspromonte conglomerates may be a suitable starting-point for paleogeographic research. This is especially true for the porphyries found in these conglomerates, which, consequently, also occur in the river-mouth sediments. We tried to use these unequivocally recycled porphyries to calibrate recycling quantitatively, but we failed, because the amount of porphyry clasts in the conglomerates varies too much (see Table 8.2 for the estimates given by Bonardi et al. 1980). We found porphyries up to 1% in the river-mouth areas of the following rivers: Bruzzano [06], Careri [09], Condoianni [10] and Portigliola [11].

Metamorphic Rocks. Gneiss and schist correlate with only $r = 0.64$* with their source rocks (Fig. 13.6, 2B), evidently worse than the granites. If we include the pegmatite clasts of the river-mouth sediments (Fig. 13.6, 4B) (see Chap. 5.2.1 for the discussion of the proper assignment of the pegmatites) this correlation improves slightly, increasing to $r = 0.71$*, the possible reason being somewhat underestimated erosion rates in the northern areas. The distribution pattern of granites is simple: rare or no granite in the Aspromonte Province, little granite in the Zomaro Province and much granite in the Serre Province. Thus, the response of the rivers is clear. This is not so for the metamorphic rocks. Their decreasing erosional budget, from south to north (Fig. 13.6, 1B) is not accompanied by a proportional decrease in metamorphic clasts in the river-mouth sediments (Fig. 13.6, 2B). There is too much metamorphic debris in the northern rivers. In this case, our hypothesis about the predicted sediment composition has to be corrected upwards, because the erosion rates in the high energy belt of the Serre Province are probably higher than expected. Obviously the data base for the calculation of erosion rates in the Serre area is too limited.

The regional appearance of pegmatite clasts (Fig. 13.6, 4B) evidently reflects their geographic relationship to the metamorphic rocks of the Aspromonte and Zomaro Provinces. The same is true for the marbles (Fig. 13.6, 3B), even if they occur very irregularly and make up only less than 10% of the pegmatites. To apply the citation of Goethe (introductory section), the extraordinary high percentage of marble clasts in the Buonamico [08] and Careri [09], about 2.5%, is a hint for those who are interested in Paleozoic stratigraphy to look at these basins.

Sedimentary Rocks. Seen as a whole, the sedimentary debris, at a disadvantage during fluvial transportation and attrition, barely reflects the distribution of the source rocks in the hinterland. Its correlation coefficient is only $r = 0.45$ (Fig. 13.6, 1C, 2C). Responsible for this is especially the sandstone (Fig. 13.6, 4C), the sources of which by no means correlate with the river-mouth debris ($r = 0.06$). This is reasonable, because only well-cemented clasts survive the fluvial transport. Thus, the content of sandstone in the river-mouth sediment is rather a function of the local resistance of this rock than of its erosional budget. The olistholitic limestone clasts (Fig. 13.6, 5C), which stem from the siltstones, the debris flow formation, correlate with only $r = 0.47$ with its source rocks. It is the only signal that this important sediment body can release into the river-mouth's gravel fraction, some very few sandstone clasts excepted. The low correlation makes it probable that this type of olistholite is not very evenly distributed over the entire siltstone area. More important is the fact that the impartial observer of the river-mouth gravels has no chance to realize that this mean of 0.9% of micritic limestone clasts is the remainder of a 18.2% feeder sediment. Even the olistholitic nature of these limestone remains undetectable.

In the case of the Jurassic limestones (Fig. 13.6, 3C), on the other hand, the river-mouth sediments fit very well with the predicted sediment composition, $r = 0.87$*. Compared to the sandstones and the debris flow siltstones, the outcrop conditions of the Jurassic limestones are very homogeneous, so the detritus of these rocks may reach the river mouths in amounts that correspond to their erosional budget.

Quartz. The quartz detritus > 16 mm stems mainly from vein quartz. An unknown, probably large amount may be recycled. The quartz distribution of the 19 rivers (Fig. 13.6, 5B) reveals no hints of special correlations with source area elements.

13.1.4 The Correlation Effective Source Rock Composition Versus River-Mouth Gravel Composition

The diagrams of Fig. 13.7 quantify the ternary diagrams (Fig. 13.5B, C) which demonstrate the relations between the effective source rock composition and the river-mouth gravel. Not the absolute values of these diagrams are important but the fact that it becomes evident which rivers pro-

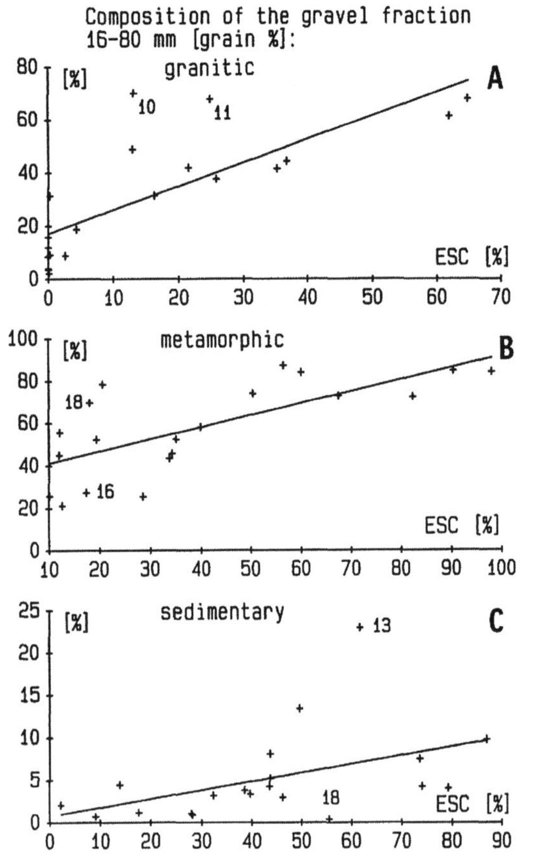

Fig. 13.7 A–C. Correlation of the effective source area composition (*ESC*) or predicted sediment composition and the petrographic composition of the gravel fraction 16–80 mm of the 19 river-mouth sediments. **A** Granitic rocks. **B** Metamorphic rocks. **C** Sedimentary rocks. For abbreviations *10, 11, 13, 16* and *18* see Fig. 13.6

duce gravel in proportion to the hinterland conditions and which rivers are unable to do this, producing more detritus or less than expected.

Granite. Figure 13.7A shows a group of basins without or with very few granites in the hinterland but with noticeable amounts of granite clasts in the river-mouth sediments. All these granites are recycled from Tertiary conglomerates or conglomeratic sandstones. A second group of basins close to the regression line reflects well-balanced proportions to the hinterland granites. This group consists of the rivers of the Serre Province. Rivers which produce granite clasts in excess are Condoianni [10] and Portigliola [11]: the granitic sources of these rivers are situated exactly in the high erosion belt. The correlation coefficient effective source rock composition versus river-mouth sediment of granites is r = 0.77*.

Metamorphic Rocks. The metamorphic rocks are more evenly distributed over the entire area than the granites, so the relationships between source and sediment are also less differentiated (Fig. 13.7B). Nevertheless, in the case of the metamorphic rocks it is much more difficult to find a satisfying explanation for an over- or under-production, because many rivers with metamorphic rocks occurring only in the high erosion belts produce very differing amounts of metamorphic clasts. The two almost neighbouring basins, Allaro [16] and Stilaro [18], for example, are characterized by the following conditions:

	Total area	Metamorphic rocks				
		Area		Erosional budget		River-mouth gravel 16–80 mm
	[km²]	[km²]	[%]	[m³/a]	[%]	[grain %]
Allaro [16]	134.66	9.8	8	2831	17.5	27.0
Stilaro [18]	98.7	8.9	11	2791	18.0	68.8

Apart from the mere size of the basins, all data on metamorphic rocks are close to identical. The amount of metamorphic rock debris in the river-mouth sediment of the Stilaro, however, exceeds that of the Allaro [16] by more than factor 2. It may be that lithological differences or local conditions are more important than the long-term erosion rates. Today huge landslides occur in the Stilaro [18] basin close to Bivongi, especially in the schists of the Torrente Pardala area, which may account for the over-production of metamorphic rocks in this river basin.

Despite these differences, the correlation coefficient effective source rock composition versus river-mouth sediments of metamorphic rocks is r = 0.71*.

Sedimentary Rocks. The relations of the sedimentary rocks between source and sediment are very unbalanced (Fig. 13.7C). Most rivers come close to the very low-gradient regression line demonstrating a production of sedimentary clasts far below the extent of sedimentary rocks in the corresponding source area, in terms of the effective source rock composition. The only over-productive river is the Novito [13] with its limestone-dominated hinterland. The Stilaro [18], however, the other important "limestone" river, is strongly under-productive.

The limestone relationships source/sediment of these rivers may serve as an example for the general dilemma calculating percentages:

	Total area	Jurassic limestone				
		Area		Erosional budget		River-mouth gravel 16–80 mm
	[km²]	[km²]	[%]	[m³/a]	[%]	[grain %]
Novito [13]	60.4	6.3	12	528.75	11.13	17.1
Stilaro [18]	98.7	5.3	6	516.25	3.33	0.6

The limestones of the two basins have more or less the same areal extent, both lie in the "high energy belt" and they display, consequently, more or less the same absolute erosional budget. Even lithologically they are identical. There is no reason why both source areas should not deliver the same amount of limestone debris into their river-mouth areas, and they certainly do. However, different size and different geological composition cause the relative erosional budget of the Novito [13] to become 11.13% and that of the Stilaro [18] 3.33%, a relation of about 3:1 instead of 1:1. Furthermore, the present-day enormous production of metamorphic debris in the Stilaro basin suppresses the limestone content of the river-mouth sediment to only 0.6%, while the Novito [13] produces 17.1%, the relation shifting to 29:1. This calculation and its background are admittedly trivial, but the widespread use of percentages, for example all ternary diagrams, may merit such a comment.

The correlation coefficient effective source rock composition versus river-mouth sediment of sedimentary rocks is only $r = 0.45$.

13.2 Conclusion

The river-mouth sediments are so perfectly mixed (Fig. 13.1), that one large sample suffices to determinate the representative petrographic composition. The dependence on grain size is strong. The most meaningful grain-size limit is the 16-mm range, where quartz starts to displace the rock fragments, a development which is successively supported, in the 4-mm range, by the appearance of single minerals of plagioclase and K-feldspar and which leads, consequently, into the realm of sand (Figs. 13.2, 13.3). Thus the proper gravel fraction reported here is the 16–80-mm fraction. It displays the best agreement with source area conditions. The evaluation of the 4–16-mm fraction often yields contradictory results hard to explain, with less agreement with source area conditions.

Ternary diagrams of the system granitic/metamorphic/sedimentary rocks (Fig. 13.5) display the development from the areal distribution of the source rocks over the effective source rock composition to the river-mouth sediment in terms of the two gravel fractions. Almost all possible combinations are realized, the picture being extremely differentiated, close to confusing. Even many years of field experience, intuitive knowledge or qualitative visual perception strongly suggest to the field geologist that geological parameters in Calabria do not change very much from river to river, but they do.

Surprisingly, the miscibility gap (Fig. 13.5), the field of non-existing source area compositions, is exactly the field occupied by the resulting river gravel compositions. The reason for this is that sedimentary clasts have only a small chance to become transmitted into the river-mouth's gravel fraction. So the cluster of the coarse gravels shifts away from the sediment corner, but the ratio granitic/metamorphic rocks, at least as a tendency, is mostly maintained or moves slightly towards more granitic compositions. Some exceptions, however, exist. This tendency, a selection to the disadvantage of metamorphic rocks, becomes even more striking in the finer gravel fraction. A certain positive feedback cannot be excluded, because recycled clasts from the conglomerates support this tendency.

Figure 13.6 demonstrates the high variability of both effective source rock composition and petrography of river-mouth gravels even between neighbouring rivers. However, the comparison of the lithologic composition of the 19 source areas with the petrography of the main components of its river-mouth gravels yields surprisingly high correlation coefficients, $r = 0.76^*$ for granites, $r = 0.71^*$ for metamorphic rocks but only $r = 0.45$ for sedimentary rocks. It is the fate of most sediments of the area that they are too weak mechanically to survive the attrition engine of fluvial transport. The Jurassic limestones, solid and lithologically very homogeneous over the entire area, are the only exception, their debris correlating very well with the source area, $r = 0.87^*$. To conclude, the broad variety of single cases makes it impossible to rely on the results of the investigation of simply one or two rivers. Only the detailed quantitative investigation of large areas yields more significant and more reliable trends of provenance.

Chapter 14

River-Mouth Sand, Silt and Clay

The number of detailed studies on modern stream sands carried out by conventional analytical procedures established in the past 20 years is limited. Recent data on the world's big rivers indicate that increased investigation of modern small streams is needed for a better understanding of sandstone genesis (Potter and Franzinelli 1985, p. 34).

As a matter of fact, research work focussing on the quantitative assessment of the variables that control sand mineralogy of short-headed streams is growing (McBride and Picard 1987). The problem of the complex interaction of the factors influencing composition makes it necessary for investigators to concentrate on stream sands derived from single sources (Tortosa et al. 1989) and to isolate the single variables that control sandstone composition, such as mixing of distinct sources (Mack 1981), effect of dilution (Whetten et al. 1969) abrasion and selective sorting (Shukis and Ethridge 1975), influence of climate (Basu 1976) and topography (Johnsson and Stallard 1989). Single-source stream sands have also been studied for testing the provenance discrimination models currently used (Girty et al. 1988). Our case study takes advantage of the data on basin configuration, relief, climate, source rocks, erosion rates and water chemistry reported in the previous chapters; the mineralogy of soil sands has been discussed in Chapter 8 and can also be compared to the mineralogy of stream sands (Suttner et al. 1981). As a consequence, the composition of Calabrian sands is a baseline that can be used as a standard of comparison. Because sample texture and counting procedure also exert some control on detrital modes (Zuffa 1985), attention is paid to the precise definition of grain categories and related petrologic parameters (Ingersoll et al. 1984), as well as to the influence of grain size on mineral proportions (Mann and Cavaroc 1973).

Summary. Calabrian river-mouth sands are texturally and mineralogically immature. Quartz grains are angular and average 0.19 (Powers). The mineralogic stability index (Q/F+Rf) is commonly < 1.0. Chemical weathering is of minor importance with respect to mechanical disaggregation. Grain proportions for medium-fine sands average Q44 F20 Rf36 and Lcr73 Lvo0 Lse27. The lithic population is dominated by crystalline phanerites, both grain size and counting procedure exerting a relevant control on detrital modes. Calabrian river sands provide compositional reference data based on the previously defined metamorphic, granitic and sedimentary source areas (Chap. 8). Granitic and metamorphic provenances can be separated simply by determining large mineral groups, such as quartzose, feldspathic and lithic grains and/or transparent, opaque and altered heavy minerals; heavy minerals, however, discriminate source provinces better than light minerals (Chap. 18).

14.1 The Mineralogical Composition of the Sand Fraction (by Renzo Valloni)

Data Base. Eighteen samples from the campaign fall 1972, and 19 samples from spring 1973. Light minerals were studied only in the 1973 campaign, heavy minerals in both 1972 and 1973 and roundness determinations were made from a total of 27 samples randomly selected from both campaigns.

Terminology. See Chapters 8.1.3.3 and 8.2.2.

Method. Light Minerals. To study the dependence of mineralogical composition from grain size, two different fractions were analyzed for each sample, 50–2000 μm and 125–500 μm. After H_2O_2 treatment to remove organic matter and residual clay, these fractions were indurated with epoxy and thin sectioned without staining the feldspars. For each sample a number of 350–400 grains was counted with a mechanical stage under polarized light. Roundness was determined on the 125–500-μm size fraction by visual estimate, using Power's (1953) comparison chart and counting 50–70 quartz grains per sample.

Heavy Minerals. Analyses were run with the 63–250-μm fraction. Minerals were separated using tetrabromoethane ($d = 2.96\ g/cm^3$) and their percentage within the 63–250-μm fraction was determined by weight. The microscopic analysis was conducted by strip counting of loose grain mounts and proceeded until at least 300 transparent non-micaceous grains were determined.

14.1.1 Texture (Roundness)

The Palizzi [04] was chosen to check the influence of grain size on roundness which expectedly increases with size:

Fraction	Roundness
125 – 250 µm	0.15
250 – 500 µm	0.18
500 – 1000 µm	0.23
1000 – 2000 µm	0.26

Roundness values of the 19 river-mouth sediments range from 0.14 to 0.26 (Fig. 14.1) and average 0.19 with a standard deviation of s = 0.03. According to Power's scale most values fall in the angular class (0.17 – 0.25) with tails in the very angular (< 0.17) and subangular (0.25 – 0.35) classes.

A moderate geographic trend is shown with higher values (0.21 – 0.26) for rivers from Bruzzano [06] to Portigliola [11]. The degree of roundness and the percentage of rounded grains should depend on the amount of recycled input into the river basin, because sand-sized first-cycle minerals have

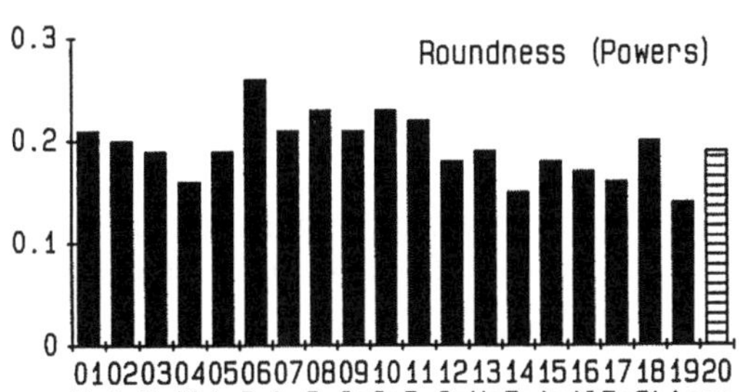

Fig. 14.1. Mean roundness of the quartz grains (125 – 500 µm) of the river-mouth sands according to Powers (1953). *01 Me*lito; *02 Am*endolea; *03 Pa*squale; *04 Pal*izzi; *05 Ar*anghia; *06 Br*uzzano; *07 La*verde; *08 Bu*onamico; *09 Ca*reri; *10 Co*ndoianni; *11 Po*rtigliola; *12 Ge*race; *13 No*vito; *14 To*rbido; *15 A*mu*sa; *16 Al*laro; *17 Pr*ecariti; *18 St*ilaro; *19 As*si; *20 xx* mean of the basins 01 – 19. Average roundness is 0.19 or angular. There is no good correlation with basins draining Tertiary sandstones as possible source of recycled and therefore less angular grains. We have no explanation for the very slight tendency of decreasing roundness values going *from left to right* in the diagram, from the metamorphic Aspromonte Province to the granitic Serre Province

no chance to become rounded during the extreme short Calabrian river transport. Therefore, sandstone- and siltstone-rich basins should yield higher amounts of rounded sand than other ones.

However, no good correlations with sandstone and siltstone are obtained, correlation coefficients not exceeding r = 0.54. Too little and too changing amounts of rounded quartz grains exist in the immature sandstones (Chapter 8.1.3.3).

14.1.2 Framework Mineral Composition

Grain categories and petrologic parameters have already been defined in Tables 8.3, 8.4 and 8.5. Results of analyses are presented below together with the discussion of the controls exerted by counting procedure and grain size on mineral composition.

Mineralogic Classification. In terms of primary parameters, analyses show similar proportions of quartzose, feldspathic and lithic grains with quartz ranging from 30 to 60%; according to the terminology of Crook (1960, 1974) Calabrian river-mouth sediments can be classified as labile or, more specifically, Q-intermediate arenites. According to the primary parameters commonly used for classification (Fig. 14.2) the sandy fraction of the river-mouth sediments occupies the feldspathic, lithofeldspathic and feldspatholithic fields in the QFL diagrams and the lithic field in the QFRf diagram.

Detrital Modes and Counting Procedure. According to the modern, texture-oriented criteria (Chap. 8.1.3.3) polymineralic grains are assigned to the lithics only if fine-grained (< 62 µm), while in the traditional, mineralogy-oriented method all polymineralic grains enter into the lithics category, irrespective of the size of the constituent crystals. Furthermore, there are grains such as lime clasts or micas which are included or excluded depending on the author (Table 8.3).

Key ternary diagrams of mineral composition (primary parameters) of the 125 – 500-µm and the 50 – 2000-µm size classes are set in a row from the most comprehensive to the most restrictive ones, from the total terrigenous grain proportions to the non-micaceous, non-carbonate terrigenous grain proportions (Fig. 14.2). Data points mostly occupy the central area of the diagrams.

In the QFRf diagrams most data points correspond to the lithic field of arenite classification,

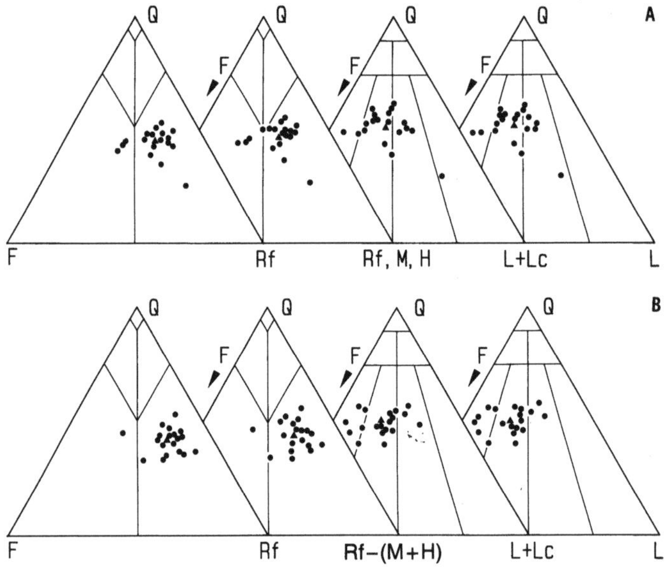

Fig. 14.2 A, B. Framework mineral composition of the 19 river-mouth sands (*triangle* is the average); *Q* quartzose, *F* feldspathic; *Rf* and *L* lithic grains (Tables 8.3 and 8.4); traditional, mineralogy-oriented *triangles on the left*, modern texture-oriented *triangles on the right*. **A** Studied size fractions 125–500 μm. **B** Studied size fractions 50–2000 μm. Single provinces are identified in the corresponding diagrams of Fig. 14.4. Shifting compositions due to different counting procedures (*rows*) are quantified in Table 14.1, those due to different grain sizes (*columns*) in Table 14.2

while in the QFL diagrams they mostly occupy the arkosic one. Indicative averages of the medium-fine grain size (125–500 μm) are Q44 F20 Rf36 and Q52 F28 L20. Spreading of data points in the 50–2000-μm and in the 125–500-μm size fraction is similar. Spread of data points is lower in the traditional diagrams. In the 125–500-μm size fraction standard deviation ranges from s = 6.0 to s = 8.0% in the traditional diagrams while in the texture-oriented diagrams (QFL) the standard deviation equals s = 7.5% for quartzose, 8.0% for feldspathic and 10.0% for lithic grains.

The switch of data clusters caused by the different counting procedures is quantified in Table 14.1 for the 125–500-μm fraction. The trend is well known in the literature with average values drifting on a line perpendicular to the QF margin (Ingersoll et al. 1984). Situations at extremes, from QFRf to QFL, cause a drift of 8% for quartzose and feldspathic grains and of 16% for lithic grains. As expected, the most relevant differences are found passing from the mineralogy-oriented QFRf to the texture-oriented QFL representations (Table 14.1).

Corresponding data from the 50–2000-μm size fraction show a similar trend with the following peculiarity: passing from QFRf−(M+H) to the QFL+Lc diagram, a much more relevant difference is shown by feldspathic (+14 instead of +7) and lithic (−19 instead of −11) grains, i.e. the drift of data points increases with grain size.

Table 14.1. Control of counting procedure on detrital modes of the 125–500-μm size fraction. Average values of quartzose (Q), feldspathic (F) and lithic (Rf, L) grains are shown according to the differing ternary diagrams of Fig. 14.2, and may be used as reference values for a mixed metamorphic/granitic source. The most evident changes are in the transition from the traditional Q-F-Rf to the modern Q-F-L diagrams of the second and third row. A shift equal to 8% for quartzose and feldspathic grains and to 16% for lithic grains is found between representations at extremes (QFRf, QFL)

125–500-μm size fraction	Quartzose	Dif.	Feldspathic	Dif.	Lithic	Dif.	Cumulative difference		
							Q	F	Rf, L
Q, F, Rf	44	–	20	–	36	–	–	–	–
Q, F, Rf−(M+H)	47	+3	21	+1	32	−4	+3	+1	−4
Q, F, L+Lc	51	+4	28	+7	21	−11	+7	+8	−15
Q, F, L	52	+1	28	0	20	−1	+8	+8	−16

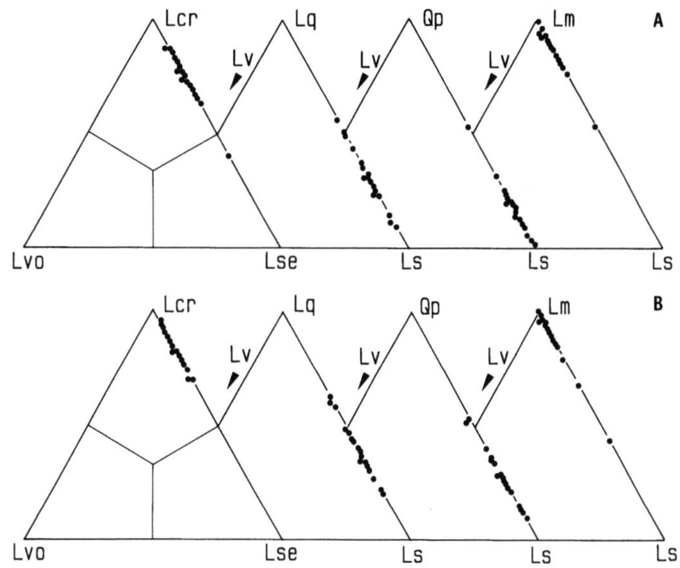

Fig. 14.3 A, B. Lithic grains population of the river-mouth sands; *Lcr* crystalline rock fragment (r.f.); *Lvo* volcanic r.f.; *Lse* sedimentary r.f.; *Lq* lithic quartzose grains; *Lv* volcanic and metavolcanic fine-grained r.f.; *Ls* sedimentary and metasedimentary fine-grained r.f.; *Qp* monomineralic, polycrystalline quartzose r.f.; *Lm* metamorphic fine-grained r.f. (Tables 8.3 and 8.4). **A** Studied size fractions 125 – 500 μm. **B** Studied size fractions 50 – 2000 μm. Single provinces are identified in the corresponding diagrams of Fig. 14.4. Averages of sample plots (125 – 500 μm) from the overlapping granitic and metamorphic sources yield: Lcr 73, Lvo 0, Lse 27; Lq 31, Lv 0, Ls 69; Qp 19, Lv 0, Ls 81; Lm 88, Lv 0, Ls 12

The data set of lithic grain proportions is reported also (Fig. 14.3). In this case, however, parameters of individual diagrams are too different and cannot be examined in a single sequence. The most striking evidence is due to the virtual absence of volcanic rock fragments, so that data points are located opposite of the Lv- (or Lvo-) pole. In the 125 – 500-μm size fraction the lithic association is dominated by crystalline rock fragments (average: Lcr73 Lvo0 Lse27) with an important role played by the coarse-grained types (average: Lq31 Lv0 Ls69). Among fine-grained lithics (average: Qp19 Lv0 Ls81) polymineralic types are overwhelming with fine-grained metamorphics dominating on siliciclastics (average: Lm88 Lv0 Ls12).

The above diagrams have also been proposed as tectonic provenance indicators; depending on the provenance scheme adopted, the data plot in fields indicative of a fold-thrusted foreland (LqLvLs, Valloni 1985), collision orogen (QpLvLs, Dickinson and Suczek 1979) and suture belt (LmLvLs, Ingersoll and Suczek 1979).

Grain-Size Control. The ternary diagrams of primary parameters (Fig. 14.2) as well as those of lithic grains parameters (Fig. 14.3) can also be cross-checked in order to determine the influence of grain size. Roughly, it can be assumed that analyses on the 125 – 500-μm and 50 – 2000-μm size fractions are indicative of sorted medium-fine and unsorted medium-coarse samples, respectively.

Whatever the diagram, data remain quite clustered and the 125 – 500-μm size fraction, which is that commonly utilized by sedimentary petrologists, largely overlaps with the "total sand" data of the 50 – 2000-μm size fraction. QFL proportions for example vary by only 2 – 6% (Table 14.2).

The medium-fine fraction of sands which is easily found in most depositional sequences is thus well representative of average grain proportions and is recommended for comparison purposes.

Table 14.2. Control of grain size on detrital modes (see Fig. 14.2). Shift of the quartzose, feldspathic and lithic grain percentages for each type of ternary diagram with increasing grain size, passing from medium-fine to medium-coarse sand. Average proportions move to the lithic pole in the traditional QFRf diagrams (5 – 6%) and to the feldspathic pole in the modern QFL representations (4%)

Transition from medium-fine to medium-coarse size fractions	Quartzose	Feldspathic	Lithic
Q, F, Rf	−2	−3	+5
Q, F, Rf − (M + H)	−3	−3	+6
Q, F, L + Lc	−2	+4	−2
Q, F, L	−2	+4	−2

The interpretation of size-controlled drifts of sample plots can be made easier by knowing that polymineralic grains are dominated by plutonic and metamorphic phanerites and subordinately represented by metamorphic aphanites (Lmet) or siliciclastics plus carbonates (Lsed). Because lithics (especially the phanerites), are known to enrich with increasing grain size (Grantham and Velbel 1988), some prediction could also be made. If one passes from the 125–500-μm to the 50–2000-μm size interval the sample cluster of the primary parameters move towards the poles containing mostly lithic phanerites (Tortosa et al. 1989, Figs. 4 and 5), i.e., the feldspathic (F) in the QFL type, and the lithic (Rf) in the QFRf type diagrams (Table 14.2). The clusters of the lithic grain proportions obviously drift toward the poles of quartzose phanerites (e.g. Lq) and crystalline lithics (e.g. Lcr) passing from the 125–500-μm to the 50–2000-μm size (Fig. 14.3).

14.1.2.1 Provenance Relations

The source area is divided into the Aspromonte Province, mainly metamorphic, basins 01–08, the Zomaro Province, mainly sedimentary, basins 09–13 and the Serre Province, mainly granitic, basins 14–19 (Chap. 2.2.3).

Many petrologic parameters of the Pasquale [03] and Precariti [17] deviate strongly from the averages of their provinces, consequently they are excluded from the ternary diagrams of Fig. 14.4.

From the key petrologic parameters of Fig. 14.4 some general conclusions can be drawn. The lithic grain population which commonly gives a clue to provenance does not cluster separately for each source area either on texture-oriented (LmLvLs) or mineralogy-oriented (LcrLvoLse) diagrams. In the latter case, data points of all provinces indicate a meta-plutonic source because they overlap and plot close to the Lcr pole (Potter 1978).

Data from granitic and metamorphic sources, on the contrary, cluster quite distinctly in terms of quartzose, feldspathic and lithic grain proportions on both texture-oriented (QFL) and mineralogy-oriented (QFRf) diagrams. Samples from sedimentary sources regularly plot between the granitic and metamorphic ones. Consequently, we disregard them and, for further simplicity, we concentrate on modern, texture-oriented parameters only in the following discussion.

The mineralogy of river-mouth sands derived from the metamorphic and granitic sources is characterized as follows (Figs. 14.4 and 14.5):

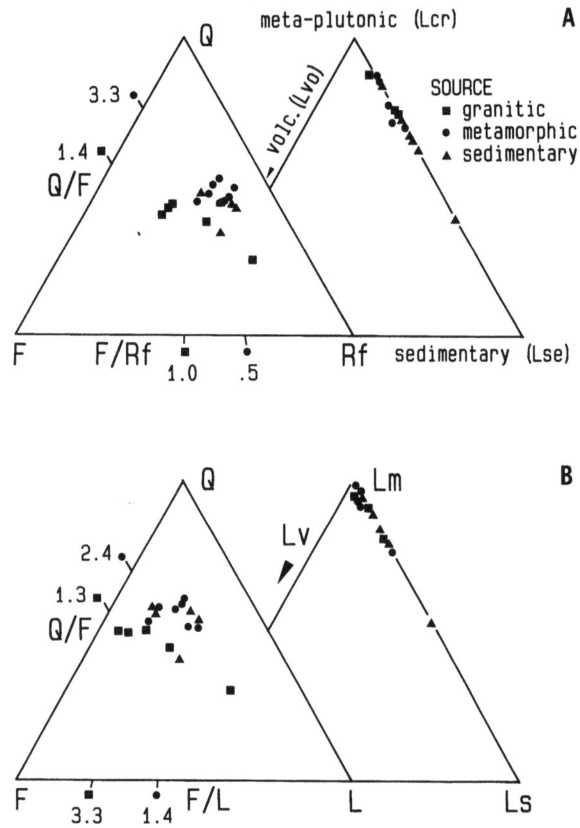

Fig. 14.4 A, B. Mineralogy-oriented (**A**) and texture-oriented (**B**) ternary diagrams showing framework composition (*left*) and lithic assemblage (*right*) of sands from metamorphic sources (Aspromonte Province), sedimentary sources (Zomaro Province) and predominantly granitic sources (Serre Province). For pole captions see Figs. 14.2 and 14.3. For provenance indexes *Q/F*, *F/Rf* and *F/L* see Table 8.5; averages for metamorphic and granitic sources are indicated ***outside the triangles***. Note that the framework grain proportions of the diagrams on the *left* better discriminate the metamorphic and granitic provenances and that sample plots of the sedimentary provenances (*triangles*) onlap the metamorphic and granitic fields

1. Average metamorphic provenances, in terms of the terrigenous framework grain proportions, plot in the upper center while average granitic sources plot in the left center areas of the QFL diagram. This trend is also evident from the literature (Basu 1976; Breyer and Bart 1978; Tortosa et al. 1989), where QFRf clusters for granitic and metamorphic sources always maintain the relative position described above, although influenced by other factors such as climate or grain size. Consequently, provenance indexes are quite distinctive, averaging Q/F = 2.4 and F/L = 1.4 for metamorphic, and Q/F = 1.3 and F/L = 3.3, for granitic sources (Fig. 14.4). The Q/F ratio, in particular, proves very efficient for discriminating granitic and metamor-

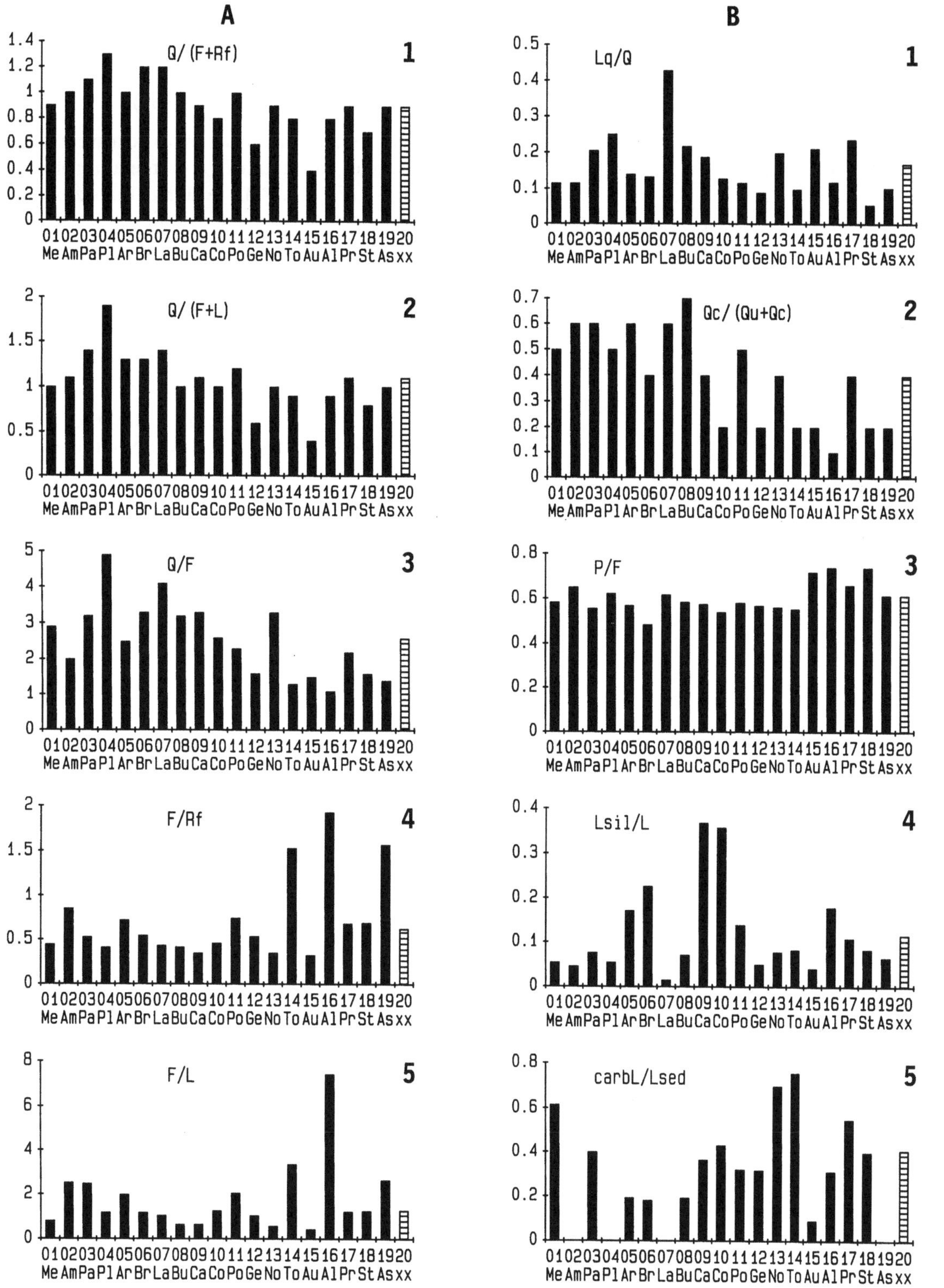

Fig. 14.5 A–C. Some mineral composition parameters of the 19 river-mouth sands (for definitions see Tables 8.3, 8.4 and 8.5). **A** Ratios between QFL populations representing Stability (*1–2*) and Provenance (*3–5*) Indexes. **B** Ratios within the quartzose (*1–2*), feldspathic (*3*) and lithic (*4–5*) grain populations. **C** Percent frequencies of single and composite grain categories. For abbreviations *01–20* see Fig. 14.1. *01–08* Aspromonte Province, mainly metamorphic; *09–13* Zomaro Province, mainly

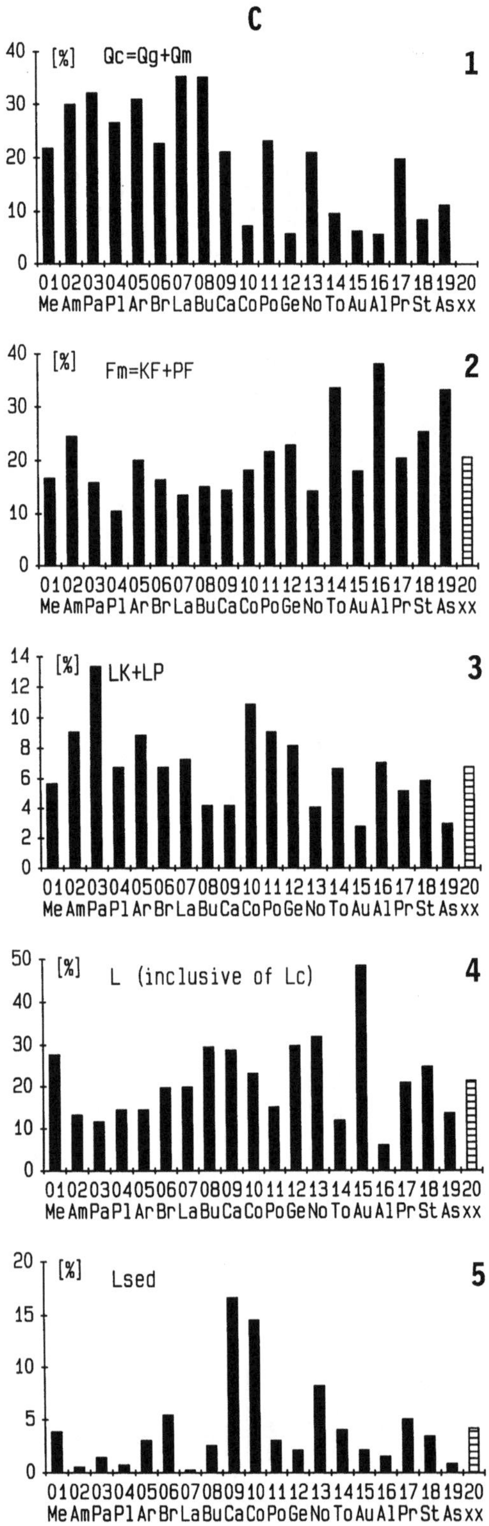

sedimentary; *14–19* Serre Province, mainly granitic. The variability even between neighbouring rivers makes it difficult to use single parameters as significant signals of single rivers. Province averages, however, are well distinct with the exception of P/F, which mostly equals around 0.6

phic sources irrespective of climatic and morphologic conditions as shown by Mann and Cavaroc (1973); Suttner et al. (1981); Breyer and Bart (1978); Tortosa et al. (1989).

2. Some grain categories have relatively different frequencies in granitic and metamorphic provenances (Fig. 14.5C). Among quartzose grains composite monomineralic grains (Qc) as well as total monomineralic grains (Qu+Qc) show relatively higher values in the metamorphic than in the granitic sources. Among feldspathic grains the granitic sources contain as much as twice monomineralic feldspars (Fm, around 30%) than the metamorphic ones (around 15%).

3. Grain ratios calculated within the quartzose types clearly discriminate between the metamorphic and granitic sources (Fig. 14.5B). In particular, metamorphic sources show relatively higher values for ratios such as lithic quartz/total quartz (Lq/Q), microgranular composite quartz/total lithic quartz (Qc-m/Lq) and composite monomineralic quartz/total monomineralic quartz (Qc/Qu+Qc), as shown also by Blatt (1967).

Reference Data. Sands from the modern Calabrian rivers draining metamorphic and granitic sources provide reference data for provenance studies that can be summarized as follows.

1. Speaking in terms of QFL grain proportion, the lithic population (L) yields 20–25% for both sources while quartzose and feldspathic grain average Q55 F24 for metamorphic and Q46 F33 for granitic sources. Thus, the frequency of quartzose grains for metamorphic plus granitic source terranes fluctuates around 50% with the stability index (Q/F+L), averaging 0.85 for granitic and 1.22 for metamorphic sources (Fig. 14.5A).

2. The abundance of lithic phanerites is similar in the two source-area types with respect to both the feldspathic types (LF, 6%, Fig. 14.5C) and the total coarse-grained lithics (LQ+LF, 10%). In both source area types the fine-grained lithic population is dominated by the metamorphic grains (Lm) with subordinate sedimentary (Ls) ones which never exceed 25% (Fig. 14.4).

3. Within feldspathic grains (F) the ratio plagioclase/K-feldspar calculated for metamorphic and granitic sources is almost identical; the classic, plagioclase/total feldspar ratio (P/F) being about 0.6, which is much higher than in the soils (Table 8.6).

Effective Source Area Composition and River-Mouth Sediment. The concept of the effective source area composition (Chap. 6.2.2) makes it possible to compare source and sediment directly. We calculated this also for the light minerals. The effective light mineral composition combines the mean content of light minerals of the soils of the eight key rock types, of the area percentages of these rock types and of their erosion rates. This value is normalized and calculated for each basin and the overall mean. In Fig. 14.9, the light minerals of both the effective source area or predicted sediment composition and the average river-mouth sediment are plotted. Main differences display Qs, Qm, PF, Lm and sL. This seems to make sense. Qs, the composite quartz, yields 3.5% in the soils and 15.1% in the river-mouth sediment. This strongly supports the argument of Harrell and Blatt (1978), who emphasize the high mechanical and low chemical durability of polymineralic quartz grains. The same may be true for Qm, the fine-grained metamorphic quartz. We take this as the mineralogic equivalent to the chemical evidence discussed in Chapter 9, that the chemical composition of the river-mouth sediment of the reference fraction 63–500 µm is closer to the solid source rocks than to the soils; in other words, fluvial attrition is the main source of sand in Calabria rather than soil decomposition. KF and PF, K-feldspar and plagioclase, are more abundant in the river-mouth sediment – newly produced by attrition, and Lm and sL, metamorphic and especially siliciclastic rock fragments, are destroyed by the same process.

14.1.3 Heavy Mineral (HM) Assemblage

Petrologic Parameters. Table 14.3 illustrates the petrologic parameters proposed for characterizing the HM fraction; they are listed according to the increasing detail of the analytic procedure passing from the laboratory preparations to the microscope analyses (Parfenoff et al. 1970). The first two parameters refer to the laboratory preparations. The study size portion is obtained by sample sieving and the heavy mineral (HM) portion is obtained by heavy-liquid separation.

Parameters which follow in the list of Table 14.3 are obtained by microscopic grain counting. A first group is calculated out of the total heavy mineral portion (minerals of the fraction 63–250 µm with $d > 2.97$ g/cm^3 in this study) and is represented by the transparent non-micaceous HM, the transparent micaceous HM, the altered HM and the opaque HM.

Table 14.3. The petrologic parameters used for characterizing the heavy mineral assemblage. Values are percent frequencies on the total indicated; only the stability indexes are taken from the literature. (Pettijohn et al. 1987)

On total sediment (weight):	
Study-size portion	= grains of the studied size fraction within sample
On total size fraction (weight):	
HM portion	= minerals with $d > 2.97$ within studied size fraction
On total minerals with $d > 2.97$ (enumeration):	
Transp. non-micac. HM	= transparent non-micaceous heavy minerals
Transp. micac. HM	= transparent micaceous heavy minerals
Altered HM	= altered heavy minerals
Opaque HM	= opaque heavy minerals
On total transp. non-micac. H.M. (enumeration):	
– Calabrian provenance indexes	
Ep	= epidote
And + Tr	= andalusite + tremolite and actinolite
St + Ky + Sil + And	= staurolite + kyanite + sillimanite + andalusite
Ath + Hbl	= rhombic amphibole + green hornblende + brown hornblende
– Stability indexes (see Table 8.7)	

Among transparent non-micaceous grains are numerous species distinctly counted. These minerals allow the calculation of some indexes related to stability already discussed (Table 8.7) and provenance. The provenance indexes are determined on the total of transparent non-micaceous HM. The minerals which define each index are found to be particularly abundant in one of the Calabrian granitic, sedimentary or metamorphic source area types.

Description of Transparent Non-Micaceous Minerals. Twenty-two mineral species have been counted separately. Specifications on relevant mineralogic varieties are:

Garnet: most grains colourless or light pink; red and yellow grains are also found occasionally

Zircon

Tourmaline: many colour varieties; two pleochroic forms are particularly common with ω = dark green, ε = hazel brown and ω = buff, ε = pale yellow

Rutile, brookite, apatite, barite

Andalusite: colourless grains very common sometimes exceeding the pleochroic ones with α = pink, β = colourless, γ = colourless

Sillimanite, kyanite, staurolite, sphene

Epidote: columnar and fibrous-lamellar crystals common; colours most frequently green and dark green and, subordinately, very light green and brown

Zoisite: contains minor clinozoisite; forms commonly irregular prismatic or prismatic with double terminations

Pumpellyite

Rhombic pyroxene: both enstatite (most commonly) and hyperstene

Diopside

Augite: contains diallage and occasional pyroxene grains having anomalous interference colours

Rhombic amphibole: only prismatic colourless forms

Green hornblende, brown hornblende

Tremolite-actinolite, the latter more frequent.

Among rare and ubiquitous mineral grains that have been assigned to the "others" category are: xenotime, anatase, monazite, olivine, piedmontite, allanite, Na-amphiboles and chloritoid.

Sampling Control. Calabrian river-mouth areas were sampled twice, 18 rivers during fall 1972, and 19 during spring 1973. This makes it possible to estimate the reliability of the data gathered. The mere amount of the HM portion does not change significantly between 1972 and 1973; the correlation coefficient is r = 0.87*. Similar evidence is given by the main mineral groups such as the transparent non-micaceous HM (r = 0.74*) and the transparent micaceous HM (r = 0.66*).

Percentages of single transparent non-micaceous minerals within the 1972 and 1973 samples of each basin are very similar even for minerals characterized by critical hydraulic forms such as hornblende, whose content varies by an average of 3%. This confirms that the "fiumara", the Calabrian rivers, provide a very efficient mixing of the available detritus (Chap. 13.1.2.1).

Frequency of Mineral Categories. Percent frequencies of mineral groups and single minerals used for petrologic parameters are plotted in Fig. 14.6. It is worth noticing that the studied size fraction (63–250 µm, fine and very fine sand) constitutes only the fine tail of the sediment size distribution. In fact, for the 37 samples analyzed the study-size portion averages 14.4% with st. dev. s = 5.9%.

HM values vary greatly as a function of source types. Ignoring an anomalous sample from the Careri [09], the HM average 4.1% with st. dev. s = 2.9%. The 37 samples studied are characterized as follows (Table 14.3, Fig. 14.6).

The opaque HM values are quite uniform and average 19.6% (st. dev. s = 9.8%). The altered HM basically consist of turbid or chloritized minerals whose original nature cannot be recognized; their content is quite variable and averages 29.9% (st. dev. s = 14.8%). The values for transparent micaceous HM change greatly probably because varying chemical composition controls mineral density; they average 14.0% (st. dev. s = 13.1%) and are represented by brown and, subordinately, by green mica and by lamellar chlorite. In the group of transparent non-micaceous HM the most common minerals, listed in decreasing order, are: garnet, green hornblende, epidote, zoisite, tremolite-actinolite and andalusite. In most samples their content varies approximately from 20 to 50%, with an average value of 36.5% and st. dev. s = 15.5%.

Transparent Non-Micaceous HM; Factor Analysis. The grain counts of the 37 samples have been subjected to a Q-mode factor analysis (Imbrie and Van Andel 1964). After grouping the 22 minerals into 17 variables (e.g. summing diopside and augite), the first four extracted factors account for 97.6% of the samples. A simplified set of nine variables, namely,

1. monoclinic pyroxenes
2. green hornblende
3. garnet
4. epidote + zoisite
5. zircon
6. tourmaline
7. andalusite
8. kyanite + sillimanite
9. sphene

accounts for 99.2% of the samples through the four factors below:

Palizzi	[04]:	green hornblende dominated.
Buonamico	[08]:	garnet dominated
Gerace	[12]:	andalusite dominated
Assi	[19]:	epidotes dominated

Figure 14.7 A shows the geographic distribution of the natural units tied to the above factors. The four units are as follows:

2–3–4–6–7–9 = Aspromonte Subprovince ([09] excepted)

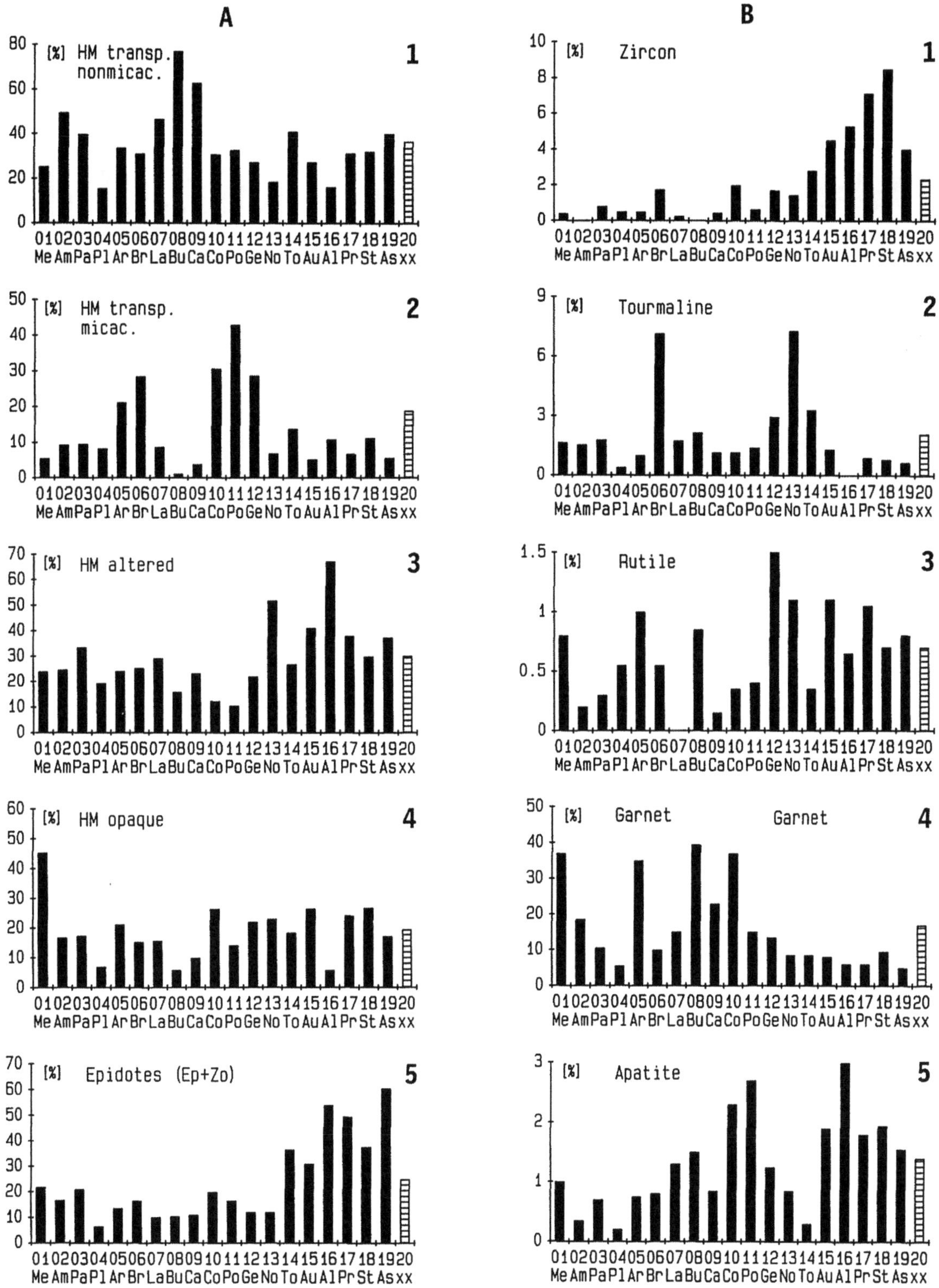

Fig. 14.6 A–D. Histograms of percent frequencies for single and key groups of heavy minerals (Tables 8.7 and 14.3) of the river-mouth sediments. For abbreviations *01–20* see Fig. 14.1

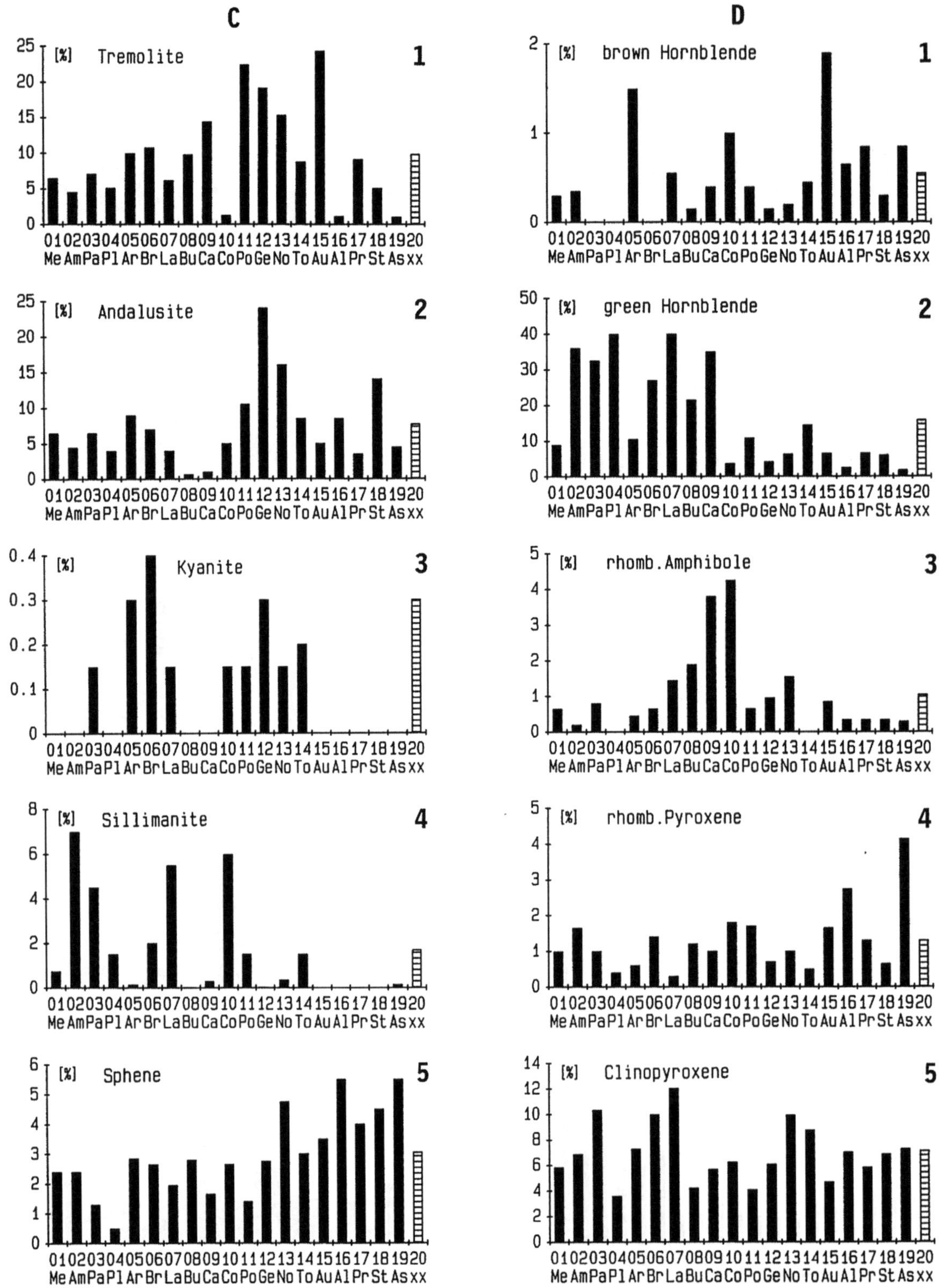
C
D
[%] Tremolite
1
[%] brown Hornblende
1
[%] Andalusite
2
[%] green Hornblende
2
[%] Kyanite
3
[%] rhomb.Amphibole
3
[%] Sillimanite
4
[%] rhomb.Pyroxene
4
[%] Sphene
5
[%] Clinopyroxene
5
01 02 03 04 05 06 07 08 09 10 11 12 13 14 15 16 17 18 19 20
Me Am Pa Pl Ar Br La Bu Ca Co Po Ge No To Au Al Pr St As xx

1 – 5 – 8 – 10	= Aspromonte Subprovince ([10] excepted)
11 – 12 – 13	= Zomaro Province
14 – 15 – 16 – 17 – 18 – 19	= Serre Province.

As a result, HM provenance units identified by factor analysis largely coincide with the physiographic and rock provinces defined in Chapter 2 (Fig. 1.3) (see Chap. 17 for a further multivariate comparison of the provinces). Calabrian provenance indexes (Table 14.3) are plotted in Fig. 14.7 B. Relative maximum values are:

Ath + Hbl	Aspromonte Province
And + Tr and St + Ky + Sil + And	Zomaro Province
Ep	Serre Province.

14.1.3.1 Provenance Relations

The heavy mineral assemblage is plotted in Fig. 14.6. In selecting the Zomaro and Aspromonte Provinces, individual mineral plots give contrasting evidence. According to the factor analysis, the Condoianni [10] is assigned to the Aspromonte (sub)Province, while on the basis of the most comprehensive parameters such as HM portion, transparent non-micaceous and micaceous HM and opaque HM as well as according to andalusite and green hornblende percentages, the same basin should be assigned to the Zomaro Province. Because of these contradictory elements the Condoianni [10] will be ignored in the following. Note that the subdivision of provinces in the next section differs slightly from that of the other chapters.

Aspromonte Province. The effective source rock composition (Chap. 6.2.2) of the basins 01 – 09 is represented by approximately 70% metamorphic and 30% sedimentary rocks (Figs. 5.2 to 5.9, Appendix B 01 – 09). Schists slightly exceed gneisses, while the sediments have similar proportions of sandstone and siltstone with only few percent conglomerates. The heavy minerals assemblage (transparent non-micaceous) is dominated by green hornblende or garnet with values up to 41 and 46%, respectively, and averages of 27.9% (st. dev. s = 11.9%) for hornblende and of 22.7% (st. dev. s = 12.5%) for garnet. Other important constituents are epidote and zoisite, averaging 14.7% (st. dev. s = 4.3%). The sillimanite content, although fluctuating, reaches its maximum with 7%. Total transparent non-micaceous grains reach their highest frequency with an average of 44.4% as well as the heavy mineral portion averaging 5.8% (Fig. 14.6).

Zomaro Province. The effective source rock composition of the basins 11 – 13 (Portigliola-Novito) is represented mostly by sedimentary rocks accompanied by metamorphic (30%) and granitic (15%) ones (Figs. 5.2 to 5.9, Appendix B 11 – 13). Abundance of sedimentary rocks in decreasing order is: siltstone, sandstone, conglomerate and limestone. The heavy mineral assemblage (transparent non-micaceous) is dominated by tremolite-actinolite averaging 19% (st. dev. s = 4.2%) and andalusite averaging 17% (st. dev. s = 6.9%). Other important constituents are garnet (average = 12.3%, st. dev. s = 3.7%) and epidotes (average 13.4%, s = 2.3%). Significant values are assumed by barite with an average frequency of 4.2% (st. dev. s = 4.3%). The mineral group of the transparent micaceous grains reaches its maximum frequency averaging 26.1% (st. dev. s = 20.8%).

Serre Province. The effective source rock composition of basins 14 – 19 is represented approximately by 40% granitic, 40% sedimentary and 20% metamorphic rocks (Figs. 5.2 to 5.9 Appendix B 14 – 19). Metamorphic rocks are almost exclusively schists, while sedimentary types are conglomerates and siltstones with subordinate sandstones. The heavy mineral assemblage (transparent non-micaceous) is dominated by epidotes which average 45.3% (st. dev. s = 11.4%). Relatively high frequencies are shown by zircon and sphene with averages of 5.3% (st. dev. s = 3.1%), and 4.3% (st. dev. s = 1.2%), respectively. Andalusite is also an important component with 7.3% average frequency while brown hornblende, although very little (around 1%) is constantly present.

The altered HM reach their maximum averaging 40.1% (st. dev. s = 14.7%) while the entire HM portion shows a minimum with 2.1% average (st. dev. s = 0.7%).

Reference Data. Table 14.4 lists petrologic variables of increasing detail which characterize source area types dominated by metamorphic rocks (Aspromonte), sedimentary rocks (Zomaro) and granitic rocks (Serre Province). These Calabrian provenances representing strong uplift, high relief, high erosion rates and consequently high sediment production should reveal useful information for both actualistic comparisons and the interpretation of ancient sandstones.

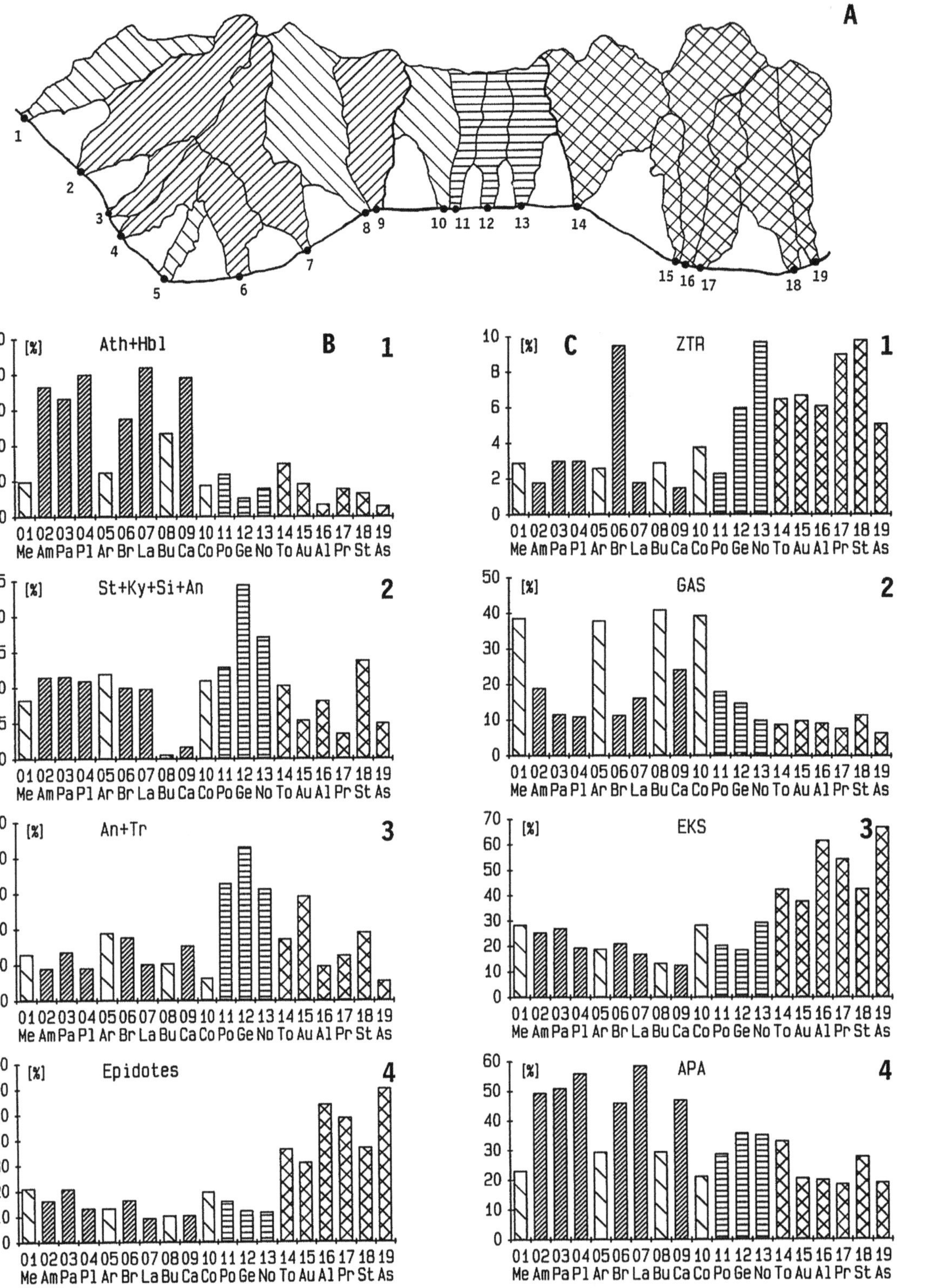

Fig. 14.7. A Geographical distribution of the four heavy mineral units identified by factor analysis of the river-mouth sands. **B** The Calabrian provenance indexes (Table 14.3). **C** Stability indexes (Table 8.7); strangely enough, these HM groups display a marked provenance discrimination power. For abbreviations *01–19* see Fig. 14.1

Table 14.4. Average frequencies (x) and standard deviations (s) of the entire set of petrological parameters (Table 14.3) and of other significant mineral grains of the HM assemblage of the river-mouth sands, for the metamorphic Aspromonte, the sedimentary Zomaro and the granitic Serre Provinces. The Condoianni [10] is omitted

Sediment yield prevalence	Average three provinces	Metamorphic Aspromonte rivers [1–9]		Sedimentary Zomaro rivers [11–13]		Granitic Serre rivers [14–19]	
		x	s	x	s	x	s
HM Portion	3.8	5.8	3.6	3.6	0.8	2.1	0.7
Transp. non-micac. HM	34.0	44.4	17.6	26.1	9.4	31.4	10.0
Transp. micac. HM	15.4	11.4	10.5	26.1	20.8	8.8	4.0
Altered HM	31.2	25.3	7.9	28.2	21.4	40.1	14.7
Opaque HM	19.4	18.8	11.9	19.6	7.2	19.7	8.3
ZTR (ultrastables)	5.5	3.3	2.4	6.0	3.7	7.3	1.8
GAS (stables)	15.6	23.5	12.6	14.3	4.1	8.9	1.7
EKS (moder. stables)	31.4	20.3	5.6	22.9	5.7	51.0	11.5
APA (unstables)	45.3	52.2	12.9	52.2	2.5	31.4	10.3
EZ (epidotes)	24.5	14.7	4.3	13.4	2.3	45.3	11.4
And + Tr	21.9	14.1	3.7	35.9	6.2	15.6	8.4
St + Ky + Sil + And	11.5	8.5	4.3	18.2	5.7	7.7	3.9
Ath + Hbl	15.0	29.4	11.9	8.4	3.3	7.2	4.5
Zircon	2.3	0.5	0.7	1.2	0.6	5.3	3.1
Garnet	14.0	22.7	12.5	12.3	3.7	7.1	1.9
Apatite	1.4	0.8	0.7	1.6	1.3	1.7	1.3
Sillimanite	1.2	2.6	2.5	0.7	0.9	0.2	0.5
Sphene	3.1	2.0	0.7	3.0	1.8	4.3	1.2
Brookite	2.6	0.9	1.4	5.7	6.0	1.2	0.7
Andalusite	9.8	5.1	3.0	17.0	6.9	7.3	3.9
Tremol.-actinol.	12.0	8.7	3.5	19.0	4.2	8.3	8.6
Rhomb. amph.	0.8	1.1	1.2	1.1	0.4	0.3	0.3
Green hornbl.	13.6	27.9	11.9	7.0	3.5	6.0	4.7
Brown hornbl.	0.5	0.4	0.5	0.3	0.1	0.9	0.6
Rhomb. pyrox.	1.3	1.0	0.4	1.1	0.5	1.8	1.3
Clinopyroxene	7.1	7.8	2.5	6.8	3.0	6.8	1.4

The amount of heavy minerals in the sands of metamorphic sources (5.8%) is three times higher than that from granitic sources (2.1%) and correlates positively with the percentage of eroded schists. Identical trends are found by Palomares et al. (1989, Fig. 2) in Spain's Central system. Transparent non-micaceous and transparent micaceous HM show opposite trends. Sedimentary sources deliver sands with the highest percentage (26.1%) of transparent micaceous and the lowest percentage (26.1%) of transparent non-micaceous grains. Altered HM show particularly high values (40.1%) in the sands from granitic sources, while percentages of opaque HM constantly average about 20% in the three source area types.

Mineralogic stability parameters are plotted in Fig. 14.7C for the source area types previously defined by factor analysis. Each stability index shows different average values in the different source area types. This means that these indexes also have an unexpected provenance significance.

Ultrastable content of river-mouth sediments is low. ZTR shows the highest average values in the sediments from granitic sources (7.3%) and subordinately in the sediments from sedimentary sources (6.0%). Low frequencies are also shown by stable grains with an index (GAS) commonly well below 20%, basins 01–05–08–10 excepted.

On the contrary, moderately stable grains (EKS) and, especially, unstable grains (APA) exhibit high frequencies; the former prevail in the sands from granitic sources (averaging 51.0%) while the latter prevail in the sands from sedimentary and metamorphic sources averaging 52.2%. The two indexes of lower stability (EKS + APA) have complementary values, summing their values within each province results between 72 and 82% are obtained.

The provenance indexes based on Calabrian sediments (Table 14.3) are compiled in Table 14.4. Epidotes (Ep + Zo) characterize the granitic source of the Serre, where its content averaging 45.3% is three times higher than in the two other provinces

(Aspromonte and Zomaro). Andalusite, tremolite and actinolite (And+Tr), and staurolite, kyanite, sillimanite and andalusite (St+Ky+Sil+And) characterize the sediment-dominated source of the Zomaro Province with values averaging 35.9 and 18.2%, respectively, which double those of the two adjacent provinces (Aspromonte and Serre). Rhombic amphibole and hornblende (Ath+Hbl) characterize the metamorphic sources of Aspromonte where its content, averaging 29.4%, is three to four times higher than in the two other provinces (Zomaro and Serre).

Among heavy mineral grains reported in Fig. 14.6 and Table 14.4 are some species which poorly reflect differences between source areas. Clinopyroxene (diopside+augite) is quantitively the most relevant and shows percentages around 7% in every province. Similar trends are shown by rutile and tourmaline with percentages constantly equal to about 1 and 2%, respectively.

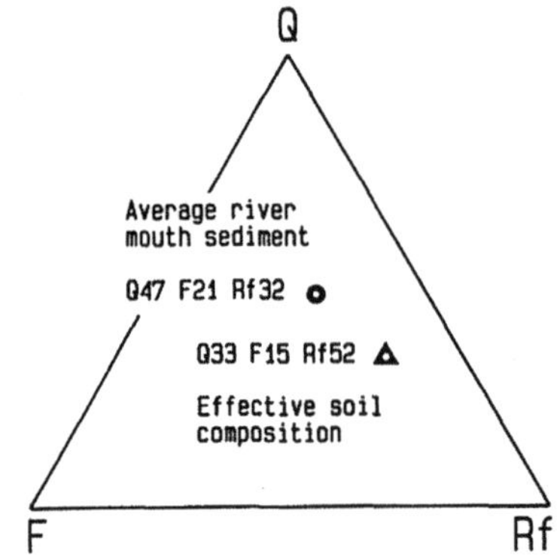

Fig. 14.8. Average QFRf grain proportions of the soil and related river sand. Both values indicate lithic composition (F<Rf). Soil and river sands have a similar Q/F ratio approaching 2/1. The higher rock fragment content of soil sands is mainly represented by the fine-grained sedimentary and metasedimentary types, *sL* and *Lm*, see Fig. 14.9

14.1.4 Discussion

A key aspect of this study is the quantitative relations between the petrography of the source area, the effective sediment composition and the composition of corresponding river sands. For the sake of the "Calabrian standard of comparison", it must be clarified that in the representative basins previously defined the metamorphic sources deliver an average 30% of sedimentary rock and the granitic sources deliver in the average about 20% of metamorphic and 50% of sedimentary rocks.

In terms of grain proportions, the contamination of metamorphic and granitic sources with sedimentary debris is of little importance because the latter mostly consist of immature quartzo-feldspathic rocks (e.g. arkose).

The mineralogic composition of soil is also fundamental. Quantitative comparisons of the bedrock with the overlying soil sand as, for instance, the sandstone and sandstone-soil of Fig. 8.13, are lacking in the literature. Conventional petrographic analyses of single-source soil sands are only a few (Suttner 1974). The data of Cameron and Blatt (1971), which provide modal compositions of Q49 F20 Rf31 for gneiss soil sand, and of Cleary and Conolly (1971), which show the selective alteration of polycrystalline quartz and rock fragments during soil formation processes, are largely coordinated with the results of this study (Chap. 8).

Calabrian river-mouth sands contain higher quantities of minerals unstable during chemical weathering than Calabrian soil sands. Feldspars are enriched in river-mouth sediment from metamorphic sources, and rock fragments in those of granitic sources. Compared to soil samples, river samples contain higher amounts of monocrystalline feldspar, lithic feldspar, polycrystalline quartz (factor 4) and have higher plagioclase/total feldspar ratios (factor 2), which indicate that river sands are mineralogically less mature than soil sands (Fig. 14.9).

In the HM population, percent differences between river and soil sands (Fig. 8.15 and Table 14.4) show that stability indexes have variable and contradictory trends from one province to another; in general, river samples contain more garnet and amphiboles and less pyroxenes and brookite than soil samples. No meaningful correlation is found between the soil versus river variation of heavy

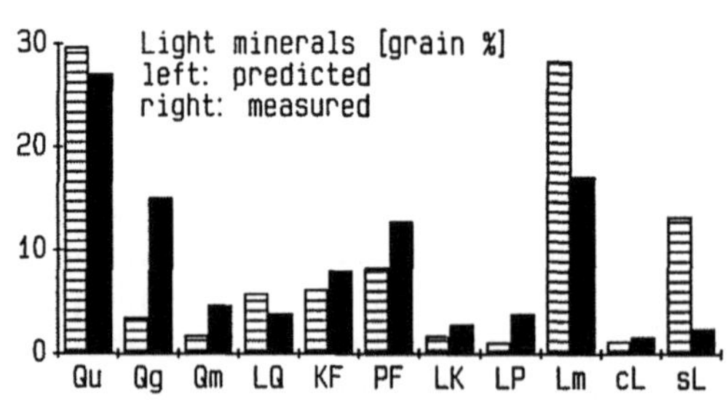

Fig. 14.9. Histogram of average light mineral grain categories; effective or predicted soil composition (*left*) and river-mouth sand (*right*). The polycrystalline grains that are particularly unstable during soil-formation processes are much more abundant in river sands (*Qg*, *Qm*, see Fig. 8.13). This indicates that sand production is essentially related to solid erosion or attrition processes and, much subordinately, to soil-formation processes

minerals and their order of mechanical stability or mineral density (Morton 1984).

The light and heavy minerals both indicate that the mineralogy of river and soil sands of Calabria show no sequential relation. This can be explained by a rapid removal of debris from the slopes with the consequent prevalence of solid erosion rates on chemical weathering as shown, for example, by Whetten et al. (1969); Ruxton (1970); Johnsson and Stallard (1989).

The Calabrian case history is characterized by rapid transport, where quartz grains show no appreciable rounding (Chap. 14.1.1). The percent of angular quartz in the Calabrian small rivers averages 96% on total single-crystal quartz and is higher than Potter's (1978) average (89%) of the world's big rivers from collision settings. Such a textural immaturity is also present in the second-cycle sands of the Zomaro Province (Fig. 14.1), which make it clear that recycling does not necessarily imply maturation of sands, as also shown by Mann and Cavaroc (1973); on the contrary, first cycle matured quartz arenites can be found in due climatic conditions (Potter and Franzinelli 1985).

Comparing the mineralogy of the Calabrian small rivers with the big rivers of the world located on collisional settings (Potter 1978, Table 9), it turns out that notwithstanding the absence of volcanics in Calabria, which average 24% on total lithics in the world's big rivers, detrital modes are very similar, with a Q44 F20 Rf36 in Calabria and a Q36 F17 Rf47 of the big rivers.

Comparable values are also shown by other petrologic parameters related to tectonic setting or mineralogic maturity; the ZTR index, for instance, averages 3% for big rivers, and 5% in Calabria. The only remarkable difference is the very high ratio Al_2O_3/SiO_2 of Calabrian rivers (Chap. 9), averaging 0.19 opposed to the 0.06 of the world's big rivers, which further demonstrates the relatively very low maturity of Calabrian river sands.

14.1.5 Conclusion

Calabrian rivers provide a standard of comparison for metamorphic (gneiss and schist) plus granitic sources of sand (Table 14.1, Figs. 14.2, 14.3 and 14.5); in terms of ternary grain proportions, averages for medium-fine-grained sands are Q52 F28 L20 and Lm88 Lv0 Ls12.

Metamorphic and granitic provenances can be distinguished on the base of both light and heavy mineral fractions (Figs. 14.4 to 14.7 and Table 14.4); within the light fraction, primary parameters (quartzose, feldspathic, lithic grains) are better provenance discriminators than the lithic grain proportions (e.g. Lm, Lv and Ls).

The provenance indexes F/L and Q/F are very efficient for discriminating metamorphic and granitic sources (Fig. 14.4); Q/F, in particular, also has small deviations from the mean within basins.

In addition to the above parameters, metamorphic sources are distinguished by the relatively high values of single-crystal quartz (Qu), composite monomineralic quartz (Qc), microgranular composite quartz on total lithic quartz (Qc-m/Lq) and of lithic quartz on total quartzose grains (Lq/Q), while granitic sources have a much higher monomineralic feldspar (Fm, around 30%) content (Fig. 14.5).

Heavy minerals of the 63–250-μm size fraction of the river sands have high provenance discrimination power and also distinguish unequivocally the sediment-dominated Zomaro Province. In terms of single mineral species (Table 14.4):

- the metamorphic source is dominated by green hornblende and/or garnet, accompanied by relevant epidote+zoisite and sillimanite;
- the granitic source is dominated by epidotes, accompanied by relevant zircon, sphene and andalusite;
- the sedimentary source is dominated by tremolite+actinolite and andalusite, accompanied by relevant garnet and epidote+zoisite.

The percent frequencies of the main categories of heavy minerals are also provenance-related (Table 14.4). Particularly abundant are the total transparent non-micaceous grains (44%) in the metamorphic, the transparent micaceous grains (26%) in the sedimentary, and the altered grains (40%) in the granitic sources.

Minerals which characterize one of the three source-area types with their high incidence on total transparent non-micaceous HM allowed the definition of four Provenance Indexes (Table 14.3):

Epidotes, granitic source
Ath+Hbl, metamorphic source
St+Ky+Sil+And and And+Tr, sedimentary source.

The transparent non-micaceous heavy minerals, grouped according to their diagenetic stability to form indexes such as the ZTR (Tables 8.7 and 14.4) prove to be provenance-dependent and have a surprisingly low value for ultrastables (ZTR), which never exceed 10%.

The mainly sedimentary sources of the Zomaro Province are difficult to recognize with the conventional petrologic parameters; in terms of ternary grain proportions, data points overlap the metamorphic and granitic fields (Fig. 14.4). In ancient orogenic and rapidly deposited terranes recycling or provenance from pre-existing arkoses may result more common than usually thought (Ferree et al. 1988).

Calabrian river sands are difficult to assign to a single provenance type using the existing tectonic-provenance models (Dickinson and Suczek 1979; Dickinson 1988) because they plot near the center of the QFL ternary diagram where provenance end-members merge. Curiously enough, Calabrian metamorphic and granitic sands can be separated by simply using primary parameters. QFRf and QFL type representations have comparable provenance discrimination efficiency (Fig. 14.4), whatever the grain size (Fig. 14.2).

The adopted counting procedure has a strong influence on detrital modes of the metamorphic and granitic source Calabrian sands; data plots drift from the arkosic (QFL) to the lithic (QFRf) composition (Fig. 14.2, Table 14.1). The notorious grain-size control on sandstone composition is also a function of the counting procedure (Table 14.2). With increasing grain size, data points drift towards the lithics in the QFRf representation and towards the feldspar pole in the QFL representation.

14.2 Mineralogy of Silt and Clay

Summary. Ternary diagrams of quartz and feldspar versus illite or chlorite plus kaolinite show that the average silt and clay composition of the effective soil, of the suspended load and of the river-mouth sediments form relatively narrow clusters.

The modern continental environments of Calabria are dominated by gravel; sand seldom exceeds 25% and silt and clay are a kind of unknown substance that never surmounts 1% of the samples, the soils or some puddles after a rainstorm excepted. On the other hand, we measured a suspended load up to 25 g/l during floods of the rivers with a mean concentration of 10% sand, 73% silt and 17% clay. This turbid water or wash load overflows the sea water and forms huge, kilometer-wide plumes off shore. On the shelf, these fines settle only below the mudline at depths beyond 50 m (Fig. 16.5E).

Thus, in spite of their scarcity in the terrestrial deposits, silt and clay are an important factor of the erosional budget of the Calabrian Massif. Analyzing the provenance by means of the gravel and sand fractions, we cannot ignore the silt and clay released from these mountains.

The accumulation of fines in the Lago Costantino, a sedimentary trap formed behind a landslide in 1973 enables an estimation of the amount of suspended load produced in Calabria (Chap. 15.1.3).

Compared to gravel and sand, little effort could be made to examine silt and clay. These analyses have to be seen as a kind of orientation, a by-product of the gravel and sand research and not as a self-reliant study.

Data Base. See Table 14.5 for the number of taken samples. Mineralogical silt analyses were made from soils, suspended load and river-mouth sediments, clay was analyzed only from river-mouth sediments. Analogously to the evaluation of the sand fraction, no siltstone and argillite soils

Table 14.5. Mineralogical composition (X-ray diffraction) of the silt and clay fraction of the river-mouth sediments and the silt fraction of the suspended load. The Palizzi [04] was not sampled

	No. of samples	Quartz	Feldspar	Calcite	Illite	Chlorite + kaolinite
		[%]				
Silt						
River 1 – 18: 1972	18	18.5	27.0	14.0	17.3	23.9
River 1 – 18: 1973	18	15.2	29.6	9.7	21.6	23.6
St. dev. 1 – 18: 1973		2.6	7.4	9.2	5.1	6.2
Mean river: 1972/1973	36	16.9	28.3	11.9	19.5	23.8
Susp. load: 1985	19	7.2	17.2	45.8	13.8	16.2
Clay						
River 1 – 18: 1973	19	6.4	17.1	15.0	22.4	39.3

were taken but these low resistant rocks were sampled themselves.

Method. The soil samples form part of the samples described in Chapter 8. The samples of the river-mouth sediments are split portions of average samples of 271 kg, so these silts and clays represent the average of a large volume of sediment instead of a possible single and tiny layer. The suspended load was sampled on the occasion of a flood, during one day, by means of simple bucket samples from highway bridges.

Analyses were made by X-ray diffraction. Strong limitations of accuracy are given by the method of measuring peak area percentages on the X-ray diffractometer plots.

14.2.1 Soils and Suspended Load

The grain-size distributions of the Calabrian soils are very uniform (Fig. 8.12). The limestone soils excepted, they form a narrow cluster on the ternary diagram characterized by a mean of 71% sand, 19% silt and 10% clay. Table 14.5 displays the petrographic composition of the silt fraction (2 – 63 µm) of the soils of our eight rock types. Feldspar dominates in soils of granite, gneiss and schist, the most frequent mineral in the soils of limestone, conglomerate, sandstone and argillite is calcite; only the siltstones show quartz as the predominant mineral.

The Effective Soil Composition. As pointed out, it is beyond the scope of this study to discuss the origin of this mineral distribution. Our problem is to find a link between the composition of the "original" silts of the source and those of the river-mouth sediments. For this, we calculated the effective soil composition (Chap. 6). Surprisingly, all minerals, feldspar excepted, proved to hold a mean of about 17%; only feldspar climbs to 31% (Table 14.5).

The chance that this fine-grained material should transmit a reliable provenance signal into the river-mouth sediment is not great, compared to sand or gravel. During flood discharge the wash load goes beyond the river mouths; because of the low settling velocities of the fines, their petrographic composition is expected to vary noticeably from deposit to deposit.

The Suspended Load. Our samples are dominated by a mean of 45.8% calcite; the single rivers, however, vary between 1 and 80% calcite. The proportions between the other minerals are comparable to those of the effective soil and the river-mouth sediment (Table 14.5).

14.2.2 River-Mouth Sediment

Table 14.5 shows the mean petrographic composition of the silt- and clay fraction of the 18 river-mouth sediments; the Palizzi [04] was not sampled. We did this campaign twice, during fall 1972 and spring 1973, after the disastrous flooding at the turn of the year (Chap. 15.2.2.2). Surprisingly, there are no significant differences to indicate, for example, a shift towards a minor maturity of the freshly deposited 1973 silts. The variability between the 18 rivers is not very pronounced, Table 14.5 shows the standard deviation of the single mineral percentages along the 18 rivers; quartz is most invariable, followed by illite. This allows it to calculate the mean of 1972/73, in order to obtain more condensed data.

Figure 14.10 compares the mineralogical composition of the silts of the river-mouth sediments; note that even neighbouring rivers may vary considerably. The same is true for Fig. 14.11, which compares the mineralogical composition of the clay of the river mouths. Most striking is the occurrence of kaolinite, limited exclusively to the Laverde [07] and the rivers 09 – 12. This does not correlate with any source area property known to us, but we know of no reason why not.

In Appendix B 01 – 20 the silt and clay mineralogy of each river-mouth sediment is plotted.

Table 14.6. Correlation coefficients of the erosional budget of schist of 18 basins versus the mineral content of the suspended load (only silt 1985) of the respective rivers and the river-mouth sediments, both silt 1972/73 and clay 1973. The Palizzi [04] was not sampled

	Quartz	Feldspar	Illite	Chlorite + kaolinite
Suspended load				
Silt 1985	0.32	0.47*	0.70*	0.60*
River-mouth sediment				
Silt 1972/73	−0.26	−0.04	0.61*	0.30
Clay 1973	0.19	0.07	0.17	−0.09

* Significant correlation (95% level).

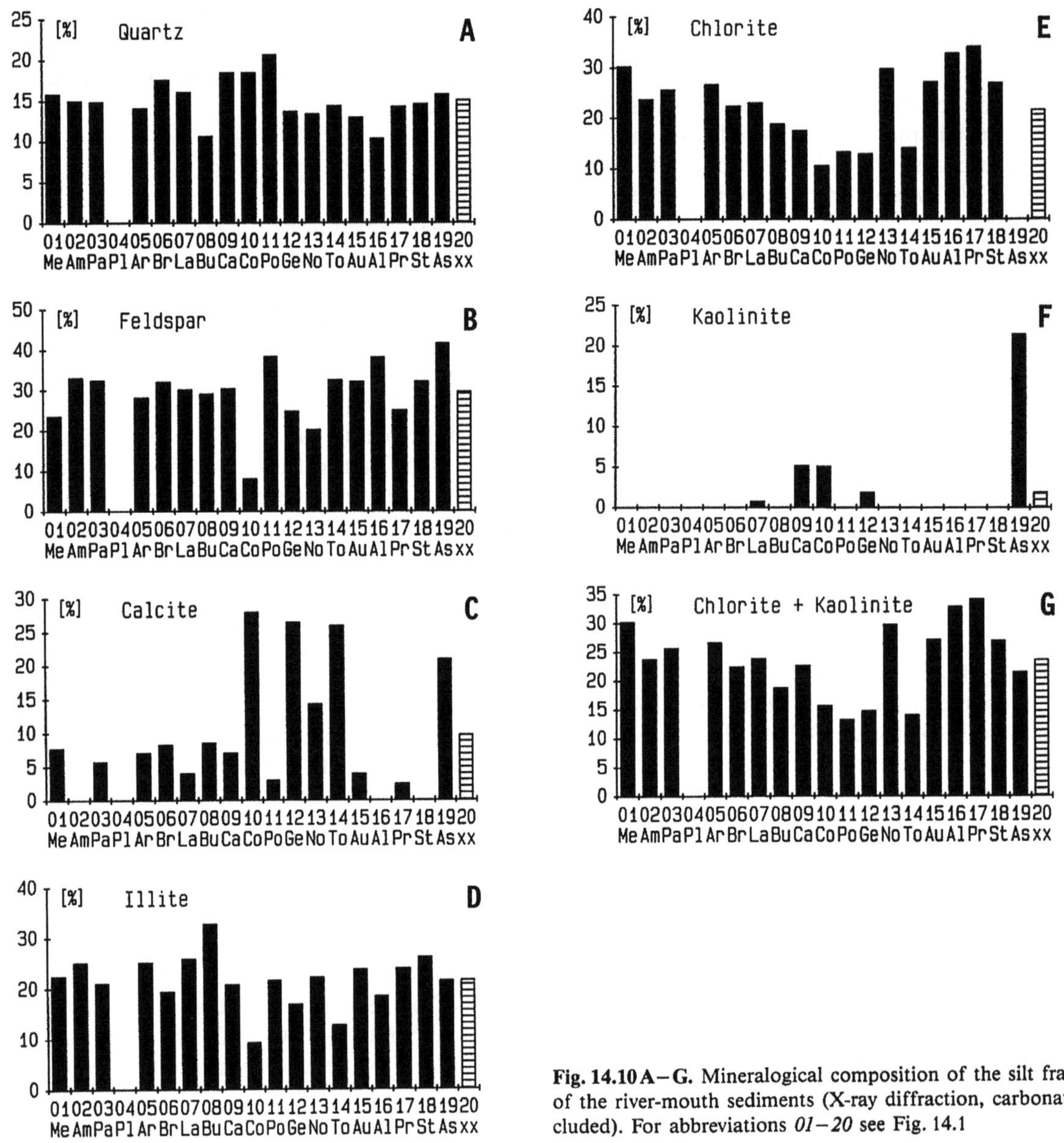

Fig. 14.10 A–G. Mineralogical composition of the silt fraction of the river-mouth sediments (X-ray diffraction, carbonate excluded). For abbreviations *01–20* see Fig. 14.1

14.2.3 Conclusion

Figure 14.12 shows two ternary diagrams with quartz and feldspar plotted versus illite (A) and versus chlorite+kaolinite (B). Because of its great variability, calcite was omitted here. The average silt compositions of the effective soil, the suspended load and the river-mouth sediments form a relatively narrow cluster, only the clay being characterized by portions of illite and especially chlorite +kaolinite higher than expected.

Thus the average mineralogical composition of the river-mouth silt corresponds fairly well to that of the effective soil composition (Fig. 14.12). This contrasts with the river-mouth sands, whose chemical composition better fits the effective source rock composition. However, we do not consider this as a contradiction. Even without differentiating the chlorite and illite of the soils – hydrolysis or disintegration – we believe that these minerals mainly stem from the metamorphic and anchimetamorphic rocks and not from soil formation. The Assi [19] excepted (Fig. 14.10), kaolinite is absent or scarce. An indirect decision as to the character of illite and chlorite, mechanical origin or hydrolysis, should be possible by correlating the

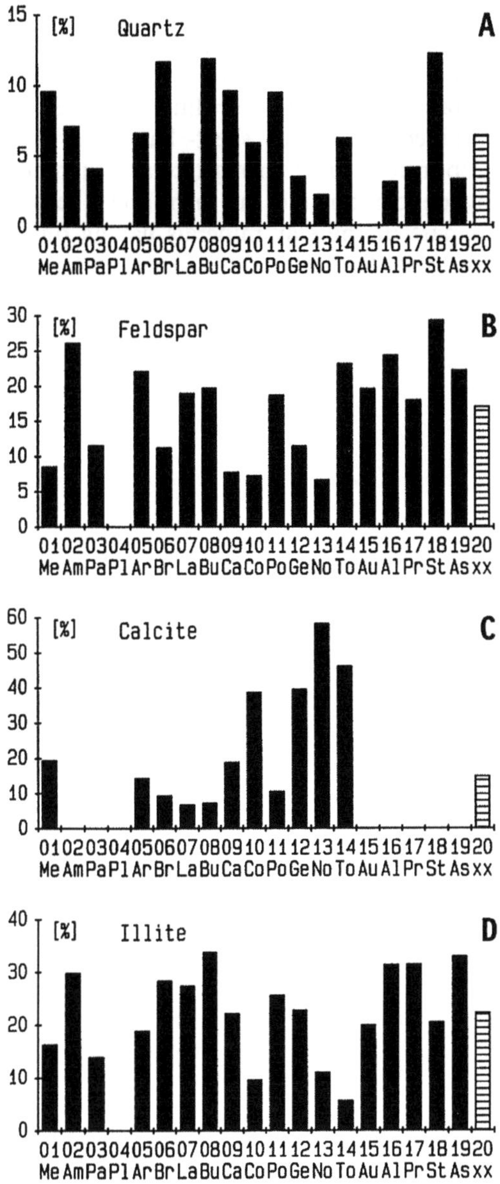

Fig. 14.11 A–G. Mineralogical composition of the clay fraction of the river-mouth sediments (X-ray diffraction, carbonate excluded). For abbreviations *01–20* see Fig. 14.1

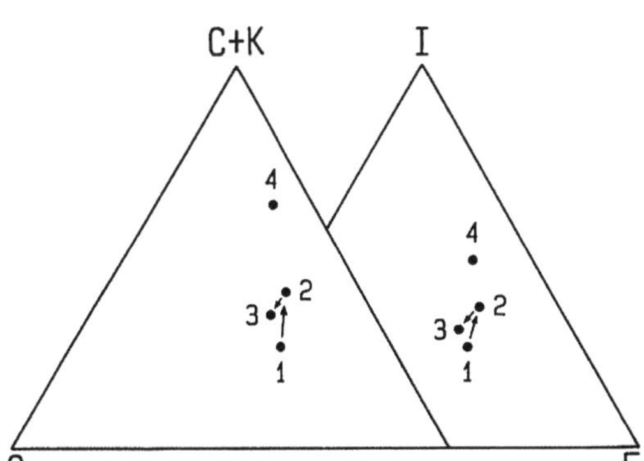

Fig. 14.12. Mineralogical composition of silt and clay, ternary diagram of quartz/feldspar/chlorite+kaolinite and quartz/feldspar/illite. *1* Effective composition of the silt fraction of the soils of the source area; *2* composition of the silt fraction of the suspended load, average of the 19 rivers (1985 campaign); *3* composition of the silt fraction of the river-mouth sediments, average of the 1972 and 1973 campaigns; *4* composition of the clay fraction of the river-mouth sediments, 1973 campaign

respective composition of the river-mouth sediments with their source rocks. On doing so, it turns out that the only promising combination is that of schist and illite. Thus we calculated the correlation coefficients of the erosional budget of schist (18 rivers) versus the mineral content of the silt fractions of these river-mouth sediments (Table 14.6).

We notice the best level of correlation between the schists and the silt of the suspended load. Here, the illite content fits best, followed by chlorite +kaolinite. The mineral content of the sediments correlates much worse, illite excepted. This is understandable, the low settling velocities of the fine-grained and mostly sheet-shaped particles being responsible for a strong lateral differentiation during deposition and, as mentioned before, the major part of the clay minerals having only little chance to settle during fluvial transport.

Part IV

Processes, Budget and Statistical Comparison

All calculations made in the foregoing chapters on effective source area composition or predicted river sediment composition are based on the long-term erosion rates deduced by means of the concept of "missing volume" (Chap. 6.1.1). However, measurements of actual mass movements and erosion rates would be helpful to support the reliability of these long-term calculations.

It is the goal of the first two chapters of Part IV to treat these actual mass balances, landside and seaside, in order to obtain a quantitative picture of the erosional removals at an active plate margin. When this is done, the presentation of the data set of our source-and-sediment study is brought to an end. Consequently, we are ready to compare the single basins and the sediment characteristics statistically. The third chapter of Part IV, Chapter 17, is devoted to such a multivariate approach.

Chapter 15

Mass Balances on Land

A huge landslide in the Buonamico [08] basin 2 years after we started research in Calabria dammed a lake, whose refill was studied over years. This presented a unique opportunity to attempt the quantification of both the amount and the grain-size distribution of the debris trapped in this reservoir and to calculate the short-term erosion rates of the respective hinterland.

Summary. Different methods yield different values of modern erosion rates:

Object: landslides; time span: 50 years. Precipitation data from 1936 to 1986 and estimates of the volumes of landslides on air photos from 1941, 1955 and 1973 result in an estimated erosion rate of about 2700 mm/ka.

Object: sediment trap Lago Costantino; time span: 12 years. The volume of accumulated sediment corresponds to an estimated erosion rate of about 1900 mm/ka.

Object: geodetic survey of the long profile; time span: 26 years. Accumulation along the river course led to an estimated erosion rate of 900 mm/ka, the amount lost to the ocean during this time not taking into account.

Object: overall mass balance of the Buonamico [08]; time span: 50 years. The input into the Buonamico is estimated at $19 \times 10^6\,m^3$, the amount still stored along the river course may add up to $7.5 \times 10^6\,m^3$ solid rock equivalents. The difference, $11.5 \times 10^6\,m^3$, is obviously exported into the Ionian Sea. This corresponds to an erosion rate of 1527 mm/ka, seven times more than calculated from the 1 Ma mean.

15.1 The Buonamico Basin

The Buonamico [08] in the Aspromonte may be seen as the most characteristic and most beautiful basin in South Calabria, its ID is compiled in Appendix B 08.

Data Base. Topographic maps 1:10000 of the Istituto Geografico Militare from 1954; air photos from 1941, 1955 and 1973; annual geodetic survey of six river cross profiles from 1972 to 1980 and survey of the long profile in 1980. Seismic data of the depth of slope fracturing.

Method. The landslides were mapped from air photos and during a field campaign 1975. Repeated geodetic survey of river cross profiles yielded positive and negative differences due to erosion and accumulation, respectively. These values were converted into volumes. The isocontour maps of paleo-valley depths beneath today's river courses stem from Jean Mouton. The development of Lago Costantino and its sedimentation were studied from a small continuously manned observatory during 1973 and 1974 (Ergenzinger group). The progradation of the lake delta was registered annually until 1985.

15.1.1 Slope Stability

It is hard to obtain a fresh and integer rock sample in Calabria. Nearly the whole rock suite is deeply fractured and loosened. Figure 15.1 A demonstrates that our Ionian research area represents the largest area of Calabria through and through exposed to the danger of landslides, >30% of the area is endangered, in the Buonamico [08] basin even 53.5% of the surface is in motion (Caloiero and Mercuri 1980).

Most slopes of the Buonamico [08] basin are steep (Fig. 15.1 B), the mode of the gradient distribution (24%) is in the 25–30° class, whereas the overall mode of the 19 basins is only 15–20°, still steep enough.

According to geophysical measurements (Jean Mouton) the fractured rocks are characterized by velocities of 0.7–2 km/s. The thickness of these bodies of low velocities yields 2–4 m at the lower parts of the slopes and up to 30 m at the middle section, decreasing slightly towards the top.

The rocks below, with velocities of 2.3 or 3 km/s and up to >5 km/s represent compact rocks, the solid ground (Fig. 15.2). It goes without saying that these bodies of fractured rocks are easily mobilized on steep slopes during heavy rainfall and shear off from their base. These loosened rock masses and landslides are the most important sources of debris for fluvial transport and deposition in Calabria.

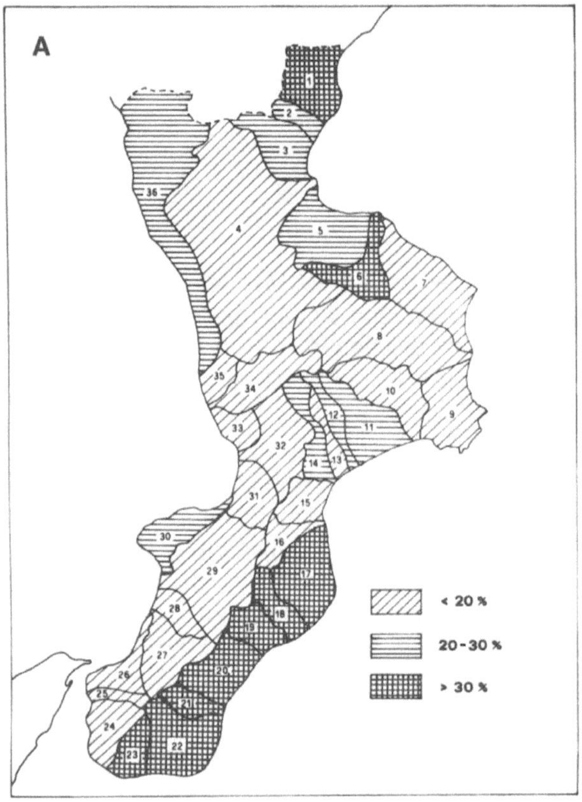

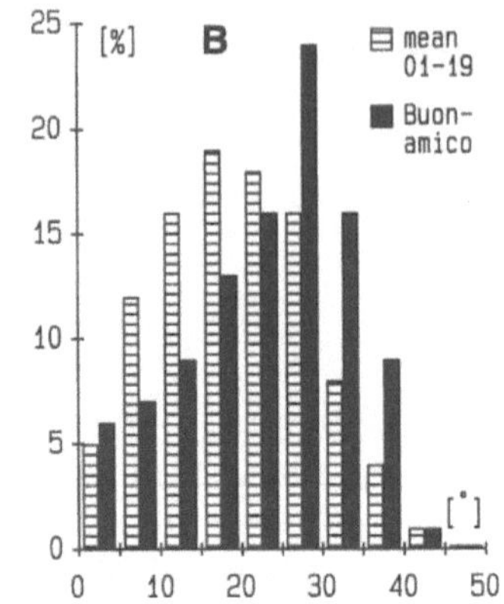

Fig. 15.1 A, B. Landslide activity and slope angles in Calabria. **A** Landslide-affected areas. (After Caloiero and Mercuri 1980). Note that the entire study area is part of the highest category of landslide risk. **B** Frequency distribution of slope angles, average of the 19 basins and Buonamico [08] basin, calculated from the geomatrix (Chap. 3). The average Calabrian slope angle mode of the 19 basins is 15–20°, the Buonamico yields 25–30°

15.1.2 Landslides

Landslides are a kind of trademark of Calabria. Landslides triggered by the disastrous 1783 earthquake dammed up 215 lakes carefully mapped by the contemporaries; one of the largest, the lake of Santa Christina d'Aspromonte, existed until 1823 (Guerricchio and Melidoro 1973).

Figures 15.3 and 15.4 show the largest landslide of the area, the Punta d'Ato slide N of Roccaforte Greco.

15.1.2.1 Occurrence of Landslides

Figure 15.5 shows three generations of landslides in the Buonamico basin, derived from 1941, 1955 and 1973 air photos. Only altitudes above 1500 m are largely exempt from mass movements; close to the river courses they accumulate. The landslides tend

Fig. 15.2. Travel times of a profile in the upper Amendolea [02] (Jean Mouton). *Numbers within the diagram* indicate travel times in km/s; *numbers at the bottom* indicate the thickness of the low velocity layer, these are weathered, fractured and partly incoherent rocks up to 30 m thick, the volumes of future landslides. The section is 2 km long

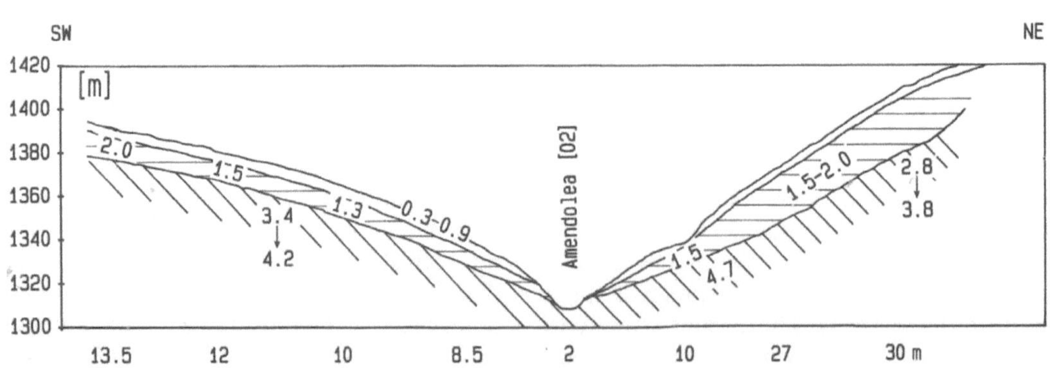

Fig. 15.3. The Punta d'Ato landslide N of Roccaforte Greco in the upper Amendolea [02] basin. Deeply fractured escarpment with rotational slides and wide, open joints extend far down the the opposite slope at the *right*

Fig. 15.4. The Punta d'Ato landslide N of Roccaforte Greco in the upper Amendolea [02] basin, same point of view as Fig. 15.3 but opposite direction. The slide bridges an altitude difference of about 800 m. The entire slope in the *fore-* and *middleground* of the picture is in motion, delivering enormous amounts of debris into the Amendolea (Fig. 2.8)

to appear in families with members of different age. There are both reactivations of the lower parts and a kind of follow-up or new sliding of the upper parts in the rear of older slides. The younger the slide, the scarcer the cover by higher vegetation (Chap. 4). Table 15.1 shows the area percentages of distinct individual landslides.

There is strong evidence that landslides are triggered by excess rainfall. The Ustra landslide

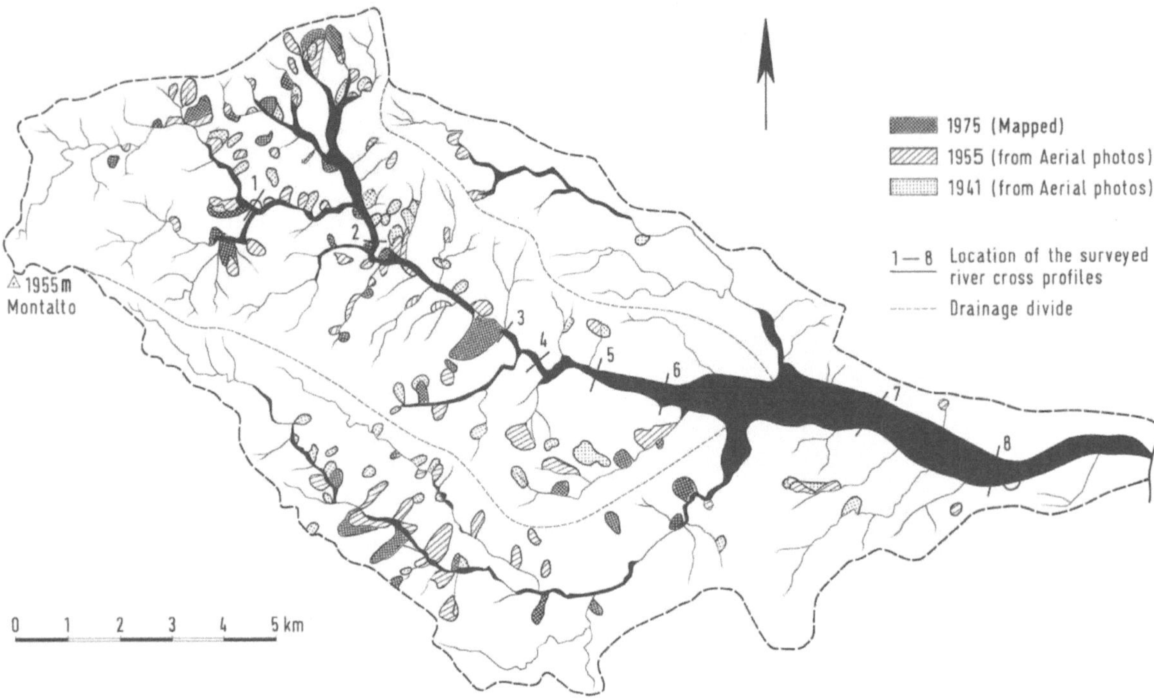

Fig. 15.5. Landslides in the Buonamico [08] basin, from aerial photos 1941 and 1955 and from mapping 1975. Location of the river cross-profiles surveyed 1972–1980

(Fig. 15.6), for example, moved after the 3-day rainstorm in October 1951, which precipitated 1495 mm (Fig. 4.3 C).

15.1.2.2 The Costantino Landslide

Best-investigated and most recent example for landslide formation in Calabria is the Costantino landslide (Guerricchio and Melidoro 1973, Figs. 15.5, 15.7, 15.8).

Before the actual landslide occurred, the slope of Costantino was 1400 m long, bridging an altitude difference of 850 m with a mean gradient of 30°.

Table 15.1. Landslide formation in the Buonamico basin

Aerial photos from	New slide area [km^2]	Volume [$\times 10^6$ m^3]	Export into the river [$\times 10^6$ m^3]
(1936) 1941	2.1	21	4
1955	4.3	43	9
1973 (1986)	3.0	30	6
50 years	9.4	94	19

The area was affected by several older slides. Three secondary valleys had been incising the slope and the Buonamico was actively undercutting the foot of the slope. At the lower parts several small fountains occurred.

The landslide happened during the night of January 3 to 4 in 1973. This was close to the end of a very wet December, and on January 2 there was finally a 400-mm rainstorm at Polsi, a sanctuary some 5000 m upstream (Fig. 15.9).

The Costantino landslide mobilized 23×10^6 m^3 of rock material, piling up a dam 80 m high (Fig. 15.8). Lago Costantino formed behind the dam, and obtained its maximum depth of 40 m after three minor flood waves. At that moment the lake contained a volume of about 3.6×10^6 m^3. On February 4, erosion started by filtration along the deepest parts of the lower slide (Fig. 15.10) and the new Buonamico canyon was eroded in only 10 days. This incision lowered the lake to a depth of about 20 m and created the first huge mobilization of landslide material downstream.

15.1.2.3 Mass Calculations of Landslides

Volume of Landslides. Most landslides move on planes close to the lower limit of the fractured rocks. This is supported by geophysical research (Fig. 15.2). The exposed central part of the Costan-

Fig. 15.6. The Ustra landslide from October 1951, triggered by a 3-day rainstorm with precipitation of nearly 1500 mm (Fig. 4.3 C). This landslide, trapped in its X-shaped upper part, as yet contributes only little to the fluvial system

Fig. 15.7. The Costantino landslide. The picture was taken in March 1973, 2 months after the event. *Left* the lower two-thirds of the slide; note that the landslide piled high up at the slope at *right*. The dam is now cut by the new valley, the lake behind is being refilled by the lake delta

tino slide also shows about 40-m-deep shear planes. Many landslides, however, are not that thick. Assuming a mean depth of 10 m for the Buonamico landslides, certainly a conservative estimate, the volume adds up to 94×10^6 m^3, corresponding to a 67.7-cm-thick layer spread over the entire Buonamico basin of 138.81 km^2.

Landslide Input Into the Fluvial System. Large parts of the landslides remain fixed on the slopes and do not enter the fluvial system for long periods (Fig. 15.6). The amount removed by re-adjusting

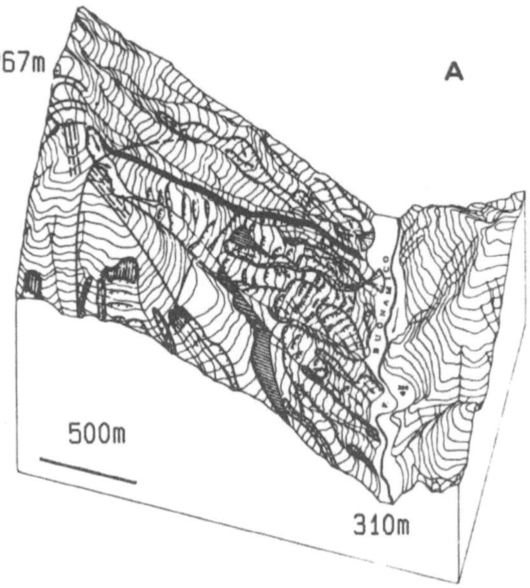

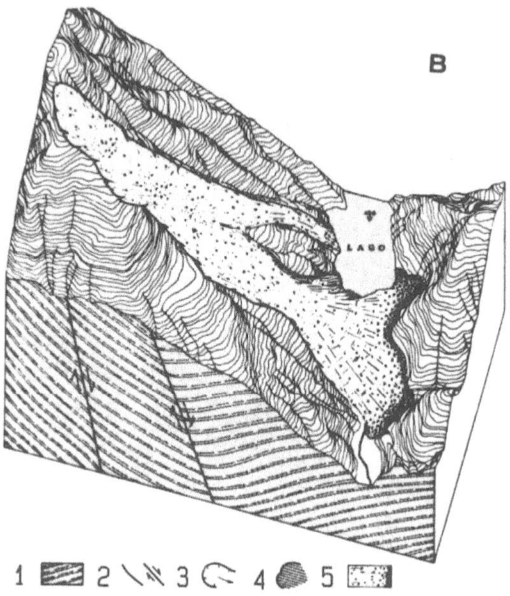

▲
Fig. 15.8 A, B. Stereo model of the Costantino area. **A** Before the event. **B** After the event. *1* Schists; *2* fractures and faults; *3* scarps and direction of movement of landslides and/or erosion; *4* slide planes of old landslides; *5* landslide debris (Guerricchio and Melidoro 1973, Fig. 9)

Fig. 15.9. Daily precipitation (mm scale) recorded at Polsi in the Buonamico [08] basin some kilometers upstream of Lake Costantino. The term "frana" means landslide, which occurred between January 3 and 4, 1973 (Guerricchio and Melidoro 1973, Fig. 17)
▼

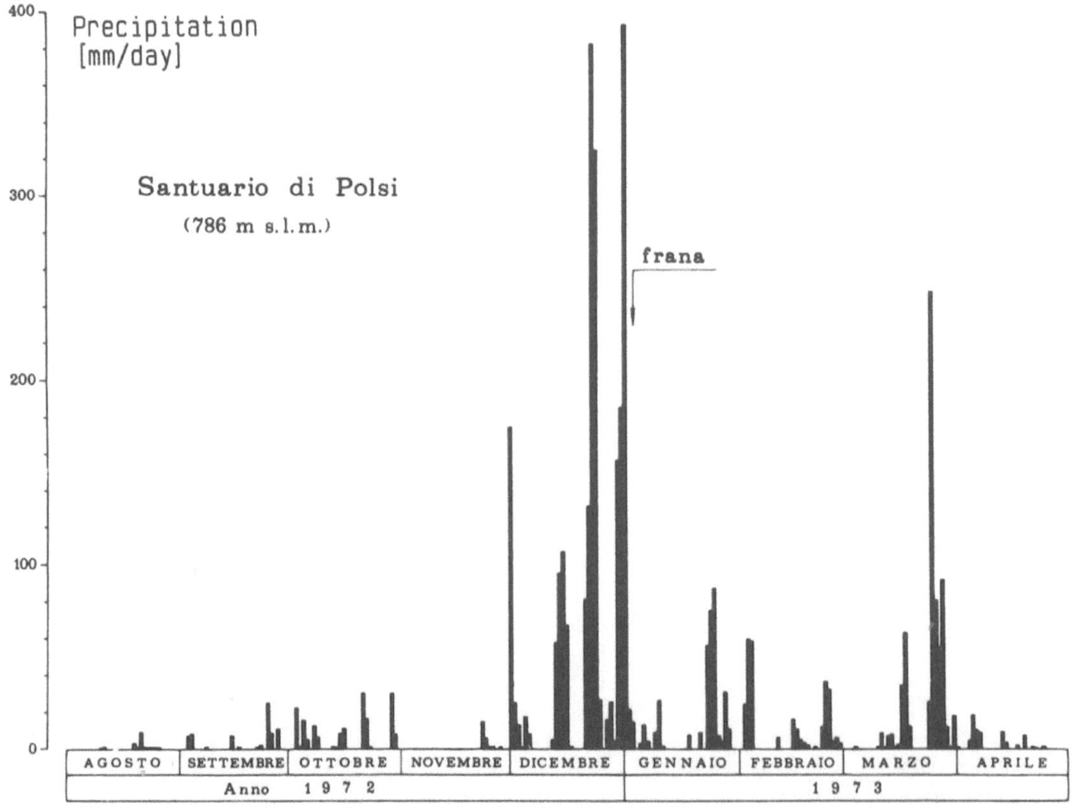

Fig. 15.10. The Costantino landslide immediately after its formation on January 4, 1973. Maximum height of the dam is 80 m. The lake level behind is still rising, note the powerful water discharge in the *lower center*. This is the place where 1 month later the new valley cut was formed, see Fig. 15.7 (Photo by Mario Stranges)

Fig. 15.11. View upstream into the outlet of the newly formed Buonamico valley. To everyone's great relief, this valley was formed slowly and quietly over a period of about 10 days and not abruptly in one catastrophic dam burst. For many years the Costantino landslide will be responsible for an over-supply of the Buonamico with sediment. Above the fall line and during floods, competence suffices to transport even larger boulders, as seen here

rivers is hard to calculate; however, it is the only way to calculate erosion rates via landslides. Our model calculation refers to the Costantino slide: the central part is drained by a deep channel and the main dam is cut by the new canyon (Figs. 15.7 and 15.11). The respective valley volumes make 10% of the entire slide. If the river course after the complete refill of the lake readjusted both the longitudinal and the cross-profile to pre-slide conditions, again 10% of the slide would be removed. This estimate is supported by similar calculations on two landslides in the upper Butramo, the main tributary of the Buonamico. We believe, therefore, the rough estimation to be justified that 20% of the landslides enter the fluvial system more or less closely after the slide event. In the Buonamico this would be 19×10^6 m^3 or a 13.5-cm layer in this basin (Table 15.1).

Time span. The datation of Table 15.1 starts with 1936 and ends with 1986. 1936 was a year of excess rainfall in the Buonamico basin (240 mm on November 15). We presuppose that this event triggered most of the fresh slides visible on the air photos from 1941. An analogous assumption is made for the new slides on the 1955 air photos (rainstorms 410 mm October 16, 1951 and 239 mm October 22, 1953) as for the slides on the 1973 photos (excess rainfall January 2, 398 mm). We know from field experience that until 1986 no new landslides occurred. This time-table makes a maximum calculation possible. During 50 years, from 1936 until 1986, 94×10^6 m^3 were moved by landslides and 19×10^6 m^3 were presumed to enter the fluvial system. This corresponds to an erosion rate of about 2700 mm/ka, an enormous value of actual erosion which surmounts the long-term (1 Ma) average erosional budget of the area, 206 mm/ka (basins only without interfluves), by factor 13.

15.1.3 The Sediment Trap Lago Costantino

Lago Costantino formed behind the Costantino landslide starting on January 4, 1973. The water level rose but never reached the coping due to strong infiltration (Fig. 15.10). In these days the lake was 700 m long with a maximum depth of 40 m, the volume being about 3.6×10^6 m^3. At the rear of the lake the inflowing Buonamico created a rapidly growing delta (Fig. 15.12), comprising about 1.1×10^6 m^3 at the end of February. On February 4 the dam gave way to a major incision, the lake level dropped by about 20 m and a new canyon cut the Costantino slide (Figs. 15.7 and 15.11). By this, the dam lost about 1×10^6 m^3 of debris downstream.

Between 1973 and 1985 the delta proceeded by 300 m, nearly half the lake, and grew by 52.5×10^3 m^2 corresponding to 0.6×10^6 m^3, thus the total volume of the delta by 1985 was 1.7×10^6 m^3 (Fig. 15.13).

The grain-size distribution of the delta sediments corresponds exactly to that of the upper Calabrian river courses (Figs. 10.2C and 10.9B), characterized by only 0.6% silt and clay and 17% sand, the gravel mode being in the 63–80-mm class.

Beyond the foresets of the Costantino delta (Fig. 15.12), in the open lake, there existed and still exists a regime of accentuated silt deposition. Here the sediments consist of 95.5% coarse silt and 4% fine silt, the remainder is clay. The amount of silt, accumulated off the delta at the lake bottom between 1973 and 1985, is estimated at 0.55×10^6 m^3, a number which equals the bed load deposits of the delta after the event, the first day's delta volume excluded. The amount of the finest wash load lost via the outlet is not known, but we do not believe it to be very important. Thus the amount of sediment accumulated in the Costantino trap during 12 more or less normal years, that is, not taking the first stage of delta formation into account, is as follows: 0.6×10^6 m^3 coarse-grained delta deposits (porosity calculated 25%) which account for 0.45×10^6 m^3 of solid rock equivalents and 0.55×10^6 m^3 fine-grained basin sediments (porosity calculated at 50%) which correspond to 0.3×10^6 m^3 of solid rocks: the sum is 0.75×10^6 m^3 solid rock equivalent. From these relations we estimate a proportion bed load/suspended load of about 3:2.

Extrapolated with regard to the 40-km^2 catchment behind the lake we are dealing with an erosion rate of about 1560 mm/ka. This is much less than the 2700 mm/ka yielded via the landslide input, but it makes sense because the lower value refers to 12 "normal" years whereas the higher one includes the peaks of landslide movement during the last 50 years.

It is a simple calculation to predict the fate of Lago Costantino. Seven additional "normal" years will suffice to put an end to this beautiful lake: 1997. One single event, however, of the 1973 magnitude will do so at one blow.

The calculated erosion rate of 1560 mm/ka is certainly still too low, because the Costantino trap did not catch all the debris moved. During the January event, probably even late December, that is, before the Costantino trap was set, huge amounts of debris must have passed this part of the

Fig. 15.12. The lake delta Costantino in September 1976. Most of this sediment body was deposited immediately after the formation of the lake

Buonamico, which have to be added to the budget. This first stage of delta formation in Lago Costantino, $1.1 \times 10^6\,m^3$, which are not included into the above calculation of normal years, may be seen as the late-comer in this event. The only witness of this early flux of debris is a large piece of the Evreo bridge, once situated 3.5 km upstream of the future slide and lake. The bridge was destroyed at some time during the excessive rainstorms at the end of 1972 (Fig. 15.9). A concrete boulder of this bridge, 1.2 m long, still with meter-long reinforcement iron, was found 7 km downstream near San Luca. Evidently there was a very effective sediment transport prior to the Costantino slide. We estimate that $1-2 \times 10^6\,m^3$ passed before the trap closed.

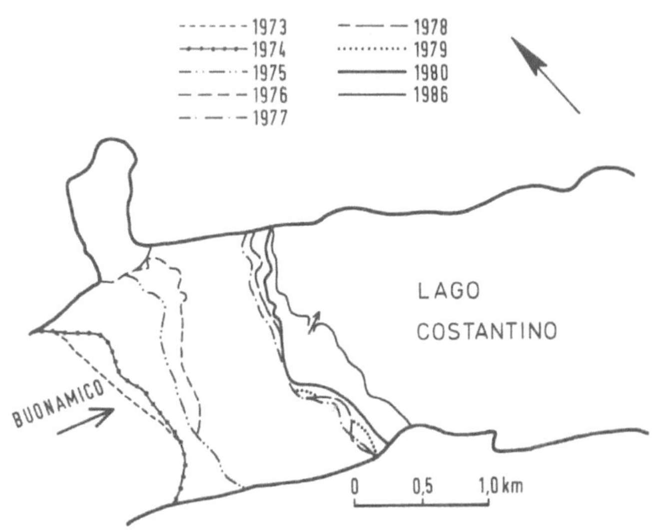

Fig. 15.13. Progression of the Costantino lake delta during the time span 1973–1986

15.1.4 Changing River Profiles

Our geodetic survey started, by good fortune, in autumn 1972, 3 months before the Costantino event. By this and repeated measurements year by year until 1980, followed by a final, single survey in 1987, the main changes of the river bed could be captured.

15.1.4.1 Cross Profiles

Figure 15.5 shows the position of the profiles, Figure 15.14 and Table 15.2 summarize the results. The uppermost profiles 1 and 2 excepted, there is a strong tendency towards accretion due to landslide activity in winter 1972/73. During the following years the re-adjusting river created more differing conditions.

The sections below the Costantino slide (profiles 3, 4 and 5) still accumulate; in profile 3, however, accretion ends in winter 1977/78, but profile 4 continues to grow whereas profile 5 is eroded. Here the older input of the Ustra slide from 1951 had to be removed. Downstream in the foreland, at profile 7 and 8, the trend turns to erosion after 1978/79. As a consequence, the huge input of 1951, 1953 and 1972/73 is not yet digested, the Buonamico was unable to return to quasi-equilibrium conditions. This certainly will take quite a while. If the Lago Costantino is to be filled up by the end of the century, the Buonamico has to remove $3.6 \times 10^6\ m^3$ in order to successfully return into its old bed.

15.1.4.2 Long Profile

In 1980 a geodetic survey of the long profile was conducted in order to quantify the departures from the long profile deduced from the topographic map 1 : 10000, which is based on an air photo evaluation from 1955.

The differences of this time span from 1955 to 1980 are summarized in Table 15.3 and plotted in Fig. 15.14A–C Here plot A shows the enormous distortion of the long profile created by the Costantino slide. Steeper slopes and the respective volumes (C) make this even more evident.

The slight erosional tendencies below the Costantino slide, at km 12 (Fig. 15.14C) stem from an alluvial cone built up by the Ustra slide (1951, Fig. 15.6) and eroded during the Costantino event. An additional, nearly 3-km-wide section of erosion is found close to the coast (Fig. 15.14C). We interpret this as an erosional remainder of the Costan-

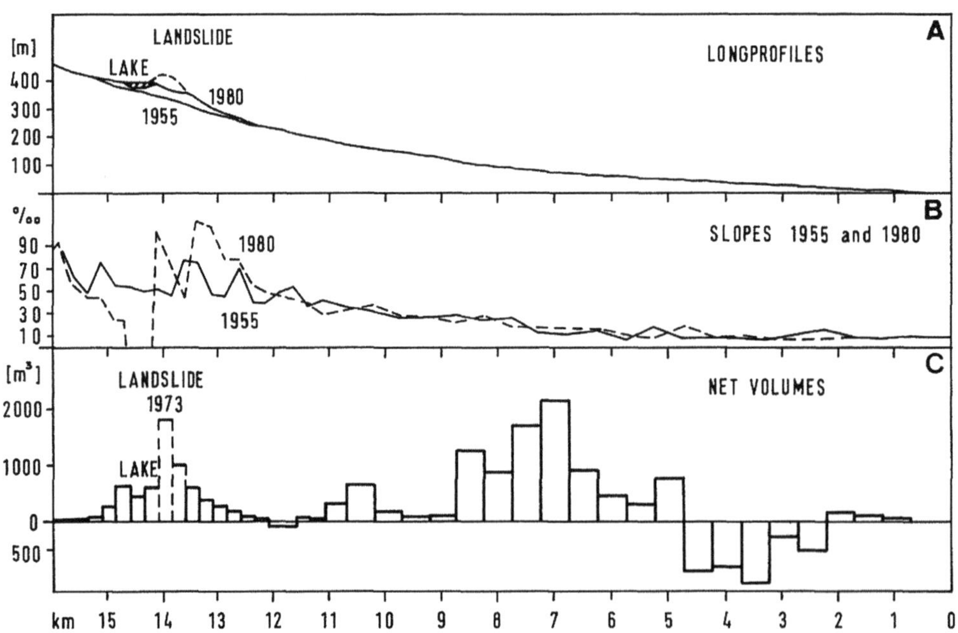

Fig. 15.14A–C. The Buonamico long profile. **A** The 1955 long profile taken from the map 1 : 10000, based on an aerial photo, and the input of the 1973 landslide, the Costanino event. **B** Slopes of the 1955 long profile and the geodetic survey carried out by Ergenzinger and Hirsch in 1980 (pers. comm.). **C** Net volumes comparing the 1955 and the 1980 long profile. The bulk of debris released by the Costantino slide is not yet digested by the system

Table 15.2. River cross-profiles 1–8 of the Buonamico [08] between 1972 and 1987; changes in cross-sectional area, (+) stands for gain, (–) for loss during the reference time. All numbers in m^2. See Fig. 15.5 for geographic positions of the eight profiles

	1972/ 1973	1973/ 1974	1974/ 1975	1975/ 1976	1976/ 1977	1977/ 1978	1978/ 1979	1979/ 1980	1980/ 1987	1972/ 1987
1	–0	–0.5	+0.2	0	0	+1.3	–0.3	0		
2	–18	+0	–15	–1	0	0		–7		
3						+80	–11	–1	–90	
4	+154	+30	+45	+43	+370	–2	+9	+69	+105	+823
5	+124	–33	+33	–5	+7	+13	–47	–1	–37	+54
6			72/75:227	+97	–24	+10	–118	+77		
7	+139	+11	+4	+24	–3	+1	+5	+2	–30	+153
8	+62	+4	+7	+3	+3	+4	+1	–17	+23	+88

tino and probably also 1951 and 1953 events not yet refilled by the approaching surplus of debris of the river sections upstream. All other sections of the river course are accreting, the degree of accumulation correlating well with the river width: the wider the river bed, the more accretion.

The accumulations of the long profile moved by fluvial transportation from 1955 to 1980 (positive differences, Fig. 15.14C), add up to $4.2\times10^6\,m^3$. This may be converted into a solid rock equivalent of $3.15\times10^6\,m^3$ assuming an average porosity of 25%. Spread over the Buonamico basin area of 138.8 km^2 this is a layer of 2.3 cm; referring to 25 years a rate of 900 mm/ka is yielded. This calculation, however, does not take into account either the bed load moved directly into the submarine canyon during this time (Chap. 16) or the wash load not captured in Calabrian river-mouth sediments.

Table 15.3. Comparison of the long profile of 1955 and 1980 of the Buonamico [08]. Width of the valley floor, difference in altitude, difference in sediment volume

No.	Distance from the coast [km]	Width of the valley floor [m]	Altitude difference [m]	Volume difference [$10^3\,m^3$]
0	0 –0.5	100	0.0	0
1	0.5–1.5	140	0.35	50
2	1.5–2.5	230	0.65	150
3	2.5–3.5	420	–0.60	–250
4	3.5–4.5	425	–1.90	–800
5	4.5–5.5	500	1.60	800
6	5.5–6.5	550	0.82	450
7	6.5–7.5	800	2.75	2200
8	7.5–8.5	780	1.20	900
9	8.5–9.5	600	0.17	100
10	9.5–10.5	400	0.38	150
11	10.5–11.5	200	1.50	300
12	11.5–12.5	150	–0.60	–100
13	12.5–13.5	75	3.33	250
14	13.5–14.5	75	26.00	1800
15	14.5–15.5	100	2.50	250
Sum of the sections below Costantino (0–13):				4200

15.1.5 Buonamico's Export into the Ionian Sea

The possible export of the Buonamico into the Ionian Sea can only be calculated indirectly, combining the gains and losses described in the foregoing sections. The total input into the fluvial system of the Buonamico by landslides during the last 50 years is estimated to add up to $19\times10^6\,m^3$. This is the gain. If we are able to calculate the amount still stored in the fluvial system, the difference should correspond to the amount definitely lost into the sea. We assume that before 1973 sediments from the upper reaches of the Buonamico were transported unimpeded. The sediment trap Lago Costantino caught $1.6\times10^6\,m^3$, including the initial delta ($1.1\times10^6\,m^3$ – 25% pore volume = $0.8\times10^6\,m^3$), the later delta and the basin sediments ($0.75\times10^6\,m^3$). The dam of the landslide still contains $3.6\times10^6\,m^3$ barricading the river profile. The lower reaches of the river course store $3.15\times10^6\,m^3$, comparing the 1955 long profile with our geodetic survey from 1980. Thus a total of about $8.4\times10^6\,m^3$ is still stored along the river course. The difference, $19-8.4\times10^6\,m^3$, is $10.6\times10^6\,m^3$, and we believe this amount to be exported into the Ionian Sea. If we take 1936 as reference time, almost 50 years, and the total of $19\times10^6\,m^3$ moved into the Buonamico's river system, an erosion rate of about 2700 mm/ka results. This amount has to be reduced regarding the amount of the effective output into the Ionian Sea, about $10.6\times10^6\,m^3$, so the effective erosion rate is about 1527 mm/ka.

15.2 Paleovalleys and Volumes of Alluvial Deposits

Beneath the surface of the lower courses of almost all larger Calabrian rivers, deeply incised paleovalleys are hidden. The depth of three paleovalleys was investigated geophysically by ARLAB and ITALPROS (Jean Mouton), the results are supported by numerous wells almost exclusively drilled for irrigation purposes. In Table 15.4 the respective volumes of the entire alluvial deposits of the single basins are listed, Appendix B 01–18 shows the depth contours of the paleovalleys. The depth of these paleovalleys below the present sea level is remarkable. The smaller rivers yield maximum depths, in the river-mouth areas, of 20–40 m, larger ones extend to 60 or 80 m and the Torbido [14] leads the 19 rivers with a maximum depth of 150 m below sea level. This is not the only exceptional quality of the Torbido [14] basin. Draining the southern rim of the Serre pluton, this basin is characterized by strong neotectonics (Ciaranfi et al. 1983). It is the only basin where an almost continuous cover of Pliocene argillites extends from the coast to the main divide (Fig. 5.9).

We do not venture to include the volumes of these paleovalleys into our mass calculations in order to complete our long- and short-term erosion rates by a kind of mid-term values. The reason for this is simply the lack of reliable time marks. It has long been known that the Mediterranean Sea is bordered by river-mouth areas with deeply incised paleovalleys cut into the coast and shelf areas during the Messinian event, "when the Mediterranean dried up" (Hsü 1977). But during the Pleistocene, low stages of the sea level again strengthened erosion which must have shaped the valleys. It is impossible to delimit these erosional processes, not to speak of the sedimentary budget, the gain and loss of alluvial deposits. From the many drillings in the alluvial deposits (Fig. 15.15) no time-stratigraphic record exists.

Table 15.4. Mountain, valley and paleovalley volumes in Calabria[a]

	Basin	A[a] [km³]	B [km³]	C [km³]	D [%]	E [× 10⁶ m³]
1	Melito	67.67	53.78	13.90	20.54	134
2	Amendolea	163.38	128.05	35.34	21.63	220
3	Pasquale	13.81	12.63	1.17	8.51	–
4	Palizzi	23.28	20.62	2.66	11.48	–
5	Aranghia	6.20	5.58	0.61	9.89	–
6	Bruzzano	26.12	18.67	7.44	28.51	63
7	Laverde	113.28	86.84	26.44	23.34	153
8	Buonamico	139.17	104.84	34.33	24.66	300
9	Careri	65.24	36.15	29.09	44.59	235
10	Condoianni	40.03	23.26	16.77	41.89	160
11	Portigliola	20.62	14.18	6.44	31.27	158
12	Gerace	20.37	14.15	6.22	30.54	141
13	Novito	32.32	26.60	5.71	17.68	164
14	Torbido	121.59	92.73	28.85	23.73	601
15	Amusa	24.79	17.80	7.00	28.23	68
16	Allaro	117.23	96.60	20.65	17.61	241
17	Precariti	41.19	31.22	9.97	24.20	161
18	Stilaro	75.16	55.65	19.51	25.95	187
19	Assi	58.58	45.17	13.40	22.88	–
20	Interfluves	75.06	57.70	17.36	23.13	–
21	Total	1245.11	942.22	302.88	24.33	–

[a] A: The presumed pre-erosional volume between z_0, the sea level plane and z_2, the upper enveloping surface (Chap. 6, Fig. 6.2), 1 Ma ago. B: Today's still existing volume between z_0, the sea level plane and z_1, the lower enveloping surface, the present-day morphology. C: The volume eroded away during 1 Ma, defined by the difference between z_1 and z_2. D: The percentage of C, the volume eroded away. E: The volume of alluvial deposits (Jean Mouton), defined by the morphology of the paleovalleys (Appendix B) and the present-day surface of the alluvial valley fills.

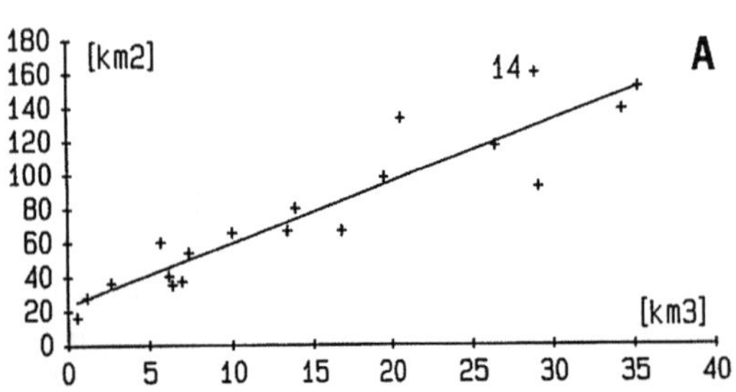

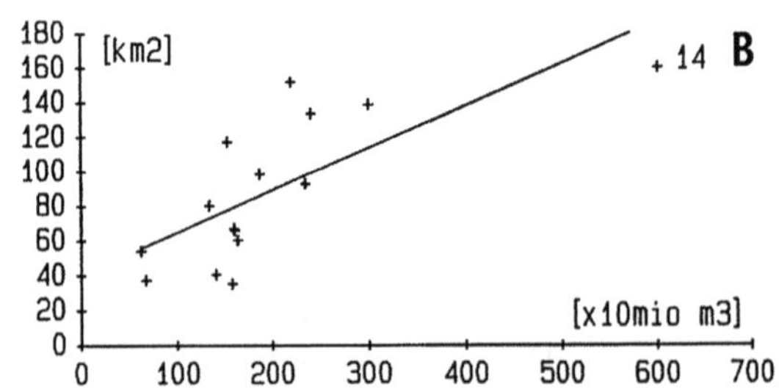

Fig. 15.15 A, B. Basin areas, present-day valley volumes and volumes of alluvial deposits. **A** Basin areas and the present-day valley volumes (Chap. 6) correlate very well (r = 0.93*), including also the Torbido [14], the largest basin of the area. Thus, the mean vertical erosion goes hand in hand with basin size. **B** Basin areas and the volumes of the paleovalleys correlate worse (r = 0.73*), evidently because the paleovalley of the Torbido and its volume filled by alluvial deposits is by far too large, compared to the other basins; a possible signal for an exceptional development of the paleo-Torbido

There are, however, some relations to other basin properties which seem to make sense: the plot basin area versus missing valley volumes (Chap. 6) displays a high degree of correlation between the two (Fig. 15.15A) including the Torbido [14]. Thus the proportion between basin size and the degree of erosion seems to be balanced. Not so in the past. The paleovalley of the Torbido, probably of Messinian age, is too large compared to the other basins. This may be seen as a signal for a special development of this basin in this time. Extrapolating today's conditions, the paleo-Torbido should have had a much larger basin. There are more arguments in this direction (Wilkes 1986).

15.3 Conclusion

It was a bitter experience that even relatively promising prerequisites such as the series of air photos and the unique occasion of the Costantino landslide and sediment trap, not to speak of the long-term and expensive operation of field stations did not suffice to yield more reliable data on erosion rates in this mountainous area. Compared to usual short-term projects in geology, the time span of 50 years bridged by the air photos and even the nearly 20 years of field work are very long, but not long enough to capture Calabrian erosion and fluvial transportation, which are so strikingly characterized by single events. The refill of Lago Costantino, which results in an estimated actual erosion rate of 1900 mm/ka, is the "hardest" value, supported by landslide calculations. Additional help comes from the shelf, where a really independent evaluation comparing the submarine morphology over the last 100 years leads to corresponding results, not only in the order of magnitude. This will be treated in the next chapter.

Chapter 16

Mass Balances in the Marine Realm

The enormous sediment production of Calabria crosses the river-mouth areas and the beaches, is dumped into the heads of submarine canyons or spread over shelf and slope. We want to present some data on submarine morphology and surface sediments which allow us to extend our calculations on mass balances from the land, from the Buonamico [08] basin discussed in the foregoing chapter into the marine realm. The comparison of our echosoundings mapped in 1974 with the original data set of the official Italian marine chart from 1877 yielded even larger differences on the shelf and in the canyon head, be it by accumulation or erosion.

According to the concept of this book, namely a study of provenance and mass balance, this chapter is not intended to make a contribution to the marine geology off Calabria as such.

Summary. The central part of the research area is bordered by the 5-km-wide Bianco shelf and the 2-km-wide Bovalino shelf, separated by the head of the Buonamico-Careri Canyon. The shelf break is between 100 and 120 m water depth; we mapped the slope to a depth of about 900 m. The comparison of our 1974 map with soundings from 1877 reveals negative deviations or erosion in the dimension of 100 m and positive deviations or accumulation up to 30 m. Loci of enhanced sedimentation are the canyon head and the outer shelf, where also the major slump scars occur. The conversion of only accumulation values on the shelf into erosion rates of the respective hinterland yields 7500 mm/ka. This huge value surmounts the 1-Ma average of the entire area (206 mm/ka) by factor 35 and the 50-year average of the Buonamico [08] basin (1527 mm/ka) by factor 5.

Data Base. Italian sea chart No. 23, C° Milazzo-Roccella Ionica, mainly from 1877. An actual marine map from 1974 (Rumohr and Ibbeken 1975). Air photos from 1941, 1955, 1973 and 1983. Ninety six surface sediment samples along ten shelf crossing profiles.

Method. Our modern bathymetric map (Fig. 16.1 A, Rumohr and Ibbeken 1975; Ergenzinger et al. 1978) is based on soundings carried out in 1973 and 1974 with a Krupp Atlas 450 echosounder installed on a 4-m boat; the profiles add up to 480 km (Fig. 16.1 B). Terrestrial navigation was done by sextant bearing. Both the Italian sea chart from 1876/77 and our modern map from 1974 were digitized using the 250×250 m grid (geomatrix). The depths of 3438 points were stored in the data bank. The differences over about one century were calculated and converted into volumes.

Ninety six sediment samples were taken by means of a small van Veen grab and a little high momentum corer using Plexiglas tubes 4 cm in diameter (Meischner and Rumohr 1974). Grain sizes were separated into the fractions <2 µm, $2-6.3$ µm, $6.3-20$ µm, $20-63$ µm and >63 µm, applying the Atterberg method.

16.1 Shelf and Canyon Area and the Delta

Figure 3.7 shows the geographic position of the marine investigation area, extending about 10 km offshore between the Laverde [07] and the fiumaras Gerace [12] and Novito [13]. The area (Fig. 16.1) is representative for eastern Calabrian shelves, including a broad shelf, 5 km wide (Bianco shelf), a narrow shelf, only 2 km wide (Bovalino shelf), a deep submarine canyon approaching the coast (Buonamico-Careri Canyon) and, unique for the entire area, a delta, the Laverde Delta (Fig. 16.8). The area comprises 215 km^2; three larger basins deliver their erosional products into this area, Laverde [07], Buonamico [08] and Careri [09] and three smaller ones, Condoianni [10], Portigliola [11] and Gerace [12].

The width of the shelf and the incision of submarine canyons obviously are dependent on the size of the respective drainage area or hinterland. We assume a causal relationship in the sense that the enormous sediment delivery of Calabrian rivers re-activates and deepens the former Messinian paleovalleys and submarine canyons: the larger the hinterland per unit of coast, the larger the canyons and the smaller the shelf. This concept was developed in Chapter 2.2.6.

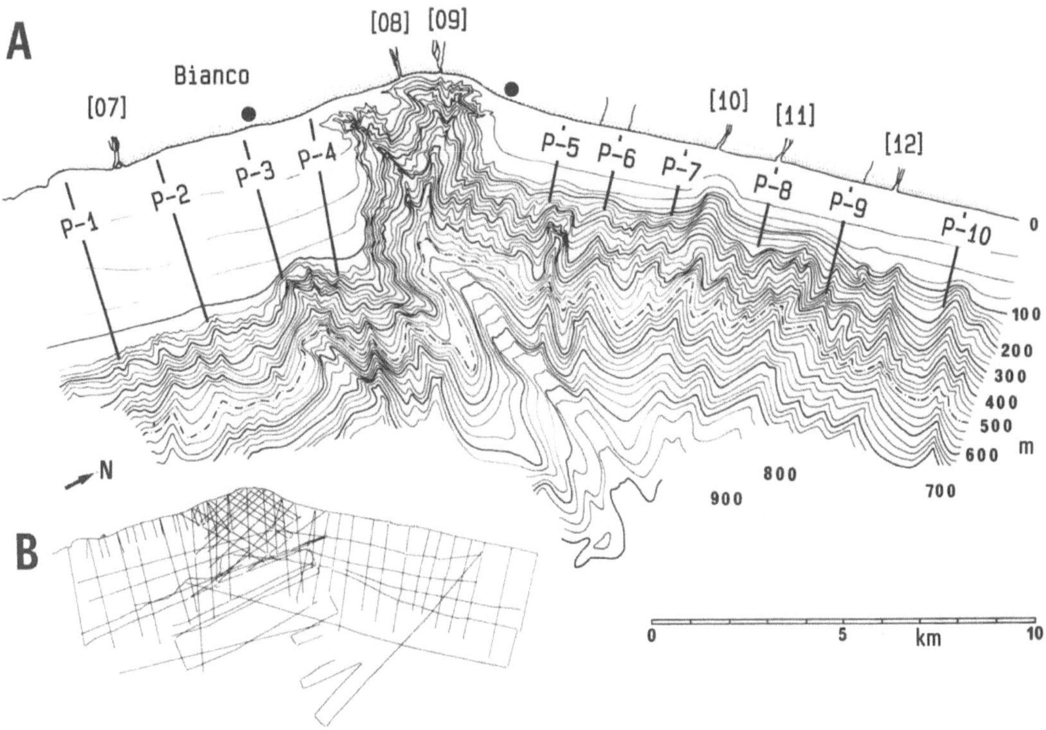

Fig. 16.1 A Bianco shelf, Buonamico-Careri Canyon and Bovalino shelf (see Fig. 3.7 for the geographic setting). A more detailed map of the proximal shelf area off Bianco is shown in Fig. 12.1. *[07]* Laverde; *[08]* Buonamico; *[09]* Careri; *[10]* Condoianni; *[11]* Portigliola; *[12]* Gerace. *P-1 – P-10* Sample profiles. **B** Main echosounding tracks (Rumohr and Ibbeken 1975)

16.2 Submarine Morphology

The Broad Bianco Shelf (Fig. 16.1). This shelf is about 5 km wide, the break is between 110 and 120 m water depth. The surface is very even. About 350 m off the coast both a distinct single longshore bar and a longshore trough down to 10 m is developed (Figs. 16.2A and 12.1). This longshore bar is fed by the Laverde Delta (Fig. 16.8). Main wave direction is from SE, provoking a net longshore sediment transport directed northward (Chap. 12.1). Mean shelf inclinations are about 1° or 1 : 50.

The Narrow Bovalino Shelf (Fig. 16.1). This shelf is only 2 km wide, the break is less distinct (Fig. 16.2B). The longshore bar is small or absent because there is no feeder river comparable to the Laverde [07]. Buonamico [08] and Careri [09] deliver their coarse debris directly into the head of the submarine canyon.

Shelf Break and Slope. Shelf break and upper slope are characterized by incisions which seem to follow neighbouring river mouths (Fig. 16.2) and by larger undulations due to slumping. The slope, on our map, extends to a water depth of about 1000 m, mean slope gradient is 8° or 1 : 7.

The Buonamico-Careri Canyon. This canyon is the most remarkable submarine morphological element of the area. Both Buonamico [08] and Careri [09] join in the canyon head which practically reaches the coastline. A southern tributary "drains" the Bianco longshore trough. The canyon extends, on the map, over 13.4 km to a depth of 900 m, which results in an average gradient of 4° or 1 : 15. The upper part displays a distinct V-shaped profile (Fig. 16.3) with slope gradients of about 7° or 1 : 8.

All slopes, not only below the shelf break but also along the submarine canyon, are dotted by

Fig. 16.2 A – C. Morphological shelf and slope profiles. **A** The broad Bianco shelf. The relatively flat shelf starts with the small but distinct couple longshore trough and bar and ends with a marked shelf break at 115-m water depth. The slope is characterized by minor slump scars. **B** The narrow Bovalino shelf. This shelf profile is curved, the shelf break is less distinct and the slope displays stronger slumping. **C** Scale of average horizontal distances and conversion graph for slope angles

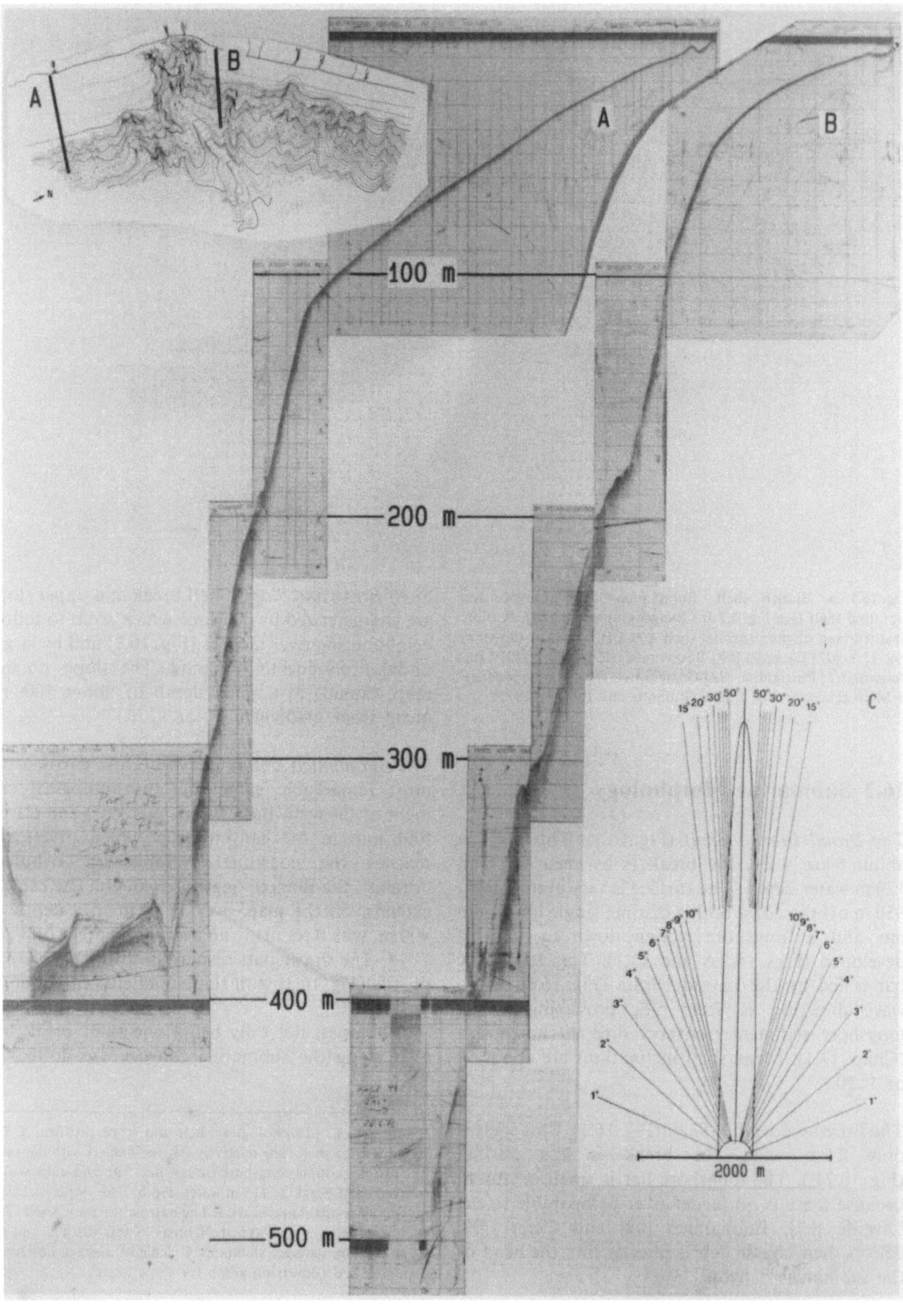
A
B
A
B
100 m
200 m
300 m
400 m
500 m
C
2000 m

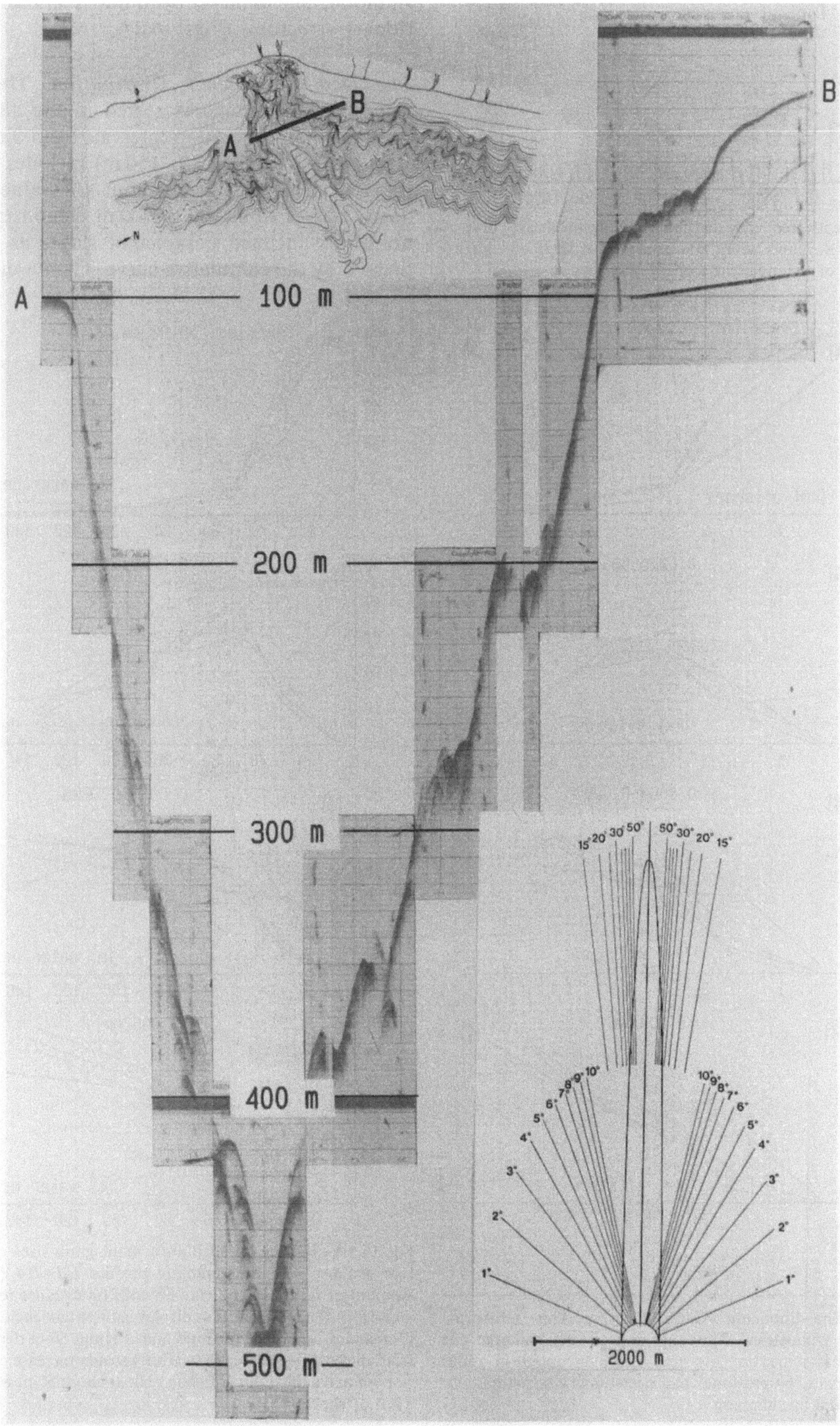
A
B
A
B
100 m
200 m
300 m
400 m
500 m
2000 m

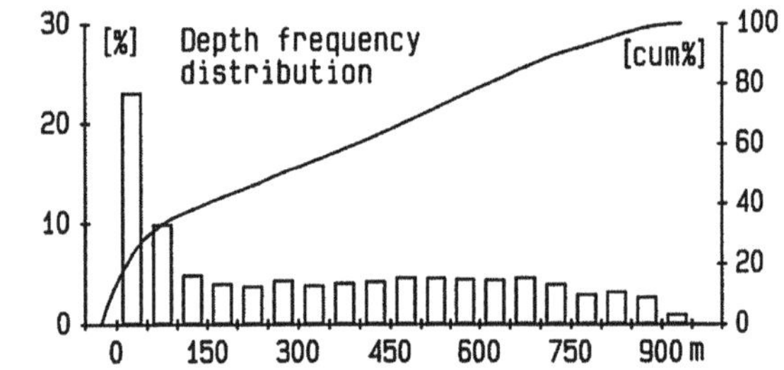

Fig. 16.4. Depth frequency distribution and cumulative curve of the Bianco-Bovalino shelf and slope. The shelf (0–150 m) covers 38% of the area; the slope is very regular

slumping phenomena, by slump scars and accretionary structures (Figs. 16.2A, 16.3).

The Depth Frequency Distribution. The depth values of the geomatrix stored in the data bank were classified into 50-m intervals from 0 to 950 m (Fig. 16.4). The shelf (0–150 m) includes 38% of the area. The plateau of about 4% values of the depth classes from 100 to 700 m demonstrates the uniformly inclined continental slope, as also expressed by the cumulative curve. The lower percen-

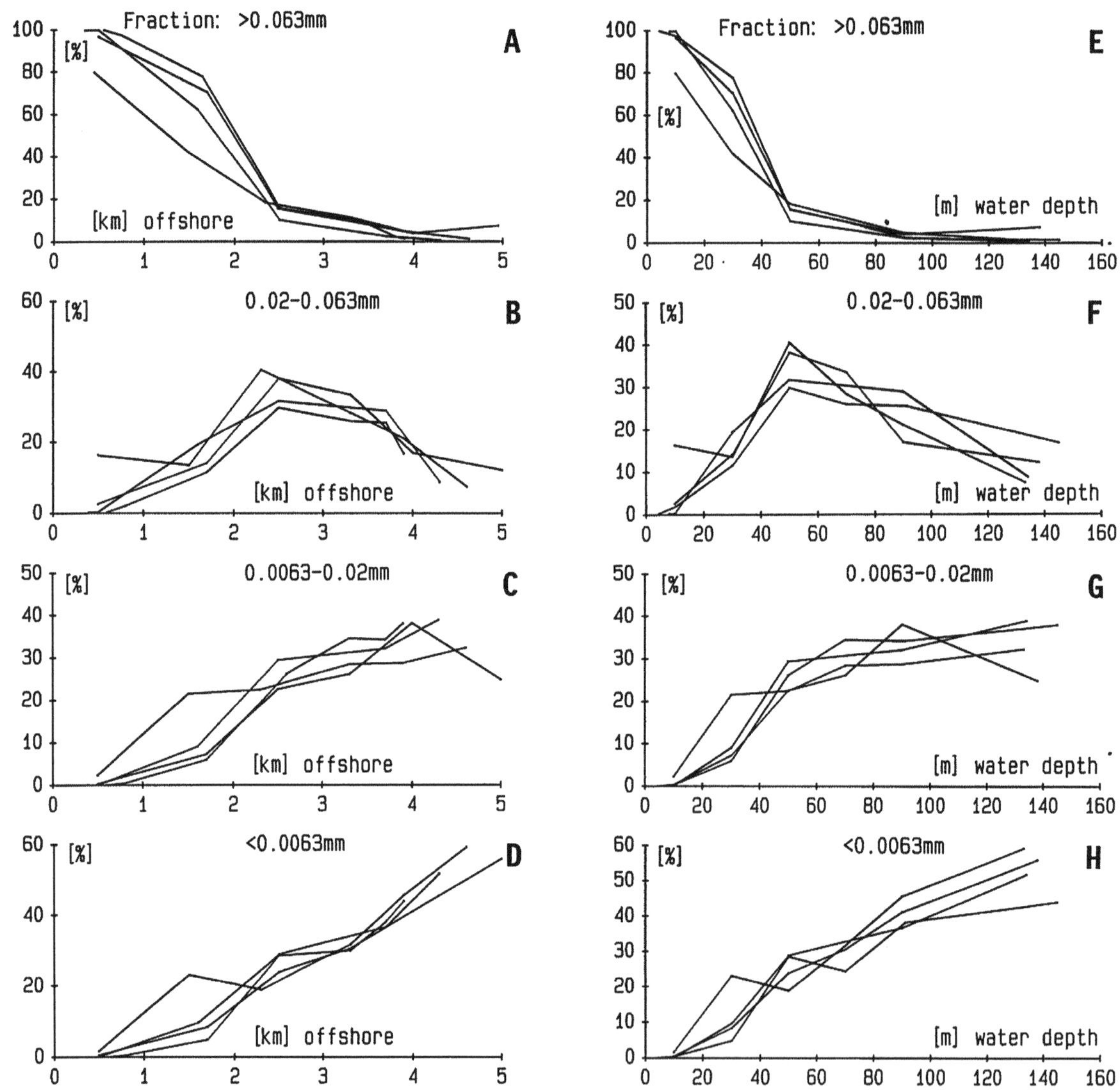

Fig. 16.5 A–H. The broad Bianco shelf, grain sizes of the surface sediments of the sampling profiles *P-1–P-4* (Fig. 16.1). Amount of sand, silt and clay depends on distance to the coast and depth. The four profiles display similar tendencies: a strong decrease of sand until the mud line at about 50-m depth 2.5 km from the coast is reached, a relatively steady increase of clay and fine silt and a maximum of coarse silt at about 60-m water depth 3 km offshore

←

Fig. 16.3. The Buonamico-Careri Canyon. This submarine valley is not symmetrical. The south slope is steep and even, the north slope is somewhat less steep and interrupted, in about 100-m intervals, by erosional and accretionary structures obviously caused by slumping tages of the last four classes are an artefact due to the irregular limitation of the area under consideration.

16.3 Sediments

There are three main pathways along which the erosional products of the hinterland are delivered to the Ionian Sea:

- The direct input of the river's bedload during floods into the head of the submarine canyon or onto the Laverde Delta.
- Plumes of light, turbid freshwater off the river mouths which distribute the suspended load over the entire area.
- The longshore transport of gravel and sand, mainly northward, which feeds both the shelf, mainly the longshore bar, and the head of the Buonamico-Careri Canyon.

The fines are redistributed mainly by wave action. Mass movements caused by overburdening of this over-supplied area or triggered by earthquakes are responsible for further transportation into greater water depth.

Sediment samples were taken along ten profiles normal to the coast, four profiles (P-1 – P-4) across

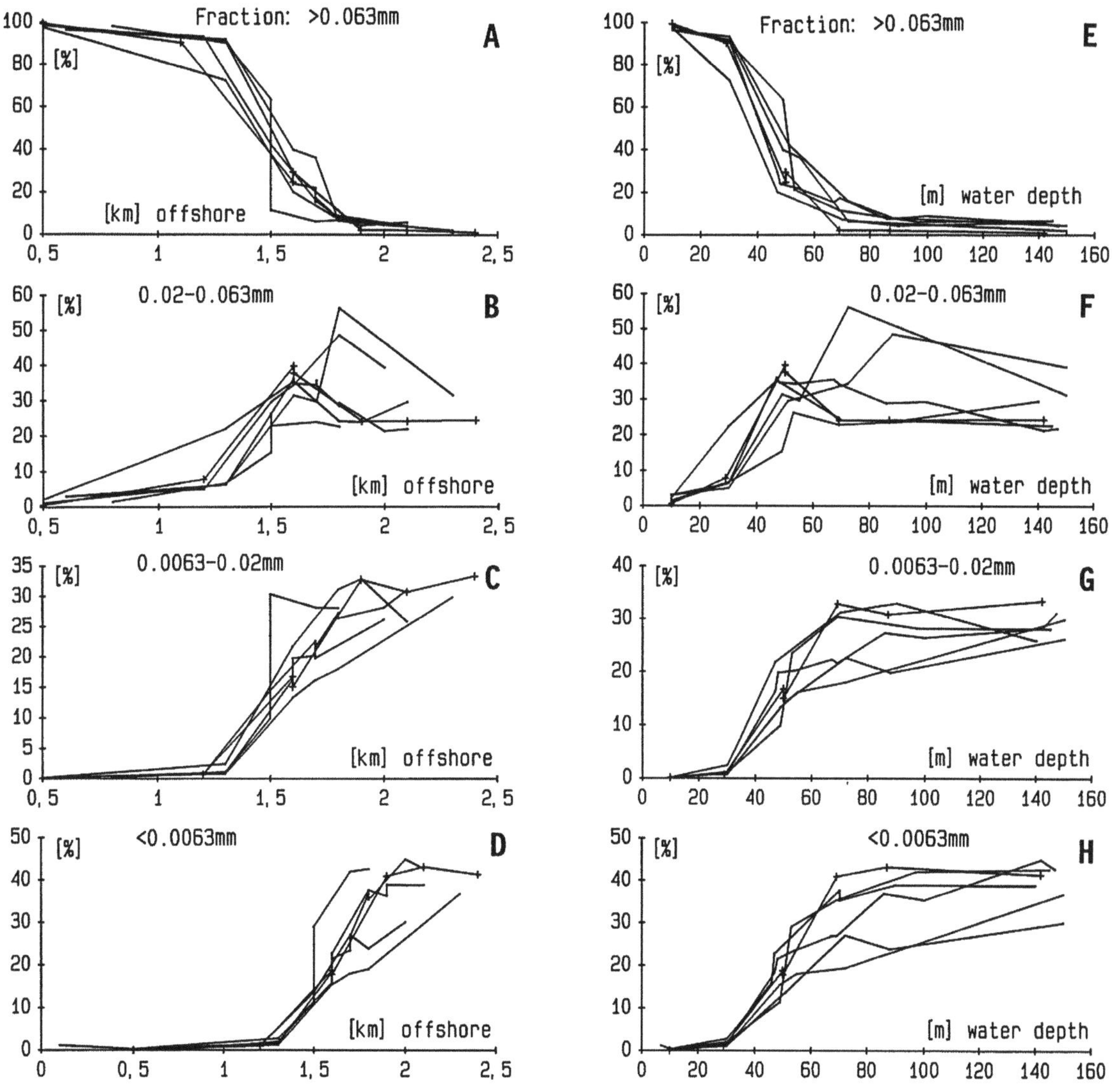

Fig. 16.6 A – H. The narrow Bovalino shelf, grain sizes of the surface sediments of the sampling profiles P-5 – P-10 (Fig. 16.1). Amount of sand, silt and clay depends on distance to the coast and depth. Tendencies of grain-size development are similar to the Bianco shelf (Fig. 16.5), but less regular. A mud line still exists at about 50 m, all other features are more variable

the broad Bianco shelf and six profiles (P-5 – P-10) across the narrow Bovalino shelf (Fig. 16.1). At each profile, six sampling sites were chosen in water depths of 10, 30, 50, 70, 90 and 140 m. The samples were taken in summer 1973 about 6 months after the exceptional flooding of the Buonamico [08] on January 4 (Chap. 15.1.2.2). For comparison, sampling profiles 2 and 3 were repeated in summer 1974.

The four surface sediment profiles of the Bianco shelf are fairly similar and show some general trends. Because the broader shelf is a slightly inclined plain, water depth correlates perfectly with offshore distance, there is only little difference evaluating grain size versus water depth or versus offshore distance (Fig. 16.5 A – D or E – H). Sand content decreases with offshore distance down to <20% in 50 m water depth, 2.5 km offshore (Fig. 16.5 A, E). This may be the depth of the mudline above which bottom feeling waves prevent fines from accumulating. Here, the sand is completely made up by mica which hydrodynamically behaves like particles <63 µm. On profile P-2 near the Laverde [07] delta, the −30-m sampling site reveals a high amount of material <20 µm (Fig. 16.5 G, H) due to the nearby river discharge. In general, the 20 – 63-µm fraction dominates within 50 – 90 m water depth 2 – 4 km offshore (Fig. 16.5 B, F).

From 1973 to 1974 a significant transformation of the fines took place, possibly due to bio-entrainment and winnowing by waves in water depth down to 70 m, that is, about 3 km offshore, as documented in grain-size percentages (Fig. 16.7 C) and the 20 – 63/6.3 – 20-µm ratios plotted in Fig. 16.7 E.

The six profiles across the narrow shelf (Fig. 16.6) reveal local differences because of near-shore shelf topography. Water depth does not correlate very well with offshore distance, thus differences occur when evaluating grain size versus water depth and grain size versus offshore distance. Sand content is better correlated to offshore distance than to water depth (Fig. 16.6 A – E).

16.4 Mass Balance

Based only on the actual submarine morphology and the data of the surface sediments presented, it is not possible to establish a mass balance in this proximal marine environment of Calabria because we do not dispose over any chronostratigraphic information. However, most data of the Italian sea chart stem from the original soundings in 1876/77.

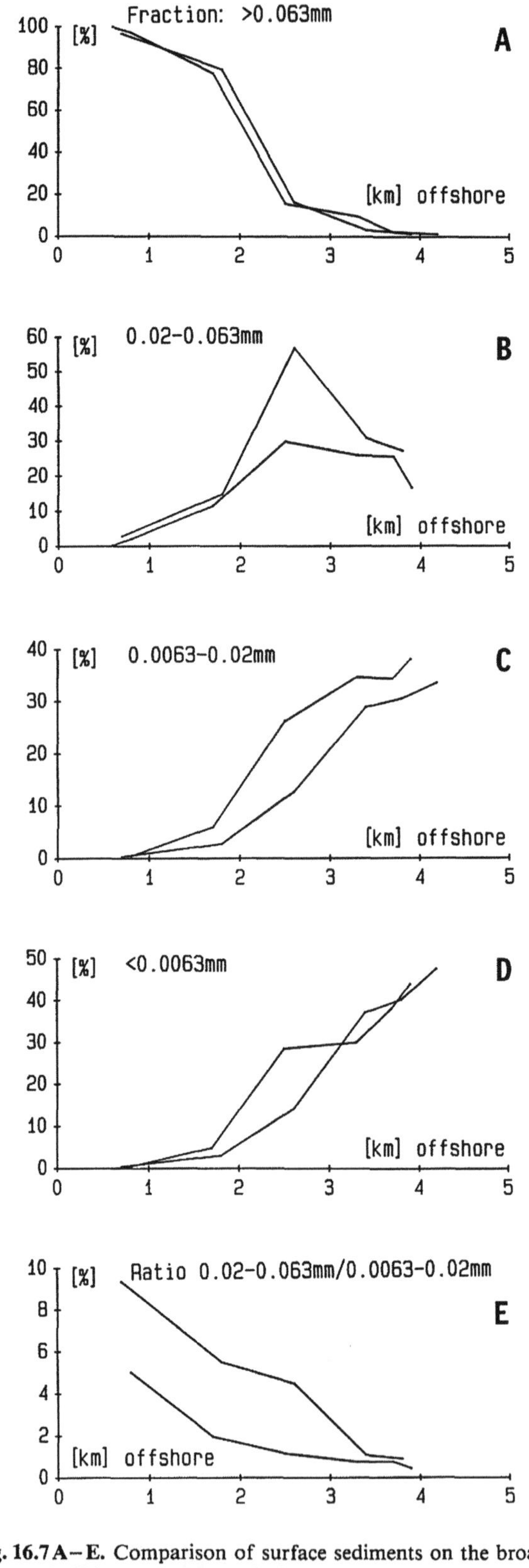

Fig. 16.7 A – E. Comparison of surface sediments on the broad Bianco shelf between 1973 and 1974, mean of the sample profiles P-1 – P-4. The sand development is identical, but fines <20 µm are depleted in water depth <70 m, this is until about 3 km offshore

We evaluated the original data of those days, still stored in the files of the Italian navy, and compared this morphology from 1877 with our own morphology from 1974, bridging nearly 100 years.

Only half a century is covered by the air photos available from the coastal areas. We evaluated a pulsation of the Laverde delta comparing air photos from 1941, 1955, 1973 and 1983.

16.4.1 The Laverde Delta 1941–1983

The Laverde [07] is an exceptional river. It possesses a paleovalley exactly like all other rivers of the area (Appendix B). However, the present-day river is the only larger one which does not feed a submarine canyon but develops a delta. We do not know why. This exceptional position of the Laverde gives rise to a special development of the coast extending northward from the Laverde river mouth, the Bianco beach (Chap. 12.1). The disastrous floodings of 1951 and 1953 created a delta whose extension is captured by the 1941 and 1955 air photos. Figure 16.8 shows this and the remainder of the delta in 1974. Ergenzinger et al. (1987) calculated the delta propagation and its destruction until 1983. From this and geodetic survey during several campaigns from 1985 until 1987 (Hurtz 1988; Fabel 1989) a mean annual net transportation (minimum) was

Fig. 16.8. *Above* The Laverde [07] delta 1974, view towards N from Capo Bruzzano. The long focal length of the photography makes the delta much more accentuated than the contour lines on the map below; *below* map of the area Capo Bruzzano-Laverde delta (after Ergenzinger et al. 1987). The shorelines from aerial photos from 1941 and 1955 are added

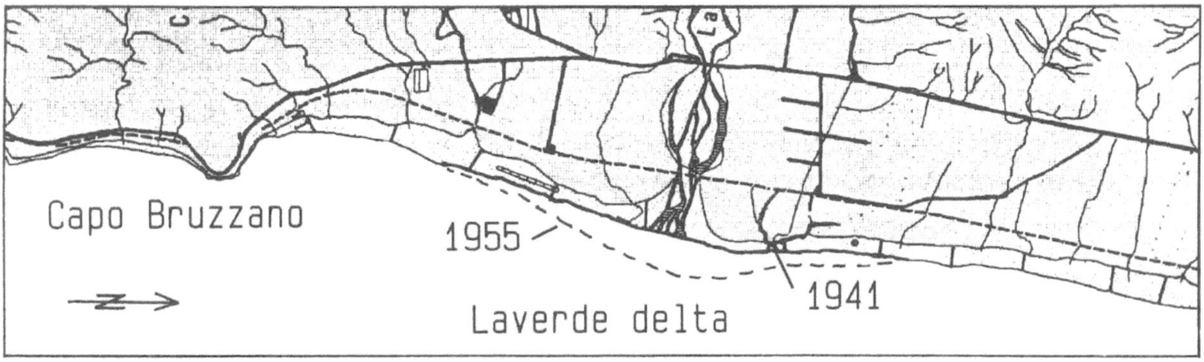

calculated of $100000 \times 10^3\,m^3$. This longshore transport is directed northward, and a large amount should enter the head of the Buonamico-Careri Canyon.

16.4.2 Shelf and Canyon 1877 and 1974

The Italian sea chart from 1877 is a very precise hydrographical survey, based on traditional plumb-

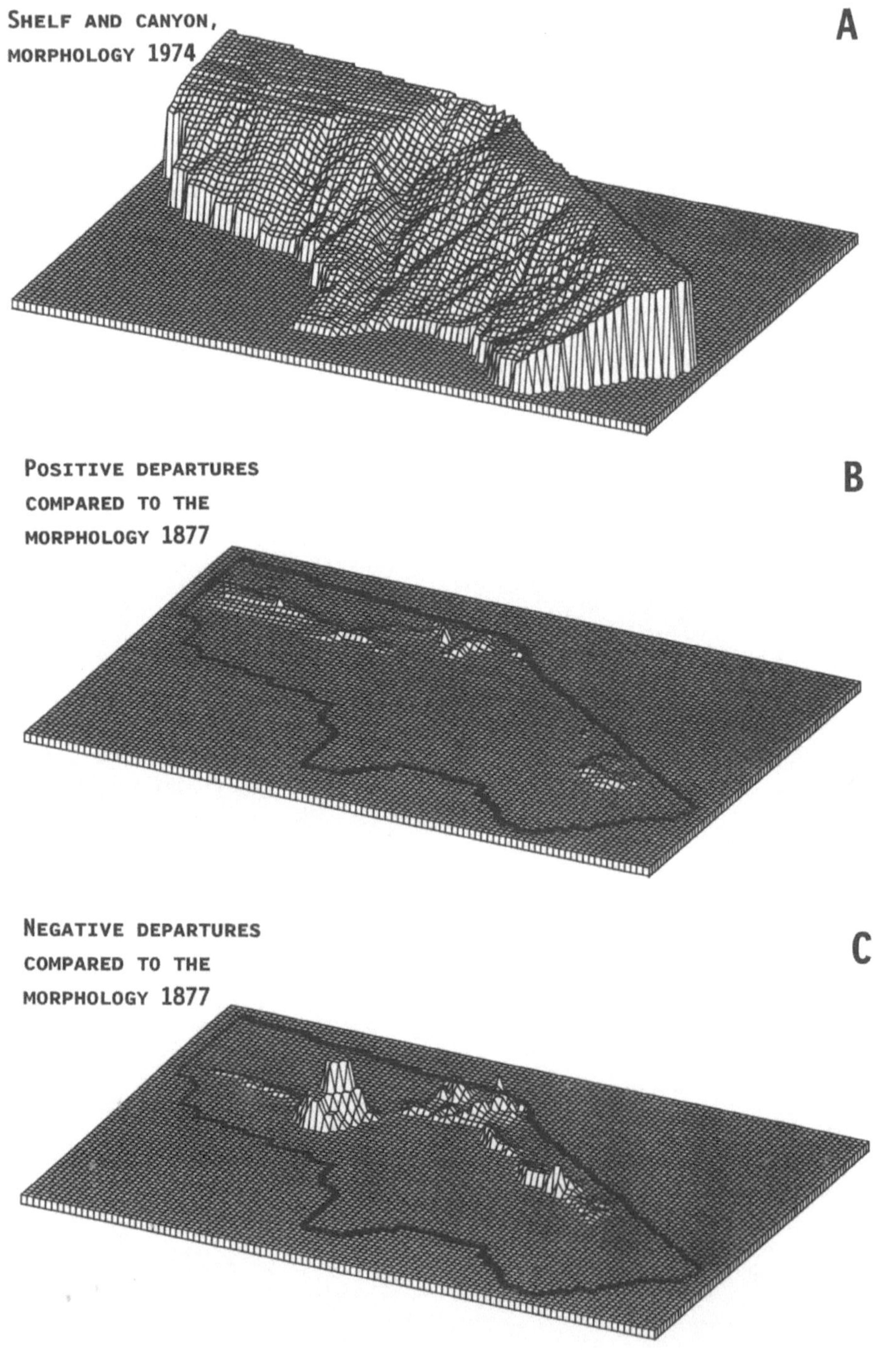

Fig. 16.9 A–C. Erosion and accumulation on the Bianco/Bovalino shelf between 1877 and 1974. **A** Stereo model of the 1974 morphology (see Fig. 16.1). **B** Positive departures or accumulation compared to the morphology of 1877. The distal Bianco shelf, parts of the canyon head and the Bovalino shelf off the Gerace [12] river mouth are loci of accentuated sedimentation. **C** Negative departures or erosion compared to the morphology of 1877. The break of the Bianco shelf, especially the "amphitheatre" close to the canyon, the canyon head itself and the Bovalino shelf off the Condoianni [10] river mouth are loci of very accentuated erosion. Neglecting the single, huge slide, the budget is more or less balanced

ing station by station, of course. The density of measurements decreases seawards. Shelf, upper slope and canyon head, however, are measured accurately enough to serve as a solid base for a contour map. This area covers 84.3 km^2 or 40% of our echosounding map, corresponding to 1349 points of the 250×250 m geomatrix. Consequently, to each matrix point of this area two depths values are assigned, the 1877 and the 1974 depth.

Comparing the two sets of data, remarkable changes of water depth can be deduced which exceed by far the possible differences caused by applying different methods. In addition, these differences are not distributed randomly over shelf and canyon head, which would be expected if we were dealing with inaccuracies of the hydrographical surveys. Consequently, the differences of the two surveys have to be interpreted as real erosional and depositional features. Calculating these differences over nearly one century there are 212 points (15.7%) with zero differences (between −1 and +1 m), 610 points (45.2%) with erosion more than −1 m and 527 points (39.1%) with sediment accumulations exceeding +1 m.

Figure 16.9A shows a stereomodel of the 1974 morphology. The positive departures (accumulations) compared to the 1877 morphology are plotted in Fig. 16.9B, while Fig. 16.9C displays the negative departures, the erosion.

Erosion takes place particularly at the shelf edge and in the two channels of the canyon head (Fig. 16.9C). The mean erosion is −33.7 m considering only those 610 points where erosion took place. Its maximum, with values up to −300 m, lies at the shelf edge south of the submarine canyon. Obviously, some time during the last 100 years a huge submarine slide happened here, forming the convex, amphitheatre-shaped niche (Fig. 16.1 A and 16.9C). The missing, mobilized sediment volume is about 700×10^6 m^3, which is 32 times more than the Costantino landslide on land (Chap. 15.2.2.2).

Submarine sediment accumulation is not as dramatic as erosion. Mean accumulation is 12.4 m, again considering only those 527 points where sediment accumulated. These are preferably situated on the outer muddy shelf plain as well as around and between the two channels of the canyon head (Fig. 16.9B).

We classified erosion and accumulation based on $\log_2$ into intervals of 1−2 m, 2−4 m, 4−8 m, 8−16 m etc. The frequency distribution (Fig. 16.10) of erosion (blank bars) and accumulation (black bars) shows regularly distributed accumulations with a mode lying in the 8−16-m range (11.2%) and maximum values up to 70 m. The erosion is bimodal with one mode in the −1- to −2-m interval and a second one in the −8- to −16-m interval. Erosion of more than −128 m is found only in the submarine slide mentioned above.

We calculate and summarize the volumes moved: a total of 1283×10^6 m^3 has been eroded and 407×10^6 m^3 have been accumulated during 100 a. If we consider the huge slumping at the shelf break as a single event and exclude this volume of about 700×10^6 m^3 from our calculation, erosion and accumulation, loss and gain of the area are more or less balanced.

16.5 Conclusion

The 100 years' comparison shows that the outer shelf with water depths of 60−100 m is the preferred locus of deposition (Fig. 16.9B). The sediments deposited here are mainly silt and clay (Figs. 16.5, 16.6). The next important place of accumulation is the canyon head. Our sampling techniques cannot provide information about the grain-size distribution of these necessarily coarser deposits, delivered directly during floods of the Buonamico [08] and Careri [09] and indirectly via longshore transportation from the Laverde [07].

We know, however, from the refill of the Lago Costantino (Chap. 15.1.3) that the proportion bed load to suspended load − fine sand may be the boundary − is approximately 3:2.

Consequently, a rough calculation of the 100 years' budget can be made. It goes without saying that this will be a minimum estimate, because nobody expects this proximal area to store the entire production of the hinterland, as the majority

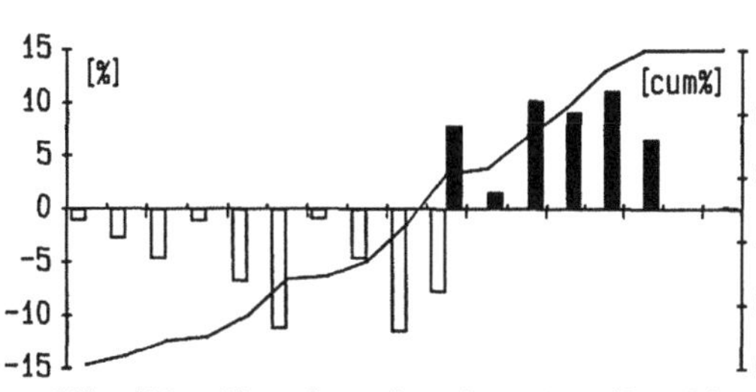

Fig. 16.10. Frequency distribution of erosion (*blank*) and accumulation (*black*). Class interval for both are geometric, 1−2, 2−4, 4−8 m etc. Maximum values of erosion in the canyon head exceed 50 and 200 m in the huge slide at the shelf break off Bianco

probably leaves the area by crossing the shelf or moving down through the canyon.

Assuming a porosity of these little consolidated silts and clays of about 60%, the accumulated volume of 407×10^6 m^3 is reduced to 163×10^6 m^3. Distributing this volume over its source area (550 km^2), an erosion rate of 3000 mm/ka results, which must be raised by 150% to about 7500 mm/ka with regard to the other part of the river's load, the bedload. This value is within the order of magnitude of the land-derived data of the Buonamico basin, but it is more than twice the amount of the highest erosion rate we calculated in the Buonamico basin, the rate of 2700 mm/ka based on landslide activity. So even this land-borne amount may be a conventional estimate.

Chapter 17

Comparison of Source and Sediment: a Multivariate Approach

In Calabria, we attempted a maximum of quantification and a minimum of statistics. We mostly used uni- and bivariate statistical methods for our interpretations, with two exceptions in the special cases of heavy minerals and grain-size distributions of gravels. The huge data set and the large number of variables, however, seem to be ideal for applying multivariate methods. It would be useful to find a quantitative model which relates the variables of the river-mouth sediment to variables which describe the source. Davis (1986) used canonical correlation for determining the relationship between petrophysical responses on well logs and formation properties on core samples from these wells.

Unfortunately, this method cannot be applied to our two data sets. Simple mean values of grain size, altitude or precipitation, for example, do not differ significantly between the basins, and other properties consist of a spectrum of variables, petrography with eight rock types, for example, which make an interpretation of factor loadings meaningless.

It may be possible to estimate the composition of a sediment if all variables contributing to this sediment are known. But this is Utopia. Even if a simulation procedure existed which makes it possible to determine the composition of a sediment from source and transport conditions, there would be no chance to trace back and find the most probable source composition: the large number of variables involved, not only in the sedimentary mill of the rivers, would lead to a "combinatorial explosion" of "valid" solutions.

Summary. The Euclidian distances between the basins in a k-dimensional space defined by k variables were used as a measure of similarity. According to this procedure, based on both source and sediment properties, the subdivision of Calabria into Aspromonte, Zomaro and Serre Provinces has to be modified. The Aspromonte basins have to be regrouped into two sub-provinces, one whose rivers reach the main divide, the large Aspromonte Province, and one whose short rivers do not, the small Aspromonte Province. Novito [13] and Torbido [14] prove to be special cases which fit neither the Zomaro Province nor the Serre Province. A t-test of the distances between different provinces and within provinces yields the following rank of source area properties: geology and altitude separate the provinces with more than 99% significance; gradient and precipitation yield between 95 and 99% significance, and erosion rates finally separate the provinces with only less than 95% significance. The same procedure ranks the sediment properties: heavy minerals, the mineralogy of the clay fraction and gravel petrography with more than 99% significance, silt minerals and geochemistry between 95 and 99% significance and, the least, grain sizes and light minerals with less than 95% significance. It is the first time a quantitative rank and evaluation of the commonly used provenance signals has been established.

Data Base. We used the following properties and their variables.

Source area properties (data available for all 103 basins):

- altitude: 20 variables
  (area percentages of 20 altitude classes from 0 to 2000 m in 100-m intervals)
- geology: 8 variables
  (area percentages of 8 rock types)
- erosion rate: 17 variables
  (area percentages of 17 erosion rate classes from 0 to 850 mm/ka in 50 mm/ka intervals)
- gradient: 15 variables
  (area percentages of 15 gradient classes from 0 to 75° in 5° intervals)
- precipitation: 9 variables
  (area percentages of 9 precipitation classes from 400 to 2200 mm in 200-mm intervals).

Sediment properties (data available for 19 river-mouth areas):

- grain size: 35 variables
  (weight percentages of 35 1/3 phi classes from 200 to 0.063 mm)
- gravel petrography >50 mm: 10 variables
  (percentages of 10 rock types)
- gravel petrography 25–50 mm: 10 variables
  (percentages of 10 rock types)

- gravel petrography <25 mm: 13 variables (percentages of 10 rock types + 3 mineral types)
- sand heavy mineral composition: 28 variables (percentages of 28 heavy minerals)
- sand light mineral composition: 16 variables (percentages of 16 light minerals)
- silt fraction mineralogy: 10 variables (percentages of 10 minerals)
- clay fraction mineralogy: 10 variables (percentages of 10 minerals)
- sand geochemistry: 10 variables (percentages of 10 oxides).

Method. The k variables of each property are used to calculate the Euclidian distance between the basins in the k-dimensional space. Each basin represents a point, its coordinates result from the values of each of the normalized k variables of a property. The formula for calculating the Euclidian distance between basin n and basin m is:

$$d_{nm} = \left[\sum_{i=1}^{k} (x_{in} - x_{im})^2 \right]^{1/2} .$$

For each of the properties listed, a matrix of distances results comprising 171 values, which is the number of possible combinations of 19 basins. In order to make the single distance matrixes comparable we normalized the 171 values putting the highest value equal to 100. By this method the value 100 means maximum Euclidian distance or maximum dissimilarity between two basins. The smaller the value, the lower the distance and the higher the similarity. Zero practically does not occur, which would mean that two basins would be identical as far as all variables of a property are concerned.

17.1 The Source

Figure 17.1 shows a dendrogram of all 103 basins, regarding all the 69 variables of the five source area

Fig. 17.1. Dendrogram of all 103 basins based on 69 variables of the five source area properties altitude, rock distribution, erosion rate, gradient and precipitation. Note the well-separated cluster of the 19 basins under closer investigation

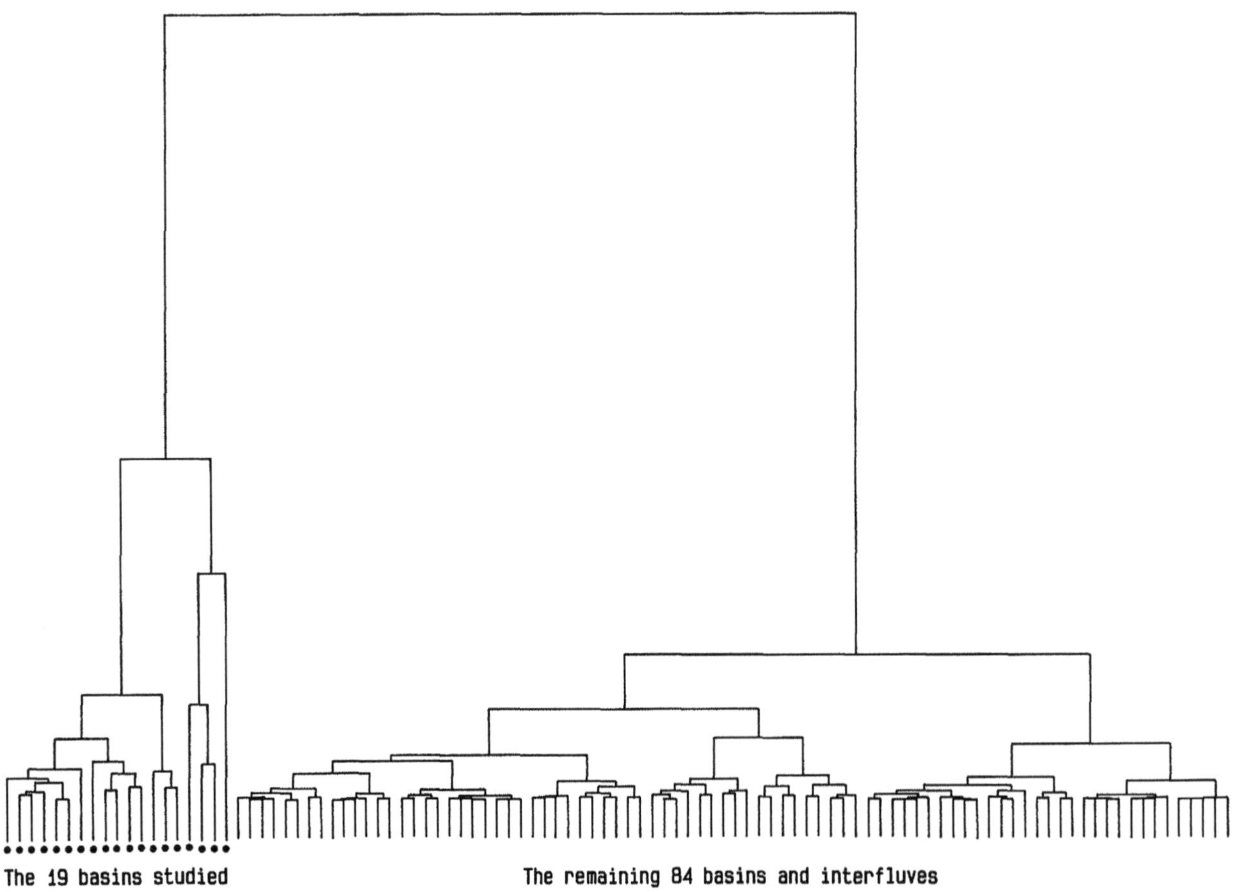

Table 17.1. Average of the five source area properties altitude, geology, erosion rate, relief and precipitation[a]

A	01 Mel	02 Ame	03 Pas	04 Pal	05 Ara	06 Bru	07 Lav	08 Buo	09 Car	10 Con	11 Por	12 Ger	13 Nov	14 Tor	15 Amu	16 All	17 Pre	18 Sti
19Ass	82.4	87.3	83.2	82.4	83.2	73.1	69.3	71.2	47.1	30.0	42.1	43.7	45.7	41.6	59.5	28.0	58.9	28.4
18Sti	83.2	87.3	73.7	83.2	76.3	65.2	68.3	71.2	42.0	24.8	45.4	38.5	38.3	45.7	69.9	47.5	38.8	
17Pre	53.3	81.2	56.0	65.4	58.7	35.7	96.4	96.4	65.5	51.9	43.8	34.8	44.1	43.8	39.3	45.3		
16All	66.7	96.4	77.2	76.4	77.2	56.3	84.1	82.6	56.8	40.8	25.9	27.0	37.7	27.8	44.8			
15Amu	37.3	69.7	57.9	48.3	56.9	54.0	88.9	85.6	70.7	67.1	41.2	52.0	53.8	44.0				
14Tor	59.1	96.4	77.2	67.4	77.2	58.0	84.1	80.7	50.0	38.2	10.8	20.1	31.6					
13Nov	69.8	96.4	68.2	60.8	68.9	67.2	84.1	82.6	59.6	44.2	27.3	36.2						
12Ger	68.6	96.4	69.3	77.2	70.4	46.9	84.1	82.6	43.0	27.5	18.5							
11Por	61.2	96.4	77.2	69.8	77.2	58.0	84.1	82.6	51.8	36.6								
10Con	83.2	87.3	75.2	83.2	76.3	63.4	71.2	69.7	28.3									
09Car	86.7	90.9	78.8	86.7	79.9	77.9	76.4	73.2										
08Buo	66.3	34.0	73.3	71.1	72.3	80.8	14.3											
07Lav	67.3	37.3	71.9	73.3	73.3	80.8												
06Bru	38.8	66.6	34.0	46.0	39.3													
05Ara	38.4	60.5	21.9	24.5														
04Pal	31.8	56.3	25.1															
03Pas	42.3	61.5																
02Ame	34.2																	

B	Large Aspromonte	Small Aspromonte	Zomaro	Serre
Large Aspromonte	42	62	81	78
Small Aspromonte		32	73	67
Zomaro			34	45
Serre				46

[a] **A:** Total Euclidian distance matrix, 171 distances between 19 basins. **B:** Euclidian distance matrix of the four provinces, calculated by averaging all distances of basins within a province and between two provinces.

properties. Strangely enough, the 19 basins investigated in detail form a distinct family differing evidently from the remaining 84 basins. This is not a matter of the absolute basin size, because all values are given in percent and three of the basins not chosen are larger than the smallest basin selected for closer research. We enjoyed this late and objective confirmation of our unexperienced selection of the 19 basins 20 years ago (Chap. 2.2.4). In the following, only the 19 basins will be treated.

In order to combine similar basins to provinces we calculated a mean distance matrix averaging the five source property matrixes altitude, geology, gradient, erosion rate and precipitation (Table 17.1 A). Lowest mean distance (14.3) is between Laverde [07] and Buonamico [08], the highest (96.4) is displayed by Amendolea [02] and Allaro [16]. Figure 17.2 shows these mean distances between neighbouring rivers in their geographic sequence. We interpret low distances between neighbouring rivers in the sense that they belong to the same province; peaks in the plot (high distances) may signal a boundary between provinces. The picture confirms the conventional division of the area into physiographic provinces (Chap. 2.2.3):

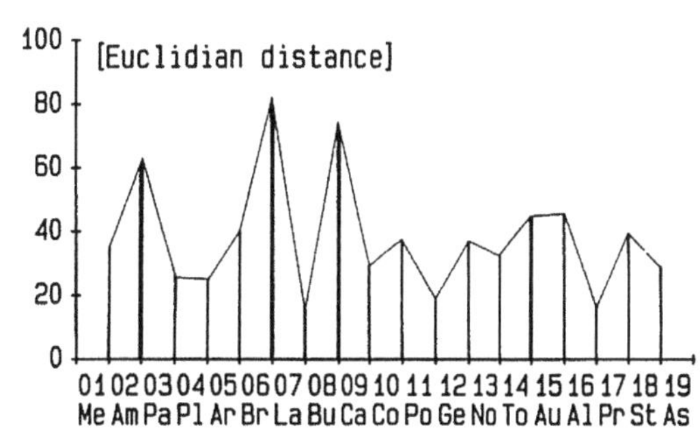

Fig. 17.2. Mean Euclidian distances between neighbouring rivers calculated by averaging the distances of the five source area properties altitude, rock distribution, erosion rate, gradient and precipitation. Peaks are marked *black* and interpreted as a boundary between two provinces. *01 Me*lito; *02 Am*endolea; *03 Pa*squale; *04 Pal*izzi; *05 Ar*anghia; *06 Br*uzzano; *07 La*verde; *08 Bu*onamico; *09 Ca*reri; *10 Co*ndoianni; *11 Po*rtigliola; *12 Ge*race; *13 No*vito; *14 To*rbido; *15 A*m*u*sa; *16 Al*laro; *17 Pr*ecariti; *18 St*ilaro; *19 As*si

Basin	Physiographic province (Chap. 2.2.3)			Cluster analysis based on source area properties			
	Aspromonte	Zomaro	Serre	Aspromonte		Zomaro	Serre
				Large	Small		
Melito	*			*			
Amendolea	*			*			
Pasquale	*				*		
Palizzi	*				*		
Aranghia	*				*		
Bruzzano	*				*		
Laverde	*			*			
Buonamico	*			*			
Careri		*				*	
Condoianni		*				*	
Portigliola		*				*	
Gerace		*				*	
Novito		*				(*)	
Torbido			*			(*)	
Amusa			*				*
Allaro			*				*
Precariti			*				*
Stilaro			*				*
Assi			*				*

However, the Aspromonte Province has to be clearly subdivided into a large Aspromonte Province comprising the four large rivers reaching the main divide and a small one which comprises only the four small rivers of this area. The border between Zomaro and Serre Provinces is not very clear and seems to slide somewhere between Gerace [12] and Amusa [15]. According to the weak peak of distance, it is drawn between Torbido [14] and Amusa [15], which means a difference to the conventional physiographic provinces, concerning Torbido [14]. Because the sediment properties of neither Novito [13] nor Tordido [14] fit very well into this picture of provinces, both basins were treated separately and are discussed later.

In order to test this division into four provinces, we calculated the mean distances of both the basins within one province and the basins of one province against those of the other provinces (Table 17.1 B). As a result, the mean distances within provinces are always lower than those between provinces. Only the Serre Province comes somewhat closer to the Zomaro Province than to itself, confirming the sliding transition between these two provinces.

According to what property of the source area are the single provinces best separated? We use the t-test to answer this question. We compare the means of both the distances within provinces and those of the distances between different provinces, separately for each property. There are 28 distances between basins within provinces and 108 distances between basins of different provinces, resulting in 134 degrees of freedom. The missing 34 distances between Novito [13] or Torbido [14] and the other basins, as well as the distance between both basins, are not treated here. One hundred and thirty four degrees of freedom result in a $t_{99\%}$ value of 2.62 and a $t_{95\%}$ value of 1.98 (Marsal 1967, p. 178).

	Distances between basins of different provinces n = 108		Distances between basins within provinces n = 28		
	mean	st. dev.	mean	st. dev.	t
Geology	79.7	28.9	19.9	15.0	10.5
Altitude	57.8	34.1	39.0	26.5	2.7
Gradient	60.7	31.0	44.7	31.5	2.4
Precipitation	66.4	39.4	46.9	40.2	2.3
Erosion rate	69.4	36.2	57.9	40.1	1.5

Note that the mean distances between different provinces in all cases reveal higher values than the mean distances within provinces, but only geology and altitude exceed the $t_{99\%}$ value and gradient and precipitation $t_{95\%}$. Erosion rate fails the t-test at a 95% significance level. The very high t-value of 10.5 of geology is interpreted that the four provinces are nearly exclusively separated by this source property. A dendrogram of the geology distance matrix is shown in Fig. 17.3.

The spectrum of rocks in the study area is reduced, volcanic and basic rocks do not occur. This is a drawback to provenance analysis, because many characteristic fine-grained rock fragments appreciated by sediment petrographers are not produced. Nonetheless, our eight-member rock suite is differentiated enough, as demonstrated by the multivariate comparison, to serve as a reliable basis for the provenance discussion of the 19 river-mouth sediments, see the "single value" maps and their histograms (Figs. 5.2 to 5.9).

The strong separation of source area geology and the weak one of source area erosion rates could be interpreted as a weakening of our concept of erosional budget and predicting sediment composition, of the weighing of source rock area percentages by erosion rates. However, this is not the case.

Because we compared only original, non-derived data, the separation of the four provinces by means of their erosional budget, a derived variable, is not known. The frequency distributions of the erosion rates of the single basins, admittedly, are so similar

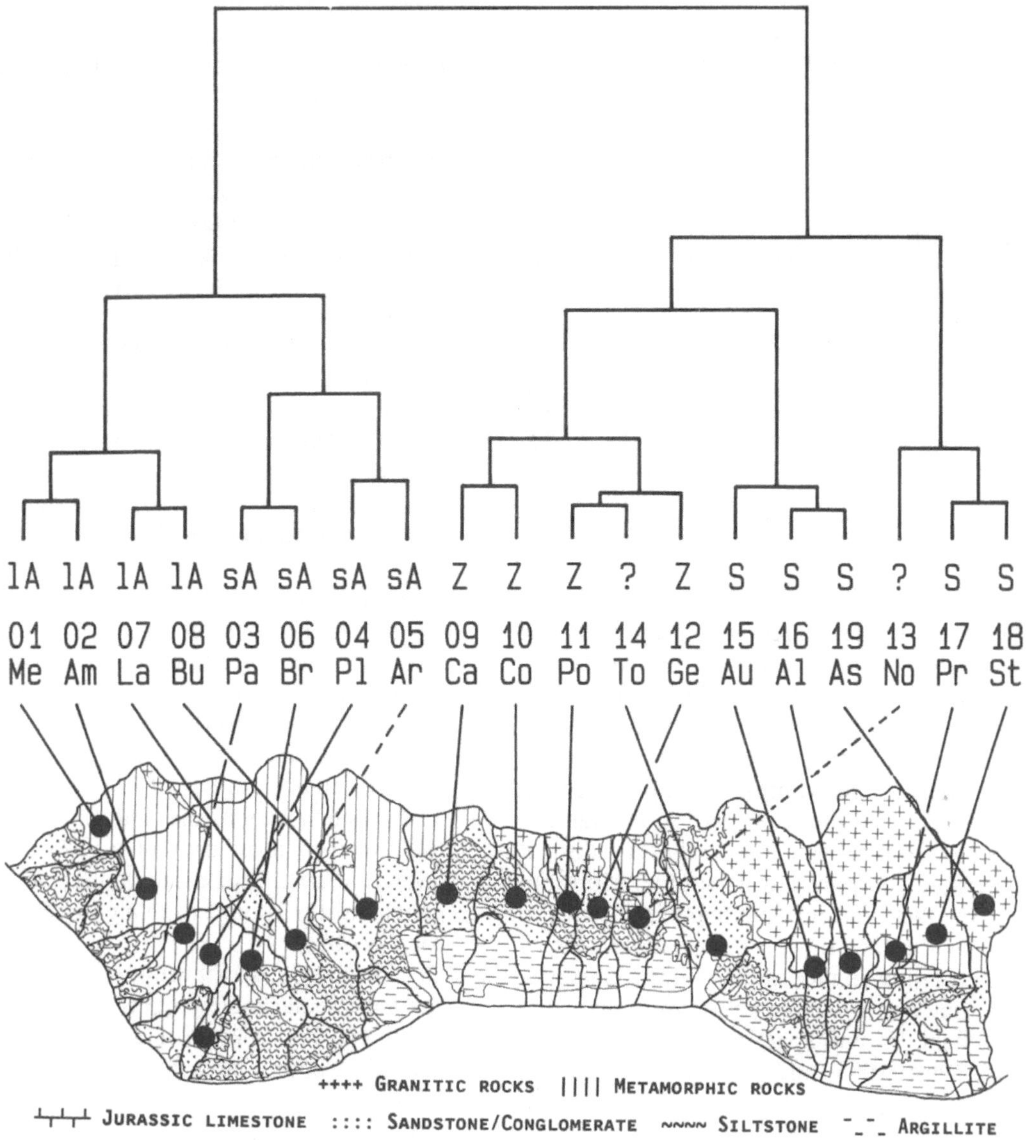

Fig. 17.3. Dendrogram of the 19 basins based only on area percentages of eight rock types. Rock distribution is the source area property discriminating the four provinces best. The *row below* indicates the four provinces large *A*spromonte, small *A*spromonte, *Z*omaro and *S*erre. For abbreviations *01–19* see Fig. 17.2

that we renounced plotting them, but this does not disqualify our findings that the area percentages of rocks change noticeably if corrected by erosion rates, as demonstrated by the direct comparison (Appendix B 01–20 and the compositional triangle, Fig. 6.11). The frequency distributions of erosion rates alone are an unsuited statistical tool characterizing these basins and provinces; the same is true for the gradients. More important, in this case, is the regional distribution of these variables, as shown by the meaningful pattern of the "energy belts" (Figs. 3.9 and 6.4).

17.2 The Sediment

We have to ask if this "provincialism" is reflected in the river-mouth sediments adequately, and if so, what property is best reflected.

Corresponding to the procedure described in the foregoing section, distance matrixes were calculated based on the sediment properties. The means, standard deviations and t-values based on the given provinces are listed below:

	Distances between basins of different provinces n = 108		Distances between basins within provinces n = 28		
	mean	st. dev.	mean	st. dev.	t
Sand heavy minerals	71.3	28.1	37.5	16.0	6.1
Mineralogy of the clay fraction	64.0	38.8	29.6	25.9	4.4
Gravel petrography:					
>50 mm	72.6	30.8	44.4	28.1	4.4
25–50 mm	62.7	21.1	45.0	20.4	4.0
<25 mm	58.9	25.8	42.1	28.2	3.0
Silt mineralogy	61.0	26.6	46.8	20.5	2.6
Sand geochemistry	52.1	21.9	40.7	18.1	2.5
Grain size	50.8	29.2	40.1	25.1	1.8
Sand light minerals	71.1	20.0	65.4	19.6	1.3

As in the case of the source area properties, the mean distances between different provinces are higher than mean distances within provinces. Again, all properties measured separate the provinces, however, the overall separation is worse; this is a mathematical description of the somewhat trivial experience that many properties or "signals" of the source areas become lost so that a sediment is impoverished compared to its source. More important is the fact that our data make it possible to establish, for the first time, a quantitative rank of the quality of provenance signals – in Calabria. This rank may disappoint some petrographers and confirm others. To anticipate a result, we have to agree with Blatt (1985) and his critical comments on sandstone petrography favouring mudrocks for provenance analysis:

- The heavy minerals of the sand fraction of the river-mouth sediments separate the four provinces best (t = 6.1); however, the well-known impoverishment of heavy minerals during diagenesis and intrastratal solution lessens their value for a provenance analysis of ancient sandstones.
- The mineralogy of the clay fraction is a side product of this study, carried out with only little effort in order to obtain a complete analysis of the river-mouth sediment. However, this variable displays an amazing high discrimination power (t = 4.4). Mudrocks are abundant, comprise about 65% of the stratigraphic record on continents and are widespread in both marine and non-marine environments (Blatt 1970). Lower permeability compared to sandstones stands for lower chemical alteration and a higher probability that provenance signals remain preserved.
- The petrographic composition of the gravel fraction separates well, but this quality decreases with decreasing grain size (from t = 4.4 to 3.0). This has certainly to be explained by the increasing percentages of quartz from everywhere and the increasing difficulty in determining the petrography macroscopically in the finer fractions. The disadvantage of gravels is their scarcity in the stratigraphical record and their geographical limitation on proximal environments. Molassic conglomerates, debris flow deposits and coarse-grained turbidites, however, are sediments that can be applied successfully in provenance calculations (Ibbeken 1970; Schleyer 1988b).
- The mineralogy of the silt fraction still separates the four provinces at a 99% significance level (t = 2.6). Here must be repeated what has been said of the mineralogy of the clay fraction. These fine fractions turn out to be a highly sensitive provenance signal which may also be valid for provenance studies of ancient mudrocks.
- The geochemistry of the sand fraction separates the four provinces only at a 95% significance level (t = 2.5). The disadvantage is that sand geochemistry is hard to relate directly to source rock geology.
- The grain-size distribution is certainly among the most remarkable properties of a terrigenous sediment. It is a kind of dogma among geologists that source area geology, with relief and hydraulics, is the most important factor controlling grain-size distribution. However, it is evident that in Calabria, within the limits of the geological and geomorphological variables, no influence is detectable of source rock geology and geomorphology on the grain-size distributions of the river-mouth sediments. Consequently, the separation of the four provinces in terms of river-mouth sediment grain-size distributions is low (t = 1.8).
- We never expected the light minerals to bring up the rear of this row (t = 1.3). The multivariate comparison is made using all light minerals evaluated, but only few show a certain provincialism (Fig. 14.5) which is a prerequisite for better discrimination. The lack of basic and volcanic rocks in the study area, usually producer of highly indicative rock fragments, may explain this low degree of separation.

We excluded Novito [13] and Torbido [14] from this collection because the province with the lowest mean distance changes from property to property:

The lowest mean distances of the Novito [13] are:

	Aspromonte		Zomaro	Serre
	Large	Small		
Sand heavy minerals			43.8	
Mineralogy of the clay fraction			43.6	
Gravel petrography				
>50 mm			63.5	63.4
25–50 mm			84.7	84.7
2–25 mm		75.3		
Silt mineralogy	29.9			27.8
Sand geochemistry				72.3
Grain sizes	57.5			
Sand light minerals	68.8			

Heavy minerals yield the best overall separation of the provinces. From this, the Novito [13] should most likely be attributed to the Zomaro Province, confirming its source properties. This is also true for mineralogy of the clay fraction, but both distances with values of about 44 are relatively high. Coarse gravel petrography attributes Novito [13], with the same high distances, to the Zomaro or Serre Province, again demonstrating a sliding transition between these two provinces. The assignment of Novito [13] by some sediment properties to provinces other than the Zomaro Province may be caused by the Jurassic limestones in its source area. Its portion in the source area is relatively small, and thus exerts only little influence on clustering source area properties, reflecting itself, however, over-proportionally in some sediment properties.

The lowest mean distances of the Torbido [14] are:

	Aspromonte		Zomaro	Serre
	Large	Small		
Sand heavy minerals	1.6	30.0		
Mineralogy of the clay fraction			43.6	
Gravel petrography				
>50 mm				41.0
25–50 mm				24.5
2–25 mm				14.5
Silt mineralogy			46.1	
Sand geochemistry	62.2			
Grain sizes		46.4		
Sand light minerals				60.0

The three best-separating sediment properties are assigned in the Torbido [14] to three different provinces; the heavy minerals to the Aspromonte, the mineralogy of the clay fraction to the Zomaro and the gravel petrography of >50 mm to the Serre Province. We have no explanation for the assignment of the heavy minerals; the clay fraction may stem from the many argillites of the Torbido [14] basin and the coarse gravel fit to the Serre.

In order to establish an overall distance matrix, the more complex data of the river-mouth sediments have to be condensed to a mean distance matrix of sediments. We felt the need to combine firstly similar properties. The three gravel matrixes were averaged (gravel matrix), the silt and clay matrixes (silt/clay matrix) and the light and heavy mineral matrixes (mineral matrix). Then we calculated the mean of the following five matrixes: gravel, silt/clay, mineral, grain-size and geochemistry matrix. The resulting overall sediment matrix is plotted in Table 17.2A.

The resumé of this overall sediment matrix, the mean distances within and between provinces, is plotted in Table 17.2B. The result comes close to the mean source-area matrix (Table 17.1B). Again, the distances within provinces are evidently better than between provinces, this time with the small Aspromonte river-mouth sediments excepted, whose distances in between do not contrast too much to the large Aspromonte river-mouth sediments.

17.3 Source and Sediment

Why not combine the mean source area matrix and the mean sediment matrix in order to obtain a mean overall distance matrix? The result is plotted

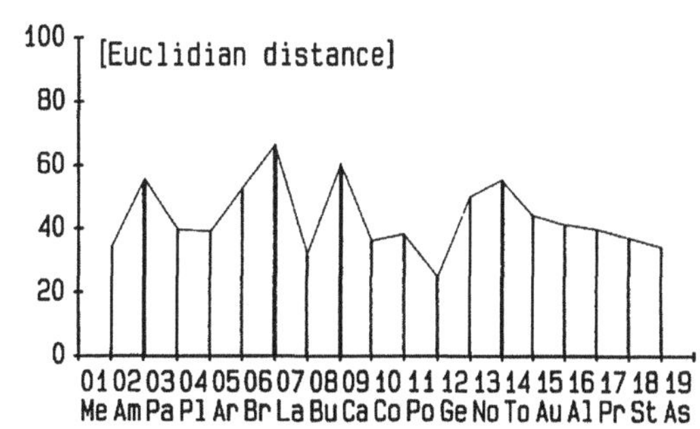

Fig. 17.4. Overall mean Euclidian distances between neighbouring rivers calculated by averaging the mean distances of all source and of all sediment properties (Table 17.3). Peaks are marked *black* and indicate limits between two provinces (cf. Fig. 17.2)

Table 17.2. Average of the nine sediment properties grain size, gravel petrography >50, 25 – 50 and <25 mm, heavy minerals, light minerals, silt mineralogy, clay mineralogy and sand geochemistry[a]

A	01 Mel	02 Ame	03 Pas	04 Pal	05 Ara	06 Bru	07 Lav	08 Buo	09 Car	10 Con	11 Por	12 Ger	13 Nov	14 Tor	15 Amu	16 All	17 Pre	18 Sti
19Ass	61.0	55.9	72.6	89.8	77.5	68.4	56.8	65.9	75.3	61.5	55.2	61.5	81.3	63.0	53.7	35.8	33.8	39.0
18Sti	44.7	47.0	58.3	83.4	69.8	51.6	46.4	57.3	68.0	71.6	56.4	51.5	72.1	66.5	47.1	43.1	34.8	
17Pre	44.2	46.5	62.0	85.4	67.2	64.4	42.7	66.3	73.7	67.9	56.3	60.7	71.9	61.7	36.9	33.3		
16All	58.4	59.0	59.2	85.4	65.5	73.5	59.6	75.4	81.9	75.4	63.7	73.5	84.8	53.6	36.7			
15Amu	52.8	58.7	45.7	74.9	58.5	65.3	57.2	64.4	74.1	69.1	55.5	58.9	83.0	43.4				
14Tor	70.3	68.2	51.5	57.5	73.8	75.8	67.4	77.2	67.4	57.5	49.7	51.3	77.4					
13Nov	61.3	66.7	86.2	69.3	71.3	54.7	69.7	72.6	66.4	75.1	62.8	62.4						
12Ger	55.7	60.0	68.5	74.4	79.7	53.4	57.5	62.4	45.0	43.1	30.1							
11Por	58.3	59.3	65.3	79.8	83.4	58.7	62.4	63.0	45.5	38.7								
10Con	69.1	70.6	71.3	78.2	84.9	61.9	71.2	68.5	43.1									
09Car	60.9	59.8	56.9	67.5	73.8	44.5	58.9	45.3										
08Buo	45.4	49.6	51.6	73.3	54.2	38.0	48.5											
07Lav	31.4	25.9	43.0	74.0	46.9	49.7												
06Bru	45.6	51.4	51.0	75.5	64.4													
05Ara	55.3	56.8	42.8	52.3														
04Pal	74.7	67.2	53.1															
03Pas	56.8	48.3																
02Ame	33.5																	

B	Large Aspromonte	Small Aspromonte	Zomaro	Serre
Large Aspromonte	39	55	61	56
Small Aspromonte		57	69	69
Zomaro			41	66
Serre				39

[a] **A:** Total Euclidian distance matrix, 171 distances between 19 basins. **B:** Euclidian distance matrix of the 4 provinces, calculated by averaging all distances of basins within a province and between two provinces.

in Table 17.3A; Table 17.3B shows the respective concentrate. It is noteworthy that the distances within the four provinces yield surprisingly similar values, 41, 44, 38 and 43. The same is true for the misfit of the non-related provinces which yield distances of 71, 70, 71 and 68. The distances of large Aspromonte rivers and small Aspromonte rivers is somewhat in between, 59, as is the distance between Zomaro and Serre Provinces, 55.

The division of provinces becomes especially evident if we plot the distance values of the overall matrix according to Fig. 17.2. The limits of provinces are characterized by distinct peaks (Fig. 17.4). The lowest overall distances are between Portigliola [11] and Gerace [12], 24.3, and between Laverde [07] and Buonamico [08], 31.4; the highest one is between Palizzi [04] and Assi [19], 86.1, a result which corresponds perfectly with our personal field experience.

17.4 Conclusion

The multivariate comparison makes it possible to rank the variables of both source and sediment which separate the provinces. Unfortunately, it is not possible to qualify the relationships of one variable between source and sediment as such, in other words, to explain why and how heavy minerals separate better than light minerals. So this procedure is a kind of black box, that works but does not tell us how. But this is the second step. The first should be to concentrate research on variables which promise to separate different provinces best. Among these, based on our Calabrian data, is the mineralogical composition of the clay and silt fractions. This confirms Blatt (1985), who comments sandstone petrography critically, favouring mudrocks for provenance analysis.

Table 17.3. Average of the mean distance matrix of the five source area properties (Table 17.1) and the mean distance matrix of the nine sediment properties (Table 17.2)[a]

A	01 Mel	02 Ame	03 Pas	04 Pal	05 Ara	06 Bru	07 Lav	08 Buo	09 Car	10 Con	11 Por	12 Ger	13 Nov	14 Tor	15 Amu	16 All	17 Pre	18 Sti
19Ass	71.7	71.6	77.9	86.1	80.3	70.8	63.1	68.6	61.2	45.8	48.7	52.6	63.5	52.3	56.6	31.9	46.3	33.7
18Sti	63.9	67.2	66.0	83.3	73.1	58.4	57.4	64.3	55.0	48.2	50.9	45.0	55.2	56.1	58.5	45.3	36.8	
17Pre	48.8	63.8	59.0	75.4	62.9	50.1	69.6	81.4	69.6	59.9	50.0	47.8	58.0	52.8	38.1	39.3		
16All	62.5	77.7	68.2	80.9	71.3	64.9	71.8	79.0	69.3	58.1	44.8	50.3	61.3	40.7	40.8			
15Amu	45.0	64.2	51.8	61.6	57.7	59.7	73.1	75.0	72.4	68.1	48.3	55.5	68.4	43.7				
14Tor	64.7	82.3	64.3	62.5	75.5	66.9	75.8	78.9	58.7	47.8	30.3	35.7	54.5					
13Nov	65.6	81.6	77.2	65.1	70.1	65.9	76.9	77.6	63.0	59.7	45.0	49.3						
12Ger	62.2	78.2	68.9	75.8	75.1	50.2	70.8	72.5	44.0	35.3	24.3							
11Por	59.8	77.8	71.3	74.8	80.3	58.3	73.3	72.8	48.7	37.7								
10Con	76.1	78.9	73.3	80.7	70.6	62.7	71.2	69.1	35.7									
09Car	73.8	75.3	67.9	77.1	76.9	61.2	67.7	59.3										
08Buo	55.9	41.8	62.5	72.2	63.3	59.4	31.4											
07Lav	49.4	31.6	57.5	73.7	60.1	65.3												
06Bru	42.2	59.0	42.5	60.8	51.8													
05Ara	46.8	58.7	32.3	38.4														
04Pal	53.3	63.3	39.1															
03Pas	49.5	54.9																
02Ame	33.8																	

B	Large Aspromonte	Small Aspromonte	Zomaro	Serre
Large Aspromonte	41	59	71	70
Small Aspromonte		44	71	68
Zomaro			38	55
Serre				43

[a] **A:** Total Euclidian distance matrix, 171 distances between 19 basins. **B:** Euclidian distance matrix of the four provinces, calculated by averaging all distances of basins within a province and between two provinces.

Chapter 18

Summary and Conclusion

After all this – what can be concluded? For clearness, we start with a synopsis, a condensed summary of the summaries of the 17 chapters.

General Summary

The Source (Part I)

Position and Geological Framework (1). Calabria makes part of the African deformed margin; the Ionian slope of the Calabrian arc is interpreted as an active plate margin, characterized by strong neotectonics, 1000 m uplift during 1 Ma, and strong seismic activity.

Basins and Channel Network (2). An order and size classification of the drainage divides and the 103 basins and interfluves did not yield meaningful results, thus remaining purely descriptive. Most basins are distinctly elongated; basin shapes (Horton's formula) do not correlate with properties of the river-mouth sediments. The arrangement of the drainage basins and the basin shapes makes us assume that uplift started with a continuous Mid-Pleistocene emergence plain. Despite maximum basin lengths of only 20 km, four rivers of the 19 main basins reach the seventh order, the other 15 rivers are all of the sixth order. Drainage density ranges between 5.1 1/km and 10.36 1/km, bifurcation ratios vary between 3.8 and 5.8.

Differences of basin configuration and channel network do not correlate with distinguishable differences of the river-mouth sediments.

Altitude and Slopes (3). The altitude standard deviation, the measure of relief, varies considerably between the single basins (213 to 481 m) as does the altitude skewness, the measure of dissection (−0.144 to 1.156). However, both variables do not exert any influence on properties of the river-mouth sediments, without regard to the fact that we are dealing with both "high" relief and dissection and consequently with steep-graded mountain rivers. The trunk slopes, the highest-order segments, vary between 1 : 30 and 1 : 110. The regional distribution of the gradients makes it possible to delineate low- and high-gradient belts of strengthened slope erosion subparallel to the coast.

Climate and Vegetation (4). Calabria has a marine mediterranean climate with a strong, altitude-dependent zonation of temperature, precipitation and consequently vegetation. From the coast to the mountains, mean annual temperatures vary between >16 and <10 °C and mean annual precipitation ranges between <600 and >2000 mm. Rainstorms with precipitations of more than 400 mm/day are the most striking climatic events of the area, the most effective driving force of erosion and transport after relief. Residence times of soils, landslides and alluvial deposits are so short that climate, at least in the sense of chemical weathering, is only of little importance in Calabria.

The Rock Units and Their Regional Distribution (5). We condensed the 55 lithological rock units of the CGDC to eight source rock types to which the provenance discussion of this book refers: granite (24% area), gneiss (7%), schist (30%), limestone (1%), conglomerate (9%), sandstone (9%), siltstone (12%) and argillite (8%). It goes without saying that source area geology is by far the most important variable controlling the composition of river-mouth sediments.

Erosion and Predicted Sediment Composition (6). The mean erosion rate, calculated as long-term (1 Ma) erosional budget instead of sediment yield is 178 mm/ka for the total area and 206 mm/ka if only the 19 basins are calculated, maximum values are 850 mm/ka. During the reference time of 1 Ma, the entire area would have been able to produce a sedimentary deposit of 100×100 km, 55 m thick, with a calculated pore volume of 50%. Corresponding to the "high gradient belt", a belt of high erosion can be established. Rock-specific erosion rates in the single basins reveal that one and the same rock type yields different erosion rates due to different positions in the mountains. The effective composition of a basin is the area percentage of its rock types weighed by its erosion rates. This effective composition may also serve as predicted sediment composition, as reference level for the provenance discussion.

Water Budget and Dissolved Load (7). The mean effective precipitation in the 50-year period between 1921 and 1970 is 698 mm/a, the mean evaporation rate 48.6%. This results in an average annual runoff of $1278 \times 10^6\,m^3$. The mean mineralization of the river waters is 322.6 ppm after the summer period and 221.4 ppm in late winter. About 47% of this dissolved load are regarded as being recycled or from atmospheric and anthropogenic input. The composition of the Calabrian river water corresponds fairly well to the worldwide mean of river water.

Rocks and Soils (8). The igneous rocks of the Serre are mostly granodiorites. Basic, ultrabasic or volcanic rocks do not occur. The majority of the metamorphic rocks belong to the transition from greenschist to amphibolite facies, sometimes retrograde. The soils are dystric and eutric cambisols, orthic luvisols and calcaric regosols; we measured a mean of 71% sand, 19% silt and only 10% clay. Soils of granite and feldspathic sandstone show an arkosic composition (F/Rf and $F/L > 1$), while the samples from gneiss and schist show a lithic composition (F/Rf and $F/L < 1$). ZTR values (zircon+tourmaline+rutile) of soil vary greatly; soils generated by a metamorphic substrate contain 12% of ZTR and those from granitic or sedimentary rocks only 4%.

From the Source to the Sediment (Part II)

Development of the Chemical Composition of Sand (9). We compared the chemical composition of the single types of source rocks and the sand fraction 63 – 500 μm of the soils, the effective composition of the source rocks and soils, the river-mouth sediments and the longshore bar. Gain and loss and constant Al_2O_3 calculations reveal that the composition of the river-mouth sands comes closer to the source rocks than to the soils. Attrition during fluvial transport is the most important source of sand at this active plate margin, chemical weathering and soil formation are less effective. The chemical composition of the average Calabrian river-mouth sand fits almost perfectly to that of active margin sandstones.

Development of the Grain-Size Distributions (10). Calabrian river-mouth sediments are both bimodal or unimodal. The gravel mode of the bimodal distributions, mostly Rosin-distributed, is interpreted as a first-order element, a primary feature inherited from the jointed and weathered source rocks, whereas the sand mode can be seen as a second-order element acquired during fluvial attrition. The shape of the gravel mode varies significantly from the jointed and weathered source rocks over the river mouths to the shoreface, the latter exhibiting log-normal grain-size distributions.

If the percentage of the fraction 1 – 20 mm exceeds about 35%, the distributions became unimodal. This fraction is thus crucial for modality and is consequently called a third-order element. It is assumed that different floods drive the "sedimentary mill" of these mountain rivers differently and that the formation of the fraction 1 – 20 mm is sensitive to these differences.

Development of Gravel Sphericity (11). Sphericity is strongly dependent from lithology and increases with grain size. Sphericity fails to be a useful provenance signal.

Special Topics of Sediment Development of Lower River Course and Beach (12). The gravel modes of the foreshore/shoreface strip are predominantly Gauss-distributed, in contrast to the chiefly Rosin-distributed river-mouth gravels. This may be seen as the only grain-size characteristic predominantly developed on the beach – in Calabria. The amount of dark gneiss, a fine grained, low resistant rock type drops from about 55 to 45% during final fluvial transportation (7.5 km) and from about 45 to 20% during longshore transportation (9 km).

Sphericity. Mean fluvial sphericity decreases from $psi_r = 0.61$ to 0.57. Along the beach, mean sphericity drops in the direction of net transportation, strongly along the backshore strip, less evidently along the shoreface and offshore strip (distances as above). The Strand Profil displays an amazing regular, unidirectional trend of increasing sphericity from land to sea. Sphericity is not a provenance signal.

River-Mouth Sediment and Provenance (Part III)

River-Mouth Gravel (13). The petrographic homogeneity of the rivers is so perfect that even after larger floods there is no change in composition. The amount of granitic and metamorphic debris in the river-mouth sediments (16–80-mm fraction) correlates well with that of the effective rock composition of the 19 feeder basins or, in other words, with the predicted sediment composition (granite: $r = 0.76^*$; metamorphic rocks: $r = 0.71^*$); but the sedimentary rocks do not.

River-Mouth Sand, Silt and Clay (14). Calabrian river-mouth sands are texturally and mineralogically immature. The mineralogic stability index (Q/F+Rf) is commonly <1.0. Grain proportions for medium-fine sands average Q44 F20 Rf36 and Lcr73 Lvo0 Lse27. The lithic population is dominated by crystalline phanerites and, consequently, grain size as well as counting procedure exert a relevant control on detrital modes. Granitic and metamorphic provenances can be separated by simply determining large mineral groups of little detail, such as quartzose, feldspathic and lithic grains and/or transparent, opaque and altered heavy minerals; heavy minerals, however, discriminate source provinces much better than light minerals.

Ternary diagrams of quartz and feldspar versus illite or chlorite plus kaolinite show that the average silt and clay composition of the effective soil, of the suspended load and of the river-mouth sediments form relatively narrow clusters.

Processes, Budget and Statistical Comparison (Part IV)

Mass Balances on Land (15). Different methods yield different values of modern erosion rates:

Object: landslides; time span: 50 years. Precipitation data from 1936 to 1986 and estimates of the volumes of landslides on air photos from 1941, 1955 and 1973 result in an estimated erosion rate of about 2700 mm/ka.

Object: sediment trap Lago Costantino; time span: 12 years. The volume of accumulated sediment corresponds to an estimated erosion rate of about 1900 mm/ka.

Object: geodetic survey of the long profile; time span: 26 years. Accumulation along the river course led to an estimated erosion rate of 900 mm/ka, the amount lost to the ocean during this time not taken into account.

Object: overall mass balance of the Buonamico; time span: 50 years. The input into the Buonamico is estimated $19 \times 10^6\,m^3$, the amount still stored along the river course may add up to $7.5 \times 10^6\,m^3$ solid rock equivalents. The difference, $11.5 \times 10^6\,m^3$, is obviously exported into the Ionian Sea. This corresponds to an erosion rate of 1527 mm/ka, seven times more than calculated from the 1 Ma mean.

Mass Balances in the Marine Realm (16). The comparison of our 1974 map with soundings from 1877 reveals negative deviations or erosion in the dimension of 100 m and positive deviations or accumulation up to 30 m. Loci of enhanced sedimentation are the canyon head and the outer shelf, here also the major slump scars occur. The conversion of only accumulation values on the shelf into erosion rates of the respective hinterland yields 7500 mm/ka.

Comparison of Source and Sediment: a Multivariate Approach (17). Multivariate comparison makes it possible to rank the variables of both source and sediment which separate the provinces. Unfortunately, it is not possible to qualify the relationships of one variable between the source and sediment as such, in other words, to explain why and how heavy minerals separate better than light minerals. So this procedure is a kind of black box, it works but does not tell us how. But this is the second step. The first should be to concentrate research on variables which promise to separate different provinces best. Among these, based on our Calabrian data, is the mineralogical composition of the clay and silt fractions.

Conclusion

Source and Sediment. Actuogeologic provenance analysis offers the possibility to quantify the relevant variables of both source and sediment. We digitized topography, geology, erosion rates and precipitation of the source area by means of a 250-m geomatrix and measured all conventional sedimentological parameters of the river-mouth sediments. The different source rocks, infinite in nature and mapped as 55 units on the Carta Geologica Della Calabria, were condensed to eight units for provenance treatment: granite (granodiorites), gneiss and schist (greenschist to amphibolite facies), reef limestone, conglomerate and sandstone (immature first-cycle deposits, marine), siltstone (debris flow) and argillite. We introduce the concept of effective source composition: these are the area percentages of the single rock types of a basin weighed by the respective erosion rates. If all rocks of a basin were able to share and share alike to produce a river-mouth sediment, sediment compositions should reflect source compositions properly. Thus the effective source composition would serve, at the same time, as predicted sediment composition. Such a prediction, however, is never verified. But the effective source composition is the only measure to compare quantitatively source and sediment in order to become aware of how far, and probably why the sediments deviate from their source. This concept is not restricted to rock distribution. The geochemistry of the source rocks and other variables such as the light and heavy minerals of the soils are used in the same way. As a result, the chemical composition of the river-mouth sands proves to be closer to the fresh source rocks than to the soils: the effective or predicted CIA (Chemical Alteration Index) of the rocks is 67.4, the soils yield 74.8 and the river-mouth sand displays 64.1 (Table 9.2). This is quantitatively supported by sand mineralogy. The effective content of Qc (composite quartz) of the soils is 3.5%, but the average river-mouth sand from the 19 basins contains 15.1% Qc (Fig. 14.9), a striking signal for the predominance of mechanical disintegration during fluvial transport compared to soil formation. The high rank of the mineralogy of the silt and clay fraction separating the provinces makes mudrocks a promising field of provenance analysis.

It goes without saying that the concept of effective source composition and predicting sediment composition is not applicable to grain-size distributions. However, we traced the shape of the gravel modes from the jointed and weathered source rocks in situ over transitory environments to the river course and river mouth and to the beach (Fig. 10.2). The Rosin distribution of the gravel modes proved to represent an inherited source rock feature preserved until the river mouth is reached. Only the beach environment symmetrizes the gravel modes. The jointed and weathered source rocks contain an average of 4% sand and the chemical signal of the soil sands is not transmitted into the river-mouth sediment. We believe from this that sand in Calabria does not come from such sources but originates mainly by attrition in the "sedimentary mill" driven by the floods of these high gradient mountain rivers. Thus the log-normal distribution of the sand modes usually found is an acquired property.

Climate and Relief. It is certainly possible to conclude from certain minerals or rock fragments in a sediment as to a qualitative minimum composition of a presumed source area. It is certainly impossible to make an equivalent statement about relief and climate, biogenic materials as climatic signal not being taken into account. We believe that even this formulation, "relief and climate", is not correct as such without specifying what is expected as a definite impact of relief and what has to be seen as a definite influence of climate on sediment formation. Basu (1985), Grantham and Velbel (1988), Johnsson and Stallard (1989) discuss this problem. From the paleogeographical point of view, the interest in learning about the trinity, paleogeology, paleomorphology and paleoclimate, is quite understandable. But what if climate and relief are linked and their forces are combined both positively or negatively in a kind of positive or negative feedback? If they intensify or counterbalance themselves? In Calabria it is almost impossible to separate the effects of these two variables.

It is evident that the geometric and morphologic differences of the 19 basins, speaking in terms of basin size, basin length or transport distance, basin shape, relief (altitude standard deviation), dissection (altitude skewness) and gradients of the trunk slopes, even if sometimes strong, are not great enough to find a reflection in differences in the river-mouth sediments, speaking in terms of compositional and textural maturity:

Variable	Range	Factor
Basin size	16 to 160 km^2	10
Basin length	10 to 26.4 km	1.5
Basin shape (Horton)	$R_f = 0.13$ to 0.44	3
Relief	213 to 481 m	2
Dissection	−0.144 to 1.156	−8
Trunk slopes	1:30 to 1:110	3.7

The same is true for climate. Mean temperatures and mean precipitation are high. The Zomaro plain is partly covered by peat bogs and the steep slopes are strongly fractured down to a maximum depth of 30 m. Consequently, chemical weathering is anything but absent, evidenced by an annual mean of 250 ppm of dissolved load of the rivers (Fig. 7.6). But these climatic conditions are barely reflected in the river-mouth sediments. Again, their chemical composition is closer to the source rocks than to the source soils. Chemical effects of weathering, even in the sand fraction, certainly exist, but they are suppressed and diluted by the overwhelming effects of mechanical disintegration, of attrition in the sedimentary mill of the mountain rivers.

It may be seen as a kind of paradox that the most important contribution of climate in Calabria, simply stated, is not chemical but mechanical. The heavy rainstorms with precipitations of >400 mm/day and recurrence intervals of about 25 years drive the sedimentary mill of the rivers, accelerating attrition. However, not only the present-day effects of climate and relief are linked and complex; their origin and history are, too.

It goes without saying that the main reason for relief, in Calabria, is plate tectonics which cause the enormous uplift of this small peninsula at an active plate margin. One result is the formation of a barrier in the center of the Mediterranean Sea, 2000 m high, which disturbs both the atmospheric and the marine circulation and which leads, over a chain of meteorologic processes, to the climatic implications sketched above.

The other result is the erosion rates, enormous in the past, close to disastrous today, which are responsible for a strong flux of sediment into the sea, again made possible by climate, by the flush during catastrophic floods.

Thus, instead of asking for "climate and relief", the appropriate question should be: what are the single factors which control the compositional and textural maturity, or immaturity, of Calabrian river-mouth sediments? The answer is: small basins, high erosion rates and high sediment production, short but steep-graded mountain rivers and short transportation paths, the absence of larger alluvial plains for effective storage in transitional deposits, and short recurrence intervals of rainstorms and floods, which make the flux of sediment not only strong but also quick. Consequently, the importance of the interaction of climate and relief consists in limiting the residence times in the continental and proximal marine environments, thus preventing maturation despite the omnipresence of chemical weathering.

A problem which cannot be solved based on our Calabrian data is the determination of the critical values of climate and relief below or above which complexes such as Calabria would produce significantly different sediments.

Mass Balance. Even if we take care not to present Calabria as a model of terrigenous sediment production, it is a prerequisite for the applicability of our results that Calabrian conditions are not a special local case but geologically relevant in a broader framework. The study area, the Ionian slope of the Calabrian Massif, forms part of an active plate margin. Long-term erosion rates are 206 mm/ka (1 Ma), present-day values climbing at about 1500 mm/ka and more, depending on the reference time.

The enormous flux of sediment, 303 000 m^3/a which corresponds to a 67-m^3 cube of solid rock, shapes the shelf and deepens the submarine canyons of the area which originate from the Messinian event.

During the presumed uplift bridging 1 Ma, the Calabrian study area lost 25% of its volume (Table 15.4). This could have led to a sediment body of 100×100 km, 55 m thick, with a mean porosity set 50%. We believe this to be a "geologic" amount, a realistic sedimentation rate verified elsewhere, but produced by a very restricted area of only 1800 km^2. Little imagination is needed to extrapolate this amount to longer geologic periods and to realize that small areas may have been responsible for larger basin fills, areas which also may quickly disappear due to more successful erosion or tectonism.

Problems of Research. The many methods applied in Calabria and the 19 basins which were compared lead to a kind of dilemma: the reliability of a single finding decreases with the increasing complexity of the research. Less complex research is better protected against such frustrations. The bimodality of grain-size distributions of fluvial sediments, for example, is attributed by many different authors to many different causes. All these single explanations can definitely be falsified in Calabria (Fig. 10.10), but at the same time we are not able to give our own satisfactory answer. Bimodality or not, this is the matter of the amount of the fraction 1–20 mm in the sediment. We believe that the formation of this fraction via attrition is sensitive to certain stages of a flood which may occur or not in space and time, making the sediments of a given river sometimes unimodal, sometimes bimodal. But this explanation was never verified (Popper 1973, p. 46), it simply exists via exclusion of others and in lack of a more plausible one.

Another example is the geochemistry of the river-mouth sand. The overall mean of all geochemical parameters of the 19 basins fits almost perfectly the worldwide mean of sandstones from active plate margins (Figs. 9.3, 9.5 A). Single rivers, however, deviate noticeably, the Amendolea [02] coming close to an island arc composition (Fig. 9.5 B), and it is hard to understand why the Buonamico [08] does not, as the cluster analysis of the geology assigns these two basins as nearly identical.

Third and last example is the complexity of erosion rates. Not only the time span involved but also the locus of research led to results which fit only in the order of magnitude, as sediment yield, erosional budget, transport rates and basin fill (Chaps. 6 and 15). As a tendency, calculated erosion rates become higher if short and more recent time intervals are chosen. In the case of Calabria, the contradiction may be resolved assuming a nonlinear uplift and/or by the fact that a relatively low area at the beginning of its uplift necessarily suffers lower erosion rates.

As a consequence, in Calabria the study of only one or even three or four basins and river-mouth sediments would have yielded misleading results.

Provenance. Provenance is a complex topic. Thus the only way to understand provenance is by complex research. Our Calabrian attempt ends with the river-mouth sediment, a transitory deposit instead of the more or less final stage of a basin fill. However, river-mouth sediments are a kind of nodal point in the development of a terrigenous sediment. All information or signals of a feeder basin which survive weathering and selective processes focus and converge towards the river mouth, where they are mixed and concentrated. The coarser grains are temporarily deposited, most fines are not. Thus a river mouth is really a suitable place for "forming an idea" about the source, in the sense of Goethe quoted at the very beginning of this study. Behind the river mouths, entering the marine realm, this flux of information diverges and is more and more diluted, information becoming more and more lost, although a single turbidite, for example, may tell us something about its provenance.

We believe from this that a satisfying provenance analysis should try to reassemble the subenvironments of a given basin, to quantify the composition of conglomerates as well as sandstones as mudrocks and to estimate their volumes and their facies interrelations. This is what still has to be done in Calabria, not taking into account the many special problems we could not solve: to carry out an equivalent study of the marine counterpart of south Calabria, of the Ionian Sea as far as it is fed by the study area, mainly by the deep submarine canyons of the Amendolea, the Torbido and the Buonamico.

Appendix A

Data Management

A comprehensive and regional study of terrigenous sediment's provenance and budget requires both the collection of thousands of data and its meaningful organization.

This quantity and our claim to these data made a conventional treatment without computer impossible. Thus a special data base was created at our university mainframe (Cyber 180) using the data base management system SIR (Scientific Information Retrieval, Robinson et al. 1980). SIR is capable of combining both hierarchical and network data structures and it is, therefore, a most convenient software tool for our Calabrian data. Moreover, SIR interfaces directly with other well-known software products, i.e. SPSS (Statistical Package for Social Sciences), a feature which facilitates the often complex statistical evaluations. Shulman (1983) gives a thorough description of a geological use of SIR.

To manage different data types such as grain size, geochemistry or precipitation data, a SIR data base is divided into several "record types" in which the data are stored as records. Each record type consists of a number of variables belonging to one another, which are defined and named during the data base creation process. In order to identify each record as member of a record type, each record begins with a variable called rec-type number, the number of the record type. The name of a second variable must be the same in all record types of a data base. Its definition is obligatory and a value must be present in each record. It is called case-ID, case-identifier, and attaches each record to a special case, that is the uppermost variable in a hierarchically structured data set, i.e. a sample. Additionally and optionally, so-called sort-IDs, sort-identifier, can be defined. Their number depends on the position of the record type data within the hierachical data structure. To give an example, if the case-ID of our data base is the sample number, a record type containing data belonging directly to the sample (e.g. grain size) needs no sort-IDs, because there is only one possible record per sample with one value for each variable, grain-size classes in this case. However, a record type with variables, which may have more than one value per case and therefore with more than one record per sample, needs sort-IDs to identify each record; this means, form data of single clasts must be additionally described by the sort-IDs "grain-size class", "rock type" and "clast number".

Finally, each record type and its records are composed of the variables rec-type number, case-ID, sort-IDs and any number of variables describing real and measured data (Table A.1).

Our Calabrian data can be divided into two main data groups:

- data of the digitized maps
- data of the taken samples

To manage these two different data groups in the same data base, a variable has to be found which is present in both data groups and which is used as case-ID in the SIR data base. This variable is simply a point in our Calabrian study area, be it a point of the geomatrix, or a sample site from which the data are available, so that the variable "point" is the uppermost variable in the data structure shown in Fig. A.1. The values for this variable are integer numbers, either the original sample number or the number of one of the 86420 digitized map squares.

The data base was named CALCOLO (*Cal*abrian *Co*mputer *Lo*g) and consists of 16 record types (Table A.1) that are closely related to the data structure, and each record type corresponds to one of the lower boxes of Fig. A.1 and to the upper right box, respectively.

With the help of the SIR data base CALCOLO, it is relatively easy to retrieve and evaluate our data in any combination.

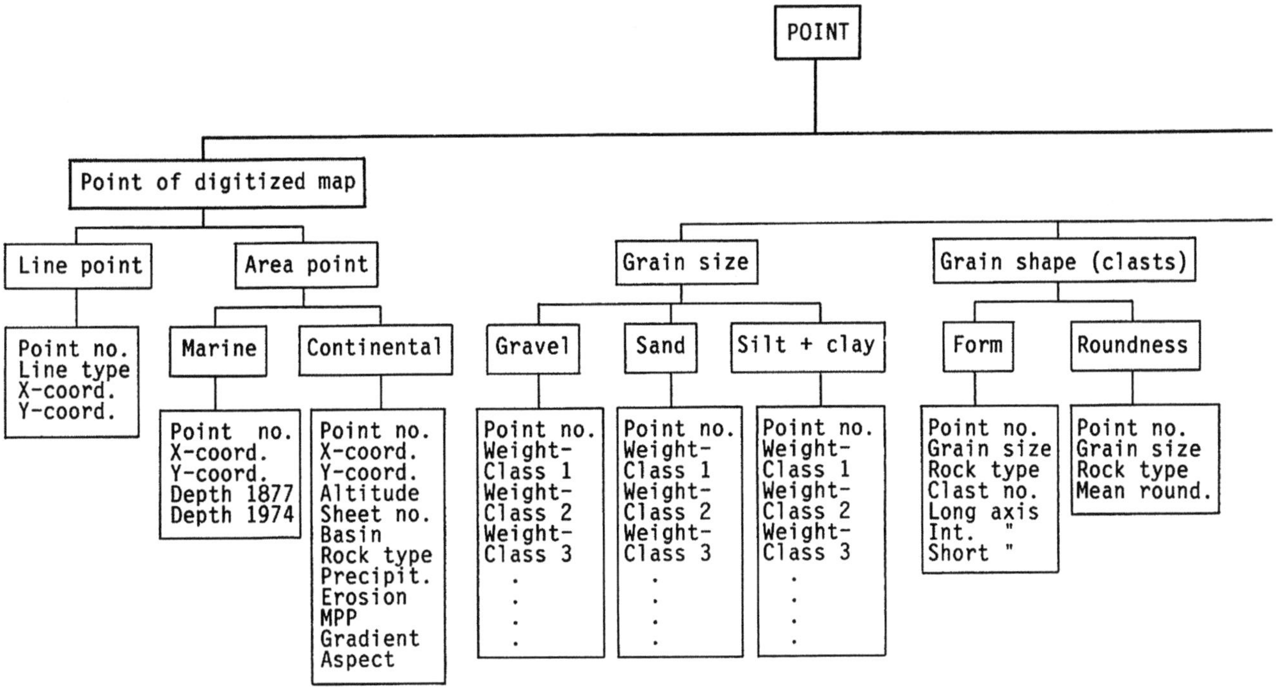

Table A.1. The record types of the data bank. Here all quantifiable data of Calabrian

Rec-type number	Description of data	Case-ID	Sort-ID
1	Coordinates of points from lines	Point number	Line type
2	Data of digitized maps; each point represents a square of 250×250 m	Point number	–
3	Data of a digitized marine map	Point number	–
4	Information on taken samples	Point number	–
5	Grain-size data gravels	Point number	–
6	Grain-size data sand	Point number	–
7	Grain-size data silt + clay	Point number	–
8	Petrographic composition of gravels	Point number	Grain-size class
9	Light mineral composition of sand	Point number	Grain-size class
10	Heavy mineral composition of sand	Point number	Grain-size class
11	Petrographic composition of silt + clay	Point number	Grain-size class
12	Petrographic composition of rocks	Point number	–
13	Form data of pebbles	Point number	Grain-size class Rock type Pebble number
14	Roundness data of pebbles	Point number	Grain-size class Rock type
15	Chemical composition of rocks	Point number	–
16	Chemical composition of water samples	Point number	–

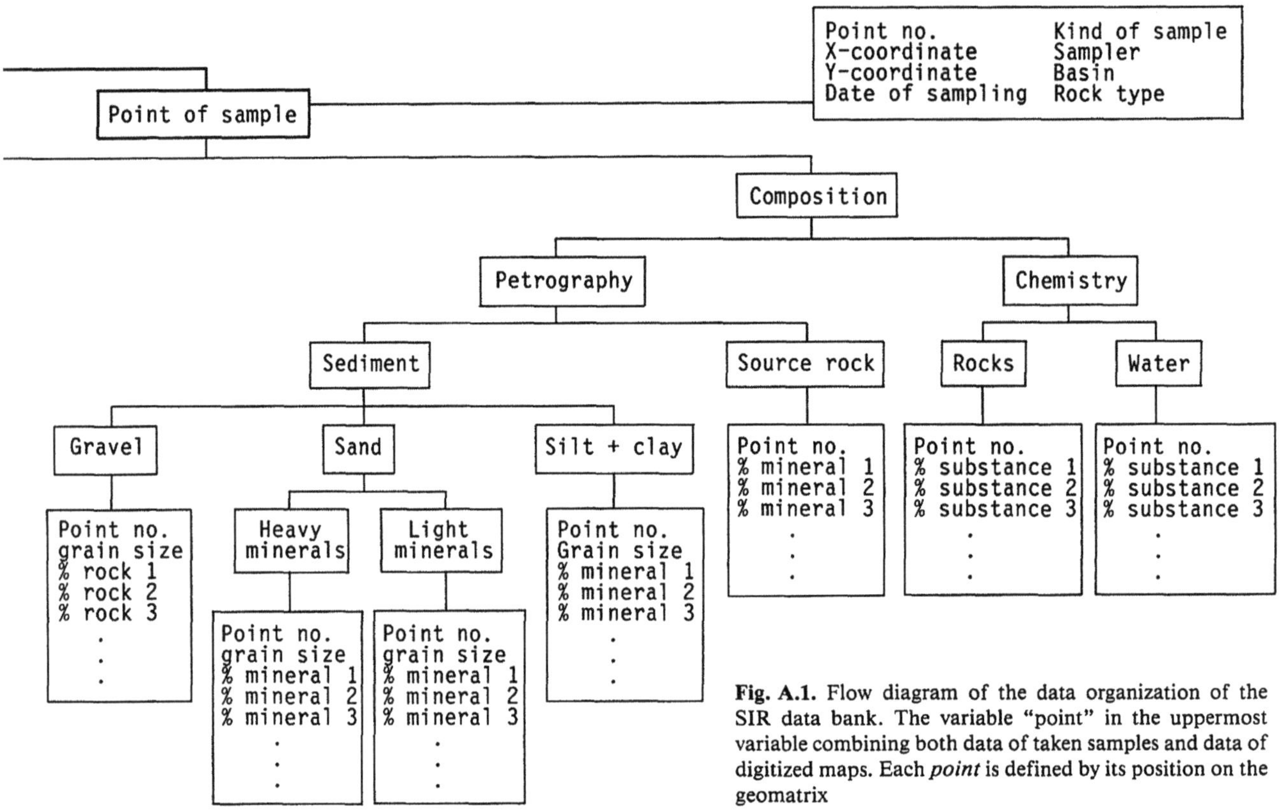

Fig. A.1. Flow diagram of the data organization of the SIR data bank. The variable "point" in the uppermost variable combining both data of taken samples and data of digitized maps. Each *point* is defined by its position on the geomatrix

maps and samples are stored, with the exception only of the digitized drainage networks

Variables	No. of records
x-Coordinate, y-coordinate	5986
x-Coordinate, y-coordinate, map number, altitude, river system, rock type, precipitation, erosion, erosion plain, relief, aspect	86420
x-Coordinate, y-coordinate, depth under sea level	7950
x-Coordinate, y-coordinate, date of sampling, sampler, river system, rock type, kind of sample	2345
From 4000 to 2 mm weights of 33 grain-size classes in 1/3-phi steps	836
From 2 to 0.063 mm weights of 15 grain-size classes in 1/3-phi steps	836
From 0.063 to 0.002 mm weights of 5 grain-size classes in 1/1-phi steps	88
Percents of maximum 22 rock types	1061
Percents of maximum 15 light minerals	71
Percents of maximum 30 heavy minerals	106
Percents of maximum 10 clay minerals	192
Percents of maximum 30 minerals	37
Lengths of long axis, intermediate axis and short axis	41211
Mean roundness	1238
Percents of 12 elements	73
Conductivity, colour, temperature, pH and percents of 10 elements	86

Appendix B

Graphic Representation of the Data Base

Appendix B presents all quantifiable data which we gathered in Calabria in condensed form, with diagrams and plots but no tables.

The intention of Appendix B is to enable the reader to picture source and sediment in Calabria himself, not filtered by our interpretation.

Basin 00, Organization and Abbreviations

Each basin 01 – 19 and its river-mouth sediment are presented on a full double page, source or producer data on the left page and sediment or product data on the right page. Basin 20 shows the mean values of the 19 basins, insofar as they are meaningful.

Source (left pages). The uppermost row shows the setting of the basin in the suite of the 19 rivers, its outline and rough drainage pattern taken from the topographic map 1 : 100000, and finally the depth contours of the paleovalley in the river-mouth area (Jean Mouton).

The lowermost row presents derived data, the hypsometric integral and three plots of basin properties plotted against their distance from the river mouth, altitude, gradient and erosion rate.

The histograms presented in the left column show the altitude frequency distribution (100-m intervals), their deviation from the mean of the 19 basins and the relations between the eight key rock types in terms of their relative areal extent and their relative erosional budget.

*Abbreviations are: Gra*nite; *Gne*iss; *Sch*ist; *Lim*estone; *Con*glomerate; *S*and*t*one; *Si*lt*s*tone and *Arg*illite.

The right column shows the regressions of the stream-segment number and the mean segment length versus their order, taken from topographic maps 1 : 10000. The remaining two histograms present the rock-specific slope angles (gradients) of the basin and the rock-specific erosion rates [mm/ka], millimeter per 1000 years or 1 kilo-year. Rock abbreviations as above.

Sediment (right pages). Above the left column is a ternary diagram which shows the changing amount of granitic, metamorphic and sedimentary rocks going from the pure areal composition over the effective composition (erosional budget) of the source basin to its river-mouth sediment represented by the gravel fractions 16 – 80 and 2 – 16 mm.

The grain-size distributions of the river-mouth sediments stem from two campaigns, fall 1972 and spring 1973, sieved in one-third phi steps.

Abbreviations of Light Minerals are: Qa single-crystal quartz, subangular to very angular; *Qr* single-crystal quartz, subrounded to well rounded; *Qg* composite monomineralic quartz of igneous origin; *Qm* composite monomineralic quartz of metamorphic origin; *LQ* quartz crystal >63 µm within crystalline, volcanic or sedimentary rock fragments; *KF* potash-feldspar; *PF* plagioclase; *LK* and *LP* K-feldspar and plagioclase >63 µm within crystalline, volcanic or sedimentary rock fragments; *Lm* fine-grained (<63 µm) metamorphic rock fragment; *Ls* fine-grained (<63 µm) siliciclastic rock fragment; and *Lc* marble, dolomite and limestone of any texture.

*Abbreviations of Heavy Minerals are: Ga*rnet, *Ky*anite, *An*dalusite, *Si*llimanite, *St*aurolite, *Zi*rcon, *Sp*hene, *Ru*tile, *Br*ookite, *Ba*rite, *Ap*atite, *EZ* epidotes, *Pu*mpellyite, *To*urmaline, *r*homb. *P*yroxene, *Dio*pside, *Au*gite, *r*homb. *A*mphibole, *g*reen *H*ornblende, *b*rown *H*ornblende and *Tr*emolite/Actinolite.

The histograms of the right row continue to demonstrate the composition of the river-mouth sediments and water.

Abbreviations of the Plot Chemical Composition are: SiO_2, TiO_2, Al_2O_3, Fe_2O_3, MnO_2, MgO, CaO, Na_2O, K_2O and P_2O_5.

The plot Al_2O_3/SiO_2 versus K_2O/Na_2O shows the broad variety of the single river-mouth sands (+); the sand described is marked (●).

Abbreviations for the Results of the X-Ray Diffraction Analysis of the River-Mouth Silt and Clay are: *Q*uar*tz*, *F*eld*sp*ar, *Cal*cite, *Ill*ite, *Chl*orite, *Kao*linite and *C+K*, chlorite+kaolinite.

The Full Terminology for the Chemistry of the River-Mouth Water is Finally: Cations Na^+, K^+, Ca^{2+}, Mg^{2+} and Fe^{2+} and anions HCO_3^-, Cl^- and SO_4^{2-}.

01 Melito

Drainage basin, the producer: Geology and geomorphology

(For abbreviations see 00 basin 00)

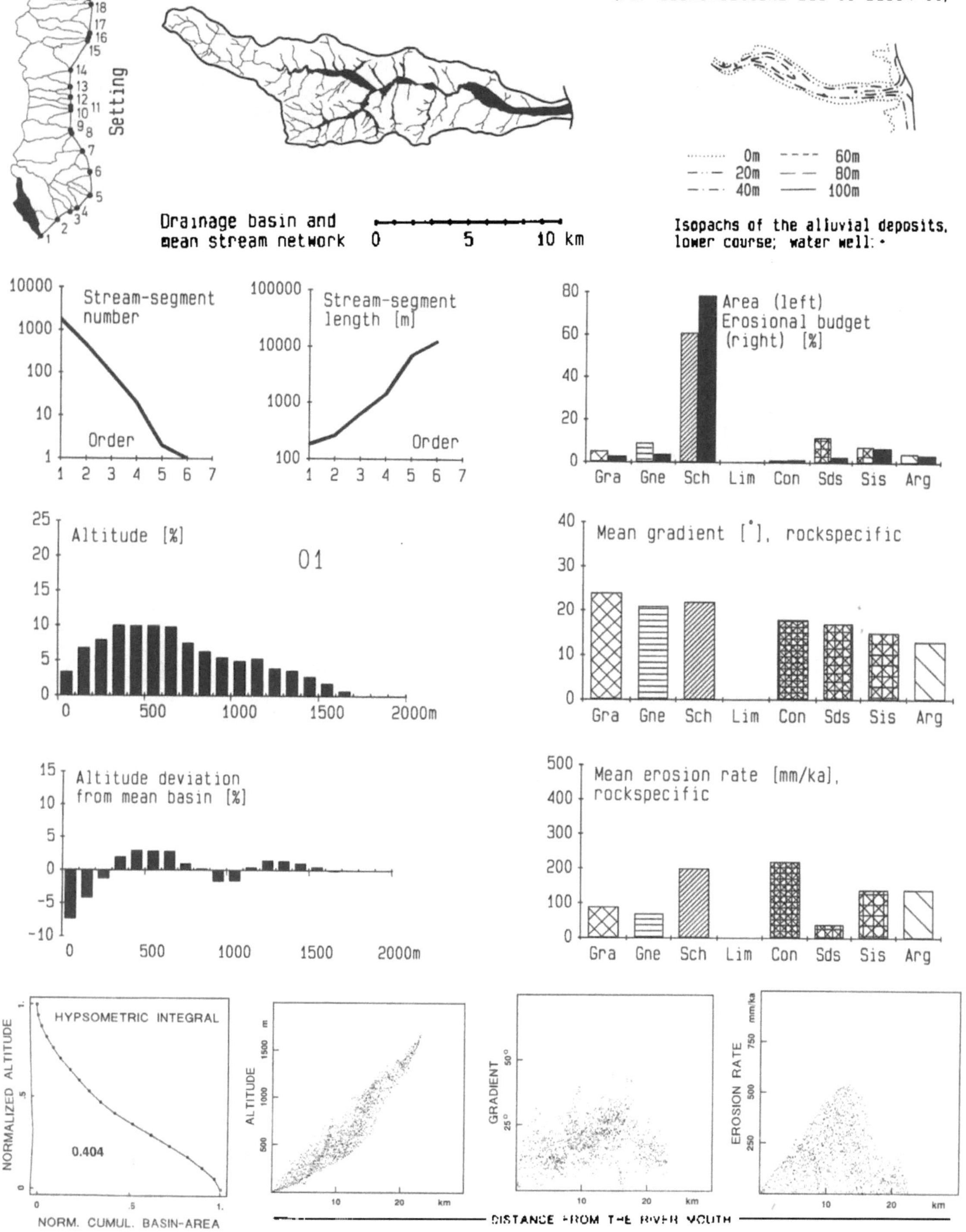

River-mouth sediment and water, the product: Grain sizes, petrography, mineralogy and chemistry

01

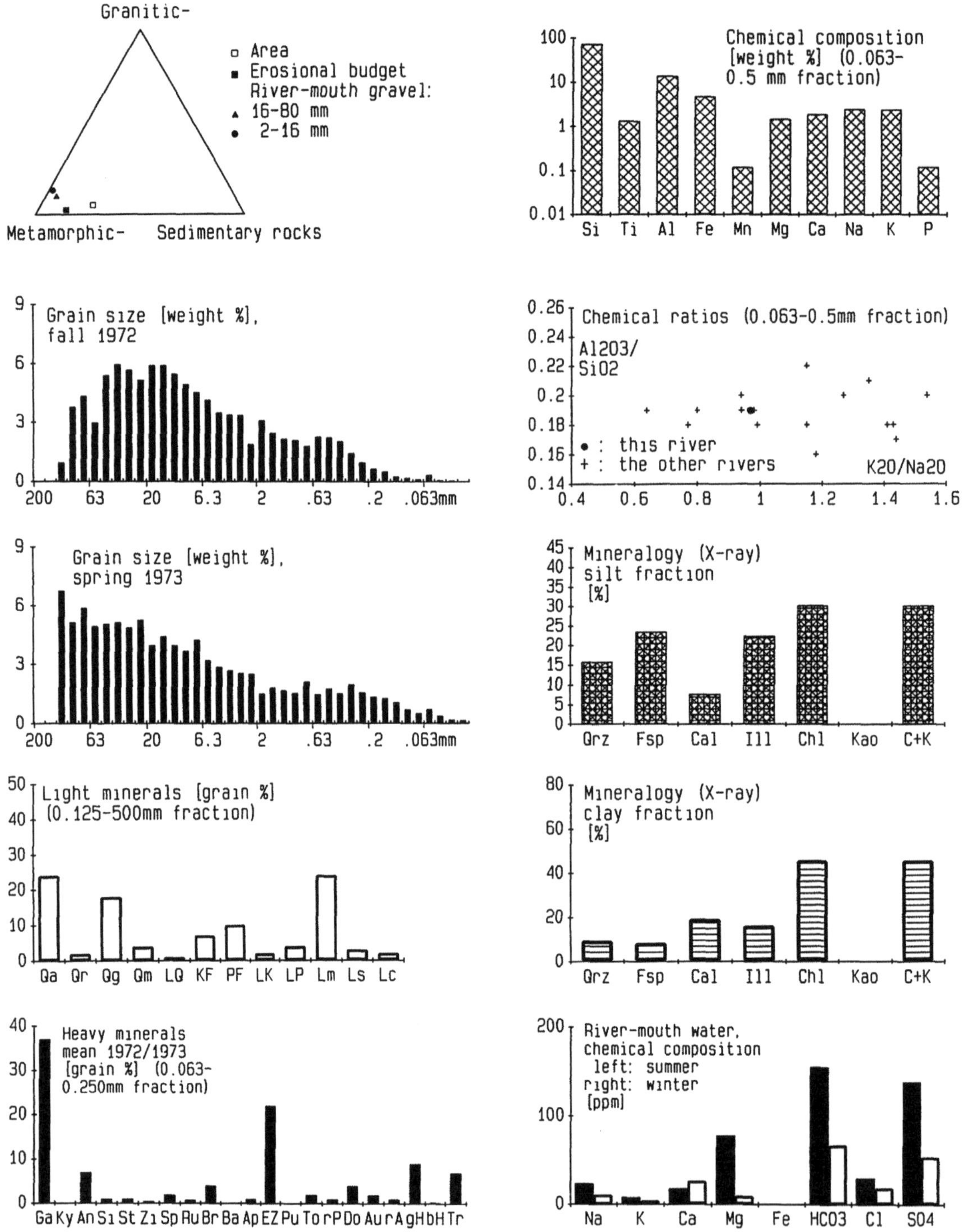

02 Amendolea

Drainage basin, the producer: Geology and geomorphology

(For abbreviations see 00 basin 00)

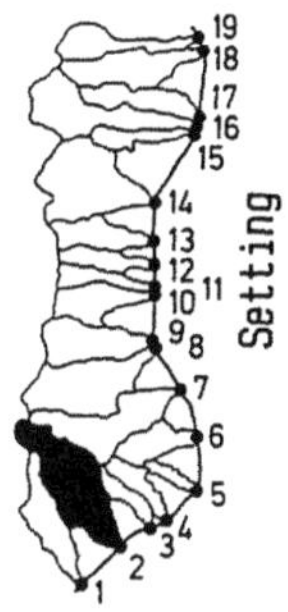

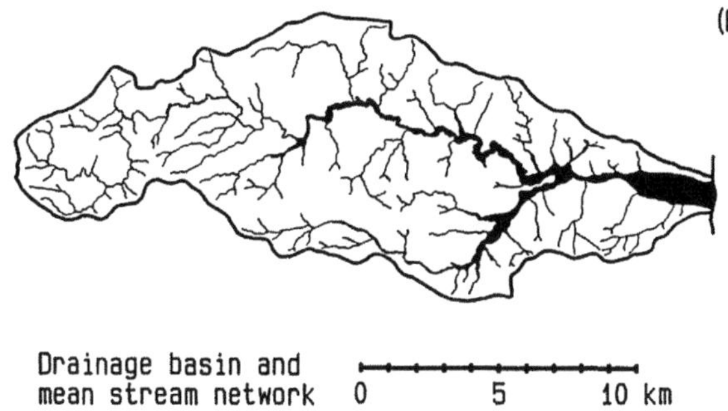

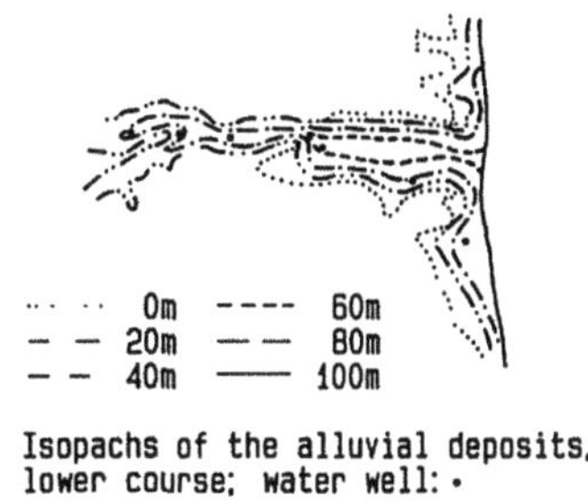

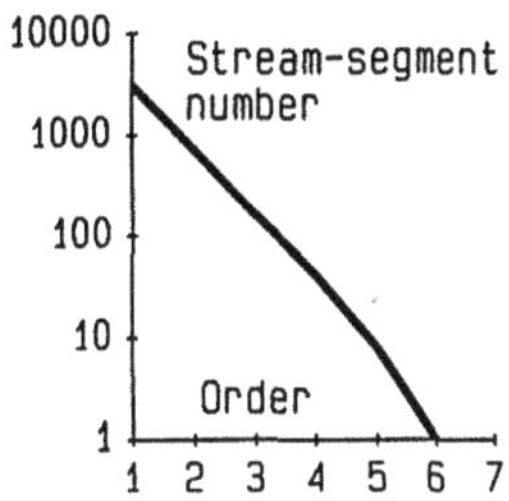

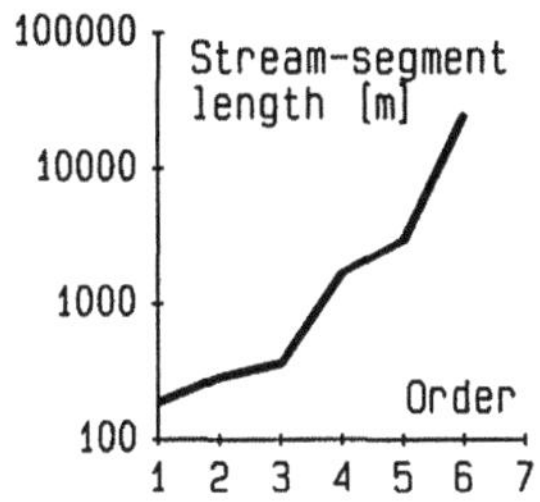

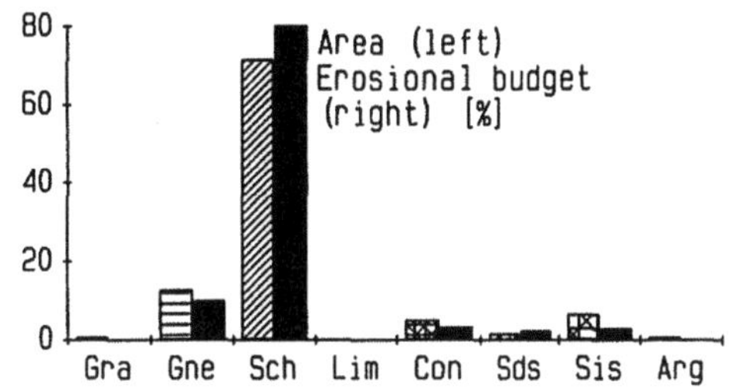

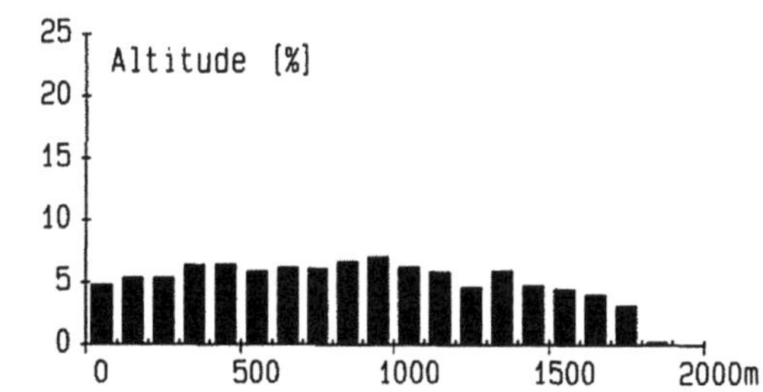

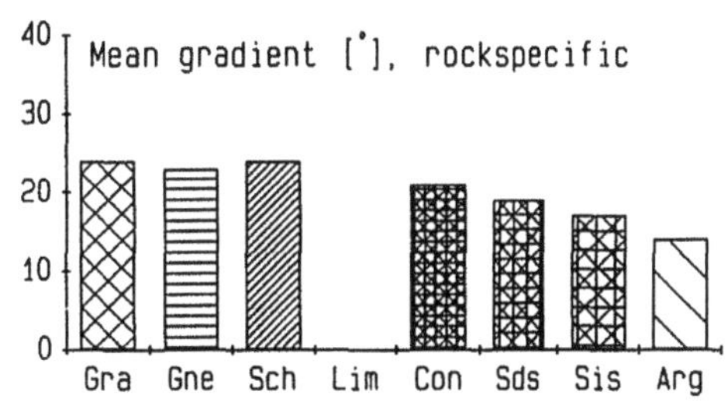

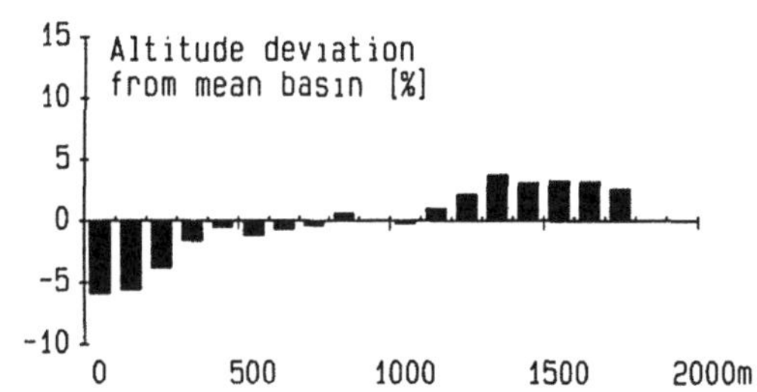

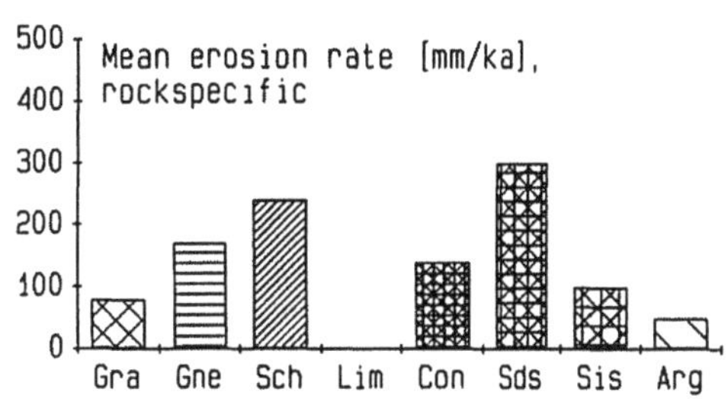

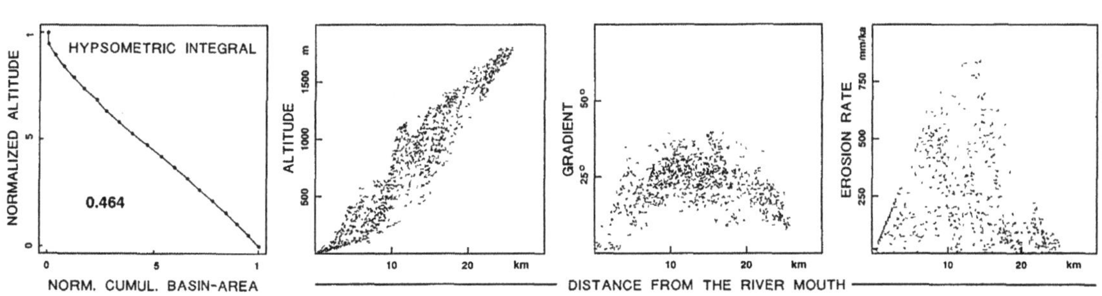

River-mouth sediment and water, the product: Grain sizes, petrography, mineralogy and chemistry

02

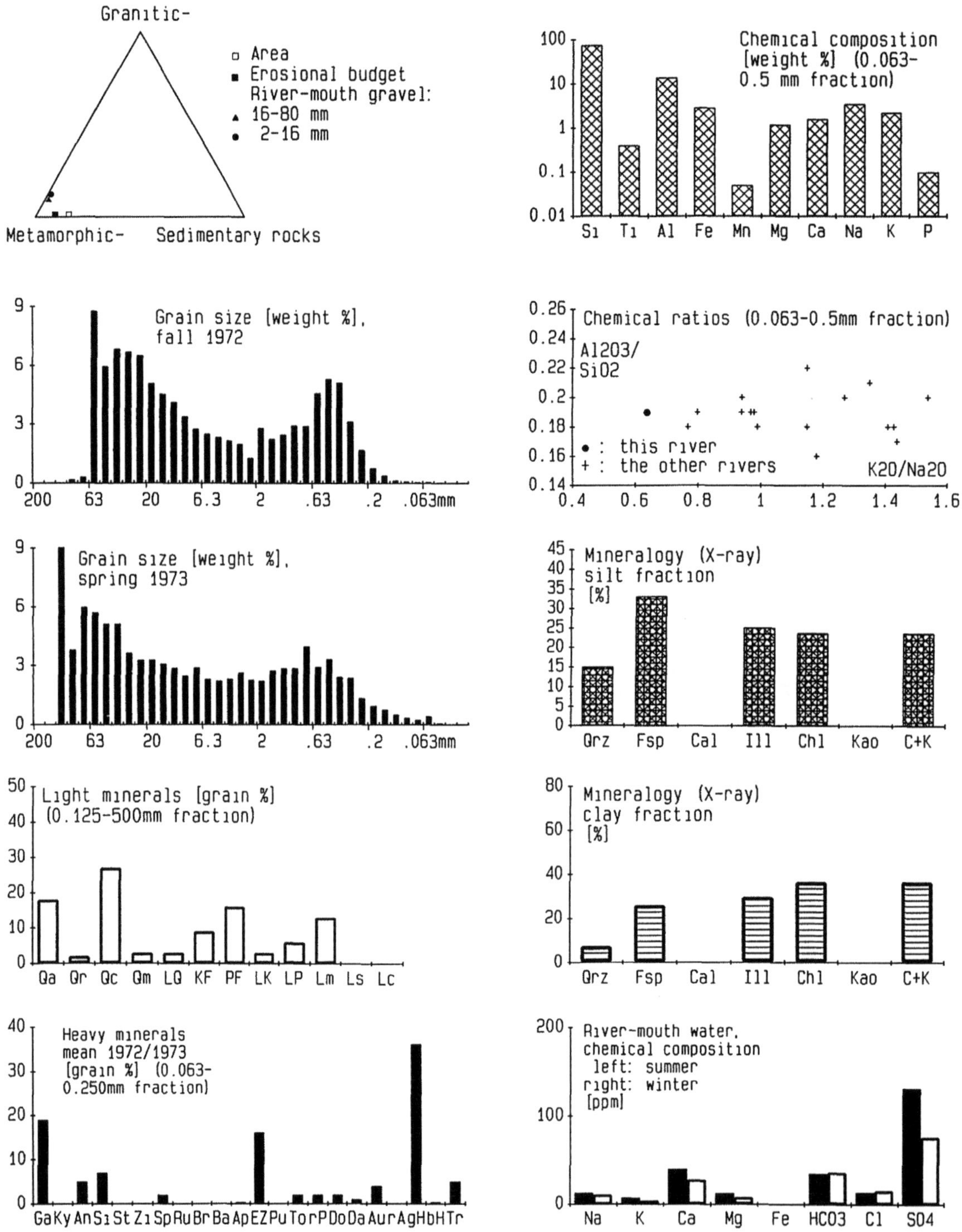

03 Pasquale

Drainage basin, the producer: Geology and geomorphology

(For abbreviations see 00 basin 00)

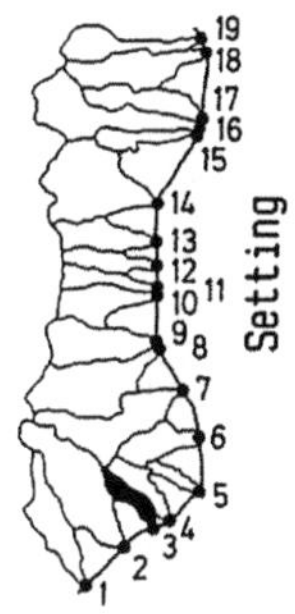

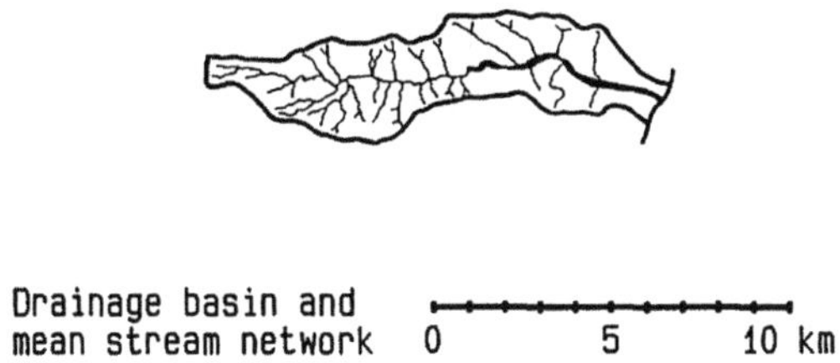

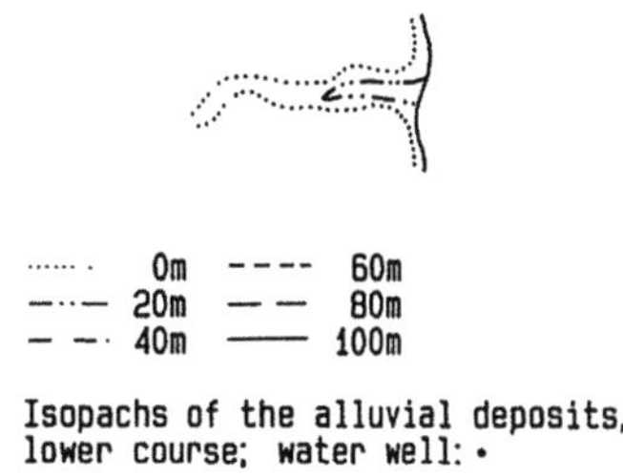

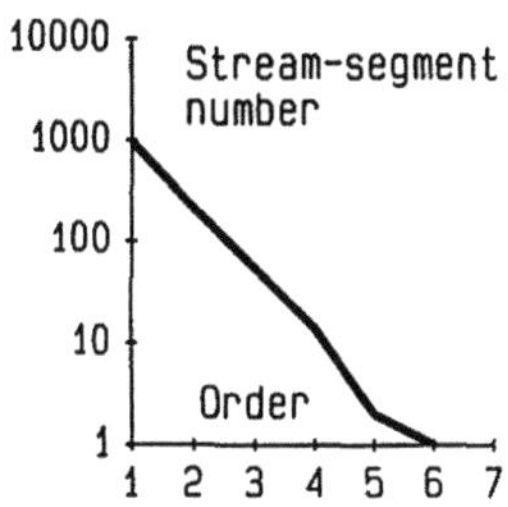

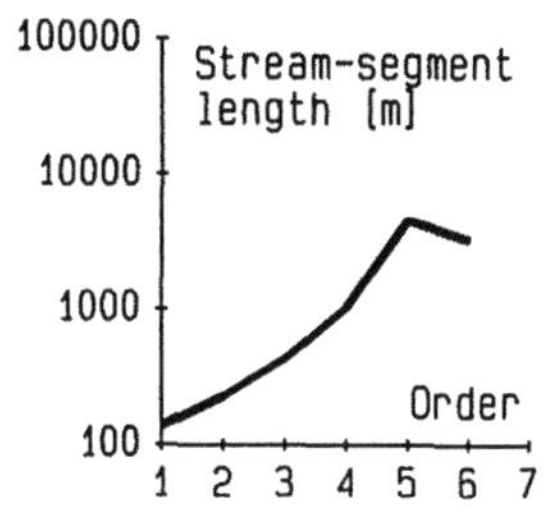

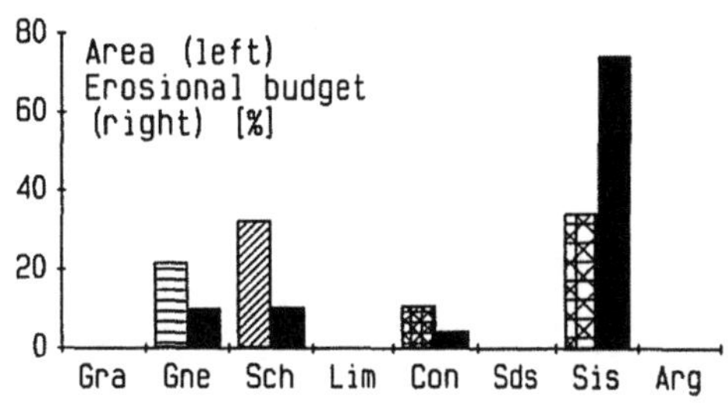

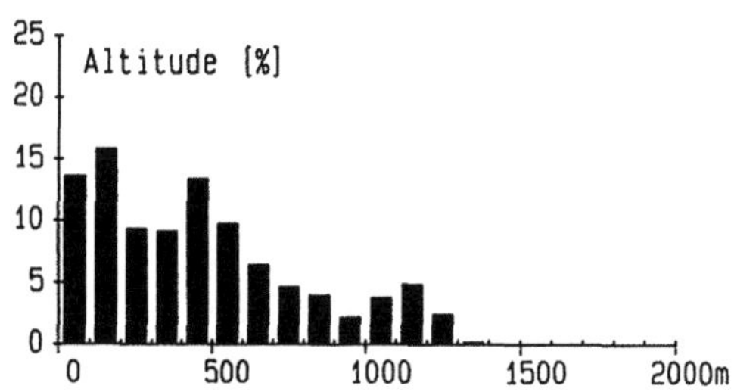

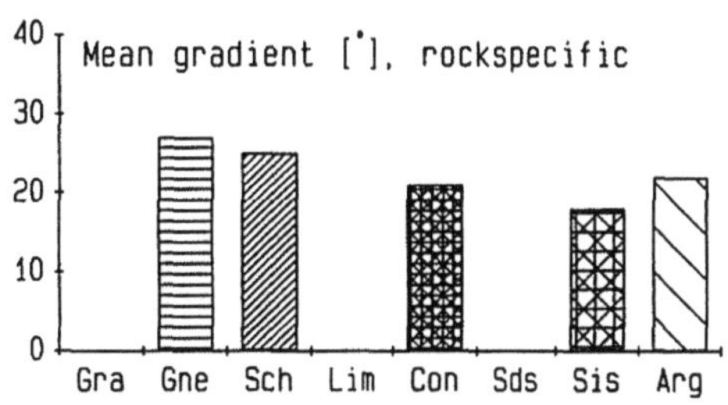

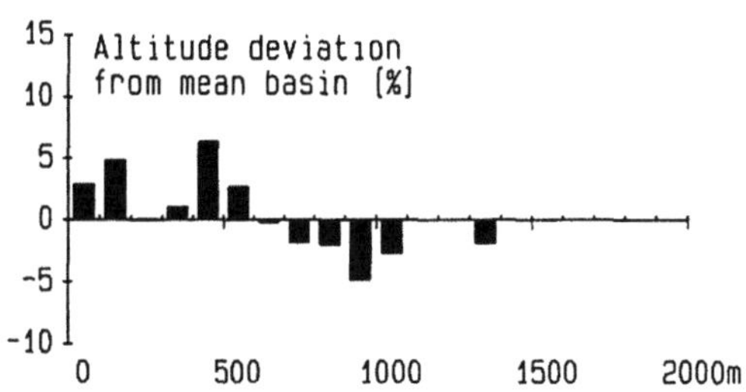

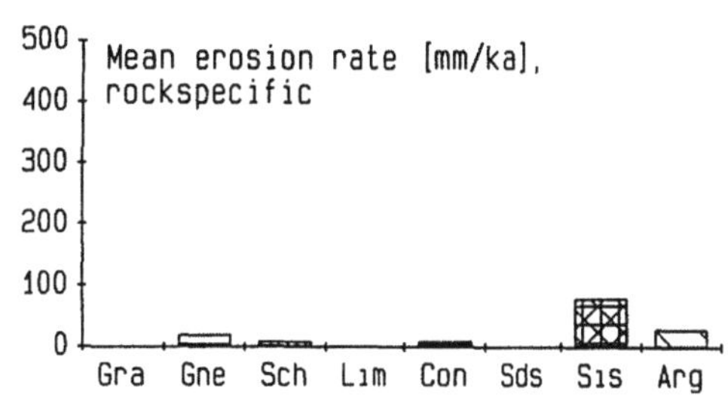

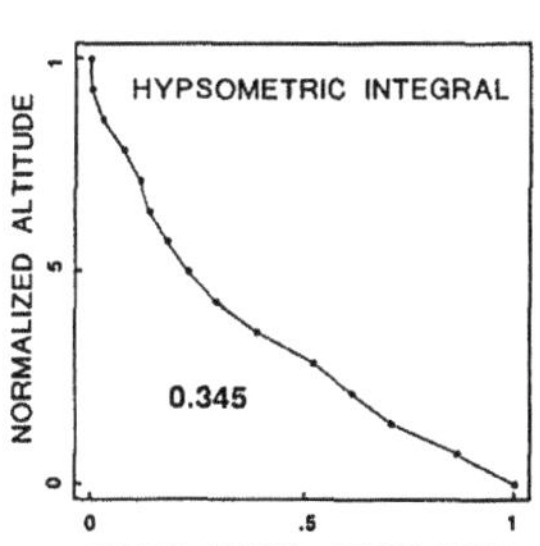

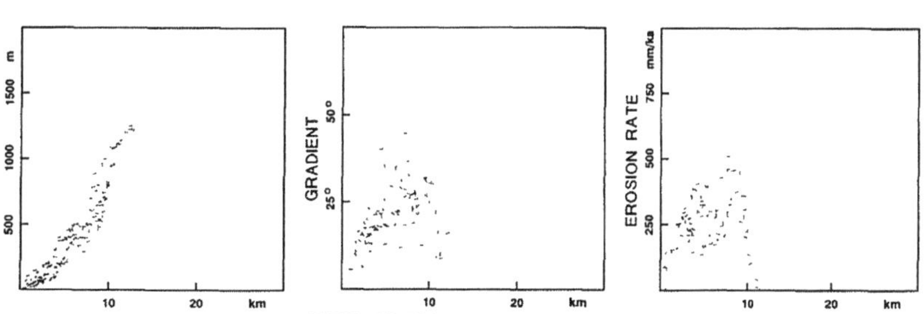

River-mouth sediment and water, the product: Grain sizes, petrography, mineralogy and chemistry 03

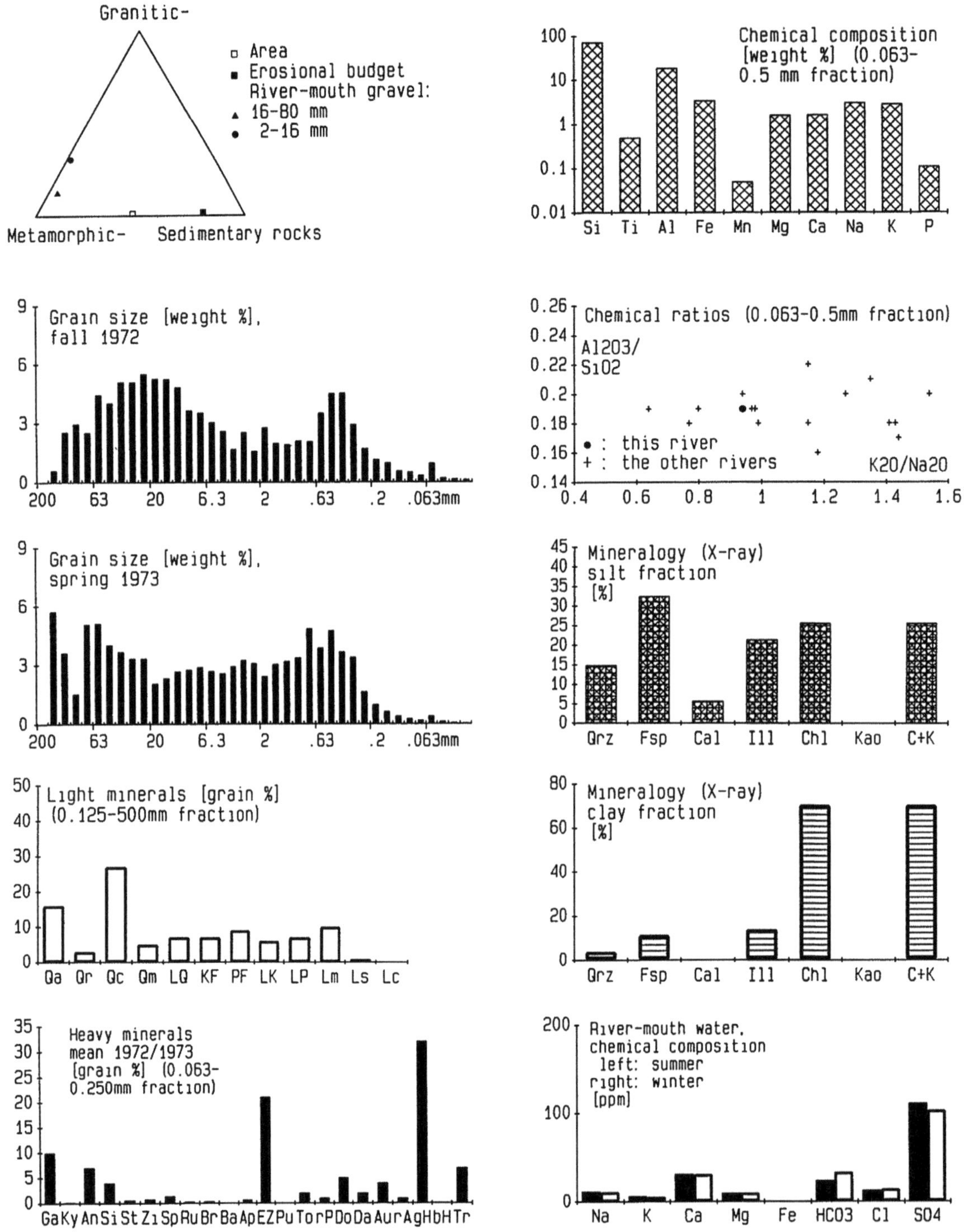

04 Palizzi

Drainage basin, the producer: Geology and geomorphology

(For abbreviations see 00 basin 00)

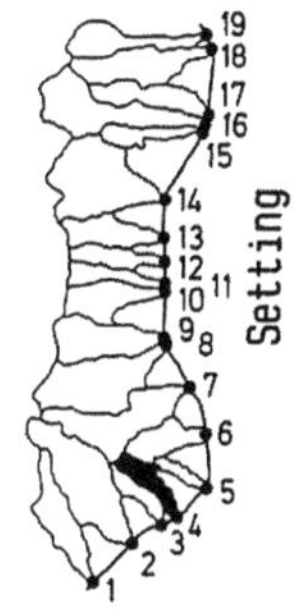

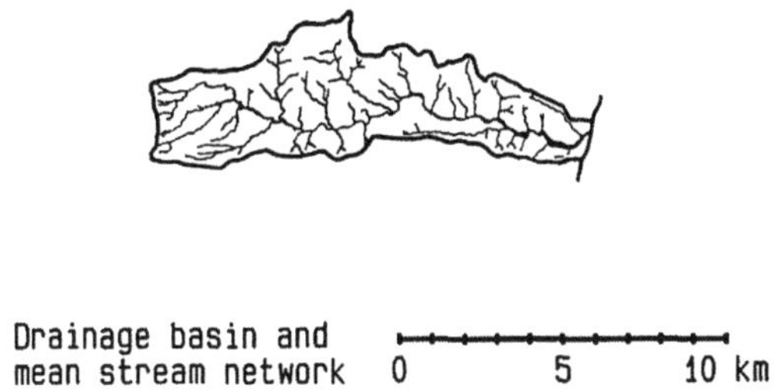

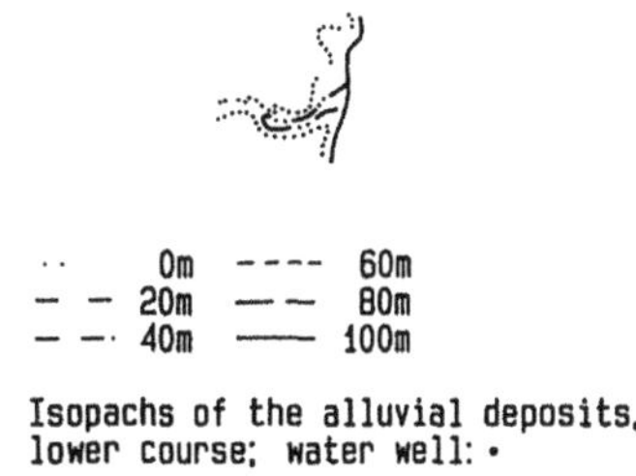

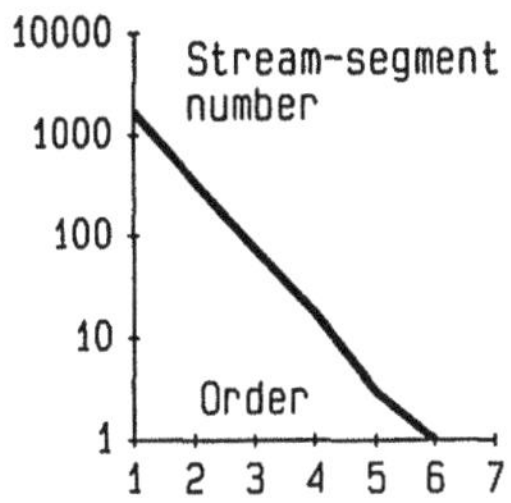

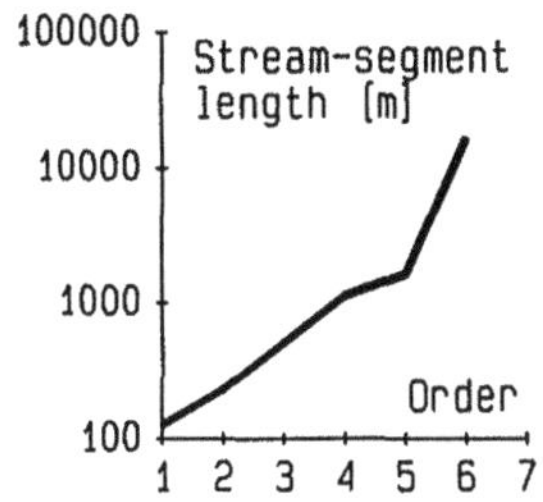

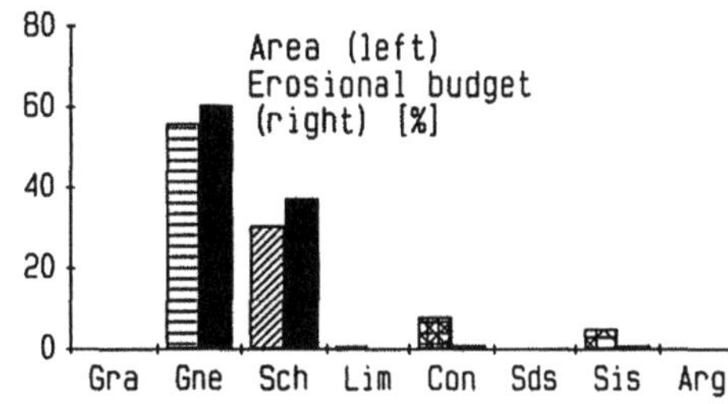

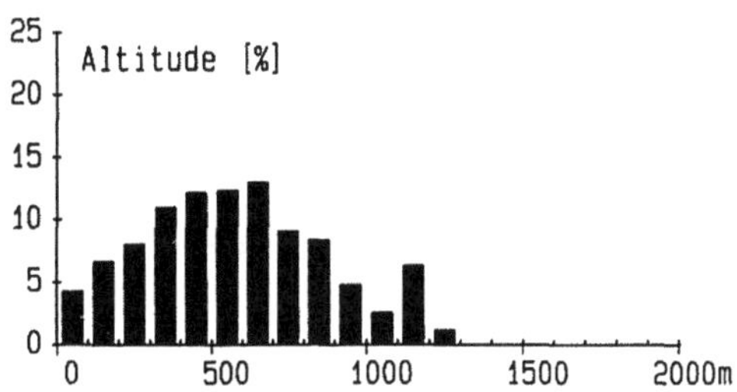

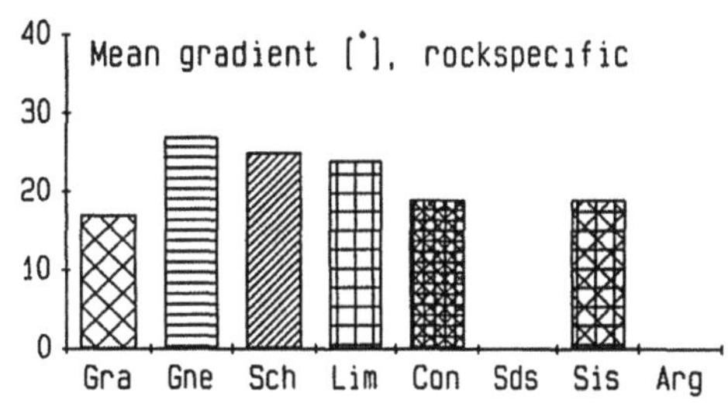

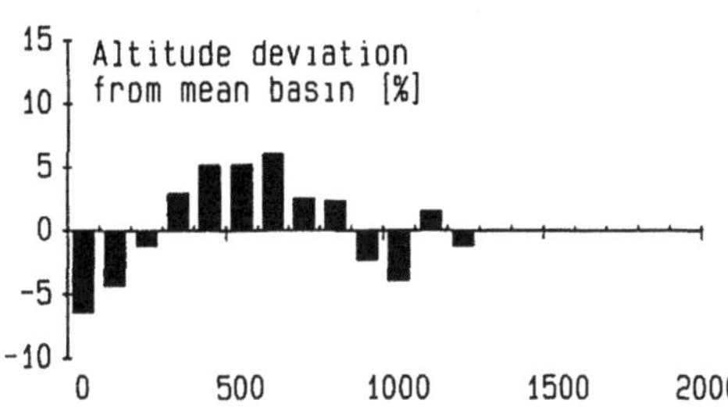

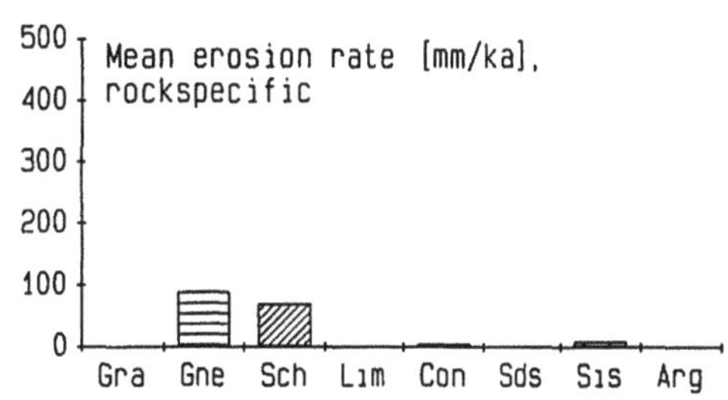

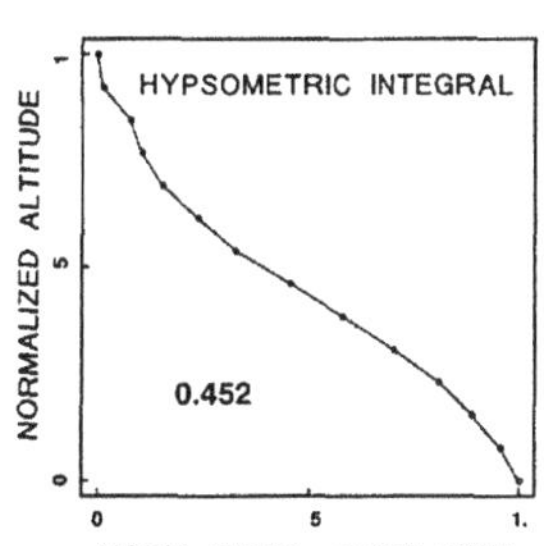

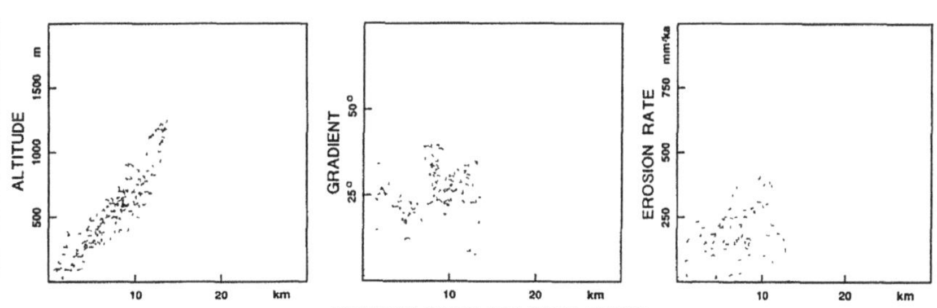

River-mouth sediment and water, the product: Grain sizes, petrography, mineralogy and chemistry

04

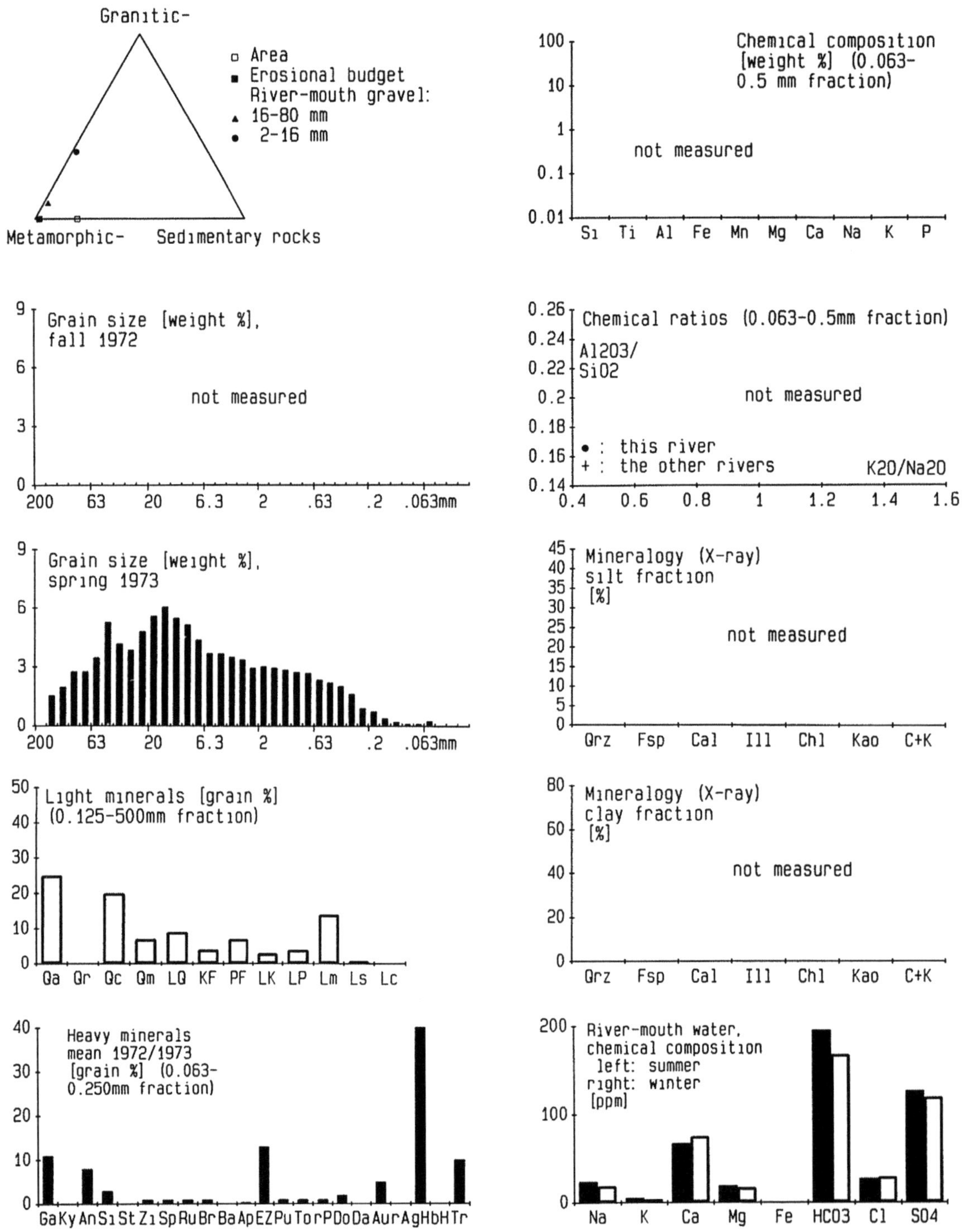

05 Aranghia

Drainage basin, the producer: Geology and geomorphology

(For abbreviations see 00 basin 00)

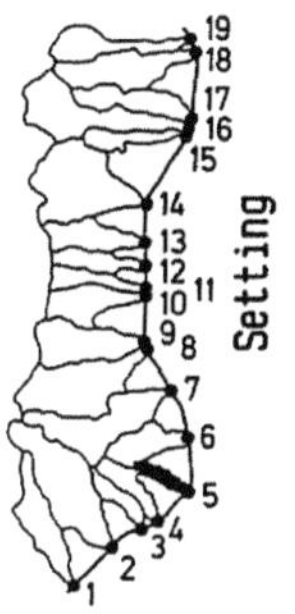

Drainage basin and mean stream network 0 5 10 km

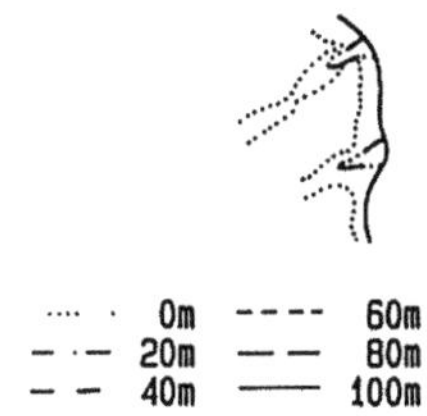

Isopachs of the alluvial deposits, lower course; water well: •

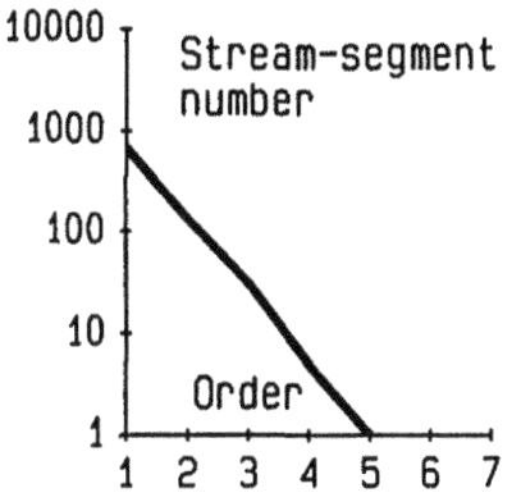

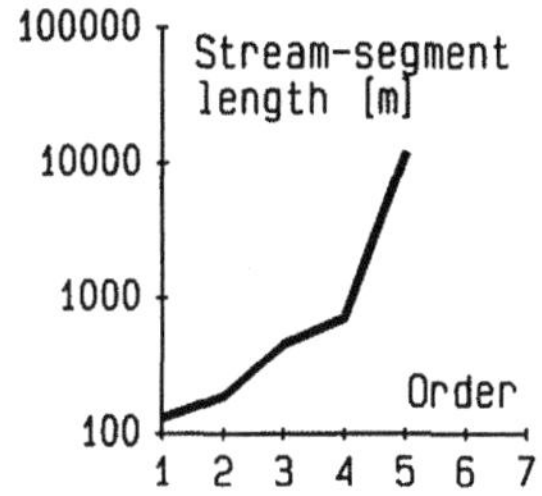

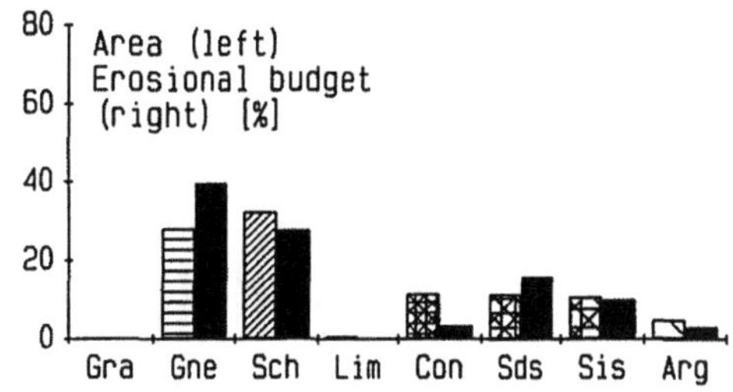

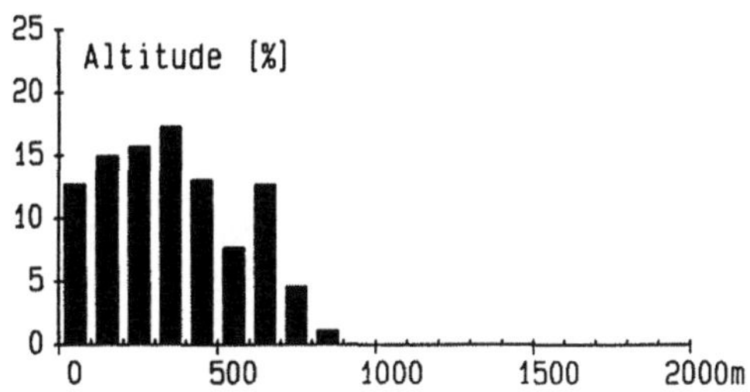

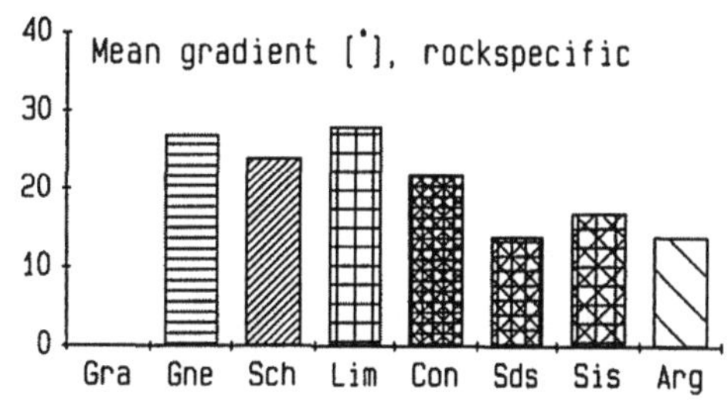

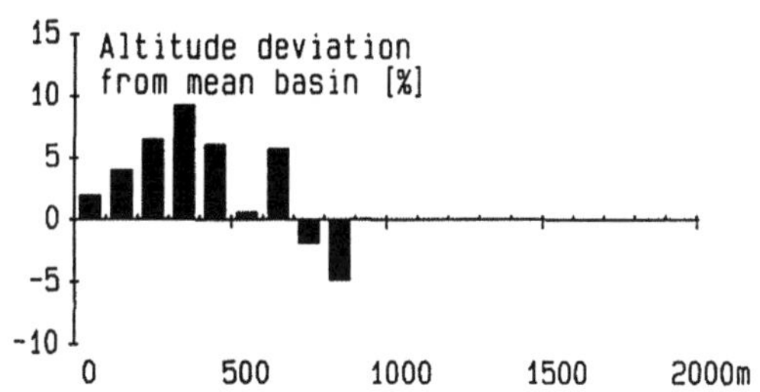

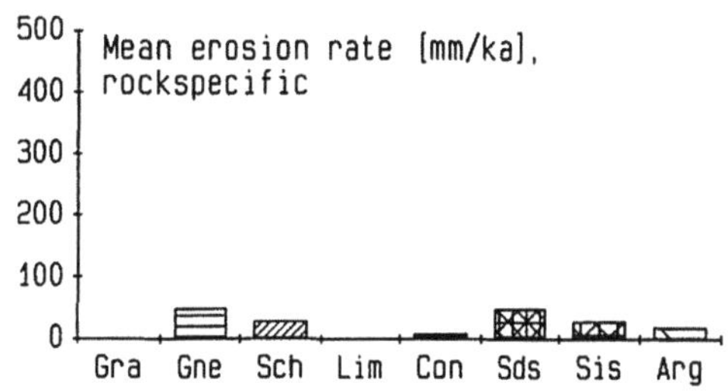

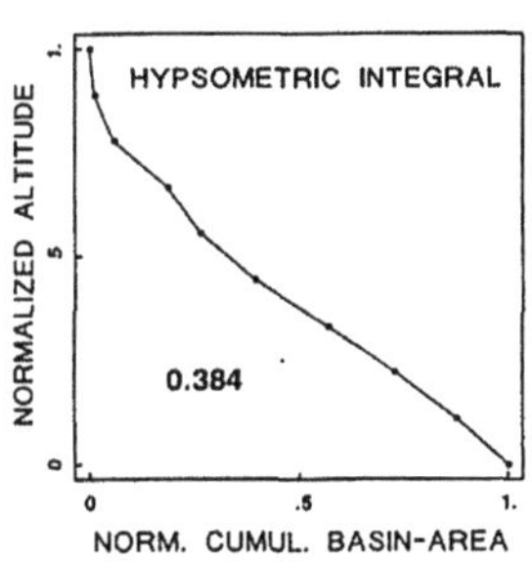

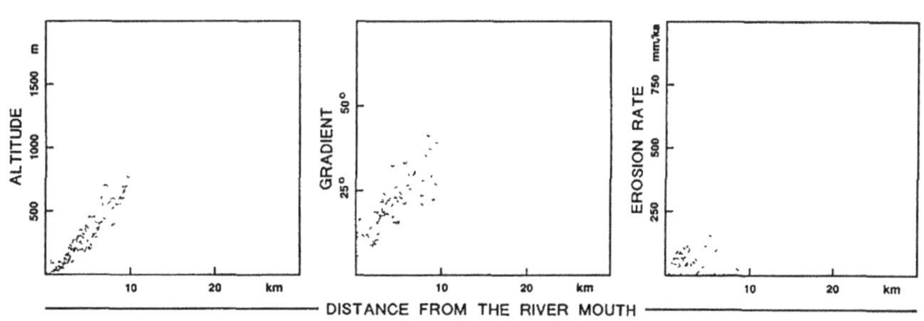

River-mouth sediment and water, the product: Grain sizes, petrography, mineralogy and chemistry 05

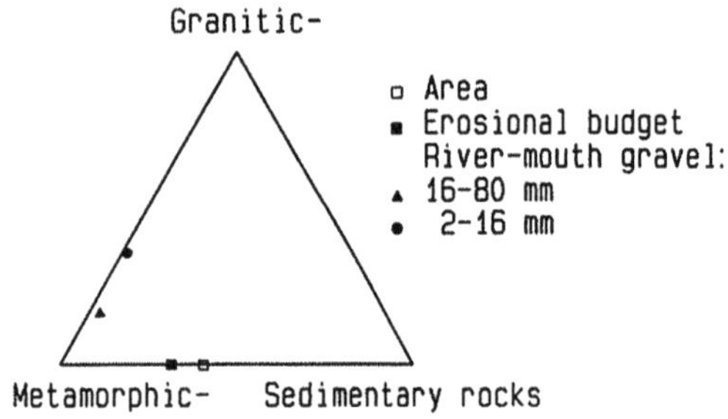

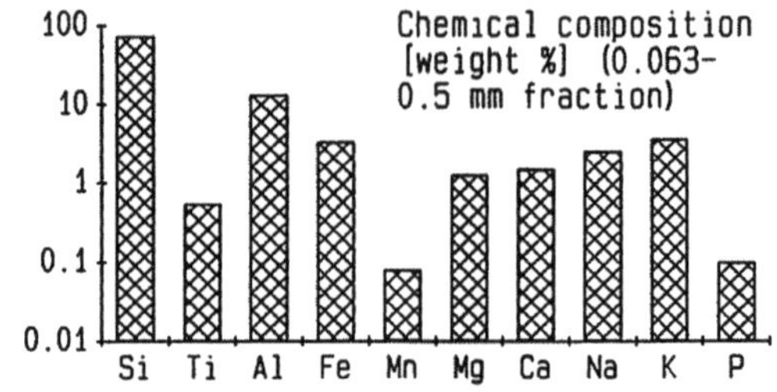

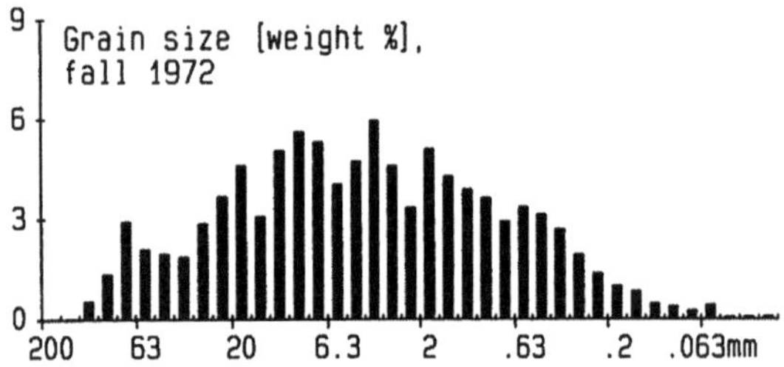

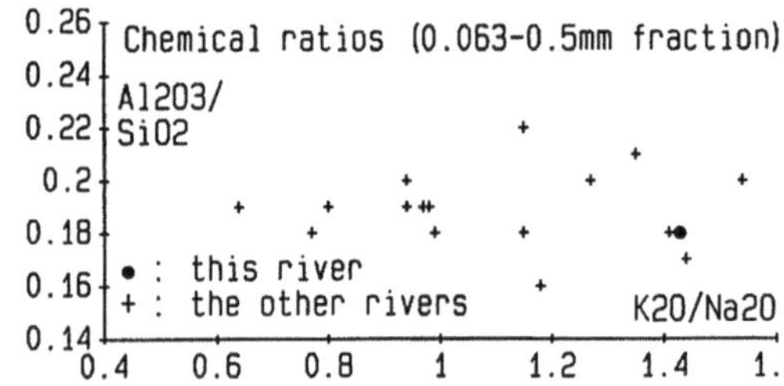

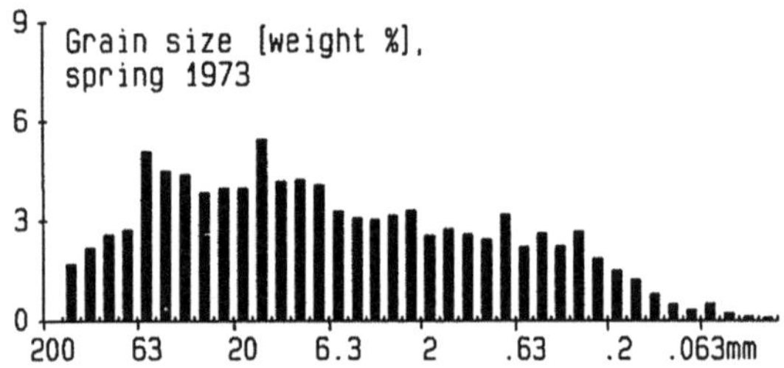

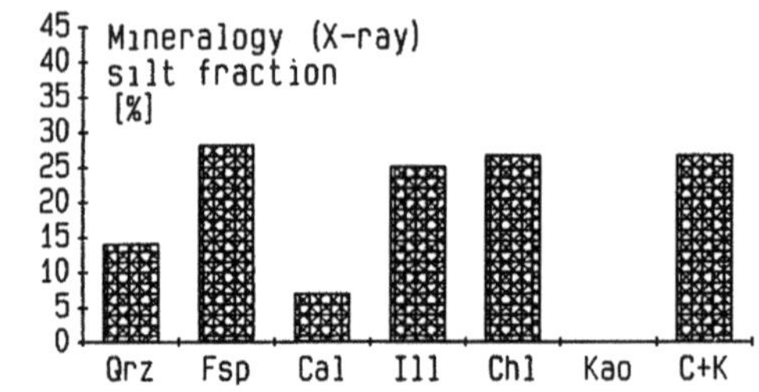

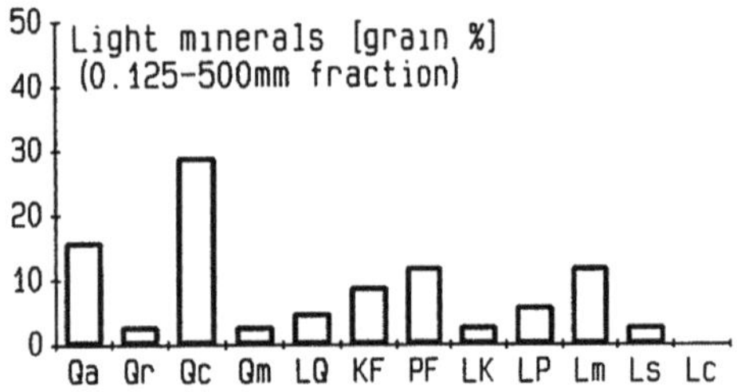

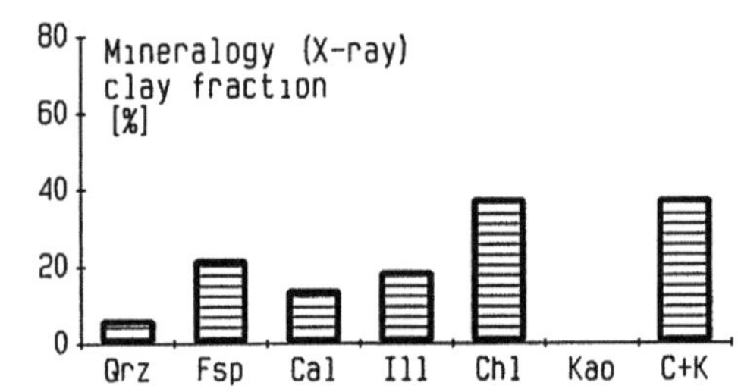

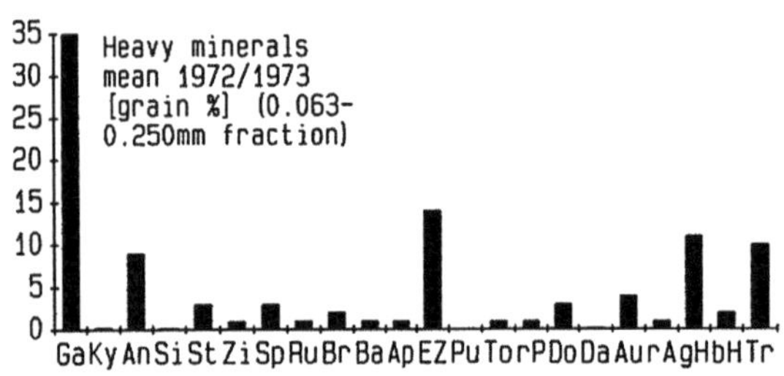

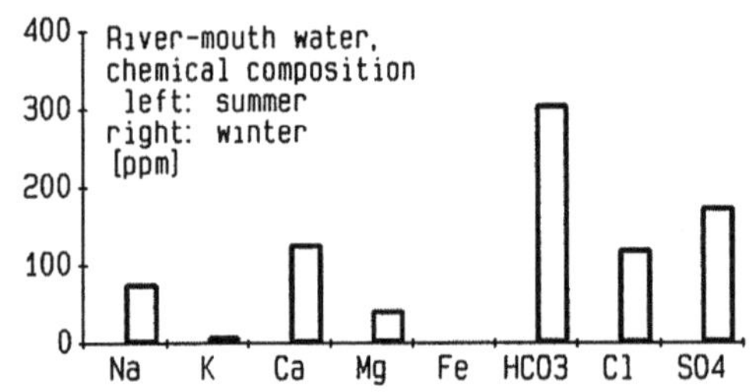

06 Bruzzano

Drainage basin, the producer: Geology and geomorphology

(For abbreviations see 00 basin 00)

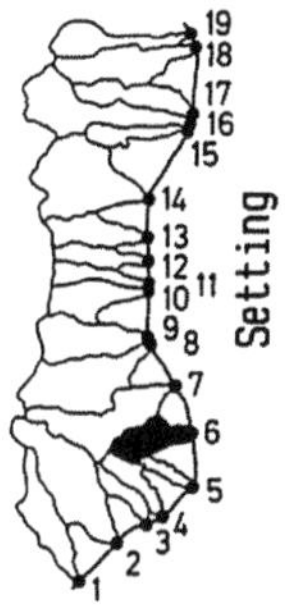

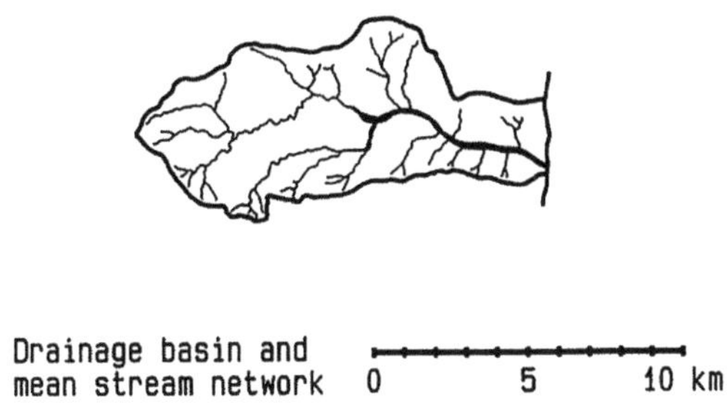

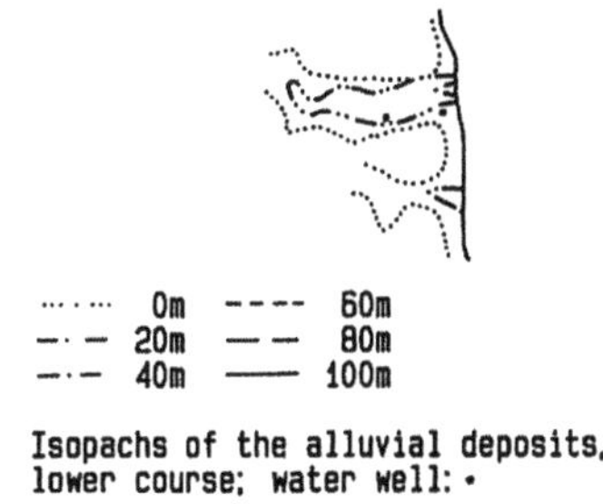

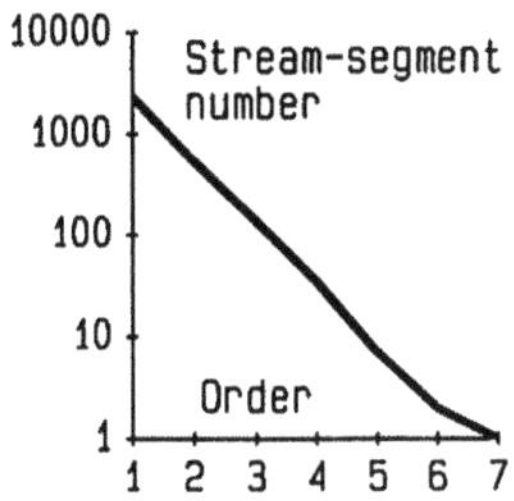

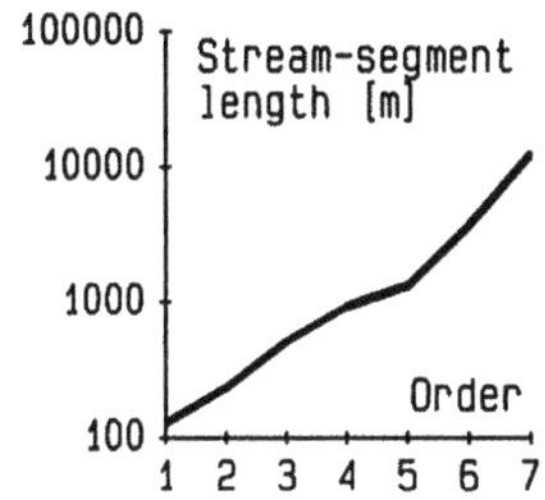

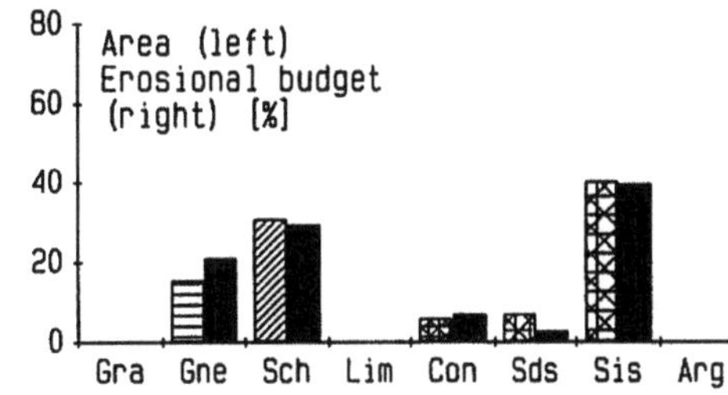

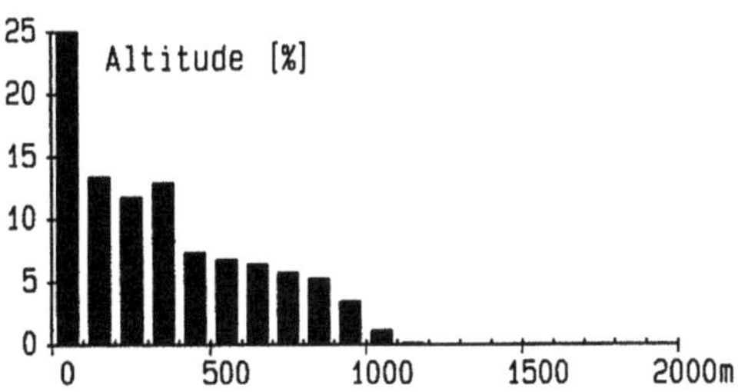

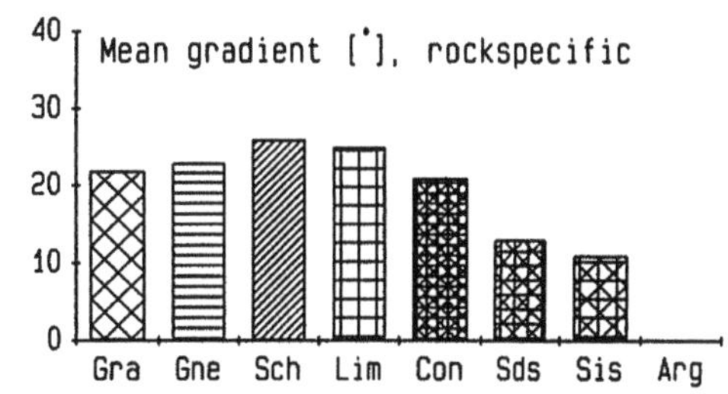

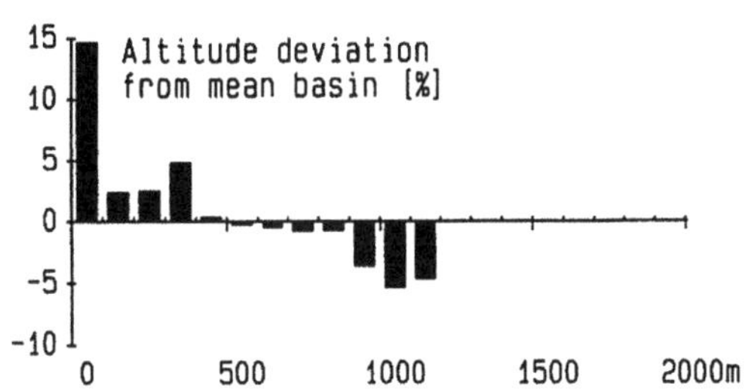

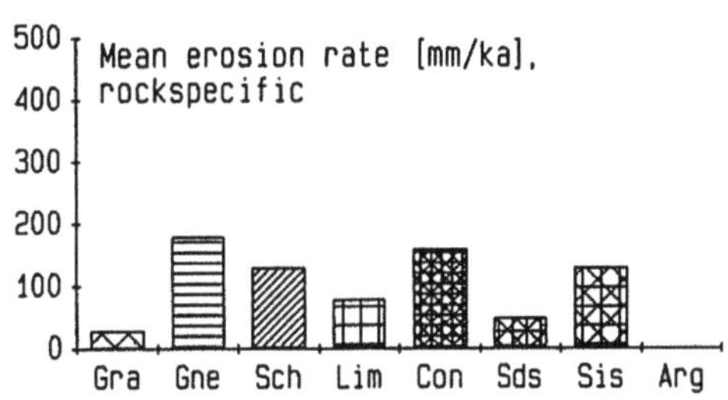

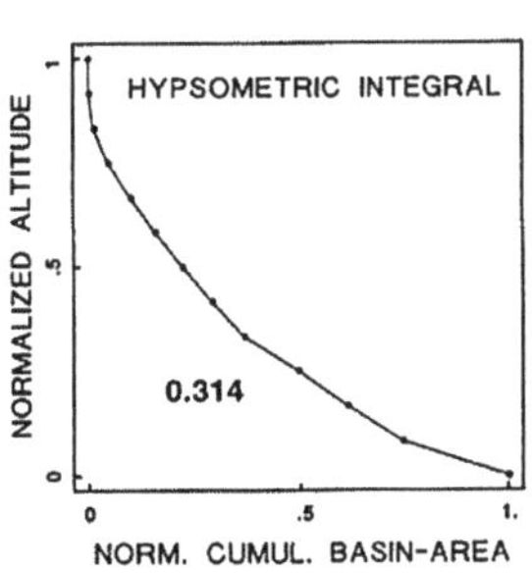

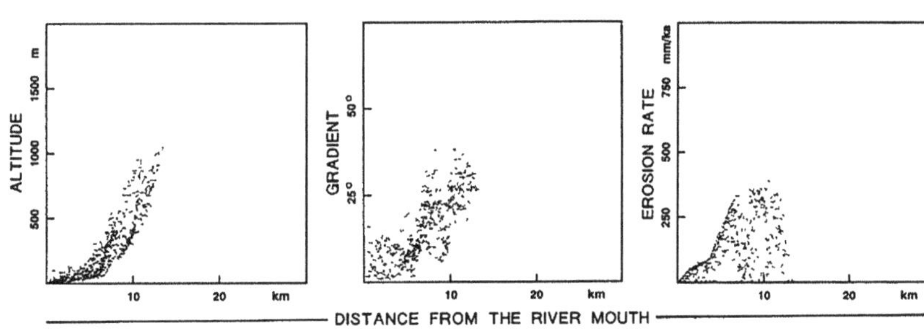

River-mouth sediment and water, the product: Grain sizes, petrography, mineralogy and chemistry 06

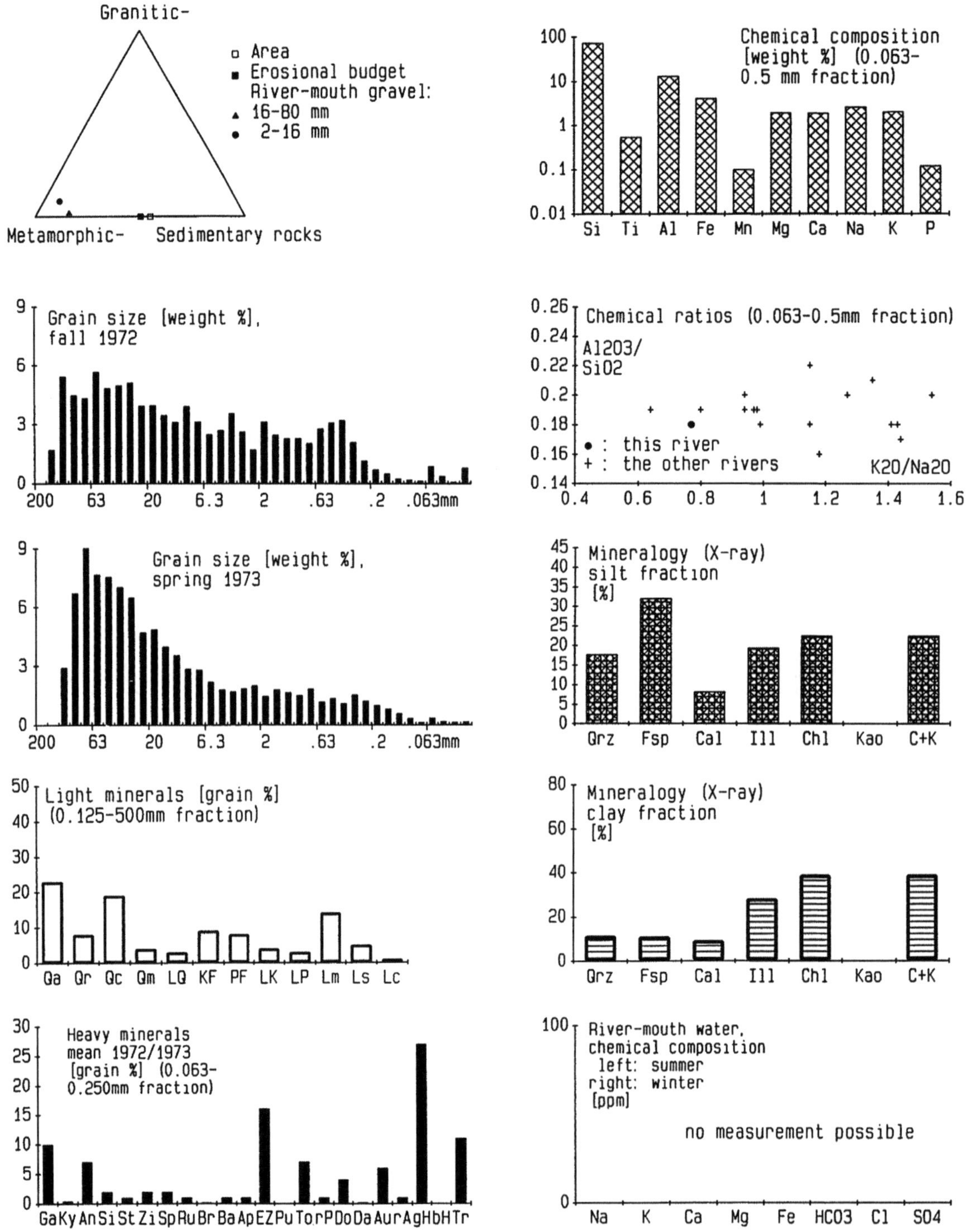

07 Laverde

Drainage basin, the producer: Geology and geomorphology

(For abbreviations see 00 basin 00)

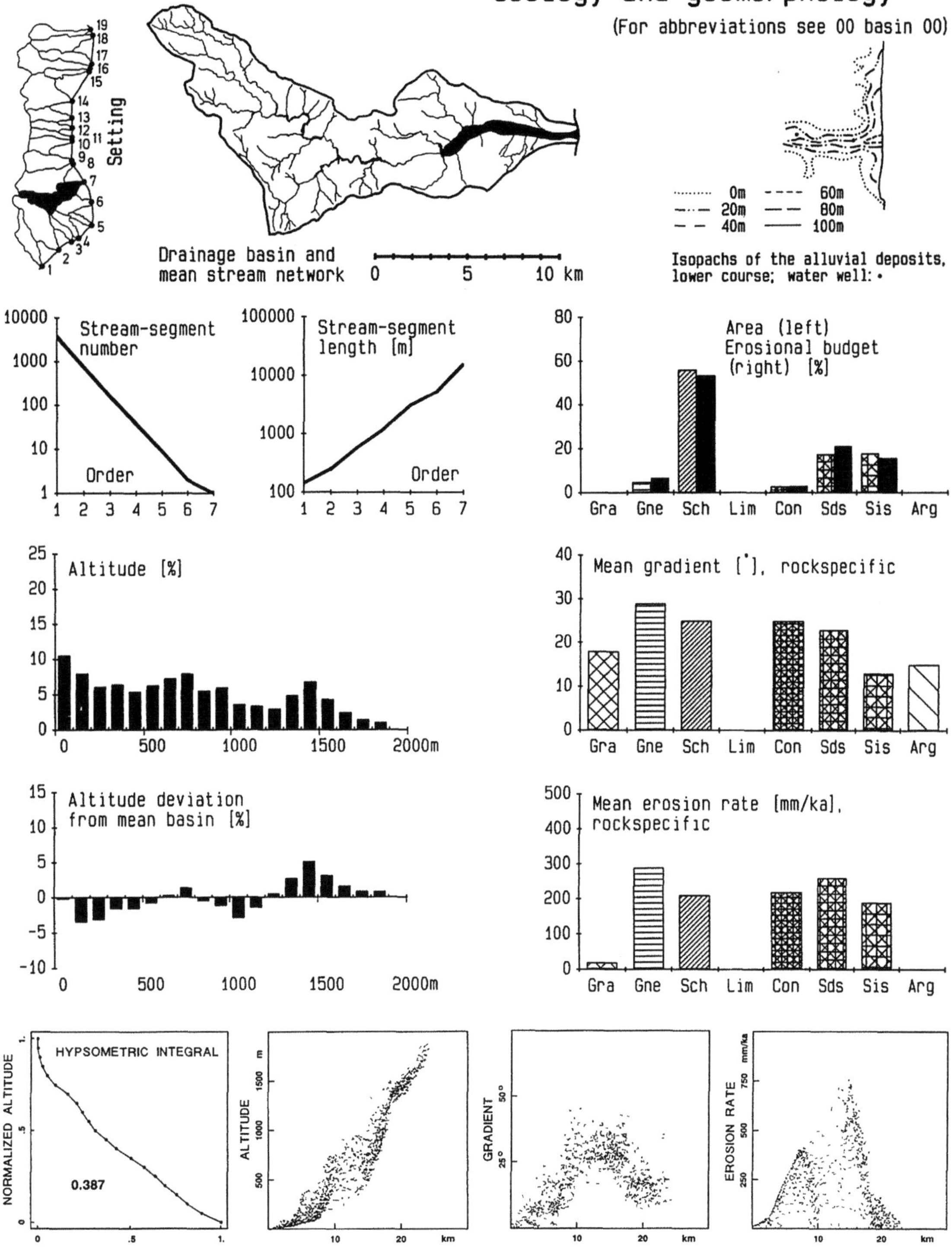

River-mouth sediment and water, the product: Grain sizes, petrography, mineralogy and chemistry 07

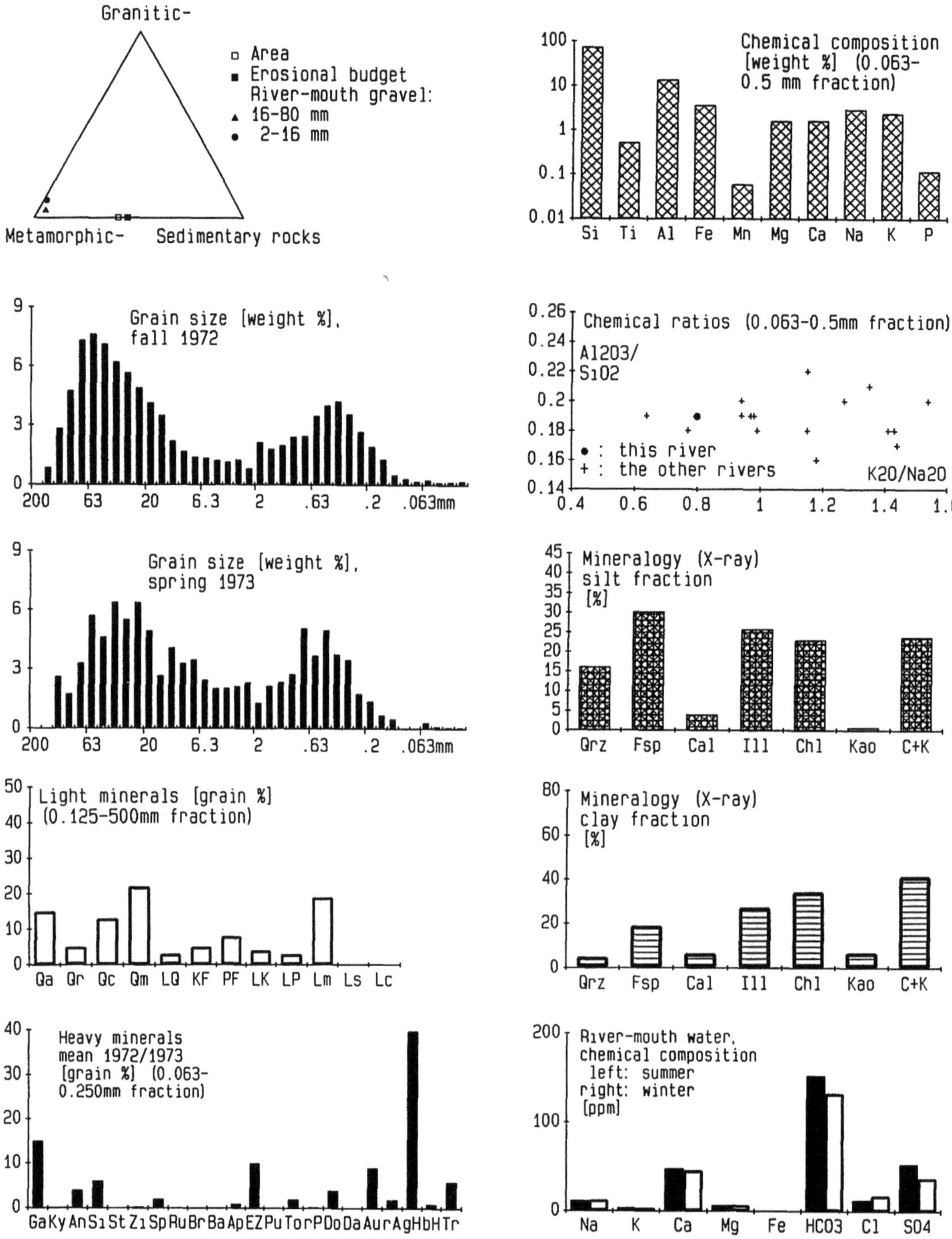

08 Buonamico

Drainage basin, the producer: Geology and geomorphology

(For abbreviations see 00 basin 00)

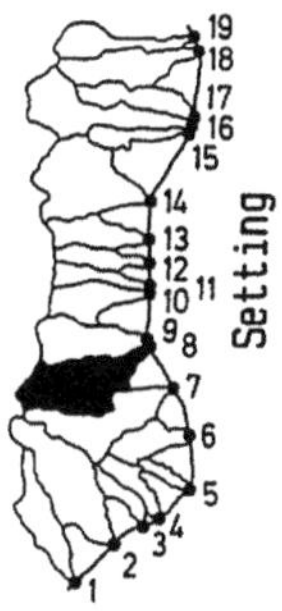

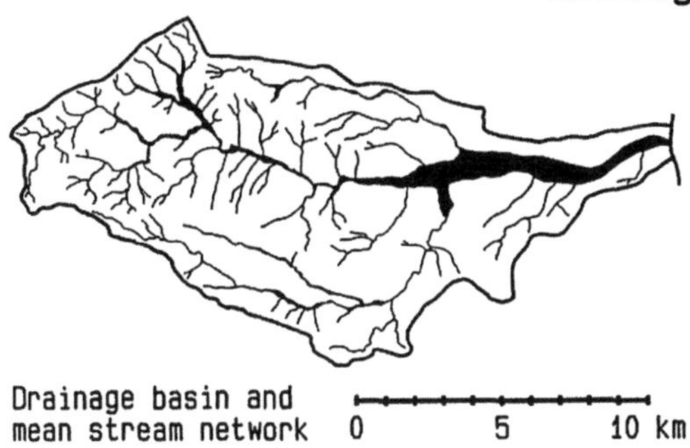

Drainage basin and mean stream network

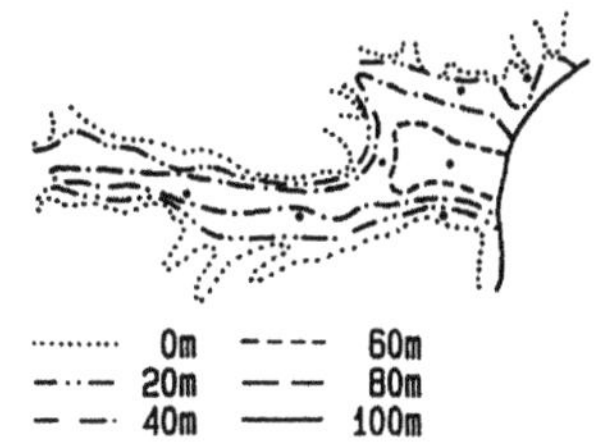

Isopachs of the alluvial deposits, lower course; water well: •

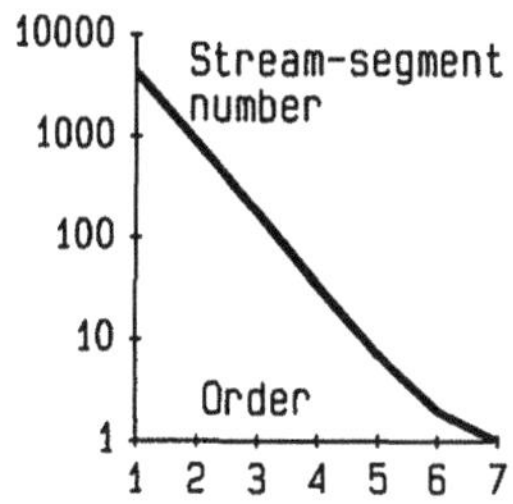

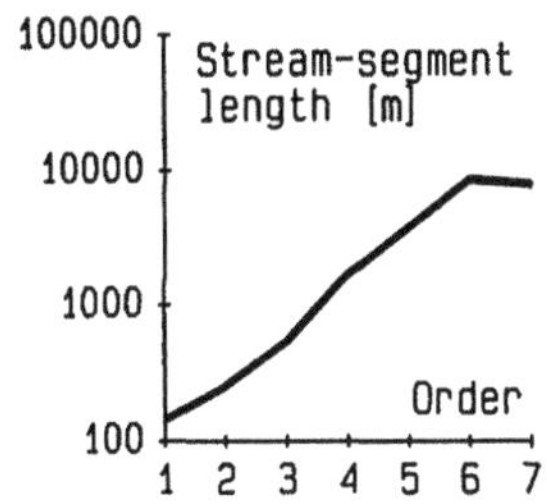

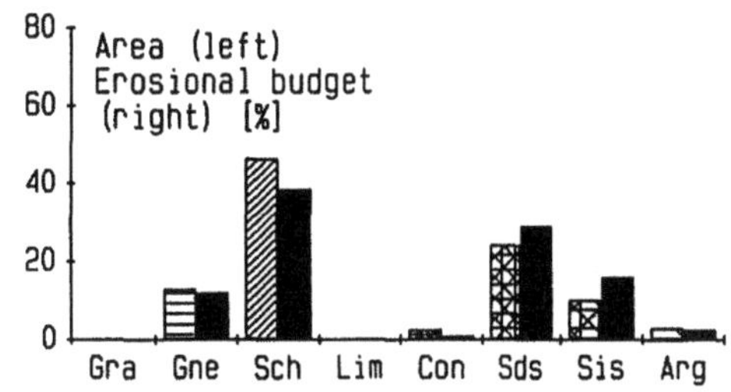

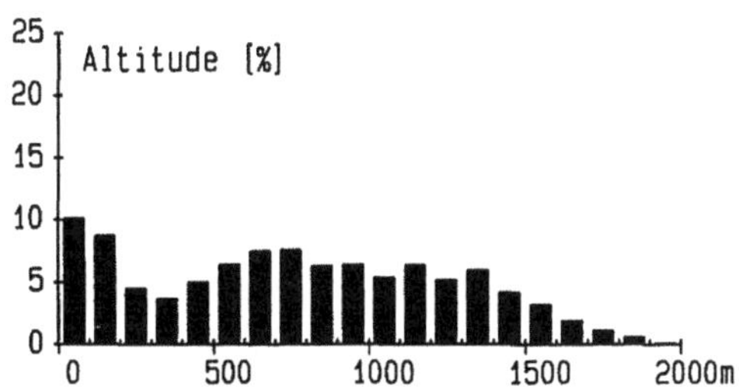

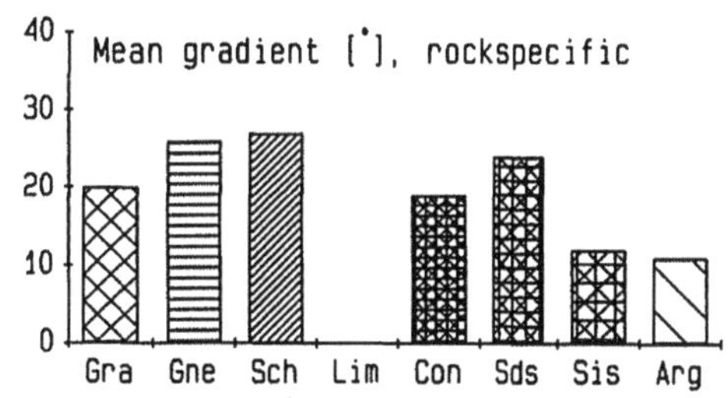

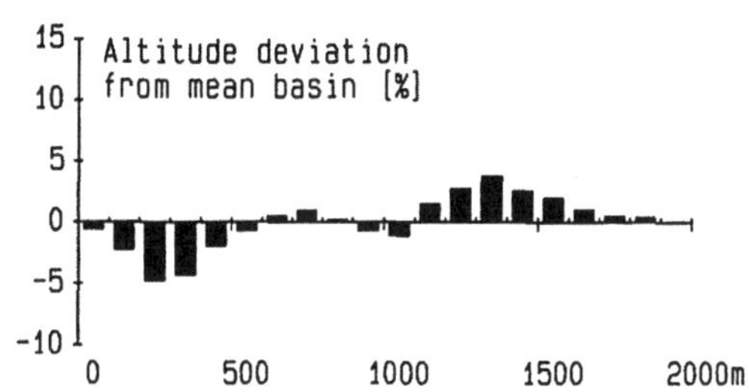

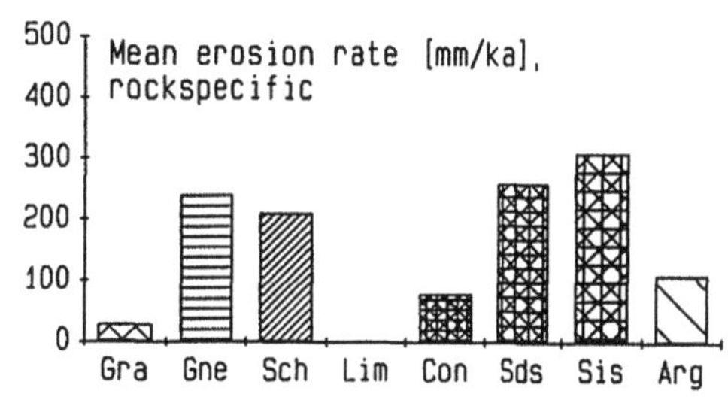

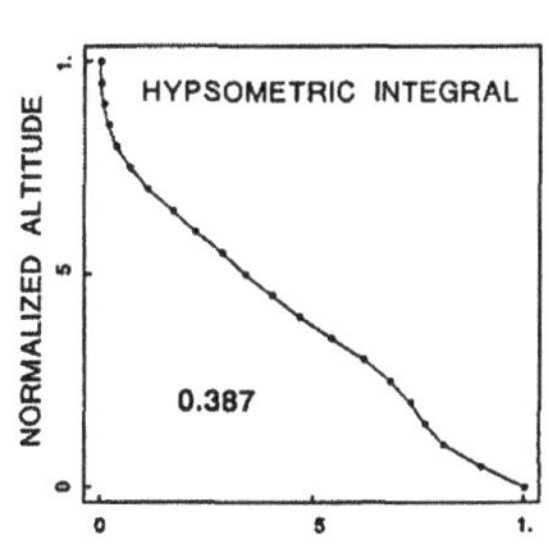

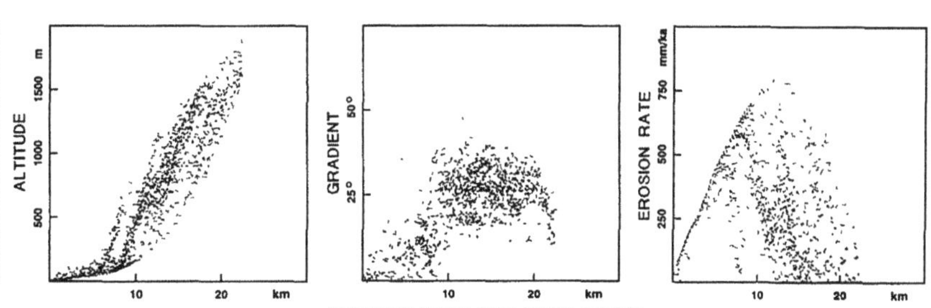

River-mouth sediment and water, the product: Grain sizes, petrography, mineralogy and chemistry 08

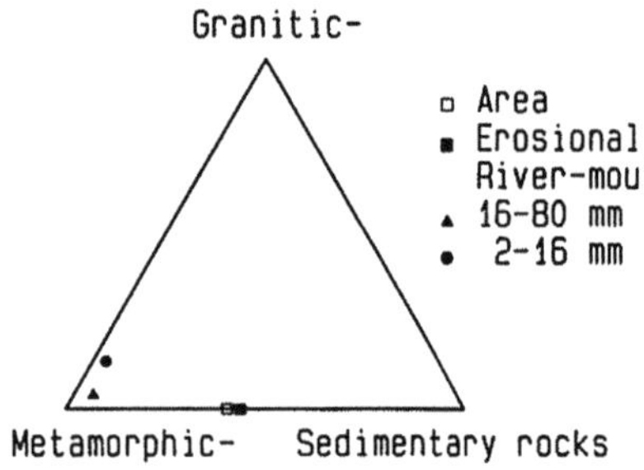

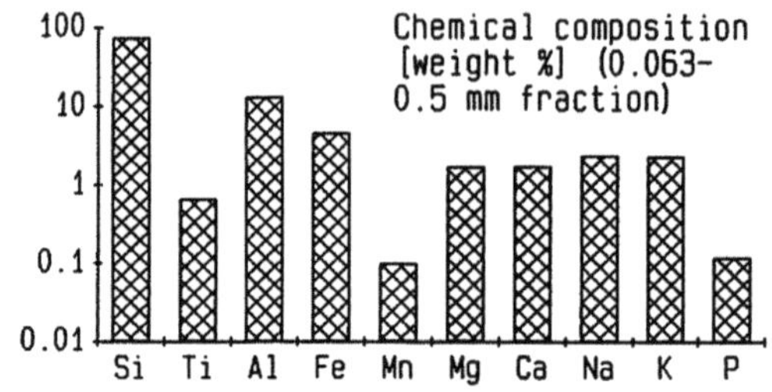

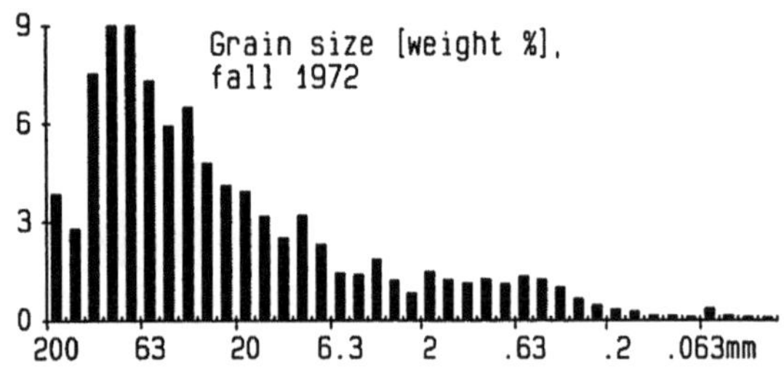

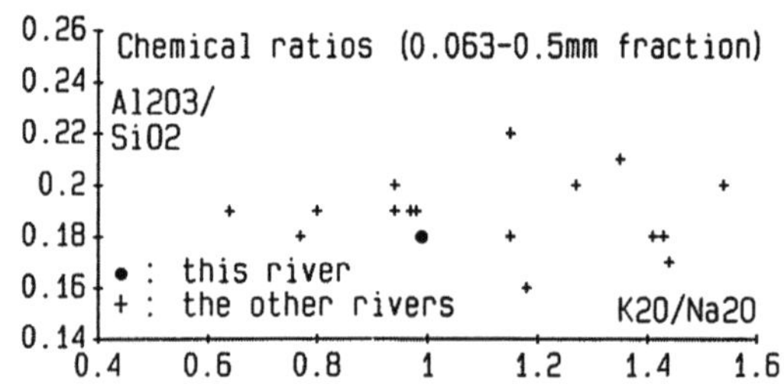

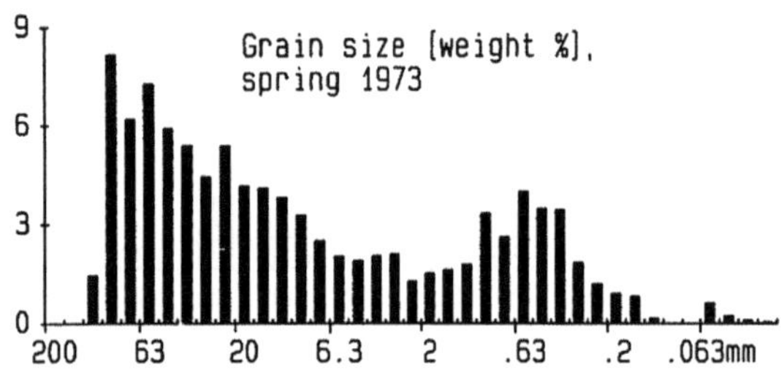

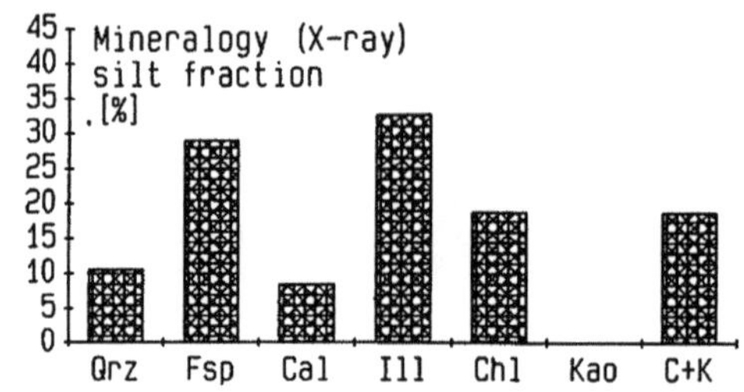

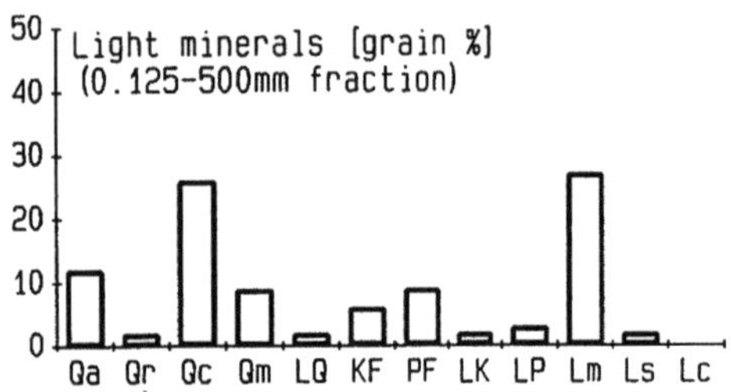

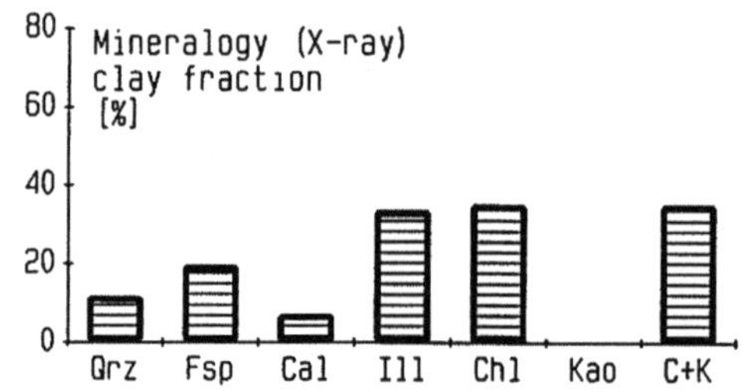

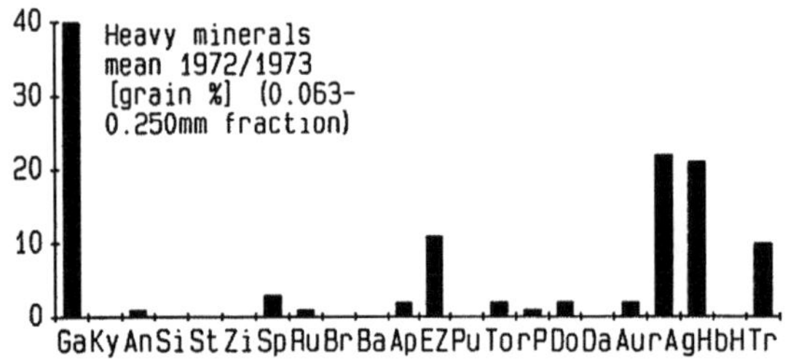

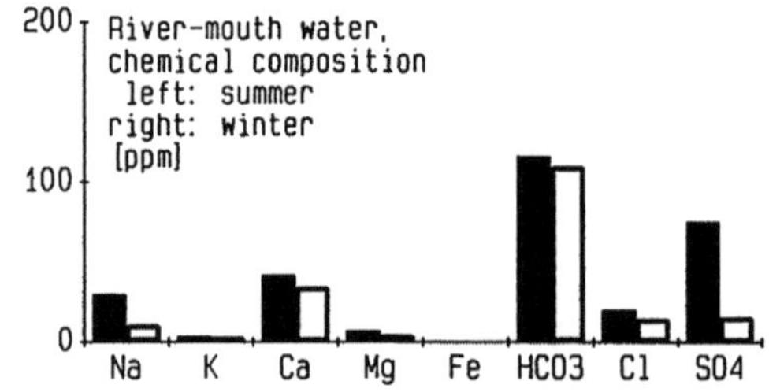

09 Careri

Drainage basin, the producer: Geology and geomorphology

(For abbreviations see 00 basin 00)

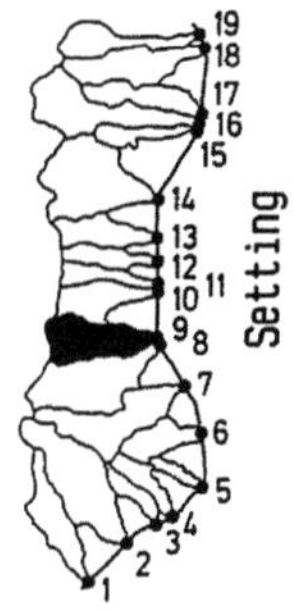

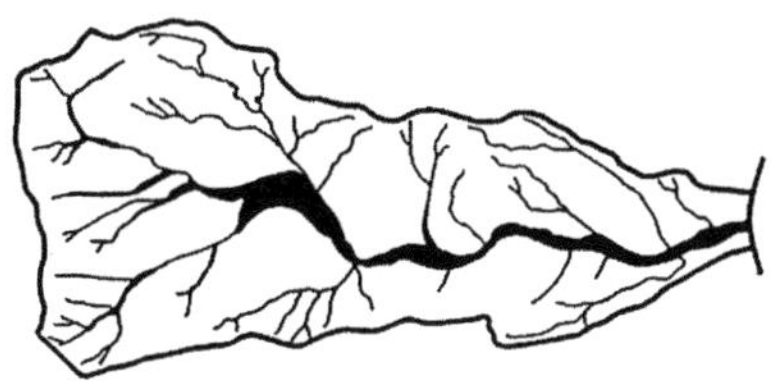

Drainage basin and mean stream network 0 5 10 km

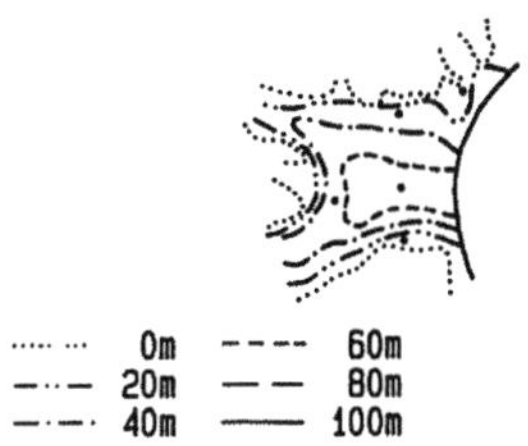

Isopachs of the alluvial deposits, lower course; water well: •

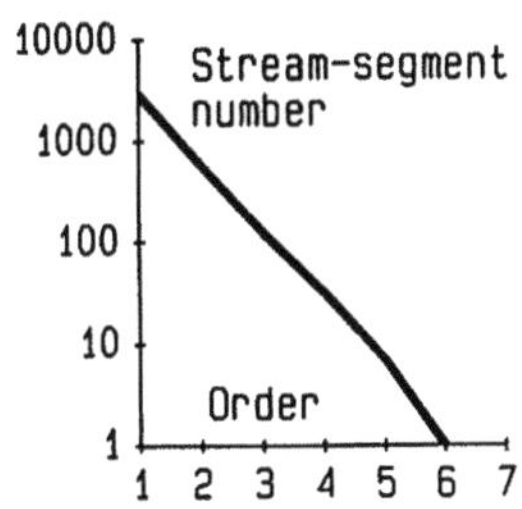

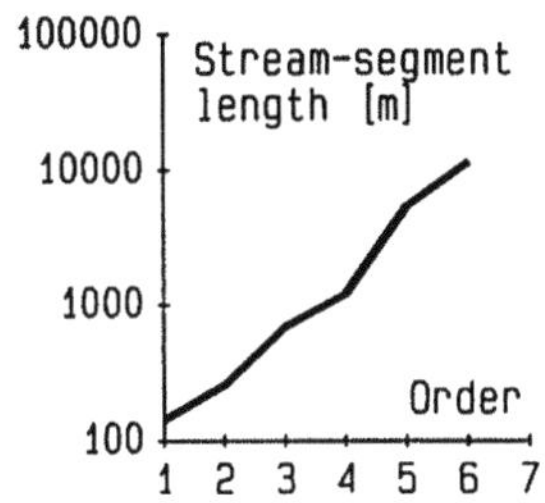

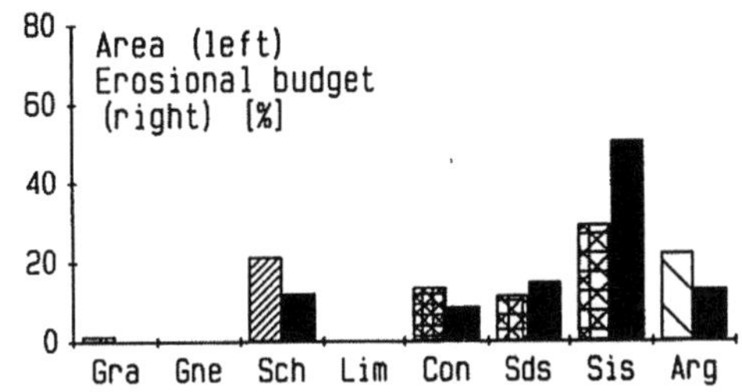

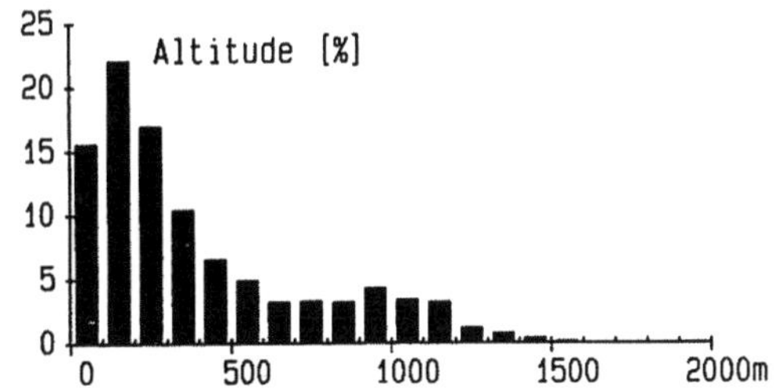

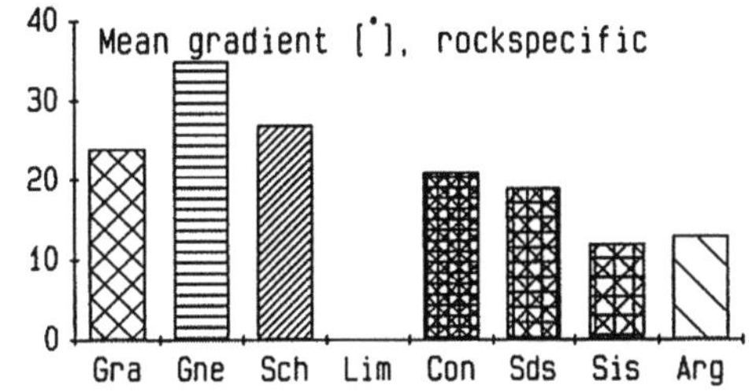

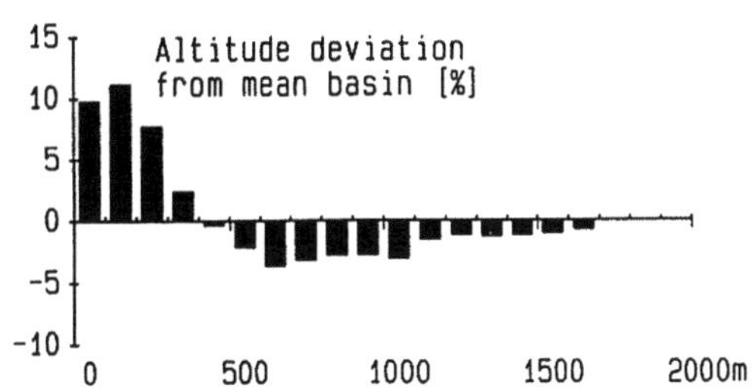

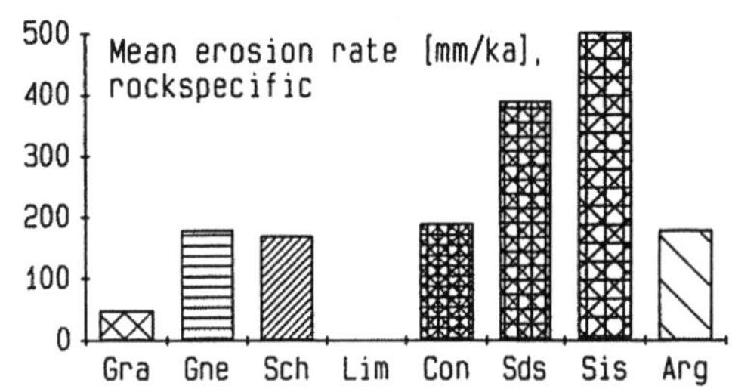

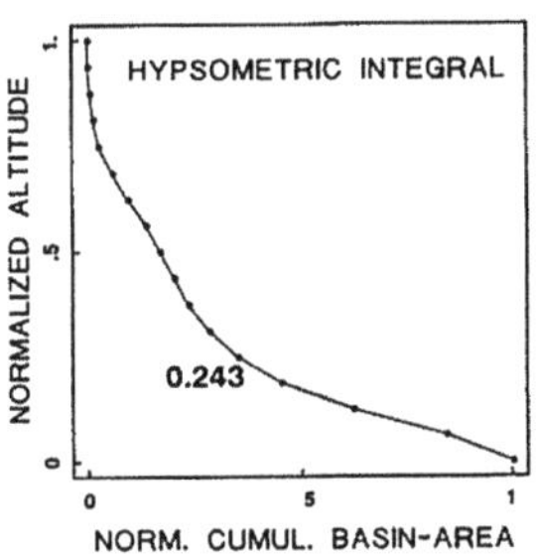

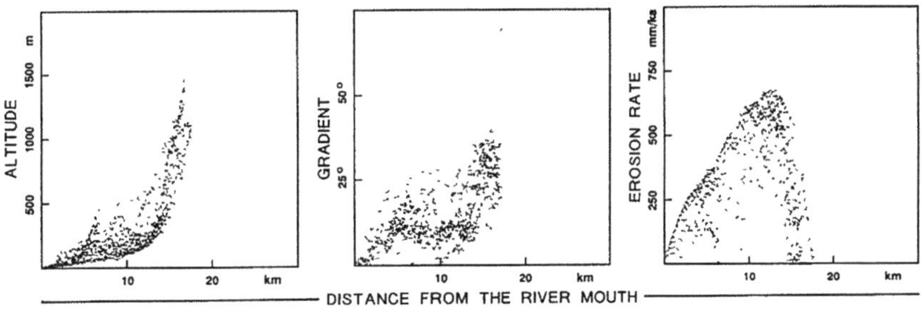

River-mouth sediment and water, the product: Grain sizes, petrography, mineralogy and chemistry 09

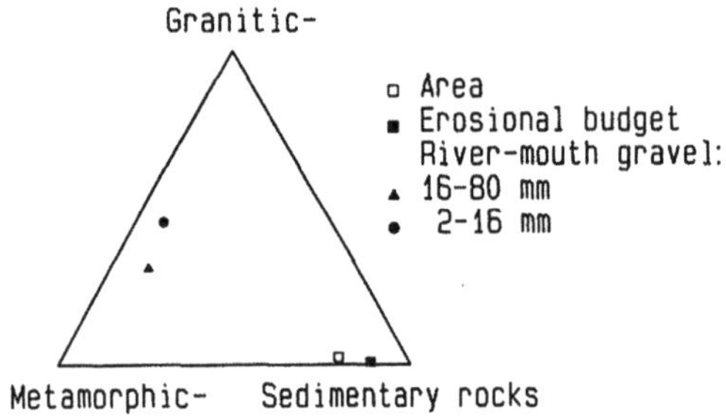

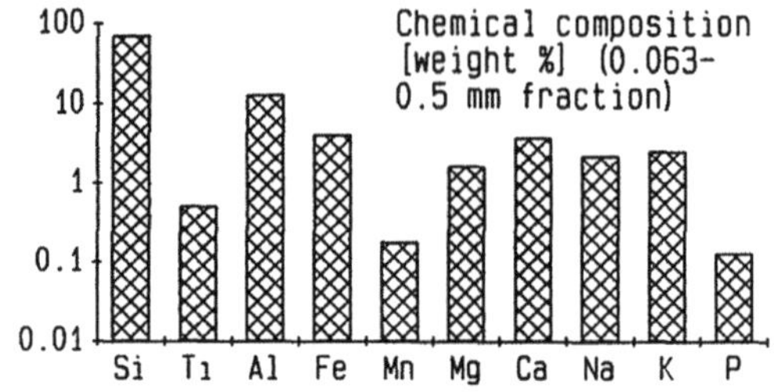

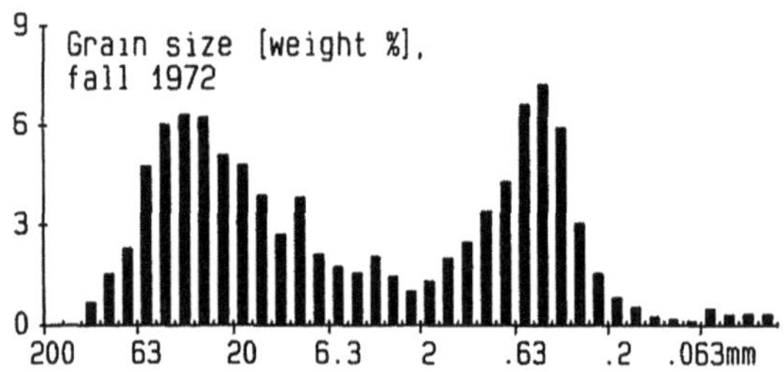

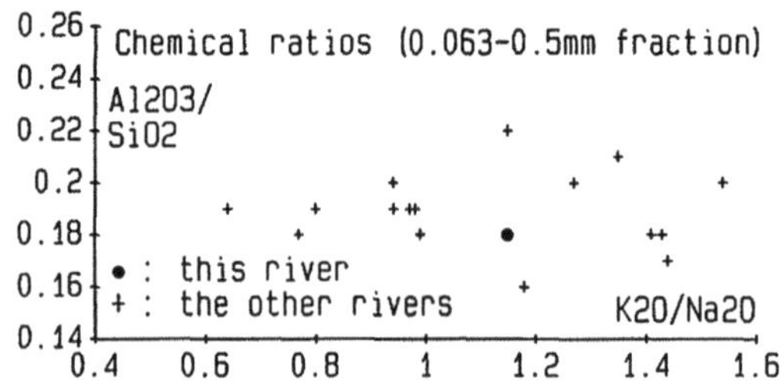

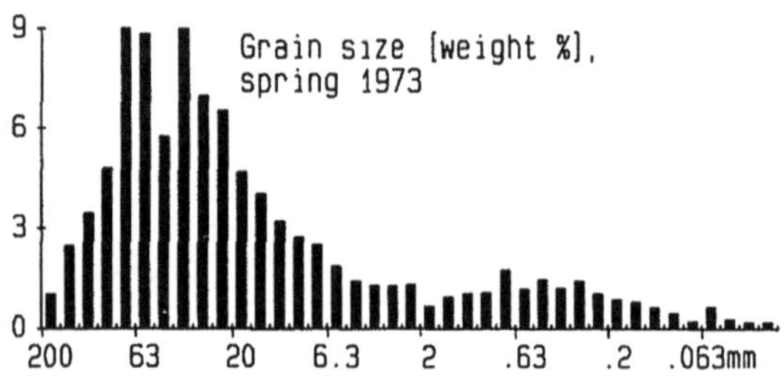

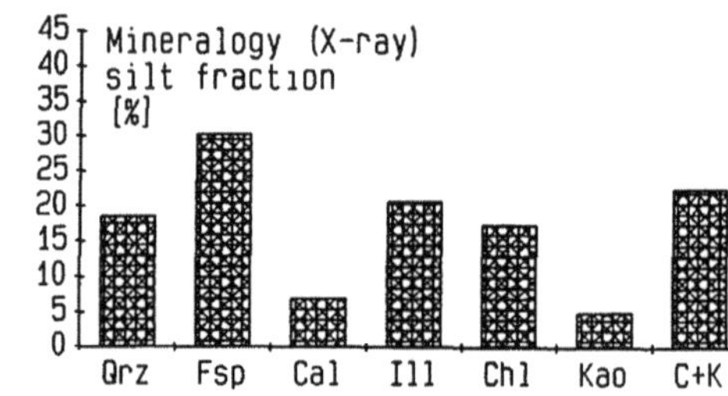

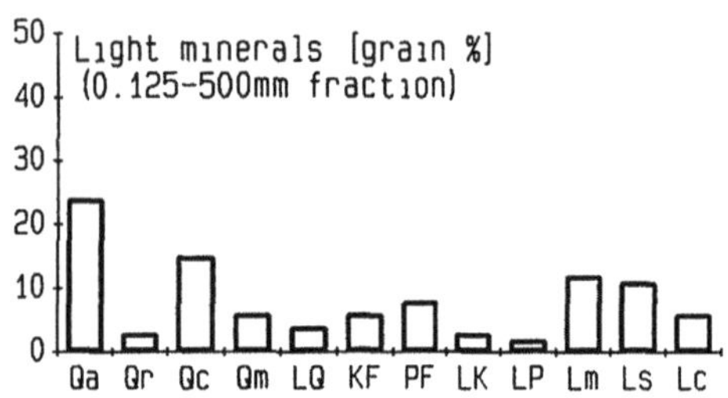

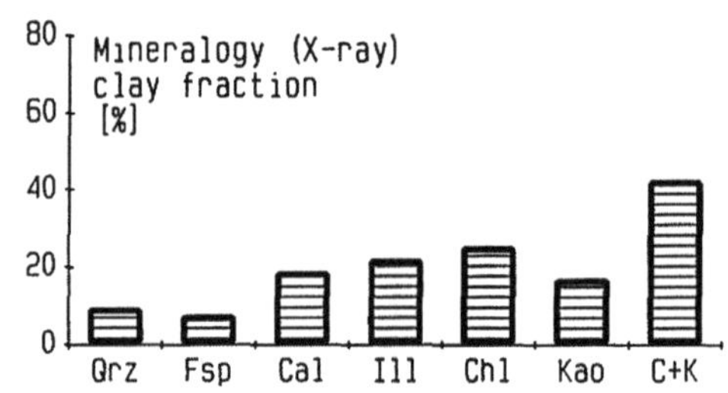

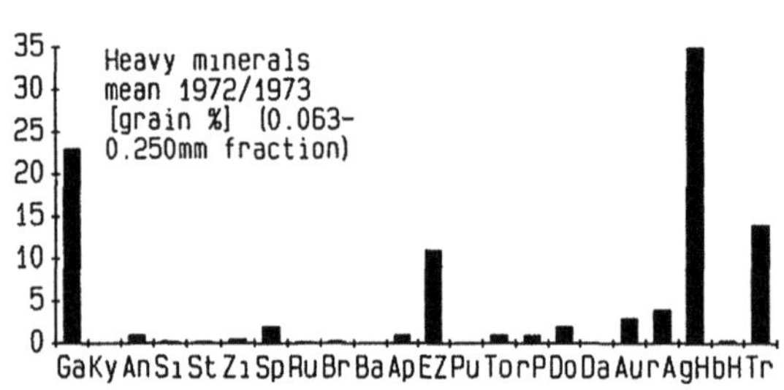

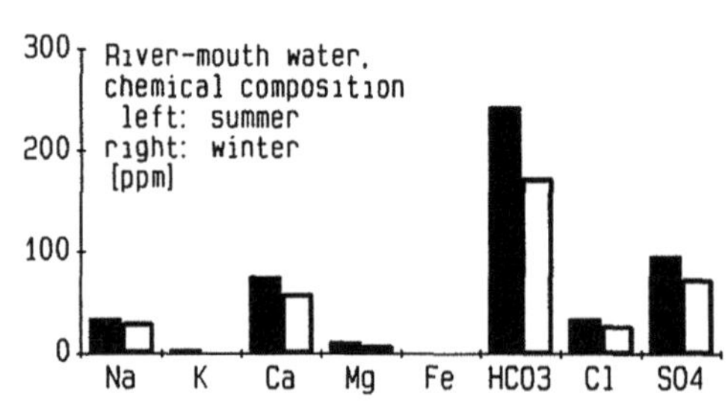

10 Condoianni

Drainage basin, the producer: Geology and geomorphology

(For abbreviations see 00 basin 00)

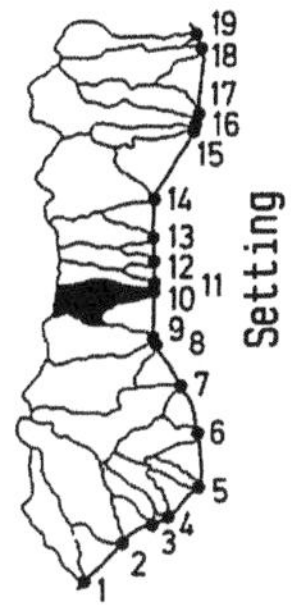

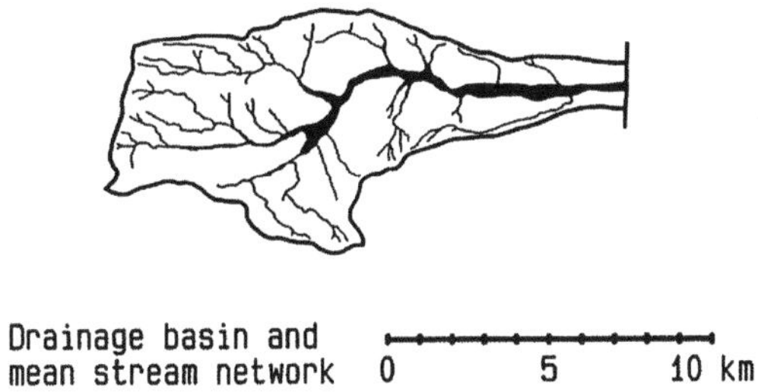

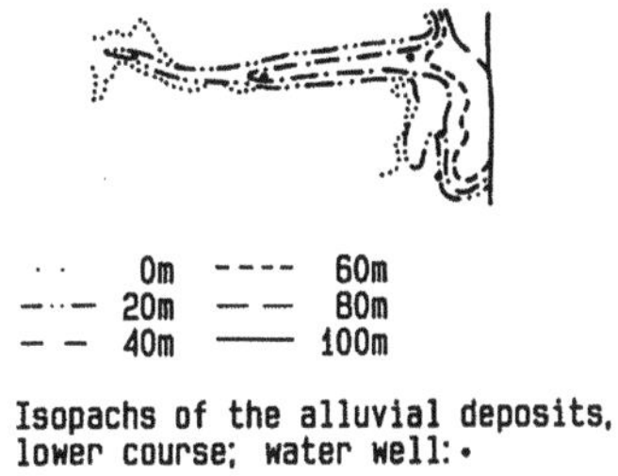

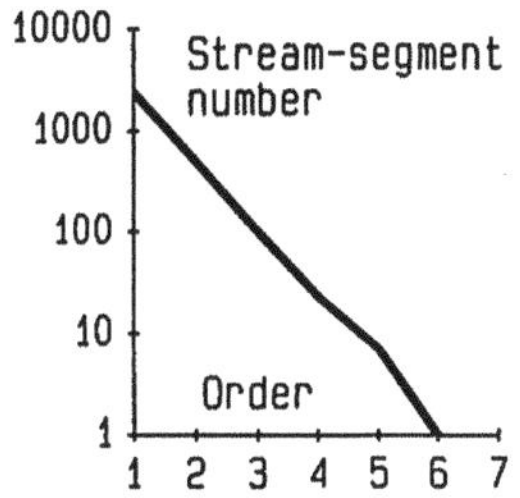

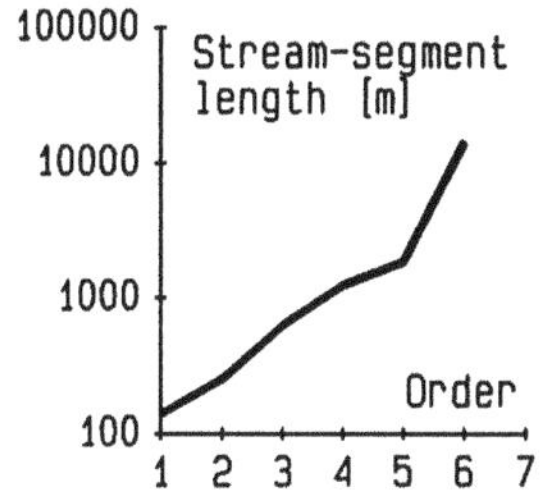

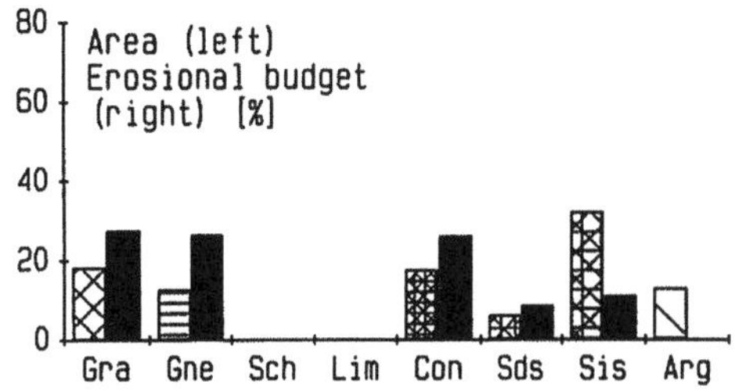

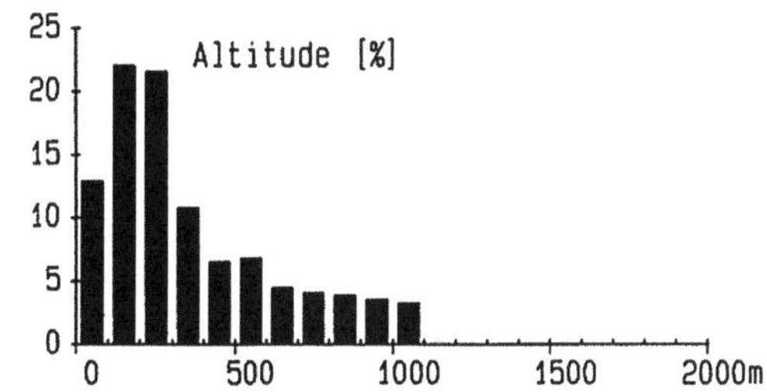

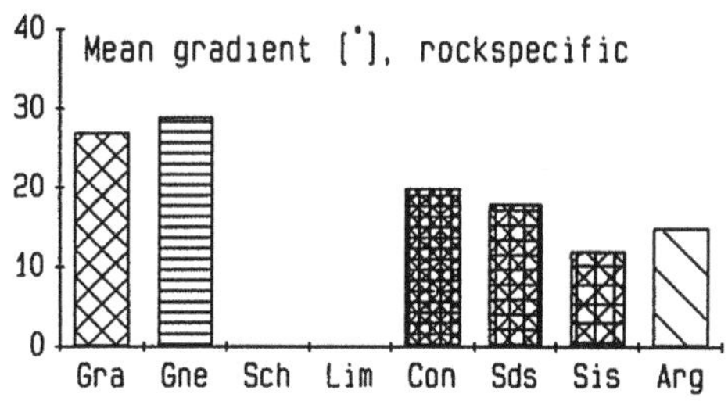

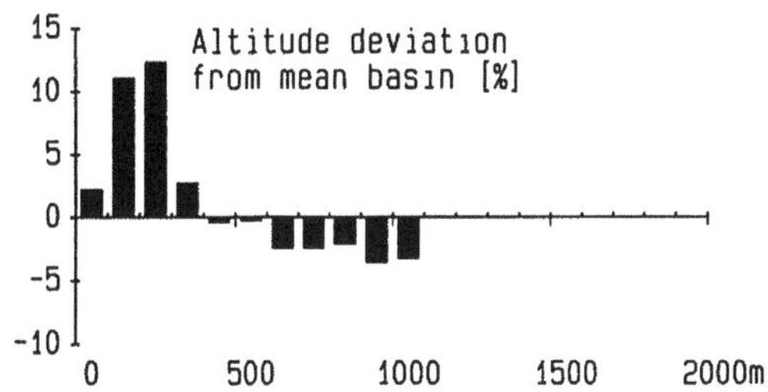

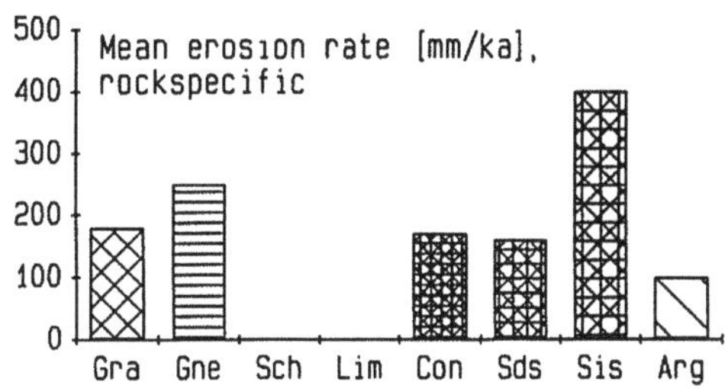

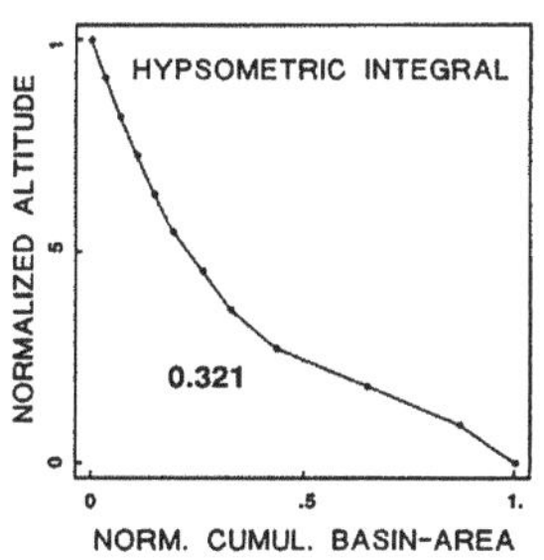

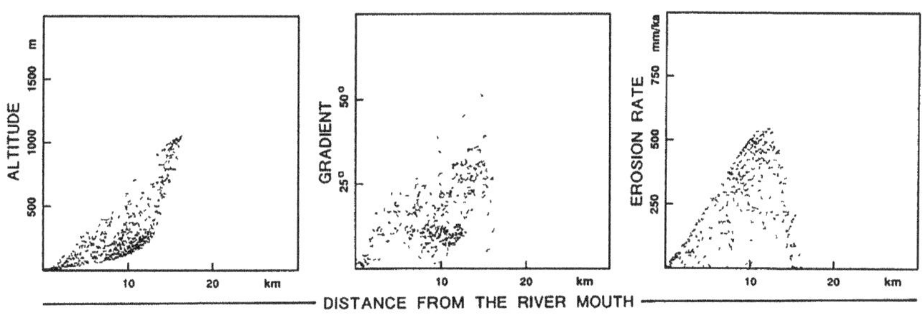

River-mouth sediment and water, the product: Grain sizes, petrography, mineralogy and chemistry 10

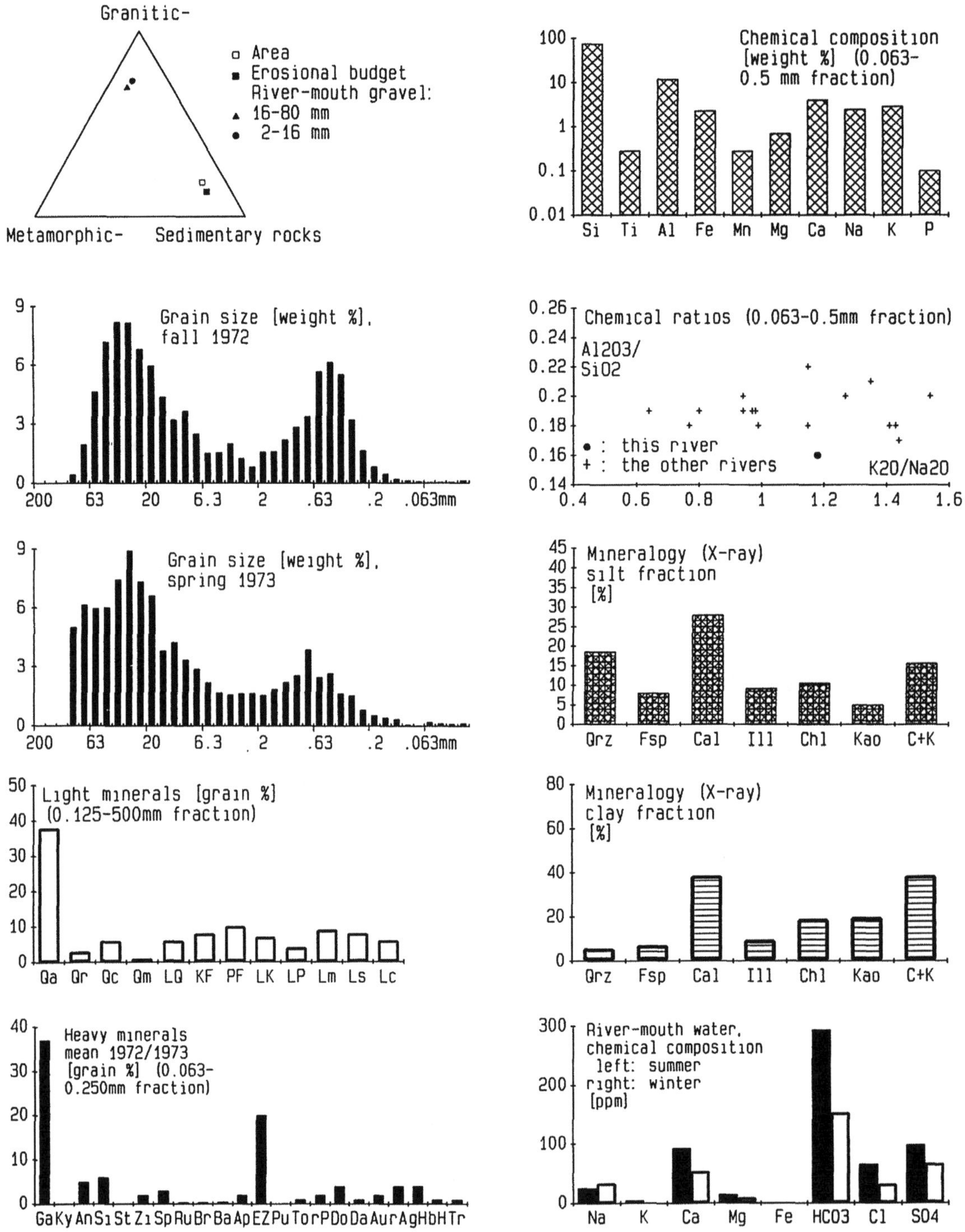

11 Portigliola Drainage basin, the producer: Geology and geomorphology

(For abbreviations see 00 basin 00)

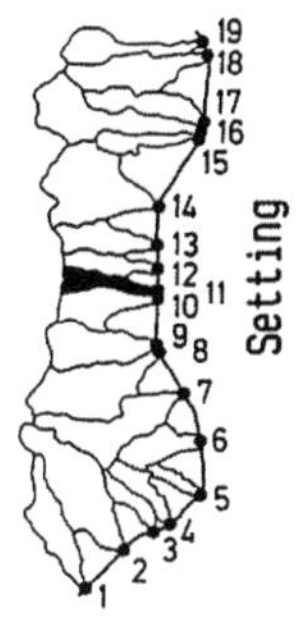

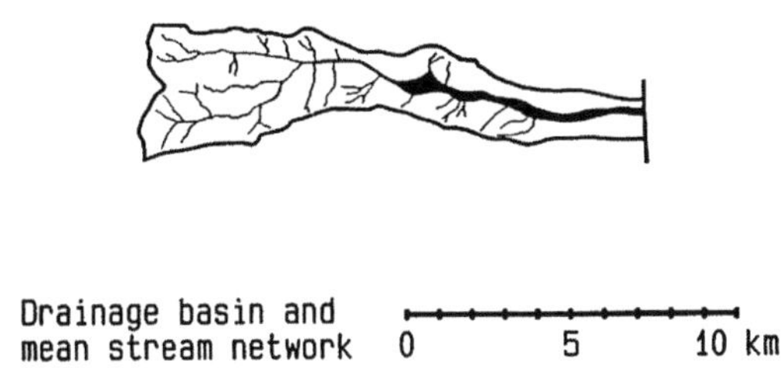

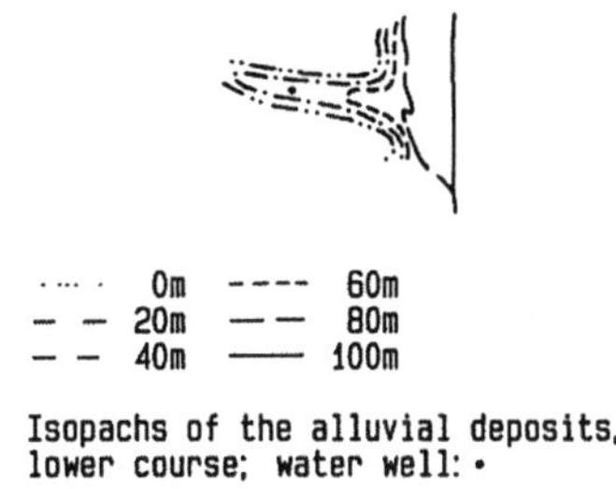

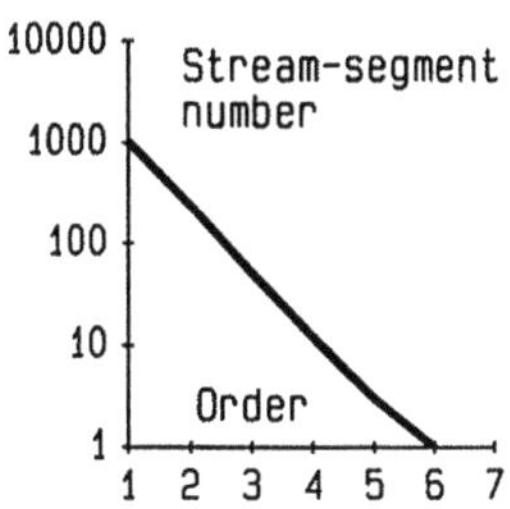

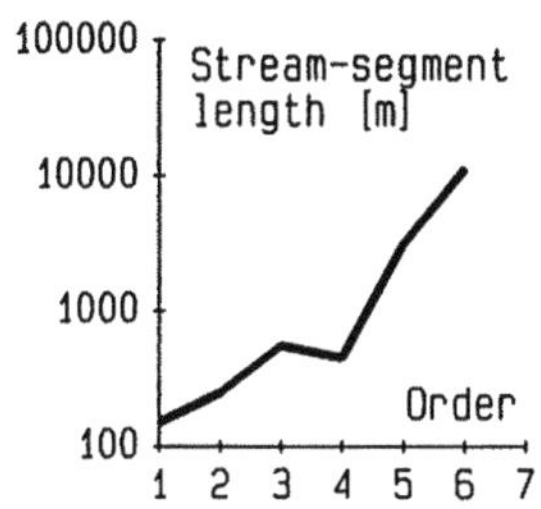

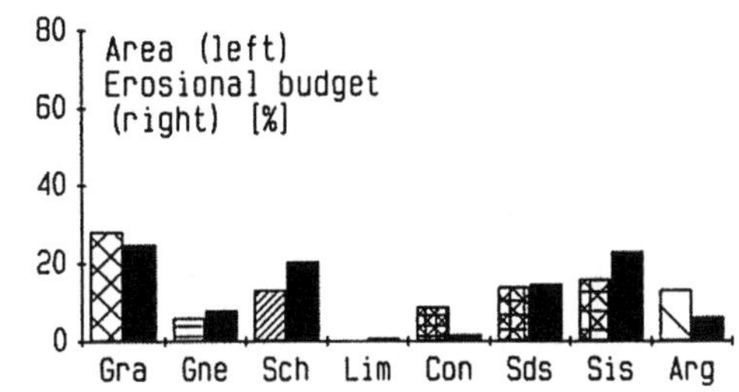

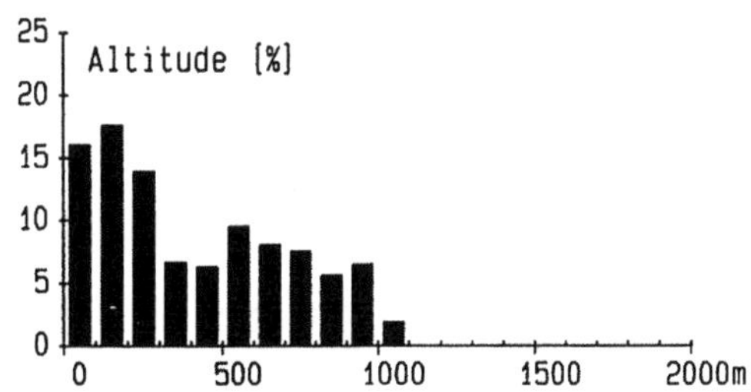

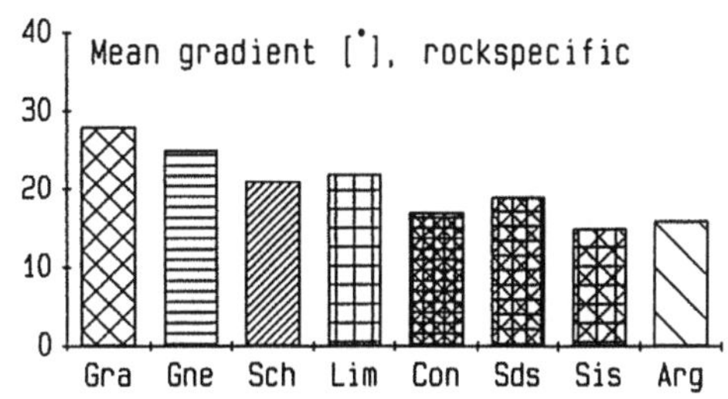

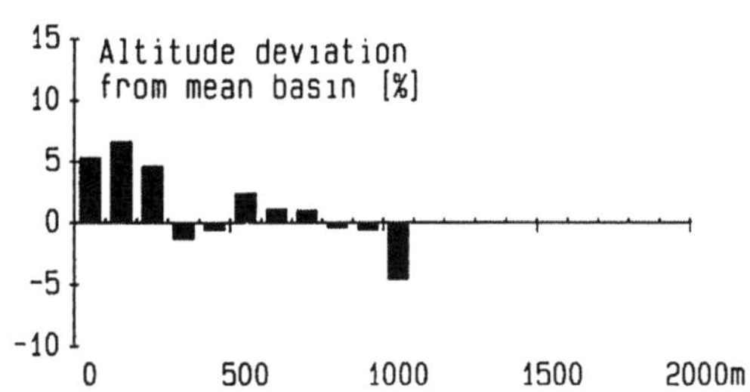

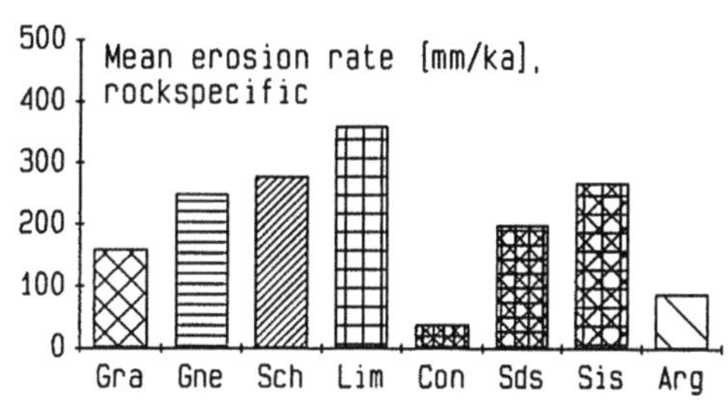

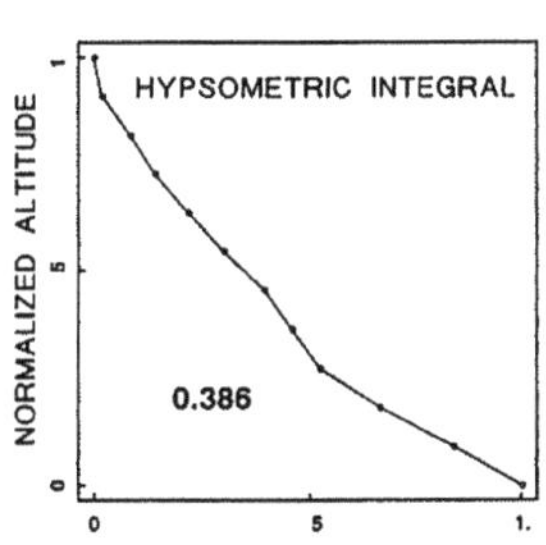

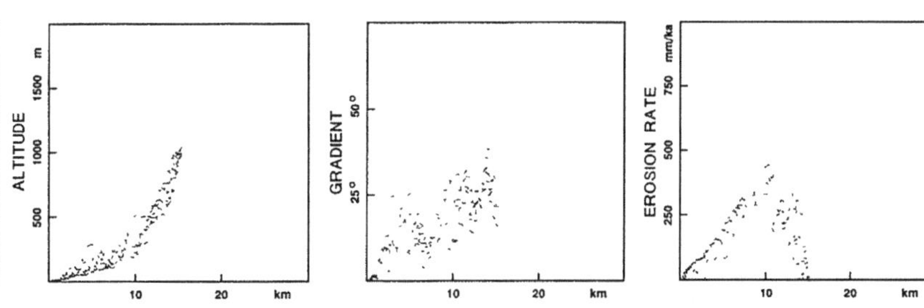

River-mouth sediment and water, the product: Grain sizes, petrography, mineralogy and chemistry 11

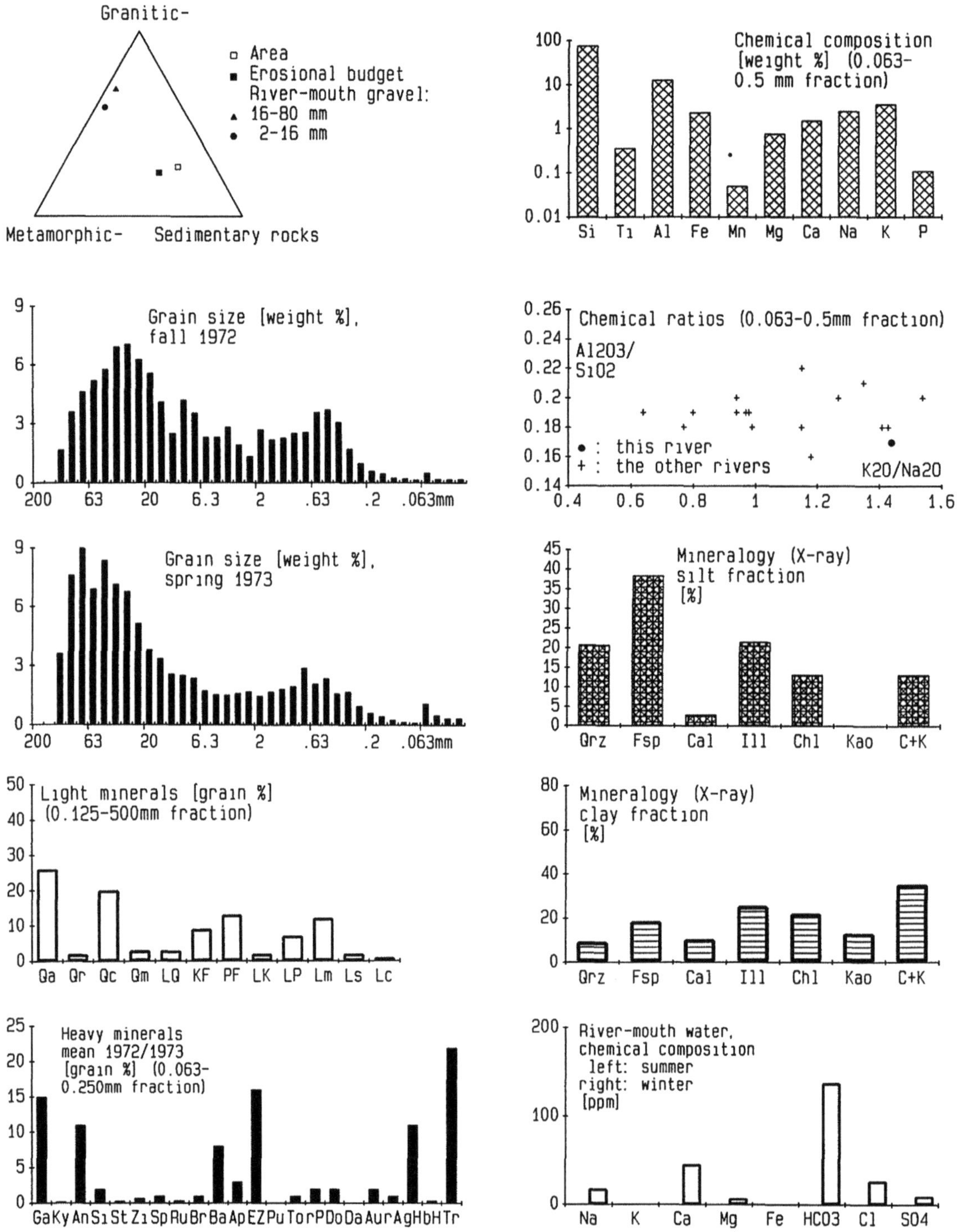

12 Gerace

Drainage basin, the producer: Geology and geomorphology

(For abbreviations see 00 basin 00)

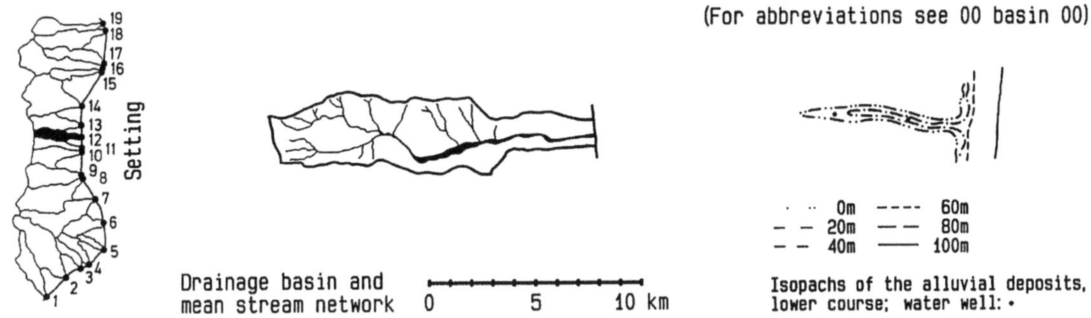

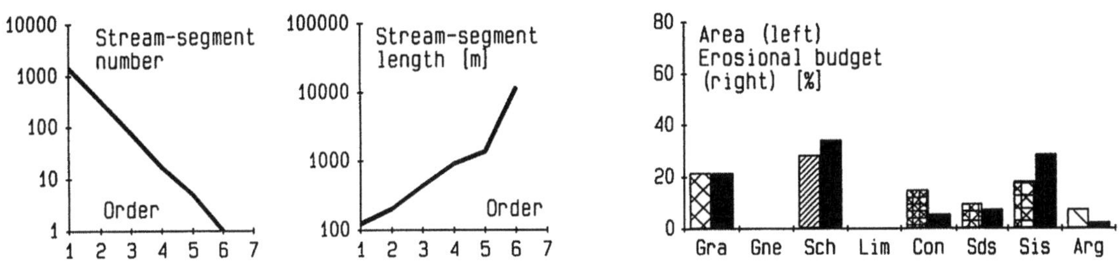

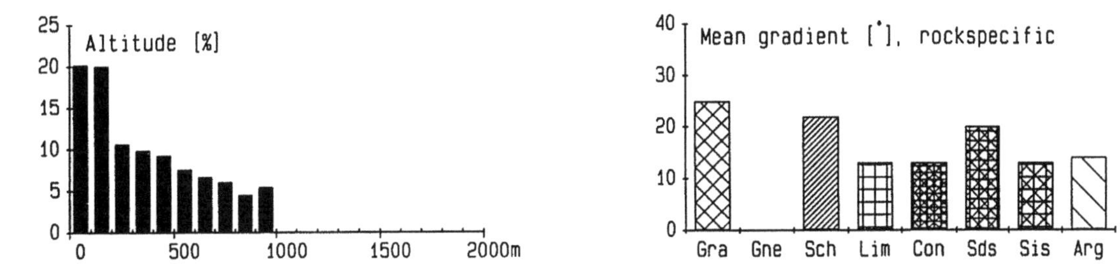

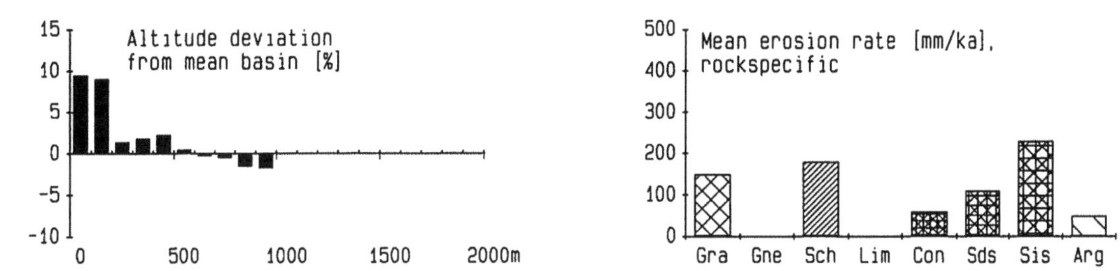

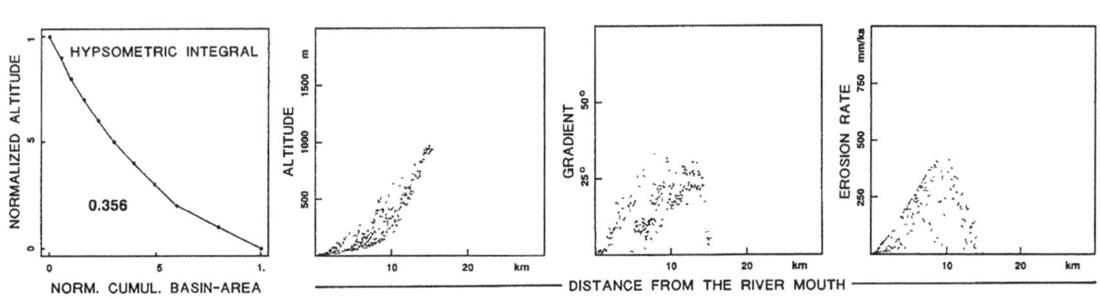

River-mouth sediment and water, the product: Grain sizes, petrography, mineralogy and chemistry 12

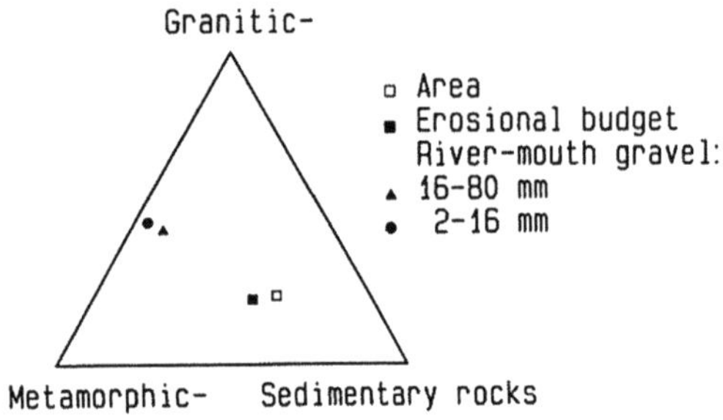

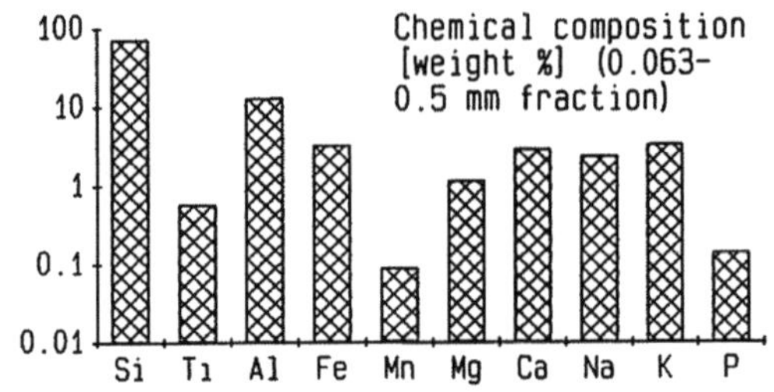

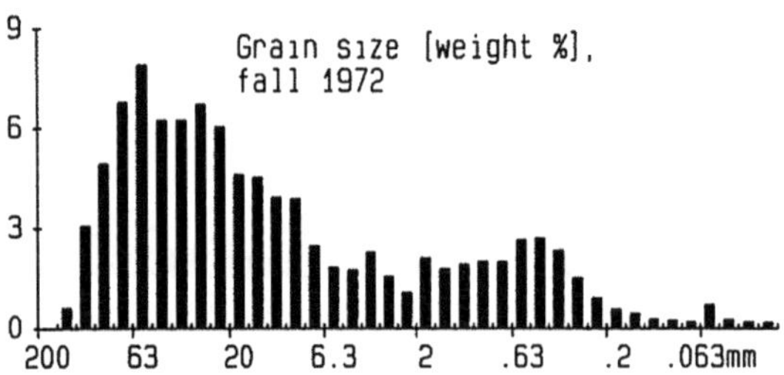

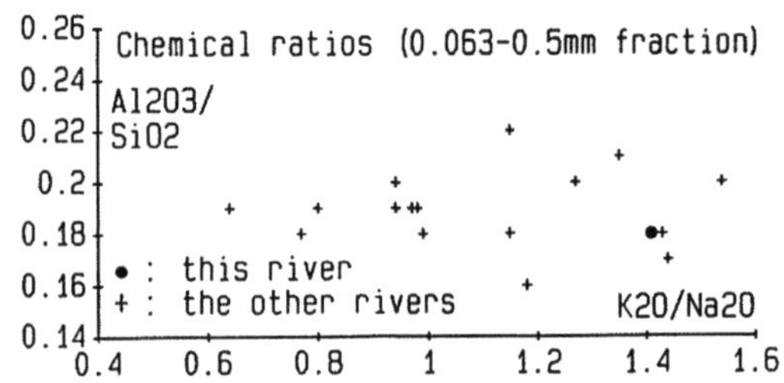

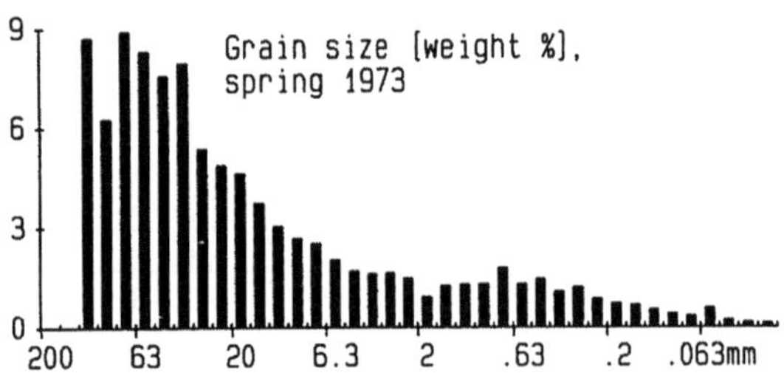

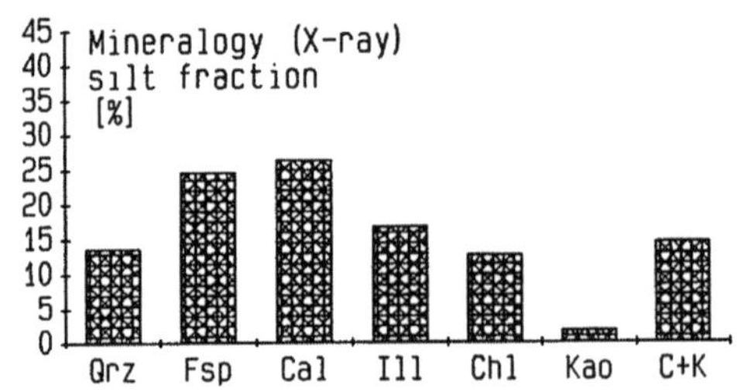

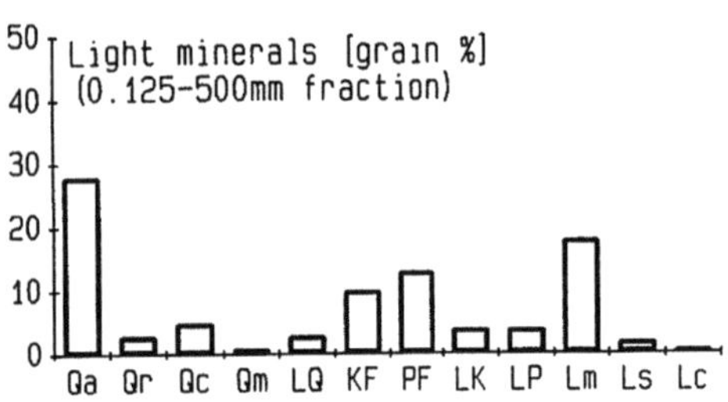

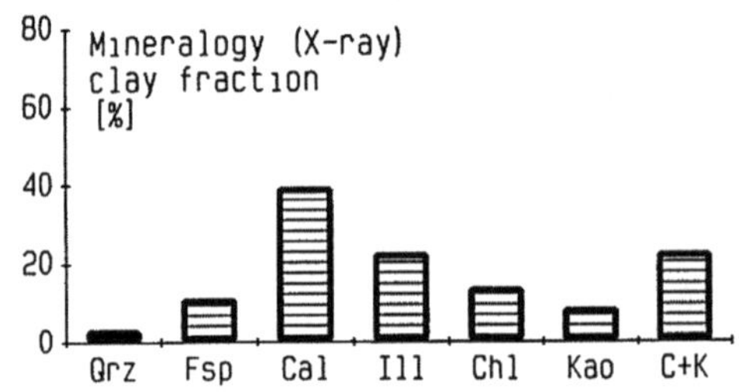

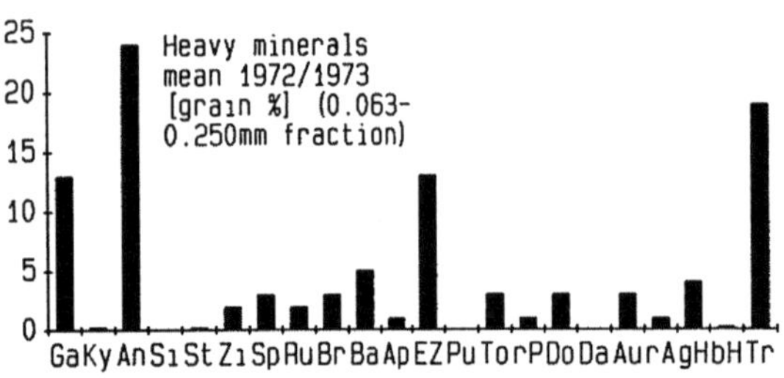

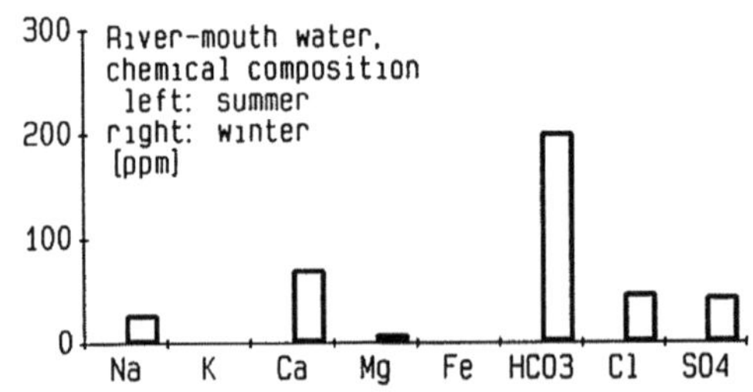

13 Novito

Drainage basin, the producer: Geology and geomorphology

(For abbreviations see 00 basin 00)

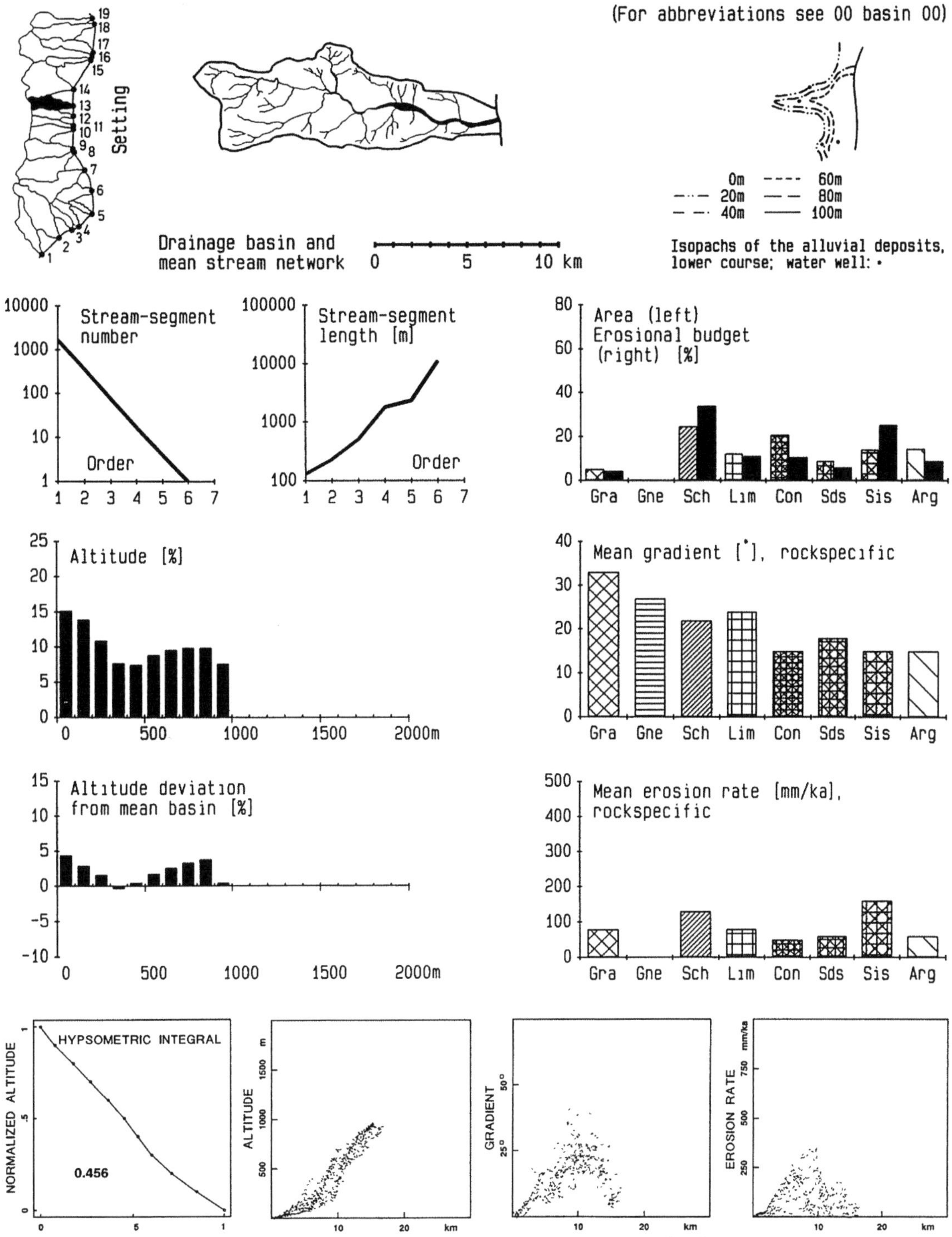

River-mouth sediment and water, the product: Grain sizes, petrography, mineralogy and chemistry 13

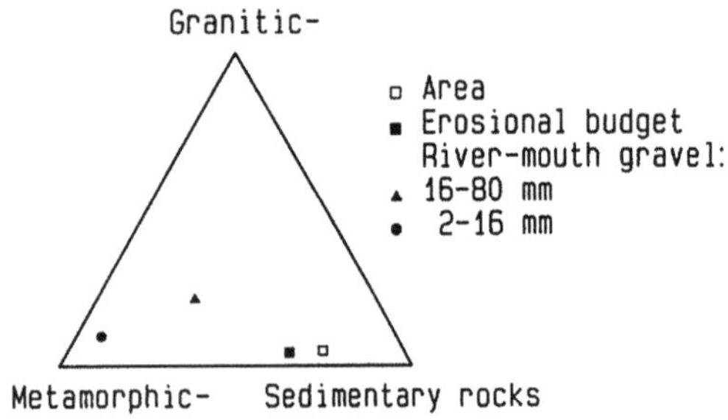

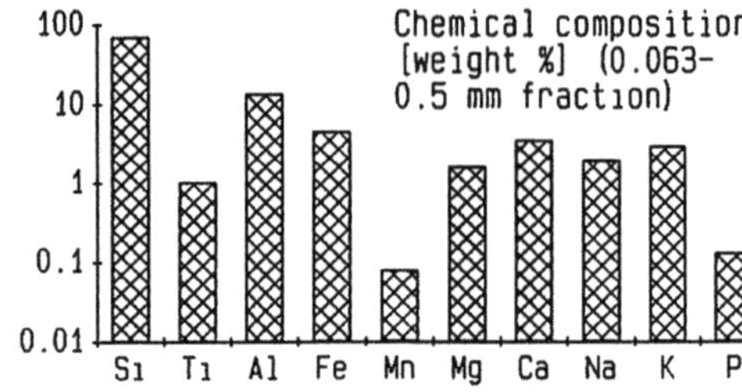

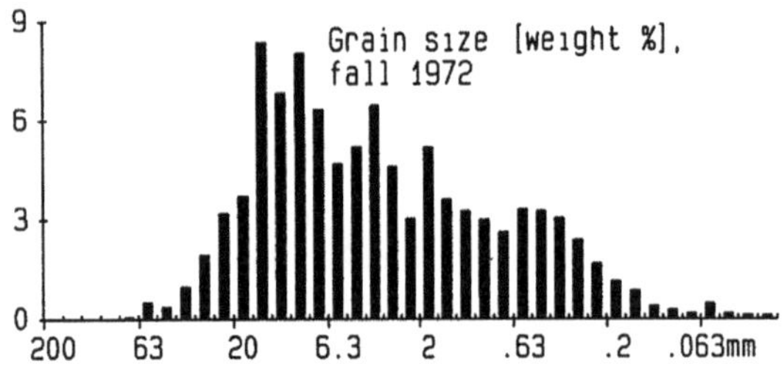

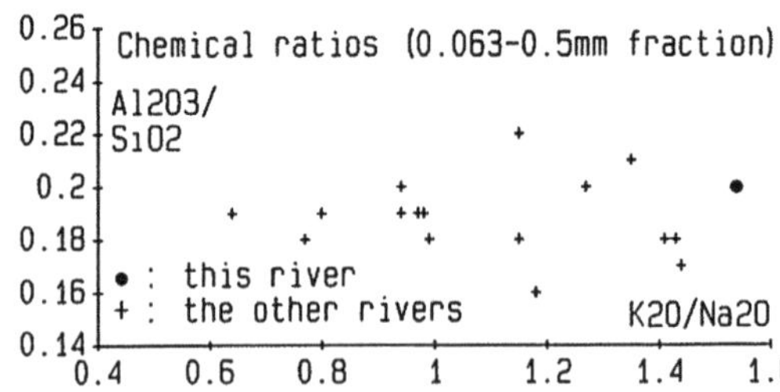

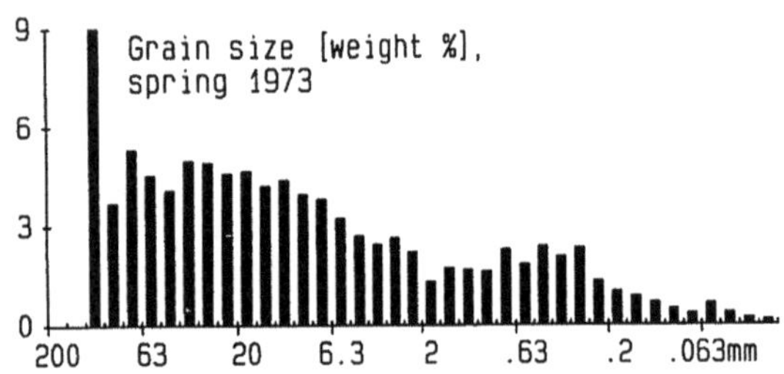

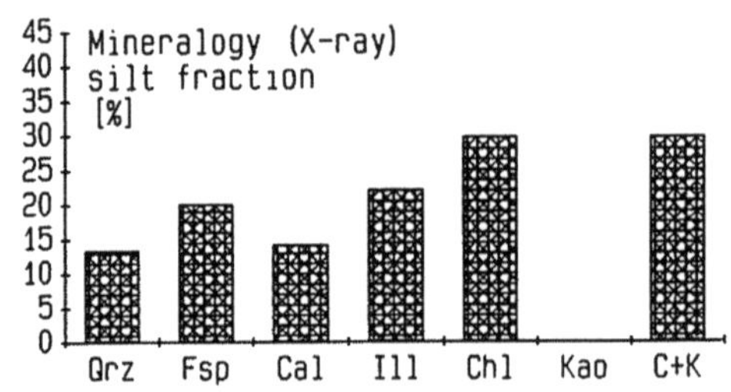

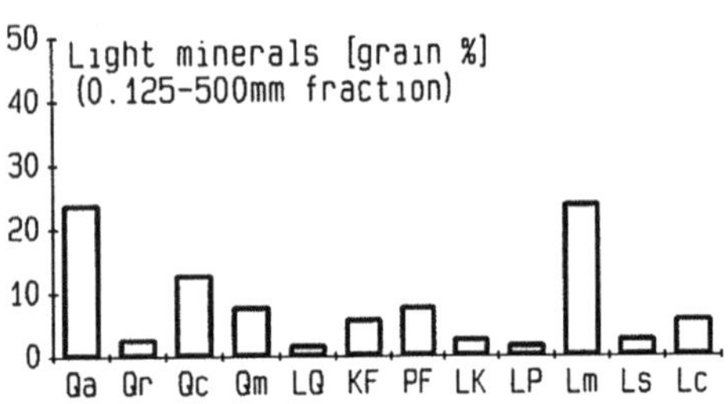

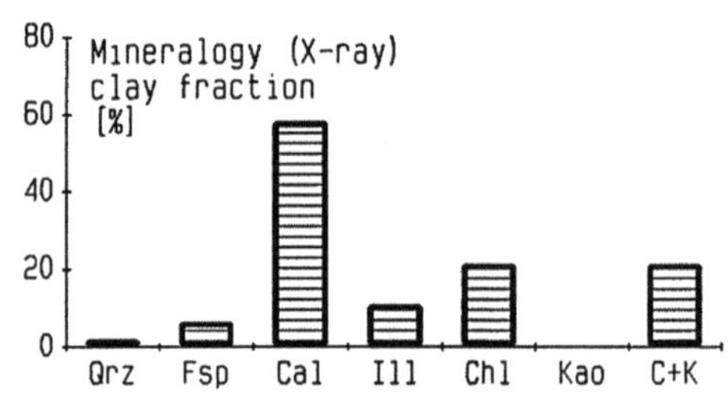

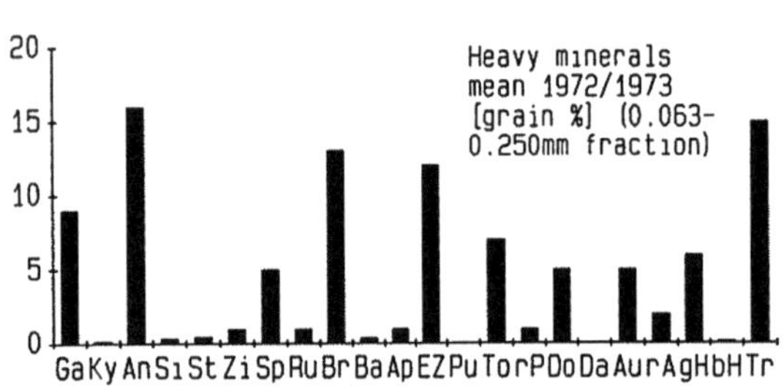

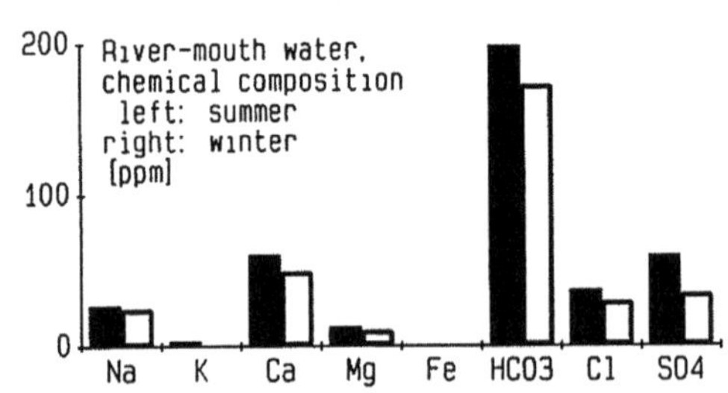

14 Torbido

Drainage basin, the producer: Geology and geomorphology

(For abbreviations see 00 basin 00)

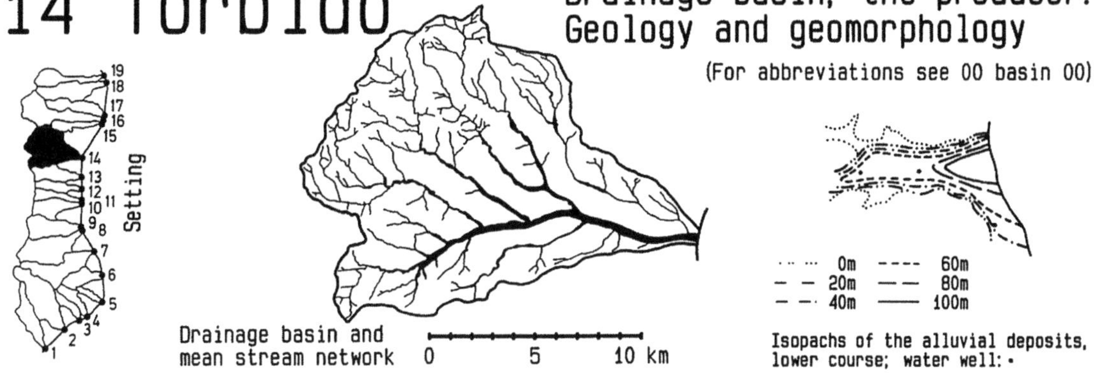

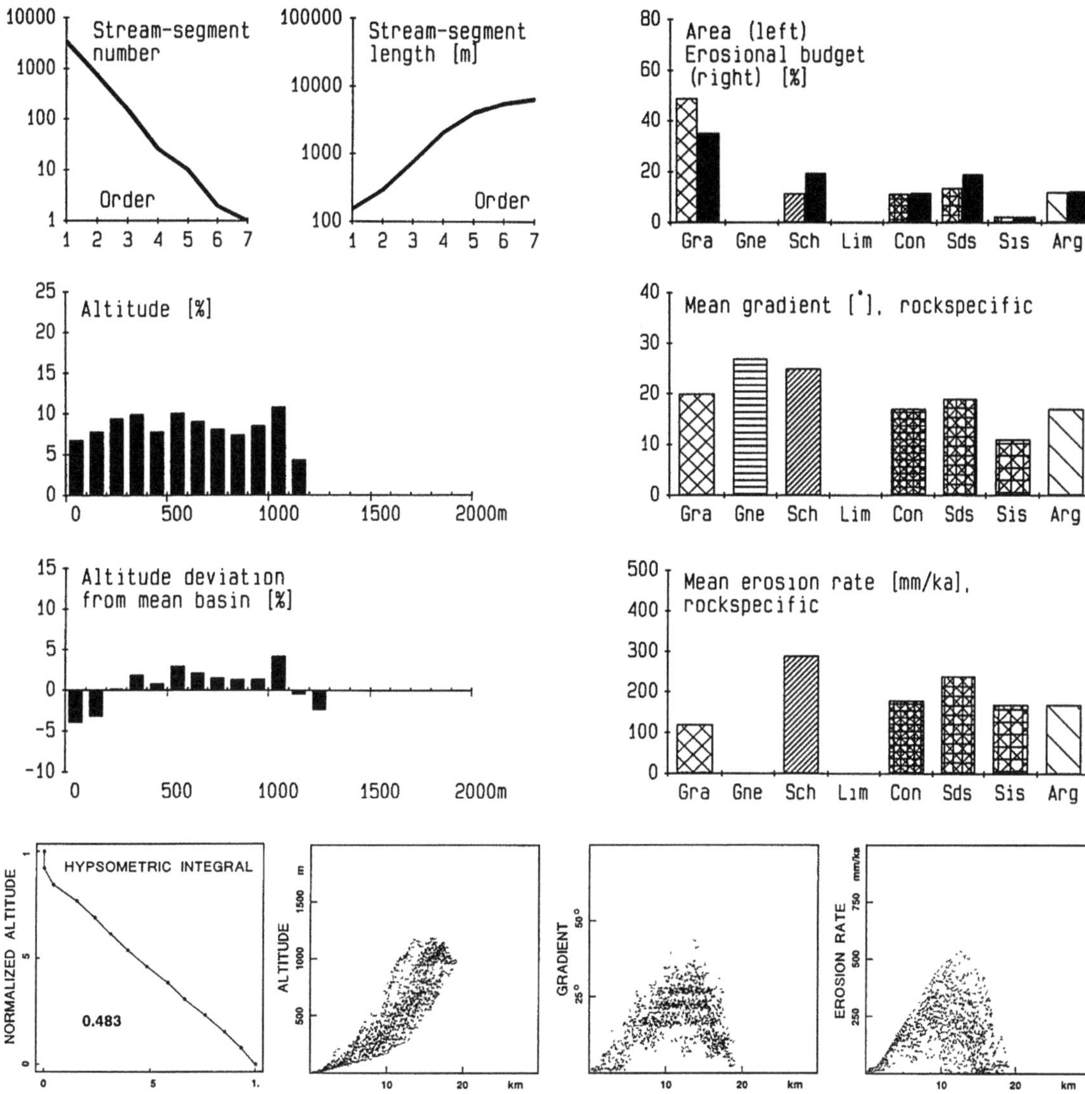

River-mouth sediment and water, the product: Grain sizes, petrography, mineralogy and chemistry 14

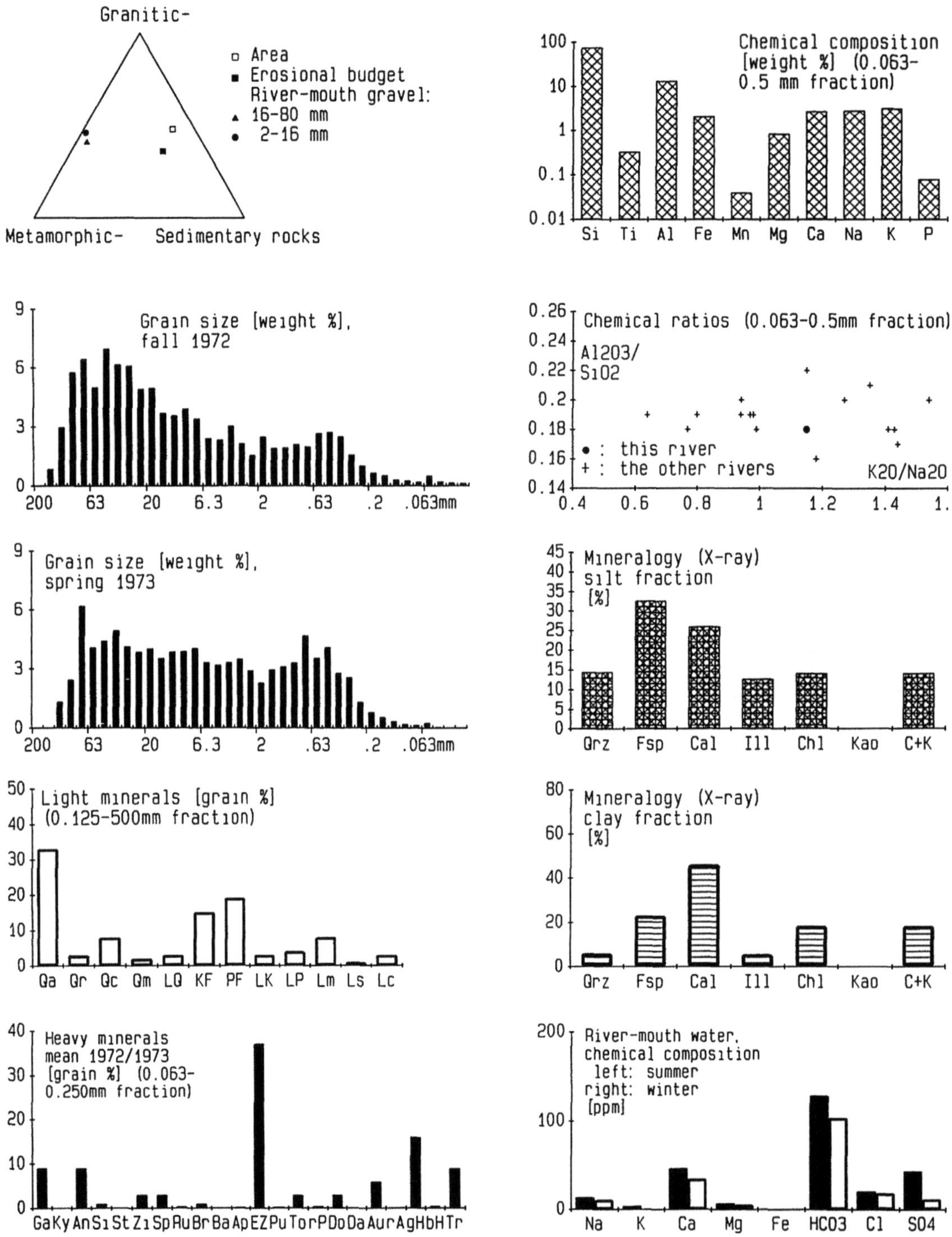

15 Amusa

Drainage basin, the producer: Geology and geomorphology

(For abbreviations see 00 basin 00)

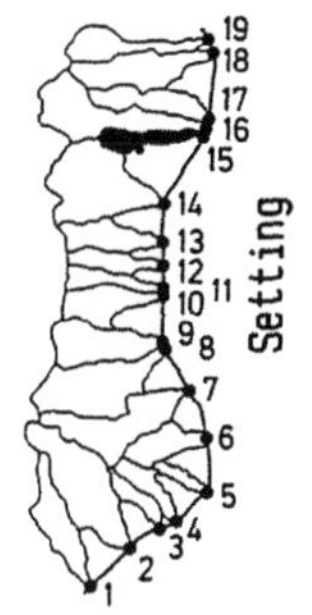

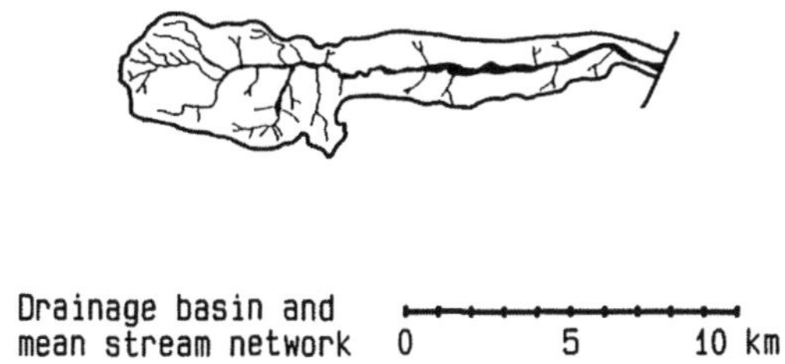

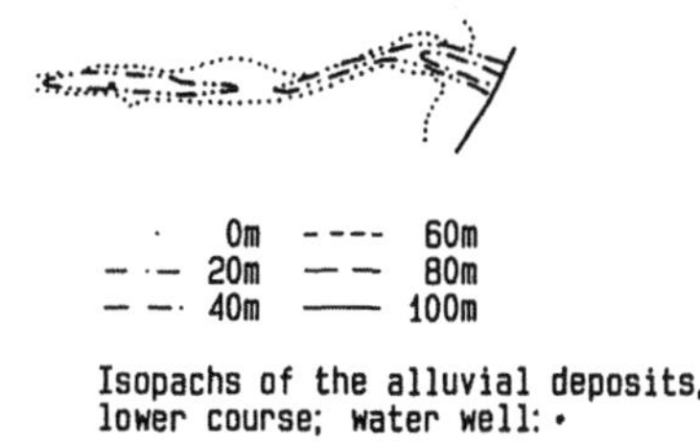

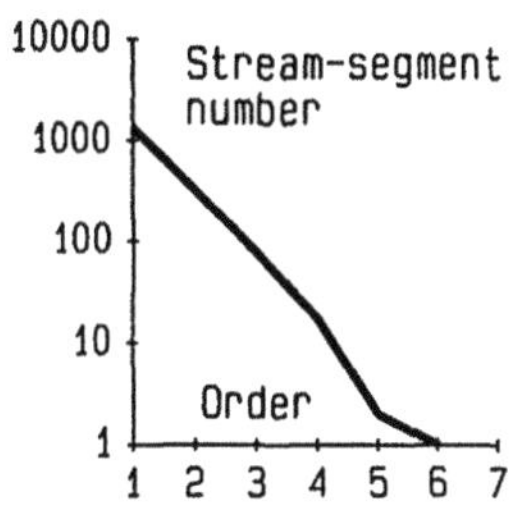

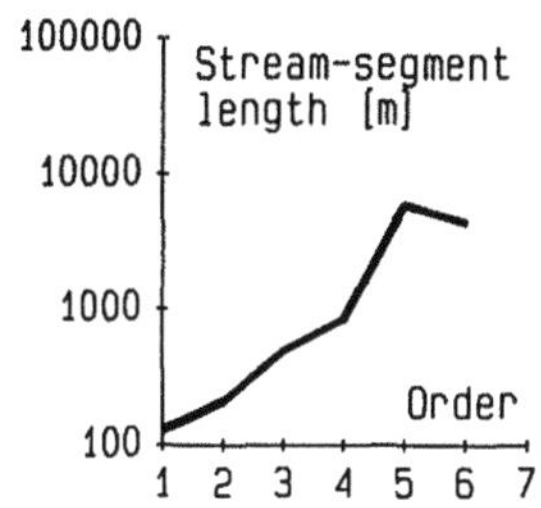

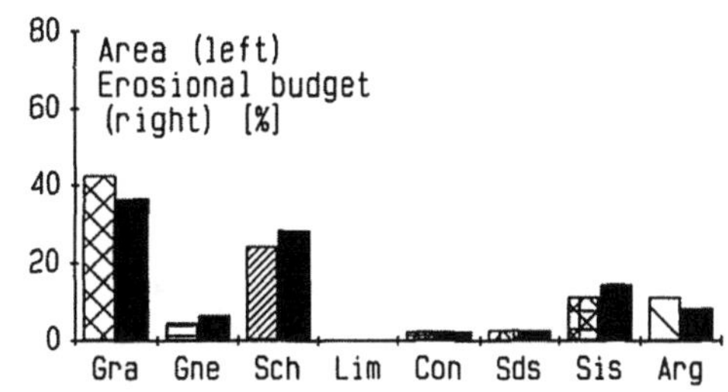

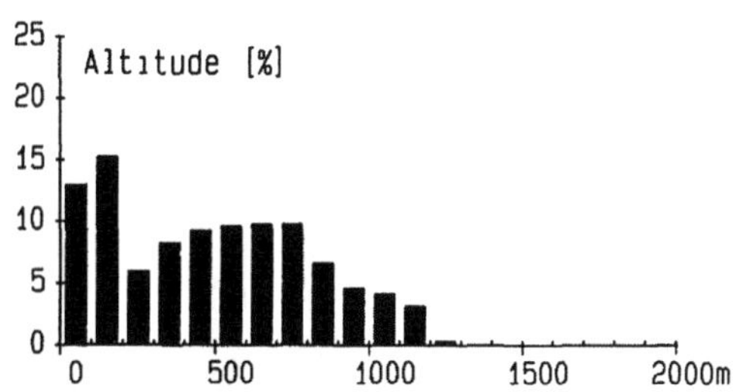

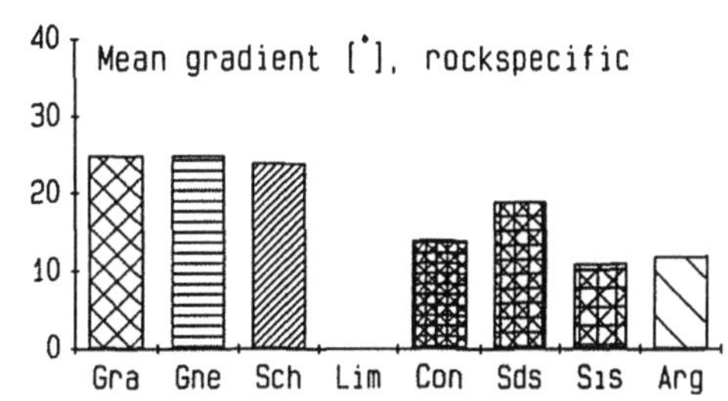

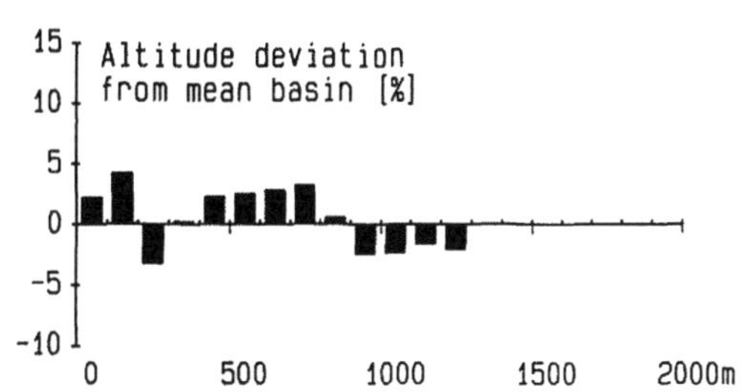

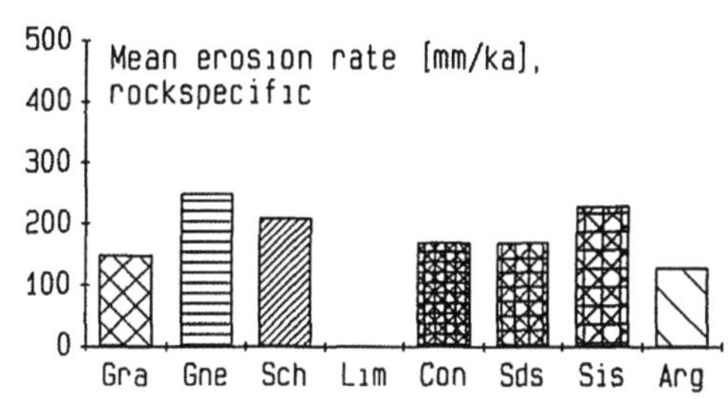

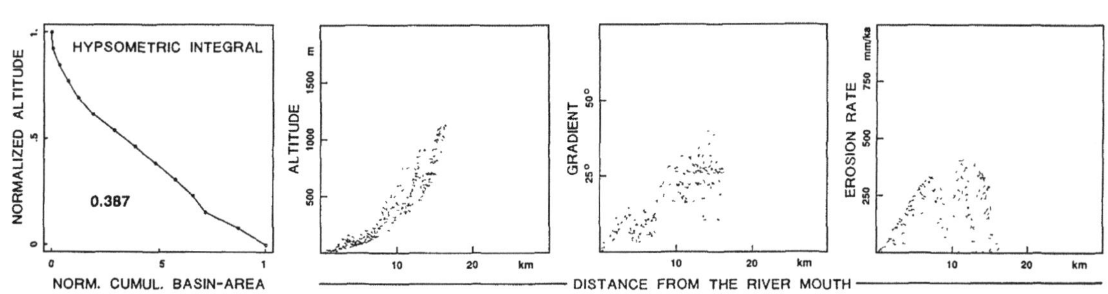

River-mouth sediment and water, the product: Grain sizes, petrography, mineralogy and chemistry 15

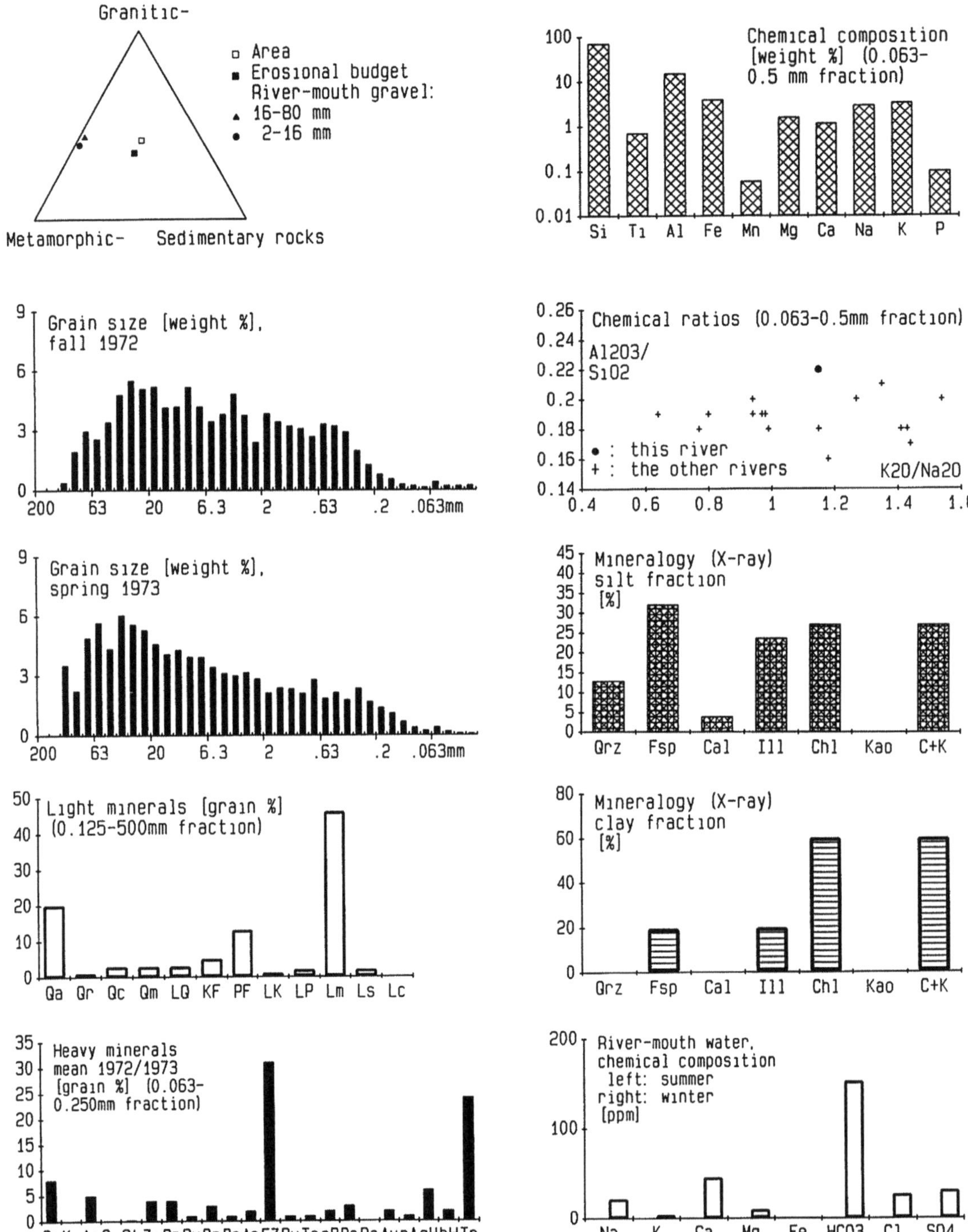

16 Allaro

Drainage basin, the producer: Geology and geomorphology

(For abbreviations see 00 basin 00)

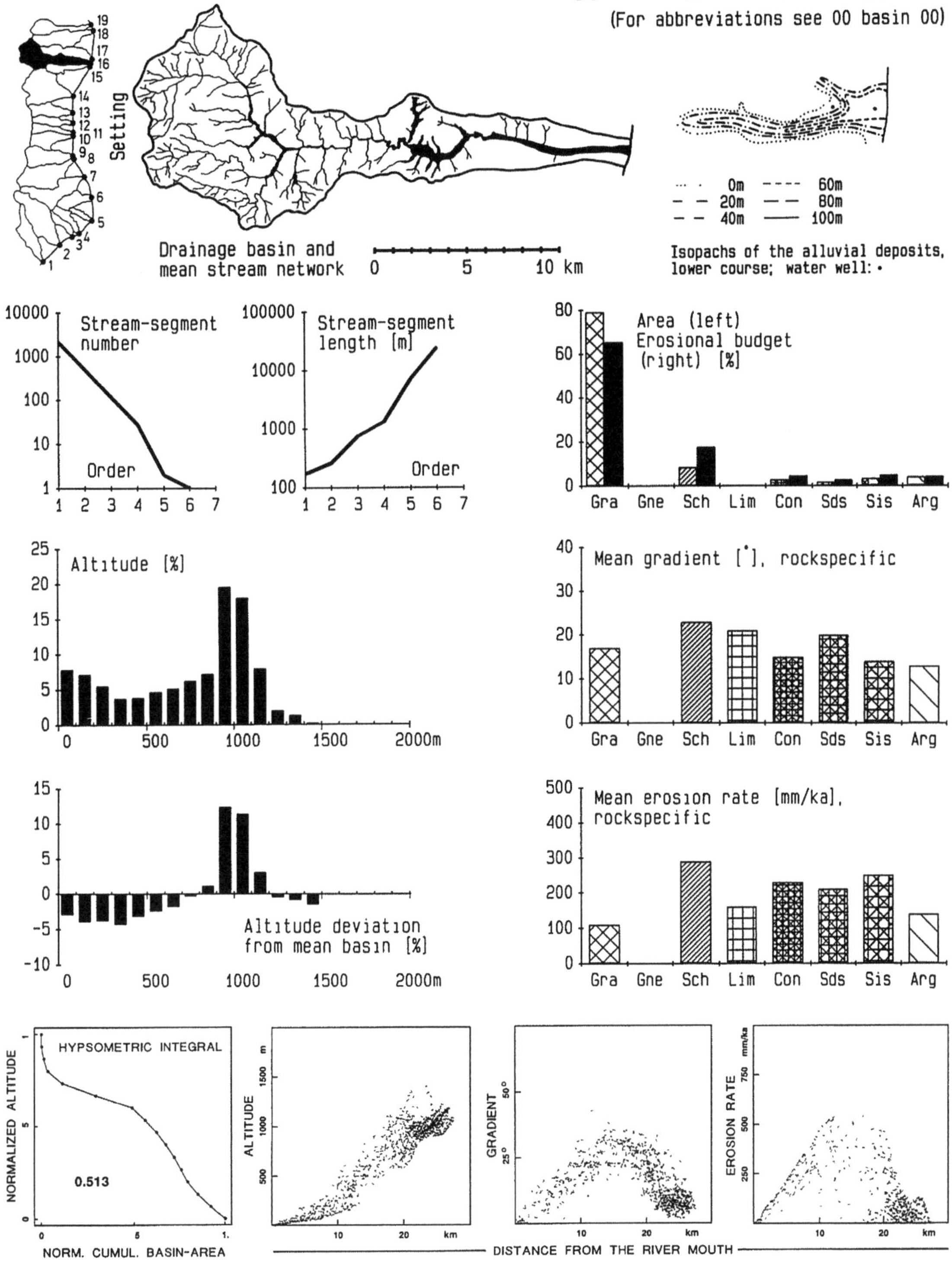

River-mouth sediment and water, the product: Grain sizes, petrography, mineralogy and chemistry

16

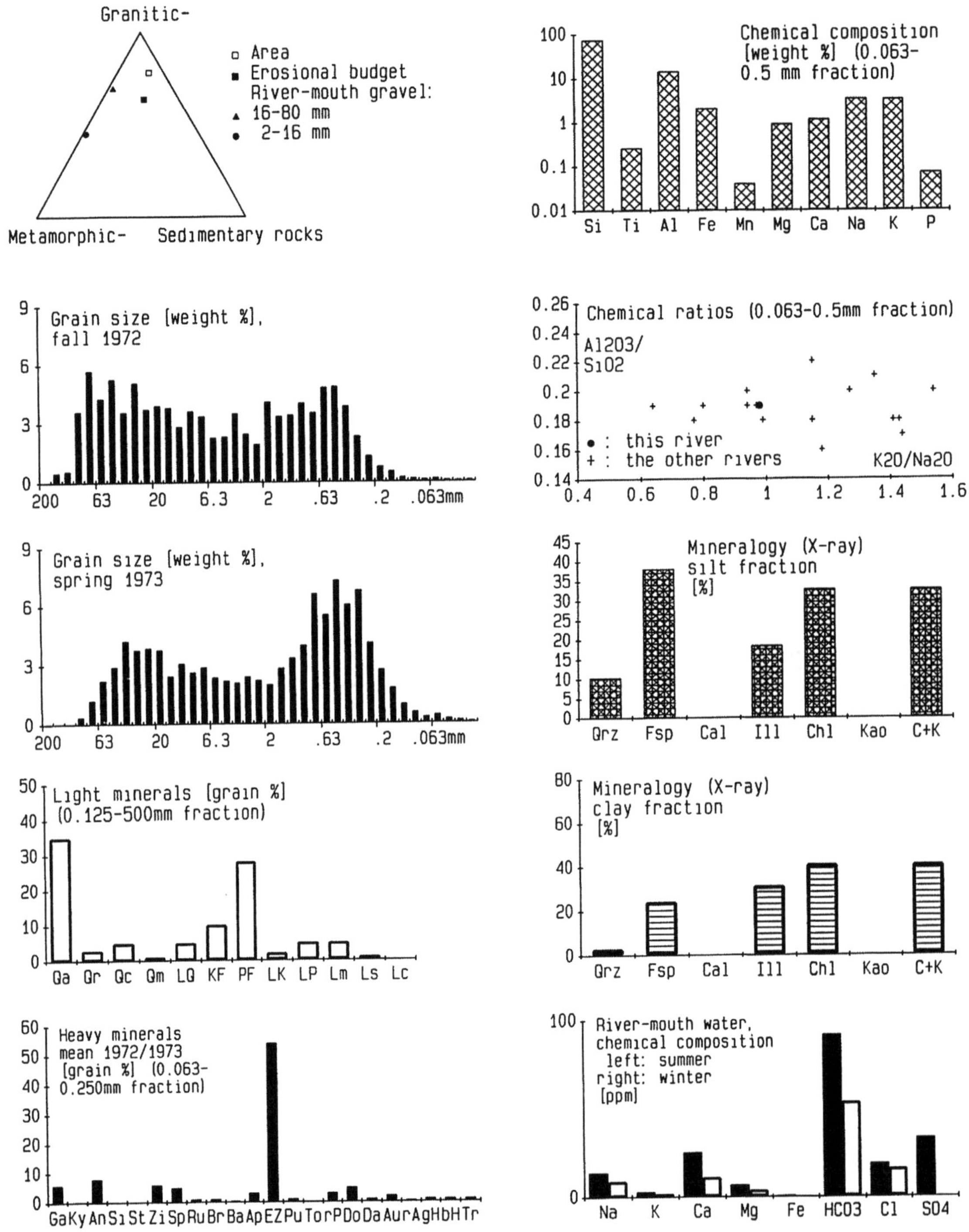

17 Precariti

Drainage basin, the producer: Geology and geomorphology

(For abbreviations see 00 basin 00)

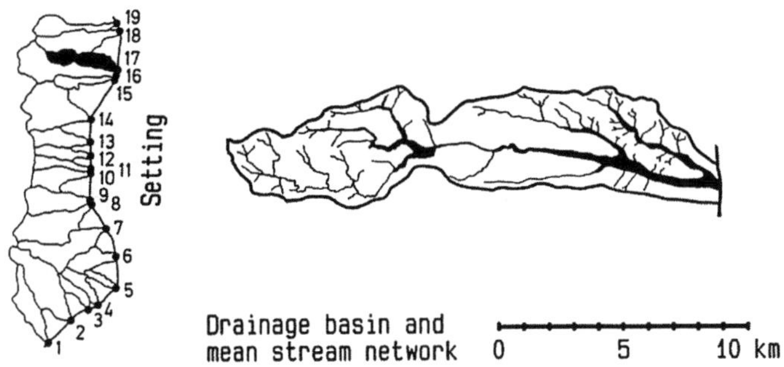

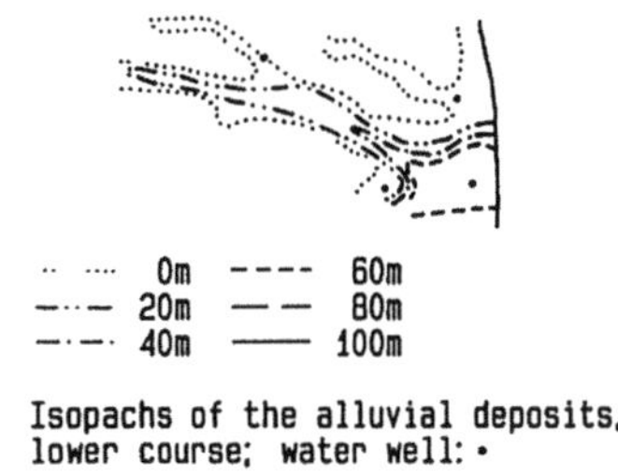

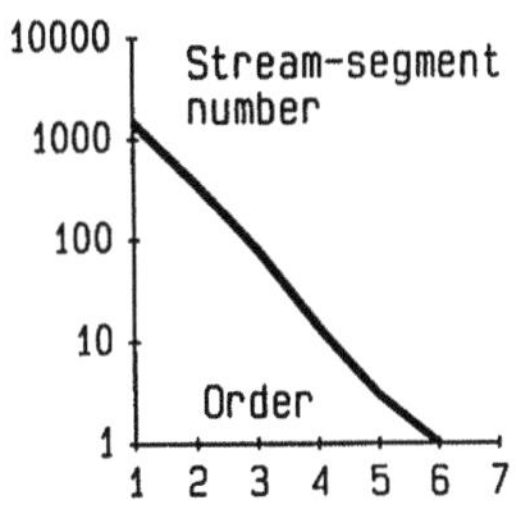

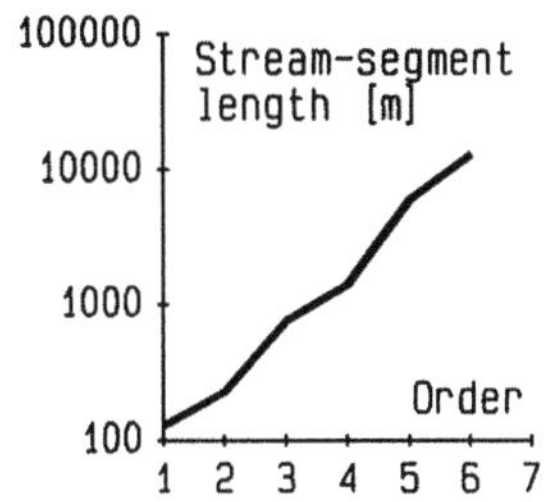

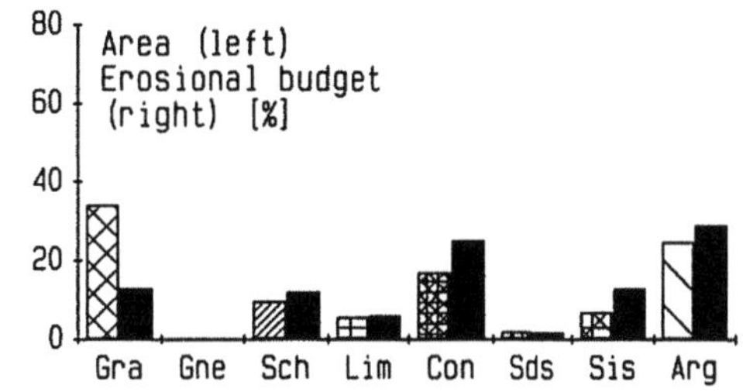

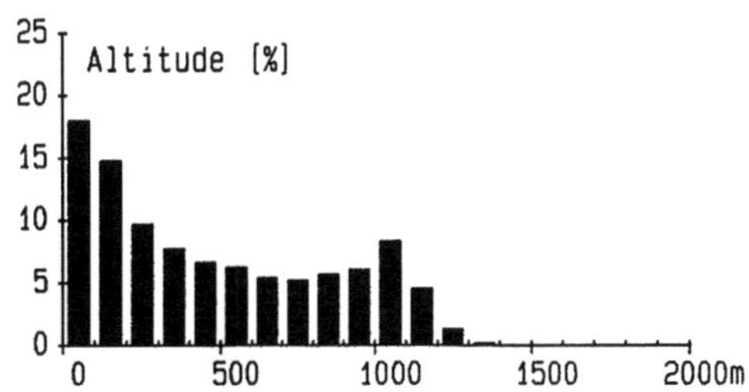

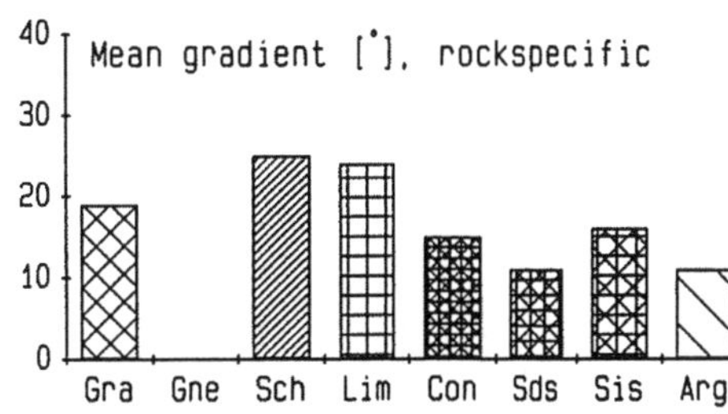

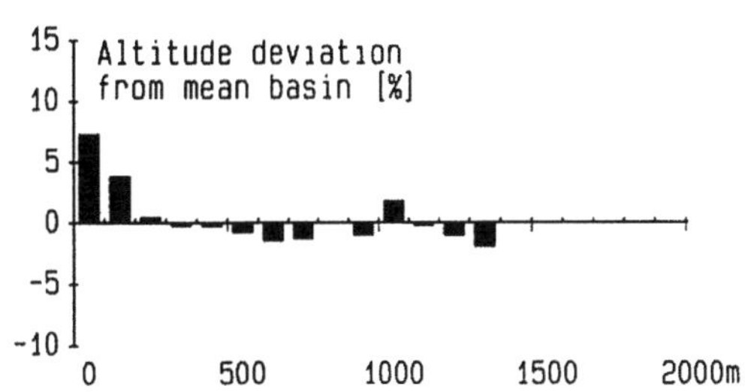

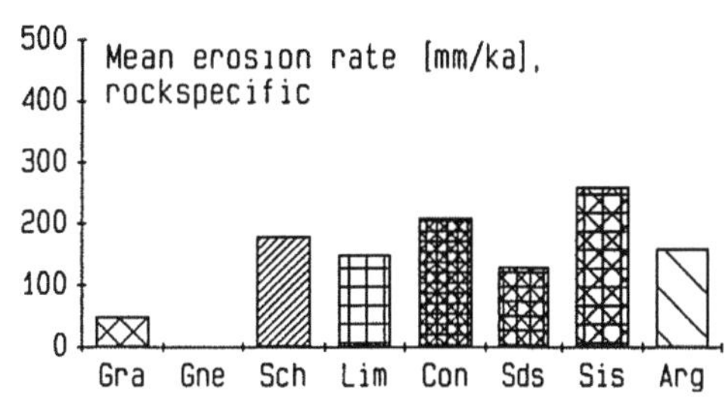

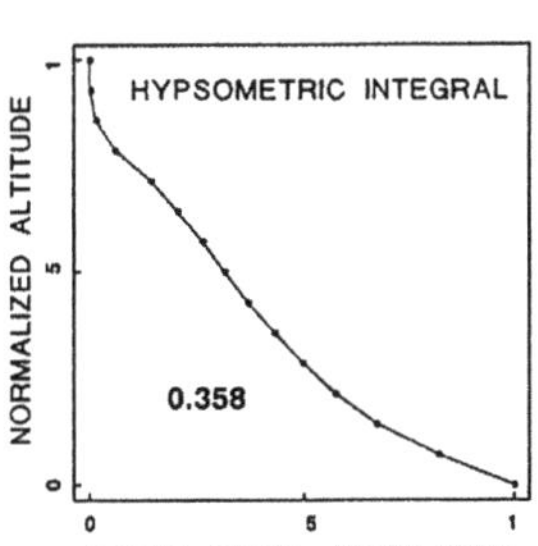

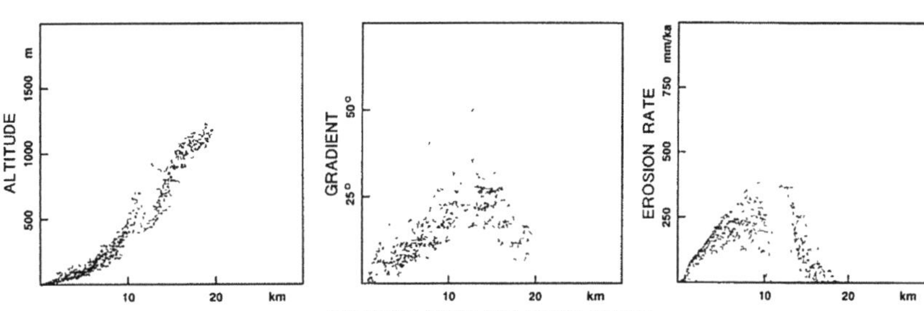

River-mouth sediment and water, the product: Grain sizes, petrography, mineralogy and chemistry

17

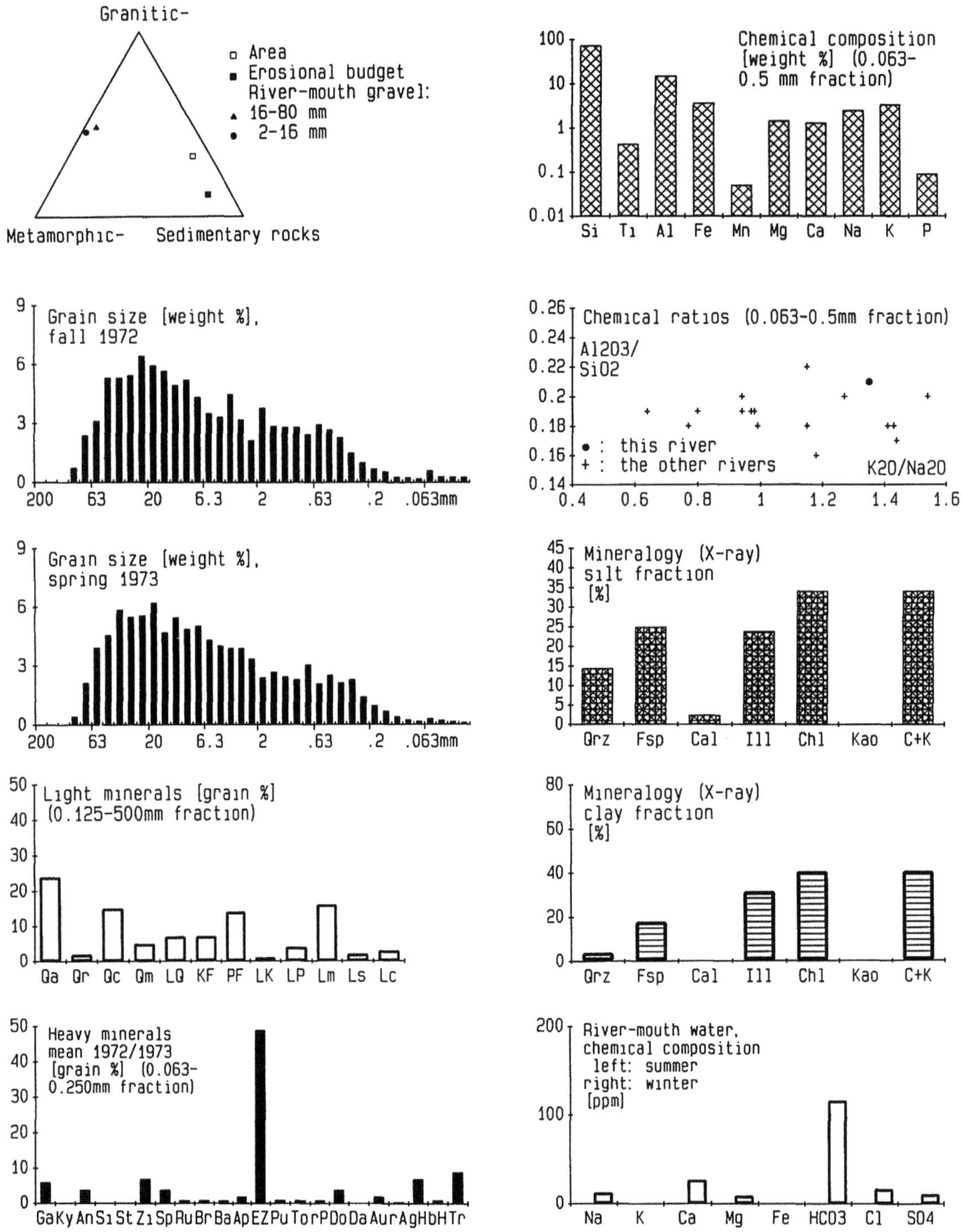

18 Stilaro

Drainage basin, the producer: Geology and geomorphology

(For abbreviations see 00 basin 00)

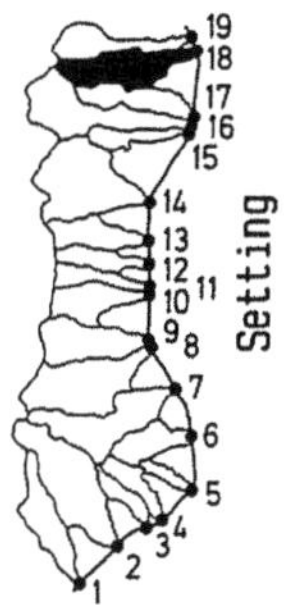

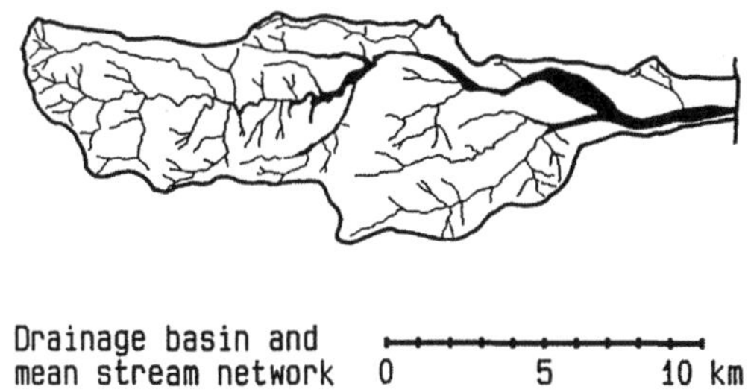

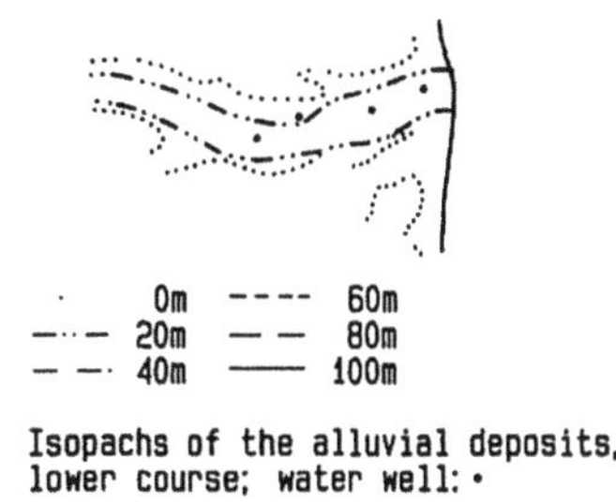

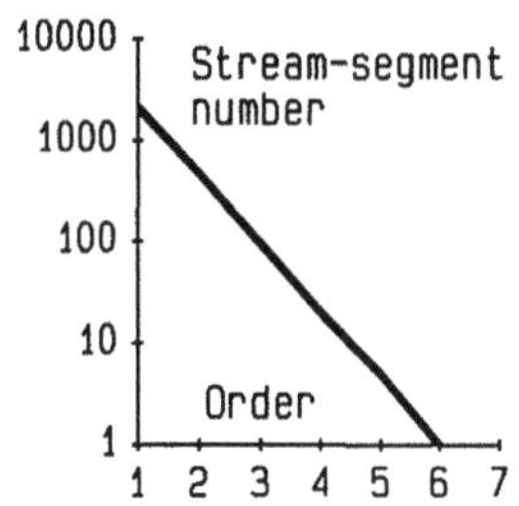

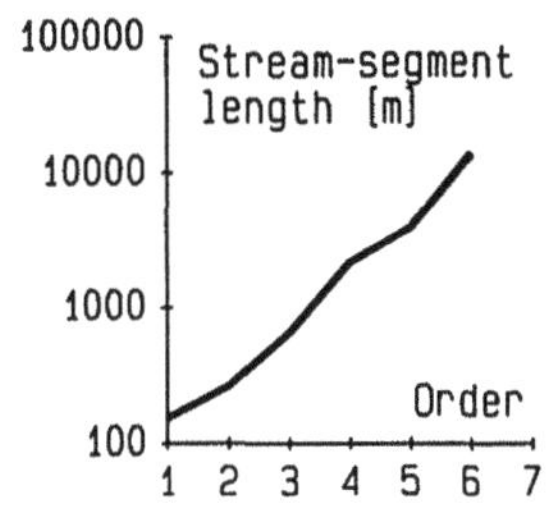

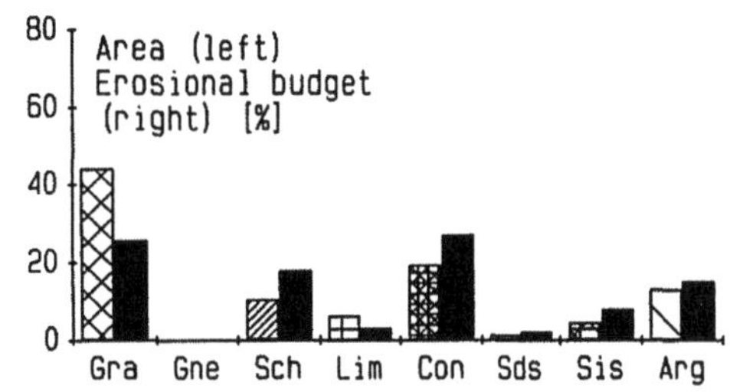

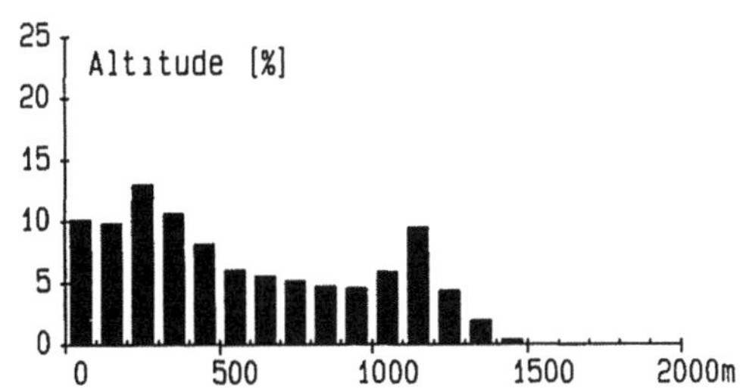

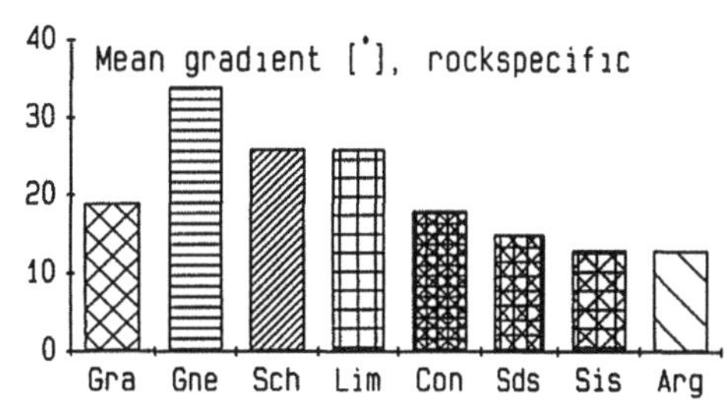

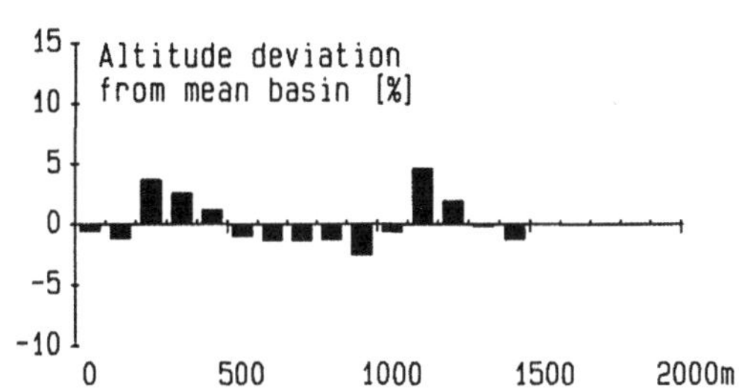

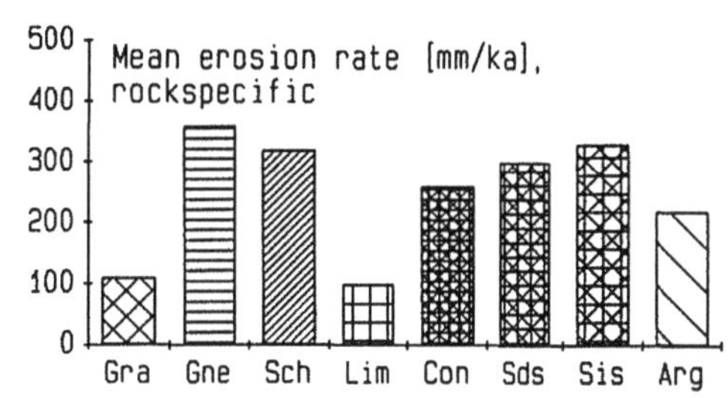

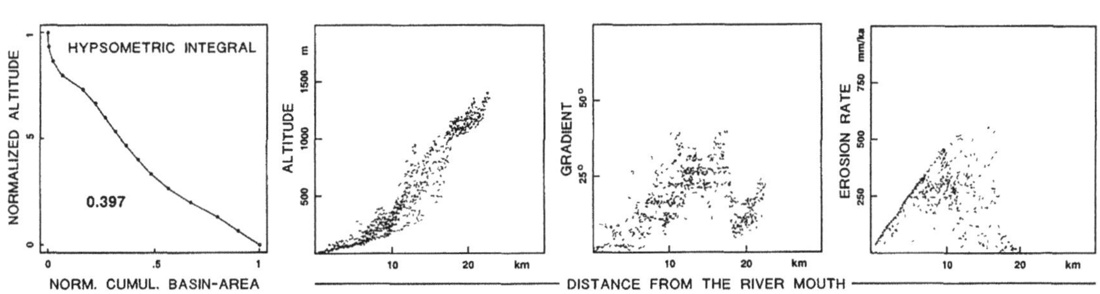

River-mouth sediment and water, the product: Grain sizes, petrography, mineralogy and chemistry 18

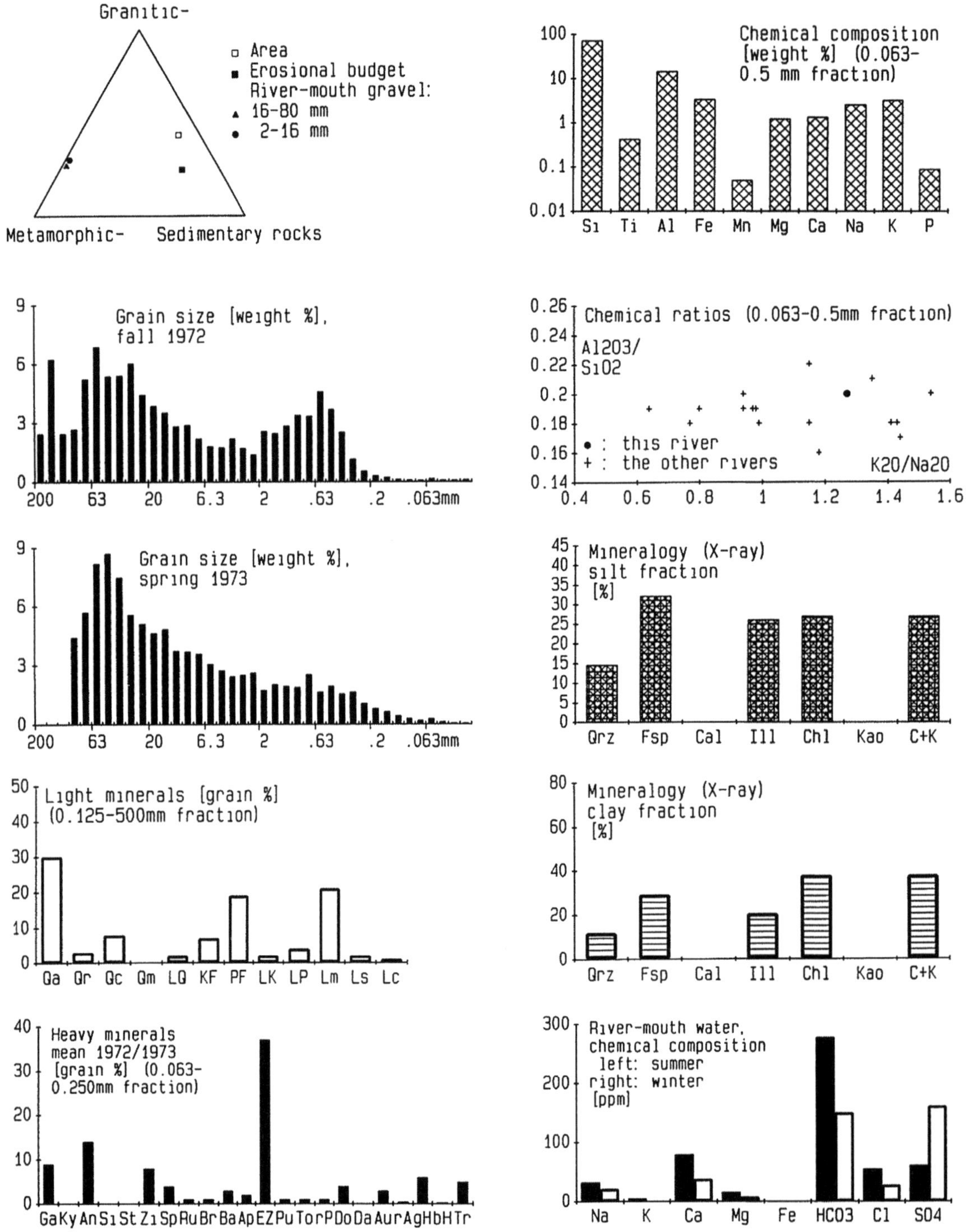

19 Assi

Drainage basin, the producer: Geology and geomorphology

(For abbreviations see 00 basin 00)

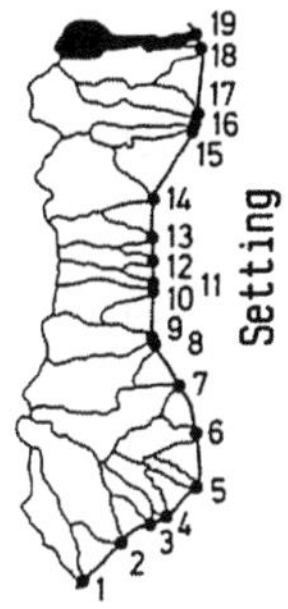

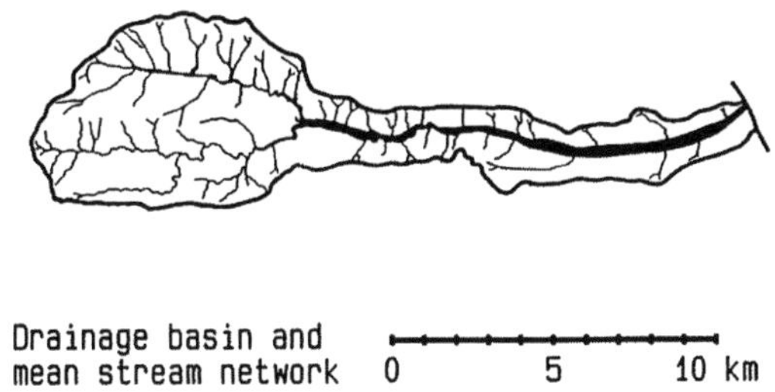

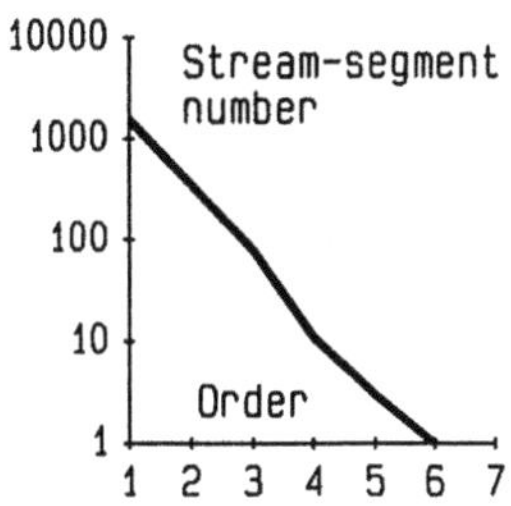

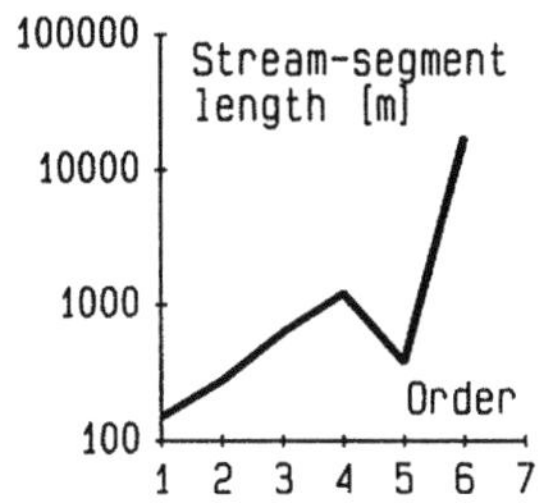

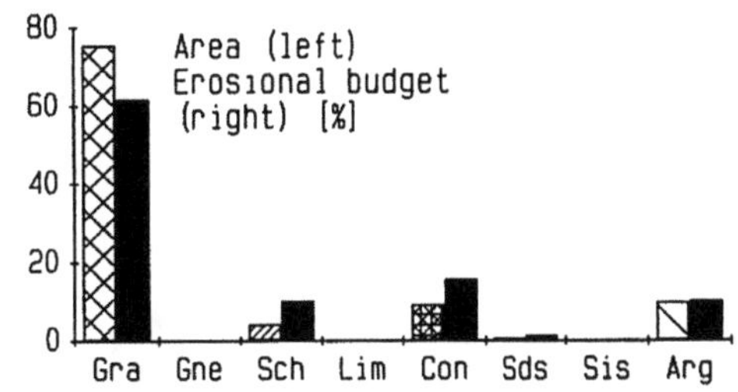

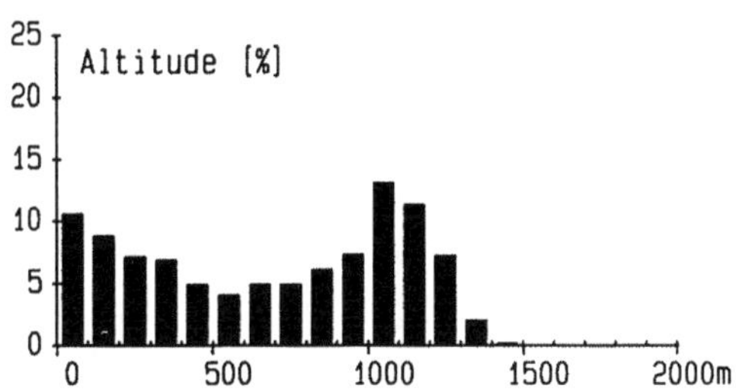

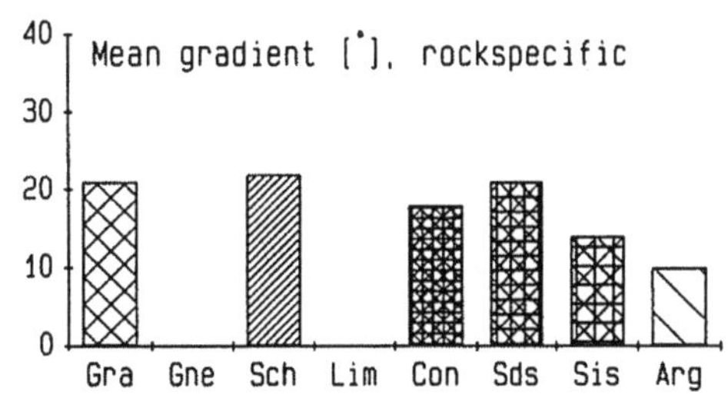

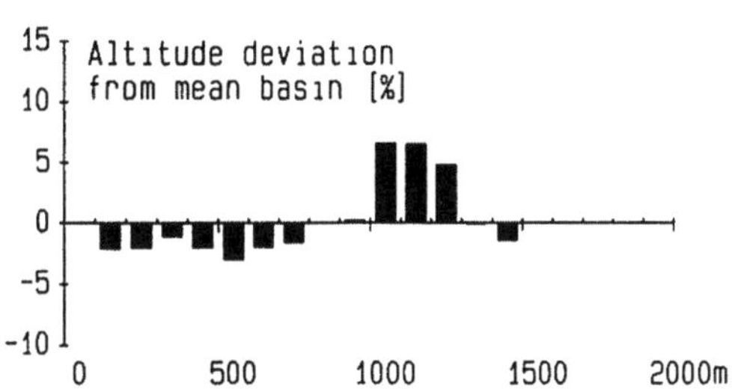

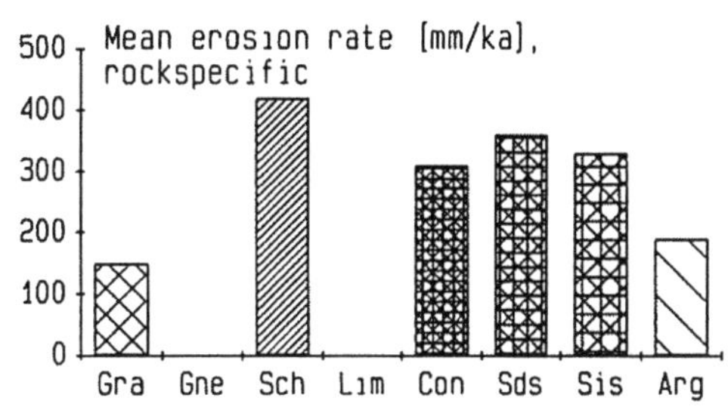

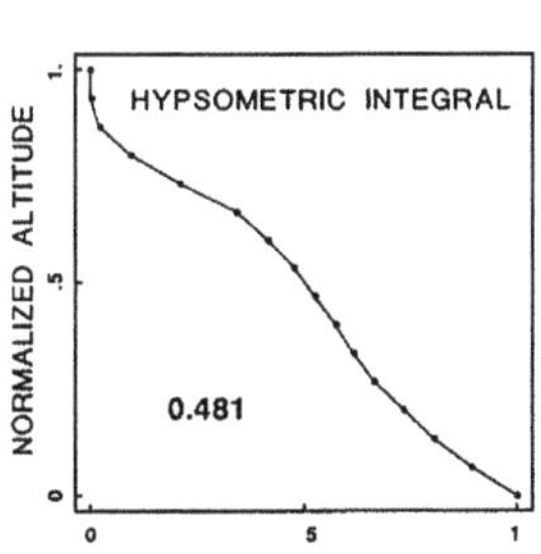

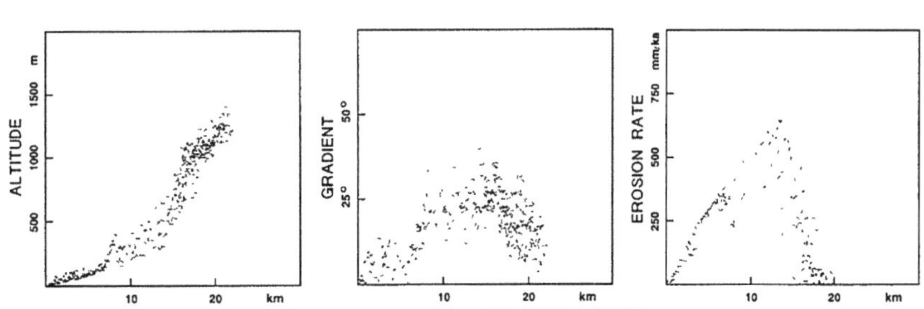

River-mouth sediment and water, the product: Grain sizes, petrography, mineralogy and chemistry

19

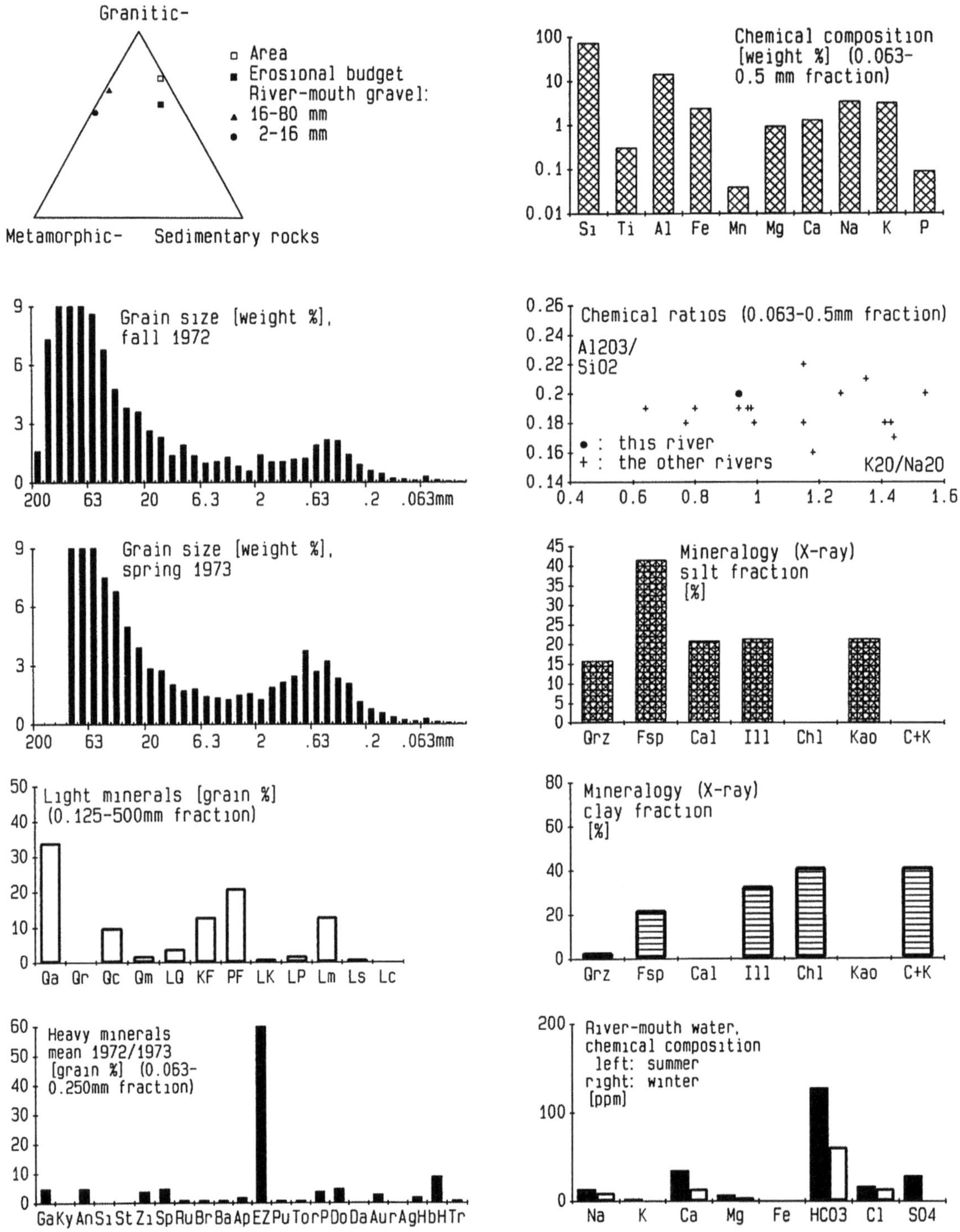

20 Mean basin 01-19

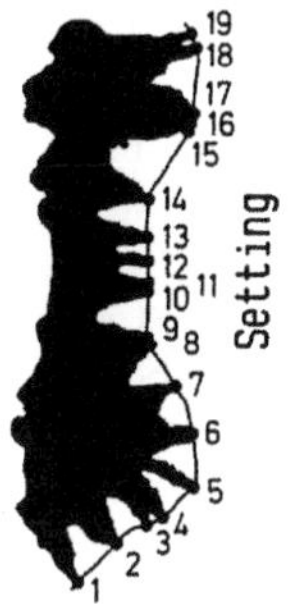

Drainage basin, the producer: Geology and geomorphology

(For abbreviations see 00 basin 00)

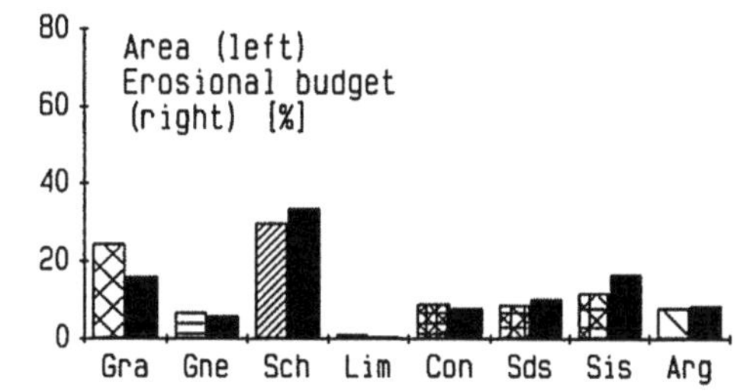

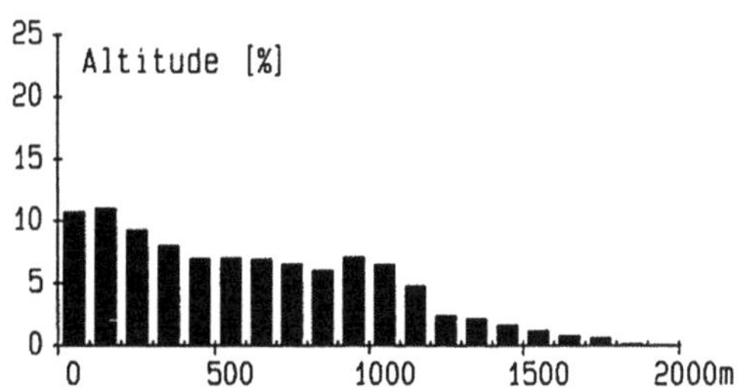

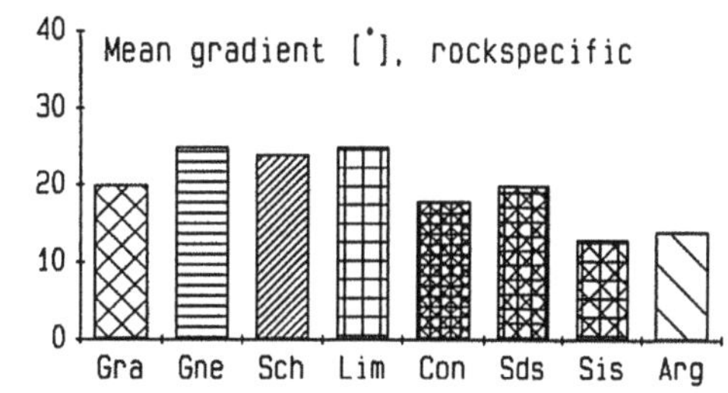

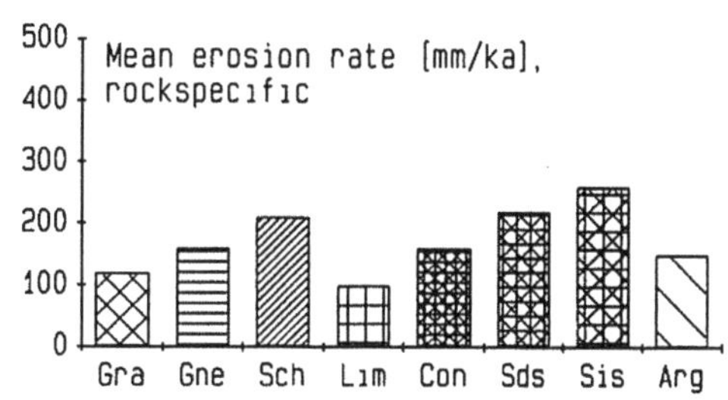

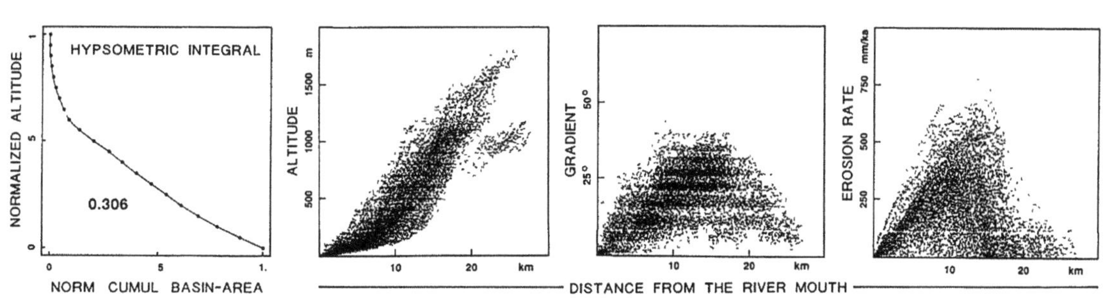

River-mouth sediment and water, the product: Grain sizes, petrography, mineralogy and chemistry

20

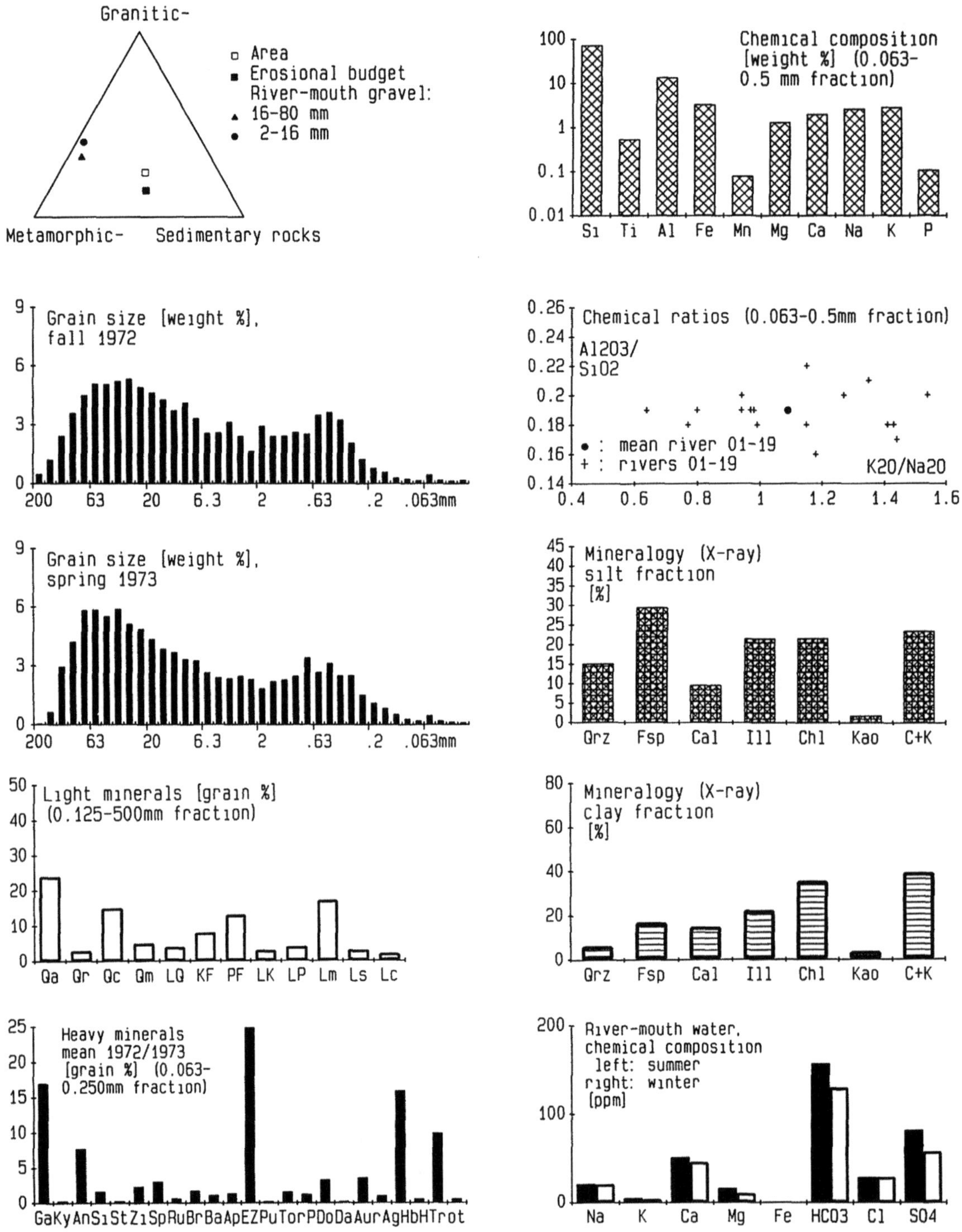

References

Ahnert F (1970) Functional relationship between denudation, relief and uplift in large mid-latitude drainage basins. Am J Sci 268:243–263

Aifa T, Feinberg H, Pozzi JP (1988) Pliocene-Pleistocene evolution of the Thyrrhenian arc: paleomagnetic determination of uplift and rotational deformation. Earth Planet Sci Lett 87:438–452

Argast S, Donnelly TW (1987) The chemical discrimination of clastic sedimentary components. J Sediment Petrol 57:813–823

Bagnold RA, Barndorff-Nielsen O (1980) The pattern of natural size distribution. Sedimentology 27:199–207

Barndorff-Nielsen O (1977) Exponentially decreasing distributions for the logarithm of particle size. Proc R Soc Lond Ser A 353 No 1674:401–419

Barndorff-Nielsen O, Christiansen C (1986) Erosion, deposition and size distribution of sands. Dep Theor Stat, Univ Aarhus, Res Rep 149:34

Barndorff-Nielsen O, Daalsgard K, Halgreen C, Kuhlman H, Moller JT, Schou G (1982) Variation in particle size distribution over a small dune. Sedimentology 29:53–65

Barndorff-Nielsen O, Blaesild P, Jensen JL, Sorensen M (1985) The fascination of sand. In: Atkinson AC, Feinberg SE (eds) A celebration of statistics. Centenary Vol Int Stat Inst, Springer, Berlin Heidelberg New York, pp 57–87

Basu A (1976) Petrology of Holocene fluvial sand derived from plutonic source rocks: implications to paleoclimatic interpretation. J Sediment Petrol 46:694–709

Basu A (1985) Influence of climate and relief on compositions of sands released at source areas. In: Zuffa GG (ed) Provenance of arenites. NATO ASI Ser C Math Phys Sci 148:1–18

Basu A, Molinaroli E (1989) Provenance characteristics of detrital opaque Fe-Ti oxide minerals. J Sediment Petrol 59:922–934

Bates RL, Jackson JA (1987) Glossary of geology. Am Geol Inst, Int Computaprint Corp, Alexandria, Virginia, 788 pp

Bhatia JC, Durst F (1988) LHPDF – a PC package for estimating parameters of the log-hyperbolic distribution, moments and mean diameters. Lehrstuhl für Strömungsmechanik der Universität Erlangen-Nürnberg No LSTM 230/T/88:67

Bhatia M (1983) Plate tectonics and geochemical composition of sandstones. J Geol 91:611–627

Blatt H (1967) Original characteristics of clastic quartz grains. J Sediment Petrol 37:401–424

Blatt H (1970) Determination of mean sediment thickness in the crust: a sedimentologic method. Geol Soc Am Bull 81:255–262

Blatt H (1985) Provenance studies and mudrocks. J Sediment Petrol 55:69–75

Blatt H, Middleton G, Murray R (1972) Origin of sedimentary rocks. Prentice Hall, Englewood Cliffs, New Jersey, 634 pp

Blatt H, Middleton G, Murray R (1980) Origin of sedimentary rocks, 2nd edn. Prentice Hall, Englewood Cliffs, New Jersey, 782 pp

Bluck BJ (1967) Sedimentation of beach gravels: examples from South Wales. J Sediment Petrol 37:128–156

Boccaletti M, Nicolich R, Tortorici L (1984) The Calabrian arc and the Ionian Sea in the dynamic evolution of the Central Mediterranean. Mar Geol 55:219–245

Boggs S Jr (1968) Experimental study of rock fragments. J Sediment Petrol 38:1326–1339

Bonardi G, Giunta G, Perrone V, Russo M, Zupetta A, Ciampo G (1980) Osservazioni sull'evoluzione dell'arco Calabro-Peloritano nel miocene inferiore: La formazione di Stilo Capo D'Orlando. Boll Soc Geol It 99:365–393

Bonardi G, Messina A, Perrone V, Russo S, Zuppetta A (1984) L'unita di Stilo nel settore meridionale dell'arco Calabro-Peloritano. Boll Soc Geol It 103:279–309

Bottari A, Lo Giudice E, Nicoletti PG, Sorriso-Valvo M (1981–1982) The Ferruzzano earthquake of 1978: macroseismic effects and slope-stability conditions in southern Calabria (Italy). Rev Géol Dyn Géogr Phys 23:73–84

Breyer JA, Bart HA (1978) The composition of fluviatile sands in a temperate semiarid region. J Sediment Petrol 48:1311–1320

Burton AN (1971) Carta Geologica Della Calabria 1:25000; Relazione Generale: Poligrafica & Cartevalori-Ercolano, Napoli

Caloiero D, Mercuri T (1980) Le alluvioni in Calabria dal 1921 al 1970. CNR, IRPI Cosenza, 161 pp

Cameron KL, Blatt H (1971) Durabilities of sand size schist and "volcanic" rock fragments during fluvial transport, Elk Creek, Black Hills, South Dakota. J Sediment Petrol 41:565–576

Cassa Per Il Mezzogiorno, Dipartimento per la regione Calabria, Ufficio tecnico lavori – Zona 1, (1986) Studi ed indagini geologiche, geofisische, idrologiche ed idrogeologiche nei bacini alluvionale delle fiumare comprese tra lo Stilaro ed il T. Melito per la correta valutazione delle resorse idriche dei subalvei nella stagione di magra e per la eventuale ricarica di serbatoi sotterranei alluvionali. Relazione Conclusiva. Cassa per il Mezzogiorno, Roma, 137 pp

Cavazza W (1989) Petrofacies and petrologic characteristics of Lower-Middle Miocene sandstones of southern Calabria. Rend Soc It Mineral Petrol 43:733–741

Cavazza W, Dahl J (1990) Relative timing of precipitation of authigenic minerals as a control on sandstone texture. J Sediment Geol (in press)

Christiansen C, Hartmann D (1988a) On using the log-hyperbolic distribution to describe the textural characteristics of eolian sediments discussion. J Sediment Petrol 58:159–160

Christiansen C, Hartmann D (1988b) SAHARA a package of PC-computer programs for estimating both log-hyperbolic grain-size parameters and standard moments. Comput Geosci 14:557–625

Ciaranfi N, Ghisetti F, Guida M, Tortorici L, Vezzani L (1983) Carta neotettonica dell' Italia Meridionale. Pubbl 515 del PF Geodinamica, Bari

Cleary WJ, Conolly JR (1971) Distribution and genesis of quartz in a Piedmont-coastal plain environment. Geol Soc Am Bull 82:2755–2766

Cortese E (1895) Descrizione geologica della Calabria. R Ufficio Geologico, Roma

Crook KAW (1960) Classification of arenites. Am J Sci 258:419–428

Crook KAW (1974) Lithogenesis and geotectonics: the significance of compositional variation in flysch arenites (graywackes). In: Dott RH, Shaver RH (eds) Modern and ancient geosynclinal sedimentation. Soc Econ Paleontol Mineral Spec Publ 19:304–310

Davies DK, Vessell RK, Miles RC, Foley MG, Bonis SF (1978) Fluvial transport and downstream modifications in an active volcanic region. In: Miall AD (ed) Fluvial sedimentology. Can Soc Petrol Geol Mem 5:61–84

Davis JC (1986) Statistics and data analysis in geology, 2nd edn. Wiley and Sons, New York

Davis RA Jr, Hayes MO (1984) What is a wave-dominated coast? Mar Geol 60:313–329

Dickinson WR (1970) Interpreting detrital modes of graywacke and arkose. J Sediment Petrol 40:695–707

Dickinson WR (1985) Interpretation of provenance relations from detrital modes of sandstones. In: Zuffa GG (ed) Provenance of arenites. NATO-ASI Ser, Reidel, New York

Dickinson WR (1988) Provenance and sediment dispersal in relation to paleotectonics and paleogeography of sedimentary basins. In: Kleinspahn KL, Paolo C (ed) New perspectives in basin analysis. Springer, Berlin Heidelberg New York, pp 3–25

Dickinson WR, Suczek CA (1979) Plate tectonics and sandstone compositions. Am Assoc Petrol Geol Bull 63:2164–2182

Diepenbroek M, Bartholomä A, Ibbeken H (1991) How round is round? A critical review of the topic "roundness" and a new approach by Fourier Grain-shape analysis (subm.)

Dobkins JE Jr, Folk RL (1970) Shape development on Tahiti-Jui. J Sediment Petrol 40:1167–1203

Dott RH Jr (1964) Wacke, graywacke and matrix – what approach to immature sandstones classification? J Sediment Petrol 34:625–632

Dumas B, Gueremy P, Hearty PJ, Lhenaff R, Raffy J (1988) Morphometric analysis and amino acid geochronology of uplifted shorelines in a tectonic region near Reggio Calabria, south Italy. Palaeogeogr Palaeoclimatol Palaeoecol 68:273–289

Durst F, Macagno M (1986) Experimental particle size distributions and their representation by log-hyperbolic functions. Powder Technol 45:223–244

Ehrlich R (1983) Size analysis wears no clothes or have moments come and gone. J Sediment Petrol 53:1

Ergenzinger P, Görler K, Ibbeken H, Obenauf P, Rumohr J (1978) Calabrian arc and Ionian Sea: vertical movements, erosional and sedimentary balance. In: Alps, Apennines, Hellenides. IUCG, Sci Rep No 38. Schweitzerbartsche Verlagsbuchhandlung, Stuttgart, pp 359–373

Ergenzinger P, Fabel E, Hurtz B (1987) Untersuchungen zu den kurz- und mittelfristigen Veränderungen des Grobgeschiebestrandes vor Bianco (Kalabrien)/Italien. Berl Geogr Studien 25:263–284

Evans IS (1972) General geomorphometry derivatives of altitude and descriptive statistics. In: Chorley RJ (ed) for British Geomorphological Research Group. Spatial analysis in geomorphology. Metanen & Co Ltd, United Kingdom, pp 17–90

Fabel E (1989) Untersuchungen zu den jahreszeitlichen Veränderungen des Grobgeschiebestrandes vor Bianco (Kalabrien/Italien). Diplomarbeit, Fachbereich Geowiss, Inst Phys Geogr, Freie Univ Berlin, 84 pp

Ferla P, Alaimo R (1976) I graniti e le rocche porfiriche calcalcaline e K-andesitiche nel conglomerato trasgressivo del Miocene Inferiore dei Monti Peloritani (Sicilia). Mem Soc Geol It 17:123–133

Ferree RA, Jordan DW, Kertes RS, Savage KM, Potter PE (1988) Comparative petrographic maturity of river and beach sand, and origin of quartz arenites. J Geol Educ 36:79–87

Folk RL (1968) Petrology of sedimentary rocks. Hemphill's, Austin, 170 pp

Fraser HJ (1935) Experimental study of the porosity and permeability of clastic sediments. J Geol 48:912–1009

Franzinelli E, Potter PE (1983) Petrology, chemistry and texture of modern river sands, Amazon River System. J Geol 91:23–39

Friedman GM (1961) Distinction between dune, beach and river sands from their textural characteristics. J Sediment Petrol 31:514–529

Friedman GM (1967) Dynamic processes and statistical parameters compared for size frequency distributions of beach and river sands. J Sediment Petrol 37:327–354

Friedman GM (1979) Differences in size distributions of populations of particles among sands of various origins. Sedimentology 26:3–32

Füchtbauer H (1974) Sediments and sedimentary rocks. Schweitzerbartsche Verlagsbuchhandlung, Stuttgart

Garrels RM, Mackenzie FT (1971) Evolution of sedimentary rocks. Norton, New York

Gasparini C, Iannaccone G, Scandone P, Scarpa R (1982) Seismotectonics of the Calabrian arc. Tectonophysics 84:267–286

Gazzi P (1966) Le arenarie del flysch sopracretaceo dell'Appennino modenese; correlazioni con il flysch di Monghidoro. Miner Petrogr Acta 12:69–97

Ghisetti F (1981) L'evoluzione strutturale del bacino Plio-Pleistocenico di Reggio Calabria nel quadro geodinamico dell'arco Calabro. Boll Soc Geol It 100:433–466

Ghisetti F, Vezzani L (1982) Different styles of deformation in the Calabrian arc (southern Italy): implications for a seismotectonic zoning. Tectonophysics 85:149–165

Giese P, Görler K (1978) The geotraverse Sicily-southern Italy. Ionian Sea (III a+b). In: Alps, Apennines, Hellenides. IUCG, Sci Rep No 38. Schweitzerbartsche Verlagsbuchhandlung, Stuttgart, pp 313–314

Gilbert CM (1958) Sedimentary rocks. In: William H, Turner FJ, Gilbert CM (eds) Petrography – an introduction to the study of rocks in thin section. Freeman, San Francisco, 406 pp

Gignoux MM (1911) Resultats generaux d'une étude des anciens rivages dans la mediteranée occidentale. Ann Univ Grenoble 23:113–133

Girty GH, Mossman BJ, Pincus SD (1988) Petrology of Holocene sand, Peninsular Ranges, California, and Baja Norte, Mexico: implications for provenance-discrimination models. J Sediment Petrol 58:881–887

Görler K, Giese P (1978) Aspects of the evolution of the Calabrian Arc. In: Alps, Apennines, Hellenides. IUCG, Sci Rep No 38. Schweitzerbartsche Verlagsbuchhandlung, Stuttgart, pp 375–388

Görler K, Uchdorf B (1980) Zur quantitativen Erfassung von Erosion und Denudation – Beispiele aus Südkalabrien. Berl Geowiss Abh (Berlin) 20:121–136

Graham SA, Ingersoll RV, Dickinson WR (1976) Common provenance for lithic grains in carboniferous sandstones from the Quachite Mountains and Black Worrior Basin. J Sediment Petrol 46:620–632

Grantham JH, Velbel MA (1988) The influence of climate and topography on rock-fragment abundance in modern fluvial sands of the southern Blue Ridge Mountains, North Carolina. J Sediment Petrol 58:219–227
Gregory KJ (1983) Background to paleohydrology, a perspective. Wiley, Chichester, 486 pp
Gregory KJ, Walling DE (1973) Drainage basin form and process – a geomorphological approach. Arnold, Norwich, 458 pp
Guerrera F, Veneri F (1989) Evidenze di attività vulcanica nei sedimenti neogenici e Pleistocenici dell'Apennino: Stato Delle Conoscenze. Boll Soc Geol It 108:121–160
Guerricchio A (1986) Il problema della difesa dalle catastrofi idrogeologiche e deformazioni gravitative profonde a Casignana, Sagata del Bianco e Caraffa del Bianco (Prov di Reggio Calabria), Sviluppo No 47. Riv Stud Ric Cassa Risp Calabria Lucania, pp 67–77
Guerricchio A, Melidoro G (1973) Segni premonitori e collassi delle grandi franelle metamorfiti della valle della fiumare Buonamico (Aspromonte, Calabria). Geol Appl Idrol 8:315–337
Hack JT (1957) Studies of longitudinal stream profiles in Virgina and Maryland. US Geol Surv Prof Pap 294B:B45–B97
Harland WB, Cox AV, Llewellyn PG, Pickton CAG, Smith AG, Walters R (1982) A geologic time scale. Cambridge Earth Science Series. Cambridge, 131 pp
Harrell J, Blatt H (1978) Polycrystallinity: effects on the durability of detrital quartz. J Sediment Petrol 48:25–30
Hartmann D (1988) The goodness-of-fit to ideal Gauss and Rosin distributions: a new grain size parameter – discussion. J Sediment Petrol 58:913–917
Heward AP (1981) A review of wave-dominated clastic shoreline+deposits. Earth Sci Rev 17:223–276
Hofmann A (1980) Die Bianco Barre-Differenzierungen sedimentologischer Parameter entlang einer Longshore bar vor der Küste SE-Kalabriens. Diplomarbeit, Freie Univ Berlin, 45 pp
Hölting B (1989) Hydrogeologie, Einführung in die allgemeine und angewandte Hydrogeologie, 3rd edn. Ferdinand Enke, Stuttgart, 396 pp
Holland HD (1978) Dissolved load and runoff of major rivers of the world: The chemistry of the atmosphere and the oceans. John Wiley & Sons, New York, 351 pp
Horton RE (1932) Drainage basin characteristics. Trans Am Geophys Union 13:351–361
Horton RE (1941) Sheet erosion, present and past. Trans Am Geophys Union 22:299–305
Horvarth F, Berckhemer H (1982) Mediterranean backarc basins: Alpine-Mediterranean geodynamics. In: Berckhemer H, Hsü KJ (eds) Vol 7, American Geophysical Union, Washington, DC, pp 141–175
Hough JL (1942) Sediments of Cape Cod Bay, Massachusetts. J Sediment Petrol 12:10–30
Hsü J, Montader L, Bernoulli D, Lita MB, Garrison RE, Kidd RB, Melieres F, Müller C, Wright R (1977) History of the Mediterranean Salinity crisis. Nature 267:399–403
Hubert JF (1962) A zircon-tourmaline-rutile maturity index and the interdependence of the composition of heavy mineral assemblages with the cross composition and texture of sandstones. J Sediment Petrol 32:440–450
Hurtz B (1988) Untersuchungen zu den mittelfristigen Veränderungen des Grobgeschiebestrandes vor Bianco (Kalabrien/Italien). Diplomarbeit, Freie Univ Berlin, 83 pp
Ibbeken H (1970) Das Ligurische Tongriano, eine resedimentierte Molmasse des Nordapennin. Beih Geol Jahrb 93:139
Ibbeken H (1974) A simple sieving and splitting device for field analysis of coarse grained sediments. J Sediment Petrol 44:939–946
Ibbeken H (1983) Jointed source rock and fluvial gravels controlled by Rosin's law: a grain-size study in Calabria, south Italy. J Sediment Petrol 53:1213–1231
Ibbeken H, Denzer P (1988) Clast measurement: a simple manual device and its semiautomatic electronic equivalent. J Sediment Petrol 58:751–752
Imbrie J, Van Andel TH (1964) Vector analysis of heavy mineral data. Geol Soc Am Bull 75:1131–1156
Ingersoll RV (1978) Petrofacies and petrologic evolution of the Late Cretaceous force-arc basin, northern+central California. J Geol 86:335–352
Ingersoll RV, Suczek CA (1979) Petrology and provenance of Neogene sand from Nicober and Bengal fans: DSDP sites 211 and 218. J Sediment Petrol 49:1217–1228
Ingersoll RV, Bullard TF, Ford RL, Grimm JP, Pickle JD, Sares SW (1984) The effect of grain size on detrital modes: a test of the Gazzi-Dickinson point-counting method. J Sediment Petrol 54:103–116
Istituto geografico militare (1957) Carta d'Italia 1:25000
Istituto geografico militare (1967) Carta Geologica d'Italia 1:100000
Jensen JL (1988) Maximum likelihood estimation of log-hyperbolic parameters from grouped observations. Comput Geosci 14:389–408
Johnsson MJ, Stallard RF, Meade RH (1988) First-cycle quartz arenites in the Orinoco River basin, Venezuela and Colombia. J Geol 96:263–277
Johnsson MJ, Stallard RF (1989) Physiographic controls on the composition of sediments derived from volcanic and sedimentary terrains on Barro Colorado Island, Panama. J Sediment Petrol 59:768–781
Kittleman LR Jr (1964) Application of Rosin's distribution in size-frequency analysis of clastic rocks. J Sediment Petrol 34:483–502
Kolmogoroff AN (1941) Über das logarithmisch normale Verteilungsgesetz der Dimensionen der Teilchen bei Zerstückelung. C R Acad Sci URSS 31:99–100
Komar PD, Li Z (1988) Applications of grain-pivoting and sliding analyses to selective entrainment of gravel and to flow-competence evaluations. Sedimentology 35:681–695
Köster H (1986) Korngrößenanalyse, Messung der Kornform und Rundungsverhalten der Küstenschotter vor Bianco RC, Italien. Diplomarbeit, Freie Univ Berlin, 55 pp
Kretz R (1983) Symbols for rock-forming minerals. Am Mineral 68:277–279
Krumbein WC (1941) Measurement and geologic significance of shape and roundness of sedimentary particles. J Sediment Petrol 16:64–72
Krumbein WC, Tidsel FW (1940) Size distribution of source rocks of sediments. Am J Sci 238:296–305
Langbein WB, Schumm SA (1954) Yield of sediment in relation to mean annual precipitation. Trans Am Geophys Union 39:1076–1084
Lembke H (1931) Beiträge zur Geomorphologie des Aspromonte (Kalabrien). Z Geomorphol 6:58–112
Livingstone DA (1963) Chemical composition of rivers and lakes. In: Data of geochemistry, 6th edn. US Geol Surv (eds) Proff Pap, 440 pp
Mack GH (1978) The survivability of labile light-mineral grains in fluvial, aeolian and littoral marine environments: the Permian Cutler and Cedar Mesa Formations, Moab Utah. Sedimentology 25:587–604

Mack GH (1981) Composition of modern stream sand in a humid climate derived from a low-grade metamorphic and sedimentary foreland fold-thrust belt of north Georgia. J Sediment Petrol 51:1247–1258

Mann WR, Cavaroc VV (1973) Composition of sand released from three source areas under humid, low relief weathering in the North Carolina Piedmont. J Sediment Petrol 43:870–881

Marsal D (1967) Statistische Methoden für Erdwissenschaftler. Schweizerbartsche Verlagsbuchhandlung, Stuttgart, 152 pp

Maynard JB, Valloni R, Ho-Shing Yu (1982) Composition of modern deep-sea sands from arc-related basins. In: Leggett JK (ed) Trench – Forearc Geology: sedimentation and tectonics on modern and ancient active plate margins. Geol Soc London, pp 551–561

McBride EF (1963) A classification of common sandstones. J Sediment Petrol 33:664–669

McBride EF, Picard MD (1987) Downstream changes in sand composition, roundness, and gravel size in a short-headed, high-gradient stream, northwestern Italy. J Sediment Petrol 57:1018–1026

McLaren P (1981) An interpretation of trends in grain size measures. J Sediment Petrol 51:611–624

McLaren P, Bowles D (1985) The effects of sediment transport on grain-size distributions. J Sediment Petrol 55:457–470

Meade RH (1969) Errors in using modern stream-load data to estimate natural rates of denudation. Geol Soc Am Bull 80:1265–1274

Meischner D, Rumohr J (1974) A light-weight, high-momentum gravity corer for subaqueous sediments. Senckenb Marit 6:105–117

Mesa OJ, Gupta (1987) On the main channel length-area relationship for channel networks. Water Resour Res 23:2119–2122

Milliman JD, Meade RH (1983) World-wide delivery of river sediments to the oceans. J Geol 91:1–21

Morton AC (1984) Stability of detrital heavy minerals in Tertiary sandstones from the North Sea Basin. Clay Miner 19:287–308

Moss AJ (1972a) Initial fluviatile fragmentation of granitic quartz. J Sediment Petrol 42:905–916

Moss AJ (1972b) Technique for assessment of particle breakage in natural and artificial environments. J Sediment Petrol 42:725–728

Mueller JE (1973) Re-evaluation of master streams and drainage basins, reply. Geol Soc Am Bull 84:3127–3130

Murray J (1887) On the total annual rainfall on the land of the globe and the relation of rainfall to the annual discharge of rivers. Scott Geogr Mag 3:65–77

Nesbitt HW, Young GM (1984) Prediction of some weathering trends of plutonic and volcanic rocks based on thermodynamic and kinetic considerations. GCA 48:1523

Nickel E (1973) Experimental dissolution of light and heavy minerals in comparison with weathering and intrastratal solution. Contrib Sedimentol 1:1–68

Okada H (1971) Classification of sandstone: analysis and proposal. J Geol 79:509–525

Ollier CD, Brown MJF (1971) Erosion of a young volcano in New Guinea. Z Geomorphol 15:12–30

Palomares M, Tortosa A, Arribas J (1989) Estudios de los fraccion de arenas actuales derivadas del Sistema Central y su aplicacion a los estudios de procedencia. Bol Geol Miner 100:695–705

Parfenoff A, Pomerol C, Tourenq J (1970) Les mineraux en grains (methodes d'étude et determination). Masson et Cie, Paris, 578 pp

Pettijohn FJ (1957) Sedimentary rocks. Harper, New York, 718 pp

Pettijohn FJ (1975) Sedimentary rocks, 3rd edn. Harper and Row, New York, 628 pp

Pettijohn FJ, Potter PE, Siever R (1987) Sand and sandstone, 2nd edn. Springer, Berlin Heidelberg New York, 553 pp

Plumley W (1948) Black Hills terrace gravels: a study in sediment transport. J Geol 56:526–577

Popper KR (1973) Logik der Forschung, 5th edn. Mohrs, Tübingen, 441 pp

Potter PE (1978) Petrology and chemistry of modern big river sands. J Geol 86:423–449

Potter PE, Franzinelli E (1985) Fraction analyses of modern river sand of Rios Negro and Solimoes, Brazil, implications for the origin of quartz-rich sandstones. Rev Bras Geoc 15:31–35

Powers MC (1953) A new roundness scale for sedimentary particles. J Sediment Petrol 23:117–119

Rehault JP, Moussat E, Fabri A (1986) Structural evolution of the Tyrrhenian back-arc basin. Mar Geol 74:123–150

Rio D, Thunnel R, Sprovieri R, Kukry D, Destefano E, Howell M, Raffi I, Sancetta C, Sanfilippo A (1989) Stratigraphy and depositional history of the Pliocene Bianco section, Calabria, southern Italy. Palaeogr Palaeoclimatol Palaeoecol 76:85–105

Ritsema AR (1972) Deep earthquakes of the Tyrrhenian Sea. Geol Min Jahrb 51:541–548

Robinson BN, Anderson GD, Cohen E, Korpel LC, Miller AH, Stein JR (1980) SIR scientific information retrieval, user's manual, version 2. SIR Inc Evanston, USA

Robinson D, Leake BE (1975) Sedimentary and igneous trends on AFM diagrams. Geol Mag 112:305–307

Roda PH (1965) Il calcare portlandiano a Dasycladacee di M Mutolo, Reggio Calabria. Geol Romana 4:259–290

Rogers JW, Krueger WC, Krog M (1963) Sizes of naturally abraded materials. J Sediment Petrol 33:628–632

Romagnoli L (1969) Osservazioni geologiche e geomorphologiche sull'Aspromonte (Calabria). Considerazioni preliminari sull'erosione e la conservazione del suolo. Boll Soc Geol It 88:245–266

Rosin P, Rammler E (1933) Die Kornzusammensetzung des Mahlgutes im Lichte der Wahrscheinlichkeitslehre. Kolloid-Zeitschrift, 16 pp

Rosin P, Rammler E, Sperling K (1933) Korngrößenprobleme des Kohlestaubes und ihre Bedeutung für die Vermahlung. Berichte des Reichskohlenrats C52, VDI-Verlag, Berlin

Rumohr J, Ibbeken H (1975) Submarine morphology of southeastern Calabria, Italy (Ionian Sea). Rapp Commint Mer Médit 23:245–246

Russel RJ (1968) Where most grains of very coarse sand and fine gravel are deposited. Sedimentology 11:31–38

Rust BR (1978) A classification of alluvial channel systems. Can Soc Petrol Geol, Calgary, Alberta, Canada

Ruxton BP (1970) Labile quartz-poor sediments from young mountain ranges in northeast Papua. J Sediment Petrol 40:1262–1270

Schleyer R (1987) The goodness of fit to ideal Gauss and Rosin distributions: a new Grain-size parameter. J Sediment Petrol 57:871–880

Schleyer R (1988a) The goodness-of-fit to ideal Gauss and Rosin distributions: a new grain-size parameter – reply. J Sediment Petrol 58:917–918

Schleyer R (1988b) Die exotischen Gerölle der Oligozän-Molassen im Grenzbereich Alpen-Apennin und ihre Provenienz. Thesis, Freie Univ Berlin, 175 pp

Schneider R, Sutter R (1982) Beitrag zur Flora und Vegetation Südkalabriens. Phytocoenologia 10:323–373
Schumm SA (1956) Evolution of drainage systems and slopes in badlands at Perth Amboy, New Jersey. Bull Geol Soc Am 67:597–646
Schumm SA (1968) Speculations concerning paleohydraulic controls of terrestrial sedimentation. Geol Soc Am Bull 79:1573–1588
Schütte KG (1978) Crustal structure of southern Italy. In: Alps Apennines Hellenides, IUCG, Sci Rep No 38. Schweitzerbartsche Verlagsbuchhandlung, Stuttgart, pp 315–321
Shea JH (1974) Deficiencies of clastic particles of certain sizes. J Sediment Petrol 33:180–190
Shih SM, Komar PD (1990) Hydraulic controls of grain-size distribution of bedload gravels in Oak Creek, Oregon, USA. Sedimentology 37:367–376
Shukis PS, Ethridge FG (1975) A petrographic reconnaissance of sand size sediment, upper St. Francis River, southeastern Missouri. J Sediment Petrol 45:115–127
Shulman MJ (1983) Data management and manipulation in mineral exploration. In: Glaeser PS (ed) Data for science and technology. North Holland, Amsterdam
Sneed ED, Folk RL (1958) Pebbles in the lower Colorado River, Texas, a study in particle morphogenesis. J Geol 27:114–150
Sorriso-Valvo M (1984) Un caso di deformazione gravitativa profonda nel bacino della fiumara di Platì (Reggio Calabria). Boll Soc Geol Ital 103:701–706
Spannagel P (1985) Korngrößenanalyse und Petrographie der Küstenschotter von Bianco (R.C./Italien). Diplomarbeit, Freie Univ Berlin, 49 pp
Strahler AN (1952) Hypsometric (area-altitude) analysis of erosional topography. Bull Geol Soc Am 63:1117–1142
Strahler AN (1964) Quantitative geomorphology of drainage basins and channel networks. In: Ven te Chow PHD (ed) Handbook of applied hydrology. McGraw-Hill, New York, pp 4–76
Sundborg A (1956) The river Klarälven, a study in fluvial processes. Geogr Ann Ser A 11:127–316
Suttner LJ (1974) Sedimentary petrographic provinces: an evaluation. Soc Econ Paleontol Mineral. SEPM Spec Publ 21:75–84
Suttner LJ, Basu A, Mack GH (1981) Climate and the origin of quartz arenite. J Sediment Petrol 51:1235–1246
Tavernier R (1985) Soil map of the European Communities 1:1000000. Directorate – General for Agriculture Coordination of Agricultural Research, Office for Official Publication of the European Communities, Luxembourg
Tortosa A, Palomares M, Arribas J (1989) Caracterization composicional de los depositos arenosos actuales generados en el Sistema Central. Estud Geol 45:205–213
Toy TJ (ed) (1977) Erosion, research techniques, erodibility and sediment delivery. Geo Abstr Ltd, Norwich, 73 pp
Valloni R (1985) Reading provenance from modern marine sands. In: Zuffa GG (ed) Provenance of arenites. NATO ASI Ser, Reidel, Dordrecht, pp 309–332
Van Andel TH (1958) Origin and classification of Cretaceous, Paleocene and Eocene sandstones of western Venezuela. Am Assoc Petrol Geol Bull 42:734–763
Van de Kamp PC, Leake BE, Senior A (1975) The petrography and geochemistry of some Californian arkoses with application to identifying gneisses of metasedimentary origin. J Geol 84:195–212
Versace P, Ferrari Fiorentino M, Gabriele S, Rossi F (1987) Valutazione delle piene in Calabria, Cartografia. CNR, IRPI Cosenza
Viparelli C (1964) Granulometria delle alluvioni dei Torrenti Calabri. Geotecnica 11:10–35
Walling DE, Webb BW (1983) Patterns of sediment yield: background to paleohydrology. In: Gregory KJ (ed) Background to Paleohydrology. Wiley and Sons, New York
Weik S (1988) Überblick über Hydrogeologie und Wasserbeschaffenheit von SE-Kalabrien/Italien. Diplomarbeit, Freie Univ Berlin, 122 pp
Wezel FC (1982) The structure of the Calabro-Sicilian Arc: result of a post-orogenic intra-plate deformation. In: Leggett JK (ed) Trench forearc geology sedimentation and tectonics on modern and ancient active plate margins. Geol Soc Lond Spec Publ No 10, pp 345–354
Whetten JT, Kelly JC, Hanson LG (1969) Characteristics of Columbia River sediment and sediment transport. J Sediment Petrol 39:1149–1166
Wilkes E (1986) Zur Rekonstruktion des obermiozänen Paläoreliefs in Südkalabrien – geologische und mikropaläontologische Methoden. Diplomarbeit, Freie Univ Berlin, 96 pp
Wischmeier WH (1977) Soil erodibility by rainfall and runoff. In: Toy TJ (ed) Erosion: research, techniques, erodibility and sediment delivery. Geo Abstr Ltd, Norwich, UK, pp 45–57
Wolcott J (1988) Nonfluvial control of bimodal grain-size distributions in river-bed gravels. J Sediment Petrol 58:979–984
Wyrwoll KH, Smyth GK (1985) On using the log-hyperbolic distribution to describe the textural characteristics of eolian sediments. J Sediment Petrol 55:471–478
Wyrwoll KH, Smyth GK (1988) On using the log-hyperbolic distribution to describe the textural characteristics of eolian sediments – reply. J Sediment Petrol 58:161–162
Yatsu E (1955) On the longitudinal profile of the graded river. Trans Am Geophys Union 36:655–663
Young A, Saunder I (1986) Rates of surface processes and denudation. In: Abrahams AD (ed) Hillslope processes. Allen and Unwin, Boston, pp 416
Zuffa GG (1985) Optical analysis of arenites: influence of methodology on compositional results. In: Zuffa GG Provenance of arenites. NATO ASI Ser, Reidel, Dordrecht, pp 165–189

Subject Index

MIX
Papier aus verantwortungsvollen Quellen
Paper from responsible sources
FSC® C105338

If you have any concerns about our products,
you can contact us on
ProductSafety@springernature.com

In case Publisher is established outside the EU,
the EU authorized representative is:
Springer Nature Customer Service Center GmbH
Europaplatz 3, 69115 Heidelberg, Germany

Printed by Libri Plureos GmbH
in Hamburg, Germany